Springer Tracts in Advanced Robotics

Volume 95

For further volumes:
http://www.springer.com/series/5208

STAR (Springer Tracts in Advanced Robotics) has been promoted under the auspices of EURON (European Robotics Research Network)

Ravi Balasubramanian · Veronica J. Santos
Editors

The Human Hand as an Inspiration for Robot Hand Development

Springer

Editors
Ravi Balasubramanian
Oregon State University
Corvallis
USA

Veronica J. Santos
Mechanical and Aerospace Engineering
Arizona State University
Tempe
USA

ISSN 1610-7438 ISSN 1610-742X (electronic)
ISBN 978-3-319-03016-6 ISBN 978-3-319-03017-3 (eBook)
DOI 10.1007/978-3-319-03017-3
Springer Cham Heidelberg New York Dordrecht London

Library of Congress Control Number: 2013957707

Printed on acid-free paper

Springer is part of Springer Science+Business Media (www.springer.com)

Foreword

Robotics is undergoing a major transformation in scope and dimension. From a largely dominant industrial focus, robotics is rapidly expanding into human environments and vigorously engaged in its new challenges. Interacting with, assisting, serving, and exploring with humans, the emerging robots will increasingly touch people and their lives.

Beyond its impact on physical robots, the body of knowledge robotics has produced is revealing a much wider range of applications reaching across diverse research areas and scientific disciplines, such as: biomechanics, haptics, neurosciences, virtual simulation, animation, surgery, and sensor networks among others. In return, the challenges of the new emerging areas are proving an abundant source of stimulation and insights for the field of robotics. It is indeed at the intersection of disciplines that the most striking advances happen.

The *Springer Tracts in Advanced Robotics (STAR)* is devoted to bringing to the research community the latest advances in the robotics field on the basis of their significance and quality. Through a wide and timely dissemination of critical research developments in robotics, our objective with this series is to promote more exchanges and collaborations among the researchers in the community and contribute to further advancements in this rapidly growing field.

The monograph by Ravi Balasubramanian and Veronica Santos is an edited collection of authoritative contributions in the area of robot hands which stemmed from a well-attended workshop organized by the first coeditor with Yoky Matsuoka as part of the Robotics: Science and Systems conference in Seattle in 2009. The 24 chapters discuss the field of robotic grasping and manipulation viewed in light of the human hand's capabilities and push the state-of-the-art in robot hand design and control. Topics discussed include human hand biomechanics, neural control, sensory feedback and perception, and robotic grasp and manipulation.

The results described in the volume are expected to lead to more robust, dependable, and inexpensive distributed systems such as those endowed with complex and advanced sensing, actuation, computation, and communication capabilities. A very fine addition to STAR!

Naples, Italy, August 2013 Bruno Siciliano

Preface

Are Robot Hands Necessary?

As identified in the 2013 roadmap for US robotics, robotics is expected to impact society on a massive scale in the coming decades economically and socially in the manufacturing, healthcare, medical, and defense sectors. In addition to the traditional use of robots for automation in factories, recent advances in the human sciences have energized the field of robotics toward the development of personal robotic assistants and brain-machine interfaces for assisting the disabled. While great strides have been made in the areas of computer vision and autonomous navigation that have enabled autonomic robotic cars, one of the biggest drawbacks with robots so far is that they cannot accomplish physical interaction tasks in everyday settings. Specifically, robots cannot grasp and manipulate objects in unstructured environments, or environments for which they have not been designed. A lack of robotic hands that are capable of robust grasping and dexterous manipulation is holding back the robotics field. Thus, there is an increased interest to solve the robotic manipulation problem. The reasons for this deficiency are many, including the lack of robust hardware, primitive sensing methods, and a limited understanding of how to integrate sensory information and motor control.

A key goal of the robotics community is to build robotic hands that can accomplish human grasping and manipulation tasks in human environments by physically interacting with humans and objects. Such robot hands will have an immediate impact on applications such as teleoperated search-and-rescue operations, semi or fully autonomous robot applications (e.g., planetary rovers), rapidly reconfigurable manufacturing, and medical and healthcare operations. In addition to automating operations in these different fields, advances in robot hands will also advance upper-extremity prostheses. Note that the most popular prosthesis to date remains the single degree of freedom, body-powered split hook, because of its robustness and the human ability to learn how to use it. There have been significant advances in myoelectric prostheses, but challenges remain in providing control signals in an intuitive manner to control numerous degrees of freedom in more sophisticated prostheses.

The Human Hand as Inspiration

The human hand has been the "gold standard" for robotic hand designers for decades. There are several reasons for this. First, the human hand exhibits tremendous dexterity and flexibility, and designers are keen to achieve such dexterity in robot hands. Second, everyday tools, objects, and environments are designed for use by a human hand (for example, where handles are placed on objects), and thus it is advantageous to mimic the human hand when designing robot hands to operate those same tools and objects in human environments. Third, the anthropomorphic form factor is highly relevant to prosthetic applications. Thus, most robot hand designs mimic the human hand.

However, the human hand is difficult to mimic since it is a complex system. In terms of "hardware," the human hand contains 22 joints driven by nearly 38 muscles through a complex web of tendons. In addition, it has thousands of embedded sensors which provide information about posture, muscle and tendon forces, contact, interaction forces, vibration, and temperature. In terms of "software," there are millions of neurons in the brain and the spinal cord that integrate information from the raw sensory signals before providing control signals through synergistic control inputs and reflex loops. Together, these different features enable the hand to perform a variety of dexterous tasks, but the roles that each component plays in different tasks is not entirely clear.

Roboticists want to understand what physical and computational features from the biological hand would benefit the design and control of highly capable robotic hands. This is the focus of this book. By bringing together the latest research on biological hands and the state-of-the-art in robotic hand technology in one book, we hope to inspire new ideas that will foster a deeper understanding of the human hand, accelerate the advancement of robotic hand research, and bridge multiple research communities through common interests in hands.

Note that some researchers are moving away from using the human hand as the template for robotic hands because of the difficulty in mimicking its compactness, form, and control. Specifically, they are designing "underactuated robotic hands" with many degrees of freedom but reduced number of actuators. These designs utilize tendon-driven systems or linkage mechanisms for creating movement and achieving human-like grasping capability. Such hands can surely address design criteria such as robustness and simplicity of sensing and control for static grasping. However, much work is still required to achieve human-like dexterity for manipulation.

How this Book Came About

The idea for a book on human and robot hands arose during a workshop organized by Dr. Ravi Balasubramanian (coeditor of this book) and Dr. Yoky Matsuoka as part of the Robotics: Science and Systems conference in Seattle in 2009. The workshop was a forum for researchers to discuss the field of robotic grasping and manipulation viewed in relation to the human hand's capabilities, and to push the state-of-the-art in robot hand design and control. Topics discussed at the workshop included human hand biomechanics, neural control, sensory feedback and perception, and robotic grasp and manipulation. Over seventy researchers attended. Since no book existed that combined research on human and robot hands, it was decided to assemble a comprehensive work that discusses the latest developments in these fields by building on the workshop's proceedings. Dr. Veronica Santos joined as a co-editor due to her expertise in bioengineering.

Book's Expected Audience

We expect that this book will benefit researchers from diverse areas such as robotics, biomechanics, bioengineering, neuroscience, and anthropologists. Together, these different fields can synergistically learn and apply each other's techniques to their problems. For example, the mathematical underpinnings of creating contact forces through a robot's motors can be applied to the analysis of using the hand's muscles to create fingertip forces. Integration of sensory data, reflex algorithms, and grasping strategies from humans can be used to develop advanced control algorithms for robots.

Book Layout

The book is divided into two Parts. Part I focuses on the human hand's anatomical structure and function and the state-of-the-art methods that are used to derive insights into those aspects. Part II provides a broad perspective on the approaches for human-inspired robotic hand design and control. Brief descriptions of the chapters in each section are below.

Part I: The Rich Complexity of the Human Hand and Its Control

The first four chapters of the book detail the neural control, kinematics, and musculotendon properties of the human hand as they relate to motion and force production capabilities. Chapter 1 by Schieber describes the cortical control of the human hand as a widely distributed network that can drive fixed synergies for grasping, act as diverse elements for individuated finger movements, and flexibly recombine elements for motor learning and reorganization after injury. Chapter 2 by Santello describes the phenomenon of common neural input as one of the mechanisms through which the central nervous system might coordinate the neural activation of groups of hand muscles acting on a single digit or multiple digits. Chapter 3 by Stillfried et al. explores the use of magnetic resonance imaging to determine the locations and axes of rotations of finger joints which can have significant impact on hand function. Chapter 4 by Lee and Kamper investigates the mechanisms of musculotendon force transmission such as finger posture, passive joint impedance, anatomical pulleys, and the tendon extensor hood for mapping muscle forces to index finger kinematics and dynamics.

The next three chapters discuss the characteristics and roles of tactile and proprioceptive sensory feedback mechanisms in the human hand. Chapter 5 by Jones presents the roles played by mechanoreceptors, or mechanical sensors, embedded in the human hand for the perception and control of finger forces. Chapter 6 by Walsh et al. describes the sense of proprioception with a focus on recent evidence that perception of posture can be affected by muscle contraction history, and that illusions of joint position and movement can be induced by simultaneous activation of slowly adapting and fast-adapting cutaneous receptors. Chapter 7 by Bensmaia and Helms Tillery reviews the combined roles of tactile and proprioceptive sensation in hand function, with a focus on the integration of multiple inputs to extract information about haptic interactions and to create somatosensory images for upper-limb neuroprosthetics.

The next two chapters address two ways in which sensory feedback is used: reactive control of fingertip forces, and active haptic perception. Chapter 8 by De Gregorio and Santos reviews how precision grip forces are affected by intrinsic object properties, anticipation, load direction, and sensory feedback, and then presents evidence that unexpected torque loads can elicit reactive, pulse-like increases in grip forces whose strength depends on orientation of the load relative to the hand. Chapter 9 by Tavakoli discusses how tactile feedback and kinesthetic feedback together influence the human ability to distinguish objects though haptic recognition of material and shape properties of objects.

Chapter 10 by Dollar discusses how human hands are used and presents classification and taxonomy schemes for grasping and manipulation behaviors based on hand-centric and motion-centric categorization. Such classifications of hand use can be applied to the fields of biomechanics, rehabilitation, prosthetics, and robotic manipulation.

Part II: Human Hand-Inspired Robotic Hand Design and Control

Chapter 11 by Controzzi et al. provides a historical perspective about robotic hand design, including simple prostheses from the Roman times to highly advanced anthropomorphic robotic hands with a multitude of joints, sensors, and motors.

The next four chapters discuss the modeling of hand motion and force production using robotics techniques, with a focus on how anatomical structures like tendons and fingertip shape and compliance influence grasping capability. Specifically, Chap. 12 by Inouye et al. describes a novel, systematic approach to analyze and optimize tendon routing and pulley size for three-dimensional force production by a tendon-driven finger. Chapter 13 by Sueda and Pai provides a novel method for dynamically simulating human hand movement, while factoring in aspects such as the sliding of tendons over bones beneath the skin. Chapter 14 by Inoue and Hirai presents an analytical exploration into how hand compliance provides robustness to grasps even if there are significant delays in conveying control input through the biological neural system. Chapter 15 by Arimoto and Yoshida describes a method for computationally modeling the stability of "blind" multifinger grasps for which there is no tactile or visual information.

The next four chapters discuss the development of robotic hardware components: haptic devices that can apply forces to human fingertips, and tactile sensors which are critical for robotic hands to sense physical interactions with the external world. Specifically, Chap. 16 by Endo and Kawasaki present the design and control of multifinger haptic devices with the goal of understanding the perception of fingertip forces in the human hand. Chapter 17 by Buttolo and Hannaford describes devices for quantifying properties of multifinger haptic interaction such as hand stiffness in pen-like grasps and sensory thresholds of multifinger versus single finger haptic exploration. Chapter 18 by Cutkosky and Ulmen present the development of miniature tactile sensors that mimic the slowly adapting and fast-adapting tactile units in the human hand with the goal of achieving dynamic tactile sensing in robots. Chapter 19 by Wettels et al. describes the development and use of a deformable, multimodal, biomimetic tactile sensor that provides simultaneous sensing of contact forces, microvibrations, and temperature.

Chapters 20 and 21 discuss two examples of the development of complete robotic hand hardware. Specifically, Chap. 20 by Varol et al. describes the challenges in designing biomimetic transradial prostheses, particularly addressing the constraints of housing many actuators and sensors in the small volume of a human-sized hand. Chapter 21 by Deshpande and Matsuoka discuss the design and control of an "anatomically correct testbed" robotic hand, in which bone shapes and tendon routing within the human hand are mimicked.

The book concludes with three chapters that focus on developing advanced grasping and manipulation strategies for robots either by learning from humans or through physics-based modeling and computation. Chapter 22 by Balasubramanian et al. presents a novel experiment to identify human heuristics for grasping tasks and the use of those heuristics to improve automatic robotic grasping performance. Chapter 23 by Chang and Pollard explores the preparatory physical interactions humans have with objects prior to grasping an object with the goal of programming robots to exploit similar interactions to improve robotic grasping capability. Finally, Chap. 24 by Allen et al. presents a unique look at advancing robotic grasping by using massive computing power to identify appropriate grasps for previously unseen and incomplete shapes by drawing relationships to grasps achieved on well-defined training objects.

Acknowledgments

We sincerely thank the many reviewers who took the time to provide substantive reviews of each chapter of this book. We also thank Dr. Thomas Ditzinger, Shine David, Ramyakrishnan Murugesan, and others at Springer for their prompt assistance in preparing this book.

Contents

Part I
The Rich Complexity of the Human Hand and its Control

Chapter 1
Constraints and Flexibility in Cortical Control of the Hand

Marc H. Schieber

Abstract The hand performs a variety of functions, from simple grasping to delicate manipulation, largely under the control of the primary motor cortex. Perhaps as a result of biomechanical interactions between the digits at the periphery, cortical control works as a widely distributed network. Whereas for grasping the cortex may drive a limited number of fixed synergies, for more individuated finger movements cortical neurons act as diverse elements to generate a remarkable degree of flexibility. In addition to providing a substrate for motor learning and for plastic reorganization after injury, this flexible network permits rapid re-combination of elements that can promptly create entirely novel movements. Such rapid flexibility enables cortical neurons to become dissociated from bodily movement during mental imagery and during closed loop control of brain machine interfaces.

Keywords Biological neural networks · Brain machine interfaces · Neural prosthesis · Neurocontrollers

1 Introduction

The human hand has evolved from an ancestral locomotor appendage to a modern organ capable of amazing dexterity. The pectoral fins of fish have boney rays like the digits of the hand, and in amphibians and reptiles the homologous parts of the distal forelimbs are used as feet for locomotion. But in mammals, including rodents, cats and monkeys, the paws of the forelimbs also can be used to grasp

M. H. Schieber (✉)
Departments of Neurology and of Neurobiology and Anatomy, School of Medicine and Denistry, University of Rochester, 601 Elmwood Avenue, Box 673, Rochester, NY 14642, USA
e-mail: mhs@cvs.rochester.edu

R. Balasubramanian and V. J. Santos (eds.), *The Human Hand as an Inspiration for Robot Hand Development*, Springer Tracts in Advanced Robotics 95,
DOI: 10.1007/978-3-319-03017-3_1,

food and other objects. Even in humans, the most frequent daily use of the hand still is simply to grasp objects [1].

In various types of grasp, multiple digits must first spread open and then close around the object. All grasping is not the same, however. Progressing from rodents to cats to monkeys to humans, shaping of the paw or hand to conform to the object being grasped becomes increasingly dexterous [2]. In monkeys and humans, a number of canonical grasps can be recognized, such as power grasp, precision pinch and hook grip [3, 4]. From monkeys to humans, the increasing dexterity of grasping reflects an increasing ability to produce differential movement of the digits, conforming more accurately to a wider variety of object shapes.

To some extent in monkeys, but dramatically more so in humans, this ability has evolved to move particular fingers relatively independently of the mass action of grasping, i.e. the ability to individuate finger movements. Monkeys extend a single digit, typically the index or middle finger, to "winkle" a piece of food out of a small hole. Macaque monkeys also have been trained to make individuated movements in which each digit is flexed or extended more than the others [5]. In humans, this ability to individuate movements of particular digits enables sophisticated manipulation of small objects as in tying one's shoelaces, as well as complex, non-manipulative performances such as finger spelling in sign language, typing at a keyboard, or playing a musical instrument.

Nevertheless, even in humans, movements of the fingers are not entirely independent or unconstrained. The biomechanical structure of the hand enables considerable flexion of the fingertips to touch the palm, for example, but does not permit equivalent extension of the fingertips to touch the back of the hand. And though the thumb can be rotated to oppose the fingers, the fingers cannot be rotated to oppose one another. Furthermore, although we can think about moving a single digit, measuring the motion of all digits simultaneously typically reveals some smaller parallel motion of other digits [2], which can be viewed as a vestige of grasping. The evolutionary process that has provided humans with individuated finger movements thus has not eliminated all the peripheral or central constraints on finger independence. The hand's dexterity, then, is largely attributable to control from the primary motor cortex, which functions as a distributed network of diverse elements, controlling the entire hand even during movements of a single digit. Such a neural substrate provides both for long-term plasticity and for rapid flexibility of dexterous manipulation.

2 Peripheral Constraints

2.1 Passive Biomechanical Coupling of the Digits

In contrast to robotic hands that typically are designed with mechanically independent digits, natural fingers are coupled significantly by the passive biomechanics of the hand's soft tissues. In humans, such passive biomechanical coupling accounts for most of the motion observed in other digits when normal subjects attempt to move a given digit by itself [6]. Some of this coupling between adjacent digits is produced by the skin and connective tissue in the web spaces between the fingers.

Additional passive coupling is produced by interconnections between the tendons of certain muscles [2]. Several extrinsic finger muscles, with bellies in the forearm and long tendons that cross the wrist and palm to reach the digits, send tendons to multiple fingers. The digital tendons of such multitendoned muscles are interconnected to varying degrees. The macaque flexor digitorum profundus (FDP), for example, sends tendons to all five digits, but the interconnections between these tendons are strong enough to cause tension exerted at one point on the proximal tendon sheet to be distributed to the distal insertions on multiple digits [7]. In humans, the homologous thumb tendon comes from an entirely separate muscle, flexor pollicis longus, rendering the human thumb substantially more independent than that of the macaque. Likewise, the FDP tendons to the four fingers in humans are more independent than those in the macaque, particularly the tendons to the index and middle finger. But even in humans, these tendons are mechanically interconnected in the palm, both by thin sheets of inelastic connective tissue and by the origins of the lumbrical muscles. Such species differences contribute to the greater ability of humans to individuate finger movements.

2.2 Active Coupling at the Periphery

When a single finger is moved actively by a normal human subject, other fingers move somewhat more than when the same finger is moved passively [6]. Part of this active component of coupling between digits may result from the structure of motor units in the extrinsic multitendoned finger muscles. If single motor units have different muscle fibers that insert on tendons to adjacent digits, then these motor units will act on both digits simultaneously. In macaque monkeys, the extensor digiti quarti et quinti has a significant fraction of single motor units that act on both the ring and little finger tendons [8]. In humans, the homologous muscle, extensor digiti quinti, has no tendon to the ring finger and therefore acts exclusively on the little finger, eliminating any coupling that might be produced by motor units acting on both tendons. But in other human multitendoned muscles, including FDP and extensor digitorum communis, the contraction of many single motor units is statistically associated with a large rise in force on one finger, and a small rise in force on one or more adjacent fingers [9, 10]. Single motor units in multitendoned muscles thus may exert force on multiple digits.

Additional active coupling arises from the divergence of last-order descending inputs among the spinal motoneuron pools. Two motor units may discharge more action potentials within a few milliseconds of one another than attributable to chance alone, indicating that both members of the pair receive a shared input that excites the two simultaneously. Such short-term synchronization of motor units in the same muscle can be attributed to axons that ramify and synapse on multiple motoneurons of the same muscle. Short-term synchronization can also be seen, however, between motor units that act on different fingers, indicating that single pre-motor input neurons make connections to motor units that act on different fingers [11, 12]. Although the branching pre-motor neurons responsible for short-term synchronization of motor units in theory might be any last-order inputs to the motoneuron pools, in humans lesions of the corticospinal system eliminate most such synchronization, implying that the responsible last-order inputs are primarily corticospinal axons that synapse directly on motoneurons [13, 14]. Both the simultaneous synaptic excitation of motoneurons that act on different fingers, and single motor units that act on multiple digits thus may contribute to some degree of active coupling between digits.

2.3 Biomechanical Interactions and Stabilizing Contractions

Because of both the passive and active biomechanical coupling between digits, muscle activity intended to move one digit will tend to move adjacent digits as well. Moreover, contraction of any of the extrinsic finger muscles will produce torque around the wrist. To move one digit more individually then, additional muscles must be activated to check the coupled motion of the adjacent digits and the wrist. As a monkey flexes its little finger, for example, extensor digiti secundi et tertii contracts to minimize simultaneous flexion of the index and middle fingers [15]. And in humans, the portion of FDP that acts chiefly on the middle finger contracts when the subject extends either the index or the ring finger [16]. The combination of contractions in some muscles to move the intended finger(s) and in other muscles to stabilize other fingers and the wrist thus requires descending neural control of the entire hand, even when a normal subject is moving a single digit.

3 Cortical Control

For many years the primary motor cortex (M1) was thought to contain a point-to-point representation of different muscles and/or different movements. Activation at a particular locus then would elicit either a given muscle contraction or a given movement, just as striking a particular key on a piano elicits a given note. M1 then was viewed as a well-ordered array of "upper motor neurons," each providing cortical output to a particular muscle. Using this array, available muscles and/or movements could be accessed as needed by higher cortical areas. More recently, the demonstration of a number of organizational principles has revised this view of M1 [17].

3.1 Convergence from Cortex to Spinal Motoneuron Pools

Rather than a given muscle being controlled from a particular locus in the motor cortex, any particular muscle now is known to receive outputs from a relatively large territory in M1. Conversely, given the number of muscles, the territories from which different neighboring muscles receive M1 output necessarily overlap. This principle of converging input on each muscle from a substantial M1 territory was defined originally using electrical stimulation of the cortical surface, and subsequently confirmed with intracortical microstimulation (ICMS) [18]. Much of the projection from M1 to the spinal motoneuron pool innervating any given muscle is oligo-synaptic, mediated via connections from M1 neurons to the red nucleus, reticular formation, and/or spinal interneurons [19]. But in macaque monkeys, apes and humans a substantial fraction of M1 neurons that project to the spinal cord make mono-synaptic connections to spinal motoneurons and hence are referred to as cortico-motoneuronal (CM) cells. Recently, retrograde transneuronal labeling with rabies virus has demonstrated anatomically that the sets of CM-cells that project to different finger muscles occupy extensively overlapping cortical territories [20].

3.2 Divergence of Output from Single Cortical Neurons

Spinal motoneurons each project to muscle fibers in a single muscle. In contrast, many individual M1 neurons each project to the spinal motoneuron pools of multiple muscles. Again, though much of this divergence may be mediated through oligo-synaptic connections, the principle of divergent output from single M1 neurons to multiple muscles has been demonstrated in macaque CM-cells both anatomically and physiologically. Anatomically, single corticospinal axons filled with horseradish peroxidase have been shown to ramify and terminate in the motoneuron pools of multiple muscles [21]. Physiologically, segments of a muscle's rectified electromyographic activity (EMG), aligned at the time of spikes discharged by an M1 neuron and then averaged, often demonstrate a post-spike facilitation (or suppression) of the motoneuron pool at a short (e.g. 5–16 ms), fixed latency, consistent with conduction from the cortical neuron to the spinal cord, a monosynaptic connection to motoneurons, and then conduction from the motoneurons to the muscle. In macaque monkeys, such post-spike effects (PSEs) are found in $\sim$27 % of CM-cells selected for a relationship between their firing rates and wrist movements [22], and in $\sim$55 % of CM-cells with a corticospinal axon and a relationship to precision pinch [23]. When an M1 neuron is found to produce PSEs, identifying it as a CM-cell with monosynaptic connections to spinal motoneurons, it often produces PSEs in the EMG activity of multiple arm and hand muscles. In humans, the short-term synchronization of single motor units in different hand and finger muscles (described above in Sect. 2.2) provides further evidence of the presence of corticospinal axons that diverge to innervate the motoneuron pools of multiple muscles.

Divergence of output from single M1 neurons to multiple muscles might be viewed as a means of creating muscle synergies, that is, a mechanism for facilitating (or suppressing) the activity of a particular set of muscles in parallel. The synaptic effect from any single M1 neuron, however, constitutes only a small fraction of the tens of thousands of synaptic inputs received by any given motoneuron. To drive a muscle synergy, many M1 neurons providing the same pattern of inputs to different motoneuron pools would be needed.

Do such cortically driven muscle synergies exist? When *spike*-triggered averages from a single M1 neuron show a pattern of PSEs in certain muscles and not others, *stimulus*-triggered averages formed using single-pulses of ICMS at the same electrode location often show much larger post-stimulus effects, but with the same pattern of muscles facilitated and/or suppressed [24]. These observations suggest the presence of local groups (or perhaps more distributed networks) of M1 neurons with outputs to the same set of muscles, as might be expected to drive a muscle synergy. Furthermore, during reaching movements, applying cluster analysis to cross-correlations between the firing rates of M1 neurons and the EMG activity of multiple muscles has indicated the presence of groups of M1 neurons with similar patterns of cross-correlation with particular subsets of muscles [25]. In contrast, during individuated finger and wrist movements, groups of M1 neurons with similar patterns of activity that might drive a limited set of muscle synergies were less evident, suggesting that the control of such individuated movements may not be achieved through fixed muscle synergies [26].

The extent to which M1 functions to control muscles via a limited set of synergies remains a topic of ongoing investigation, however. M1 neurons might drive a limited set of muscle synergies which, when combined in different proportions at different times, could be used to generate a wide range of movements, from grasping to typing [27]. Alternatively, M1 neurons might drive different sets of synergies for particular tasks: one set of synergies for grasping and another for typing. A third possibility would be that M1 neurons with dissimilar sets of outputs might be activated in a wide variety of combinations to produce a continuously varying repertoire of muscle activations that vary from power grasp to precision grip to typing movements. These three possibilities, of course, are not mutually exclusive. For example, power grasping might be driven as a fixed synergy, precision pinch as a task-specific synergy, and typing movements as various combinations of M1 neurons with dissimilar outputs. Moreover, M1 neurons might act sometimes to activate, and sometimes to fractionate muscle synergies organized in the brainstem or spinal cord [28–30].

3.3 Distributed Activation in the Hand Representation

Whereas the classic view would predict that distinct patches of M1 cortex would be active for movements of different fingers, active neurons are found throughout the M1 hand representation when any single finger is moved, and similar territories are activated for movements of different fingers. In monkeys, single M1 neurons typically discharge in relation to multiple different individuated finger and wrist movements [26, 31]. Often, a given neuron discharges in relation to movements of non-adjacent digits. The distribution of neurons active during movements of particular digits gives little if any evidence of somatotopic segregation of neurons controlling different digits, or of functional groups of neurons that might represent particular movements or movement primitives. Horizontal intracortical axon collaterals that interconnect the entire M1 hand representation therefore may coordinate the necessary pattern of outputs to multiple muscles simultaneously [32]. Similarly in humans, functional magnetic resonance imaging (fMRI) shows a base of widespread activation throughout the hand representation no matter which digit is moved [33]. In humans however, subtraction of the widespread activation common to all finger movements leaves a remainder of specific activation for each digit; and this remainder shows some degree of somatotopic segregation for movements of different digits [34].

Recent studies have shown that long trains of intracortical microstimulation (ltICMS) at different cortical locations in alert monkeys can elicit movements of the upper extremity to particular end postures suggestive of canonical actions, including reaching, bringing food to the mouth, or defending the head from a blow [35]. Similarly in cats, concurrent stimulation at two cortical loci elicits a forearm movement that is the vector sum of the movement evoked by stimulating each locus separately [36]. The repetitive electrical stimulation of the cortex at a constant frequency used in these studies, however, is quite unlike the natural discharge of cortical neurons, where various neurons discharge asynchronously and at different frequencies that change differently in time. Furthermore, most of the effect of repetitive ICMS is not achieved by stimulating local neuron somata, but rather by stimulating axons that produce trans-synaptic activation of more remote neurons. Consequently ltICMS probably is sufficient to activate neuron populations in other cortical areas, the brain-stem and the spinal cord. Nevertheless, these ltICMS studies suggest the possibility that different basic movements of the extremity may be organized from different cortical locations. A considerable repertoire of movements then could be generated from the cortex by producing different combinations of such a basic set of movements. Widespread activity during a given individuated finger movement then might reflect the combination of a few basic movements (or "movement primitives"), each represented at a different cortical location.

The concept of neural representation of only a limited set of basic movements is meaningful, however, only if: (i) combining a small number of basic movements can account for a much larger set of observed movements; (ii) neuronal activity correlates more strongly with the basic movements than with other features of the observed movements; and therefore (iii) neuronal activity also shows a limited number of basic patterns. Recordings of neuronal activity suggest, however, that as the repertoire of observed movements increases, the diversity of activity observed in the neuron population increases. Studies of one-dimensional wrist flexion/extension movements, for example, identified oppositely acting M1 neurons related either to wrist flexion or to wrist extension [37]. One might have expected then, that two-dimensional reaching movements would be controlled by two such sets of neurons, each set controlling movement in one of the two dimensions. But instead, studies of two-dimensional reaching movements consistently have shown that individual M1 neurons are broadly tuned with respect to reach direction, and that the preferred directions of different M1 neurons are distributed across all sampled movement directions [38]. The same principles extend to three-dimensional reaching movements [39]. Rather than suggesting control by a few sets of neurons with similar preferred directions, reach-related M1 neurons show a wide distribution of preferred directions.

Likewise, recordings during individuated finger movements suggest a great diversity of single-neuron activity patterns. Single M1 neurons tend to be active during a number of different finger movements, often movements of non-contiguous digits, and often movements of digits in opposite directions [26, 31]. Little evidence has been found for distinct groups of M1 neurons related to particular finger movements or sub-sets of finger movements, either in the general M1 population, or in the sub-population (including CM-cells) that produces relatively direct effects on spinal motoneuron pools [40]. Even the sub-population of M1 neurons that provide input to a given muscle's motoneuron pool show diverse patterns of activity across different finger movements, in part because different M1 neurons provide additional inputs to various other motoneuron pools [41]. Rather than representing a limited number of basic patterns, M1 neurons thus appear to control individuated finger movements by contributing a wide variety of patterns of muscle facilitations and suppressions that combine to the pattern needed for a particular movement, and then recombine flexibly for a different movement. Such flexible recombination of diverse facilitative and suppressive actions would provide a repertoire even more extensive than that provided by a limited number of muscle synergies or basic movements. A challenge for future study is to understand how the combination of M1 neurons used for a given movement is activated at one time, and the combination for another movement is activated at another time.

3.4 Advantages of Distributed Organization

Classically the motor cortex was viewed as an ordered somatotopic representation of movements or muscles. The principles of convergence, divergence and distributed activation combine to suggest a revised view, illustrated for the M1 hand

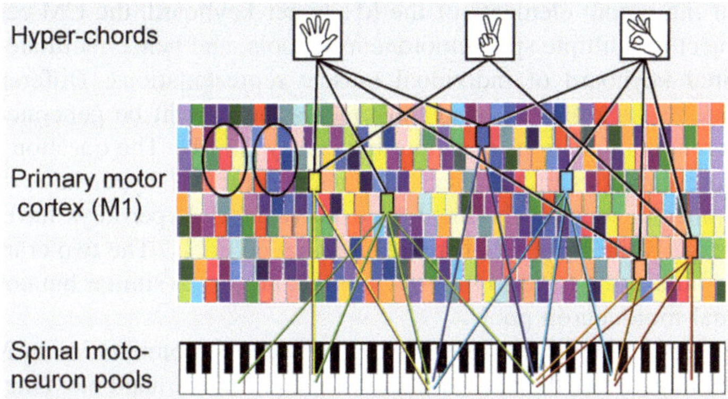

Fig. 1 The primary motor cortex hand representation as a hyper-keyboard. For many years the primary motor cortex (M1) was thought of as a well-ordered cortical representation of the body's muscles, likened to a conventional piano keyboard. The motoneuron pools of the spinal cord, however, already provide the CNS with an orderly somatotopic representation of the peripheral musculature. In contrast, the "hyper-keyboard" of M1 contains corticomotoneuronal (CM) cells with individual projections that diverge to innervate multiple motoneuron pools, depicted here as *colored rectangles*, with the connections to particular motoneuron pools shown for some CM-cells as lines descending to keys on the classic, spinal keyboard. Any given spinal motoneuron pool then receives input converging from a wide territory in M1, which overlaps extensively with the territory innervating the motoneuron pools of other hand muscles. Consequently, M1 neurons projecting to different hand muscles are intermingled, and neurons projecting to particular muscles or muscle combinations are re-represented in multiple locations. This re-represented and intermingled organization brings into close proximity a wide variety of combinations. The *two ovals* indicate two similar *red* "hyper-keys," for example, that each lie in close proximity to different combinations of other hyper-keys. In the musical analogy, a given CM-cell facilitates a "chord" of activation in a particular combination of muscles. Different hand movements then might be organized as different "hyper-chords," each of which involves activation of a particular combination of M1 neurons such that their outputs converge to facilitate different levels of activation in various spinal motoneuron pools. (Modified with permission from Fig. 10 of [17].)

representation in Fig. 1. The motoneuron pools of the spinal cord already provide the central nervous system with a detailed, highly-ordered somatotopic representation of the peripheral musculature. Another well-ordered somatotopic representation in the cortex with a similar level of detail therefore would be superfluous. Rather than being organized as a second classic keyboard, the M1 hand representation might be viewed as a "hyper-keyboard." The M1 hyper-keyboard has a wide variety of different output elements, indicated by different colors. Different outputs are re-represented at multiple, distributed locations, and intermingled in proximity to a variety of other outputs, as depicted in Fig. 1 by the two ovals marking two red hyper-keys each surrounded by different combinations of other hyper-keys. Compared to any more highly-ordered arrangement, the hyper-keyboard vastly increases the number of output combinations represented in close spatial proximity to one another, which may offer some advantages for neuronal computations.

Many of the output elements of the M1 hyper-keyboard, the CM-cells in particular, project to multiple spinal motoneuron pools, and hence facilitate "chords" on the spinal keyboard of individual muscle representations. Different muscle activity patterns for different hand movements then might be generated by activating different combinations of output hyper-keys in M1. The question of muscle synergy representations in M1 then becomes: Do the M1 hyper-keys represent a limited number of muscle chords? Or do no two M1 hyper-keys have identical outputs to exactly the same set of spinal motoneuron pools? The two orange hyper-keys highlighted at the lower right, for example, project to similar but not identical sets of spinal motoneuron pools.

Some pattern of inputs must selectively activate the combination of M1 hyper-keys needed to generate the muscle pattern for a particular hand movement. Such a combination M1 outputs can be viewed as a "hyper-chord." Neural activity from decision-making and movement-planning processes, vision and somatic sensation, all may participate in creating, selecting and driving the right hyper-chord at the right time. These inputs arrive in M1 from premotor and supplementary cortical motor areas, from the primary somatosensory cortex, and from the basal ganglia and cerebellum via the motor nuclei of the thalamus. But as yet we know relatively little about how these cortico-cortical and thalamo-cortical inputs converge on individual neurons in M1, and whether the neurons that represent hyper-chords lie in M1 or in the structures from which M1 receives input.

Why should the M1 hand representation have such a distributed, intermingled organization? As described above in Sects. 1 and 2, the various digits do not move independently of one another due to both passive and active biomechanical coupling. Consequently, in addition to muscle contractions that produce the intended motion of certain digits, M1 must drive muscle activity that stabilizes other digits. Biomechanical interactions in the hand thus require M1 to control the entire hand during any movement. Even movement of only a single digit requires the proper forces be generated by multiple muscles that act on that digit, on other digits and on the wrist. The distributed and intermingled organization of the M1 hand representation presumably is optimized for controlling the whole hand all the time.

Though less direct, the elbow and shoulder will also have biomechanical interactions with the digits, as when a reaching movement transports the hand to grasp an object. Hence M1 outputs to elbow and shoulder muscle also are intermingled to a considerable extent with outputs to finger and wrist muscles [42, 43]. As body parts become more completely biomechanically independent, such as the face and the hand, their representations in M1 become more completely segregated.

Finally, the distributed array of multiply re-represented and highly varied output elements offered by the M1 hyper-keyboard may facilitate the initial selection and generation, as well as the subsequent refinement of new combinations of outputs, i.e. new hyper-chords. Such a distributed organization provides a substrate better suited to produce both the long-term plastic changes that are sustained over many days, weeks and months, and the rapid flexibility that enables outputs to be changed within minutes. Such a substrate might provide both for normal learning and for recovery after injury.

4 Plasticity of the Motor Cortex

4.1 Motor Learning

Because no territory is entirely devoted to a particular muscle, body part or movement, the distributed organization of the motor cortex provides a substrate that can change during motor learning. For example, when squirrel monkeys practice hand movements by repeatedly extracting pieces of food from small holes, the M1 territory from which ICMS evokes movement of the distal forelimb and hand increases [44]. Similarly in humans, repeated practice of a thumb movement in a particular direction changes the direction of movement evoked by transcranial magnetic stimulation (TMS) toward the practiced direction [45]. Practicing particular movements thus leads to plastic changes which increase the M1 output representation of the practiced movement(s). ICMS delivered conditionally in M1 also can change M1 output. When ICMS at a wrist extensor site was triggered by spikes from an M1 neuron that discharged during wrist flexion, the effect of ICMS at that site changed progressively from wrist extension to wrist flexion [46]. The net output evoked by stimulation at a given M1 site thus is not fixed, but can be changed depending on recent conditioning and practice.

The activity of M1 neurons also changes during motor learning [47]. As monkeys adapt to reaching movements in a novel force field, the directional tuning of some M1 neurons changes as the monkey practices in the novel field, and then de-adapts to the prior tuning when the force field is turned off. Other M1 neurons, however, show no change in tuning as the monkey adapts to the novel field, but then do change when the force field is turned off. These observations suggest that the activity of individual neurons is being adjusted continually, such that the total M1 population can best produce the required movements [48].

4.2 Reorganization After Injury

Plastic reorganization of M1 occurs after many types of central or peripheral nervous system injury. In non-human primates, for example, if the M1 representation of the distal forelimb is infarcted experimentally, and the animal afterwards is required to obtain food using the impaired hand, territory that had provided output primarily to proximal upper extremity muscles comes to provide more output to the distal musculature [49]. Similar flexible reorganization of M1 output may underlie the improvement in functional recovery after stroke obtained by requiring patients to make use of their impaired hand.

Studies of amputees have provided valuable insights into the extent to which cortical output can be plastic while also retaining a permanent native organization. After amputation of an extremity, the M1 region that previously had represented the distal, amputated body part comes to provide output to remaining, proximal body parts. Such plastic reorganization, in which representation of the proximal stump muscles "expands" into cortical territory that previously had primarily represented the distal amputated part, has been observed in human upper extremity amputees using either TMS [50] or fMRI [51], and in non-human primates using ICMS [52].

Amputation can be simulated, from the point of view of the nervous system, by nerve block. For example, inflation of a blood pressure cuff above arterial pressure produces ischemia that leads to temporary block of both motor and sensory conduction in nerves distal to the cuff. The distal extremity is then both paralyzed and numb, as when one's arm "falls asleep." Within minutes, TMS over M1 then elicits larger EMG responses in the proximal musculature. Although this might suggest that territory previously devoted to distal musculature now provides output to the proximal musculature, intra-neural recordings have revealed that the cortical output to muscles both proximal and distal to the block is increased [53]. The expansion of output to proximal muscles into the territory that normally represents distal muscles thus does not require elimination of distal representation in the cortex.

Similarly in humans with actual amputations, representation of movements of the amputated body part remains resident to some extent in the CNS. Many amputees have the impression that they still can move their phantom limb. TMS of M1 can elicit the perception that the phantom has moved [54]. And EMG recordings from stump muscles during volitional "movement" of the distal phantom show patterns of activity that are distinct from those used to move the proximal stump [55]. All these observations suggest that representations of distal movements remain in the cortex long after amputation of the extremity.

Furthermore, when the truncated ends of nerves (radial, median and ulnar) that had innervated the distal musculature are surgically repositioned to innervate remaining proximal musculature, volitional movement of the distal phantom produces specific patterns of EMG activity in the re-innervated muscle [56], and touch of the overlying skin elicits the perception that the distal phantom has been touched in a nerve-specific location [57]. Hence long after amputation, the motor output to the amputated muscles continues to be sent down the severed nerves, and activation of the sensory fibers elicits sensations of touch in the phantom.

Some human amputees have received hand transplants. Within a few months of the transplantation, fMRI studies indicate that the expanded representations of proximal muscles have receded to their pre-amputation size, and representation of the hand in M1 has re-emerged [51]. So although the cortex reorganizes to a considerable extent after amputation, native representation of the amputated part may never be completely eliminated.

5 Rapid Flexibility in Cortical Control of the Hand

5.1 Ability to Perform Novel Movements

The motor cortex enables normal subjects to perform an extremely wide repertoire of dexterous hand and finger movements. While subcortical descending pathways such as the reticulospinal system may contribute to basic opening and closing of the hand, the variation in hand shapes used to grasp different objects (such as power grip versus precision pinch), and especially then to manipulate them with dexterity, requires control from M1. Beyond grasping objects of different shapes, the individuated finger movements used for typing or playing a musical instrument, as well as for haptic manipulation of an object hidden in one's pocket or purse, depend heavily on control from M1.

The flexibility of control from M1 enables normal human subjects to perform a wide variety of what might otherwise be considered 'unnatural' finger movements. The complex finger gestures used to communicate "OK" or to spell words in sign language, have no clear relationship to manipulative use of the hand. While these examples might be considered highly practiced hand movements, humans also can rapidly produce novel finger movements that have neither manipulative utility nor communicative purpose. Perhaps you have never flexed your index and ring fingers, and held their tips together in front of your palm by flexing your thumb, while simultaneously extending your middle and little fingers, but you can do it right now if you try.

Humans also can rapidly learn to make use of such novel movements. When required to use the 22 degrees of freedom in the hand to move a cursor in an arbitrarily defined two-dimensional space, normal human subjects learned quickly to move the cursor linearly in the two-dimensional space, although this required the fingers to move through postures uncommon in daily hand movements [58]. Similarly, human subjects can rapidly learn to contract muscles in arbitrary combinations to control a cursor [59]. Interestingly, performance is better with hand muscles than with more proximal arm muscles, suggesting that control of the hand may be particularly flexible. Such flexibility enables humans to adapt their hand and finger movements to an extraordinarily wide variety of task constraints.

5.2 Dissociation from Movement

The extreme consequence of flexibility in motor cortical output is the ability to dissociate cortical neuron activity from particular movements per se. In seminal studies, Fetz and colleagues found that monkeys could be conditioned in single sessions to modulate the firing rate of an M1 neuron voluntarily, dissociating the activity of the conditioned M1 neuron from that of the muscles with which it correlated during normal movements [60, 61]. These observations recently have been extended to show that such voluntary modulation can be used by the monkey for closed-loop control of wrist muscles, even when the controlling neuron(s) had no prior relationship to wrist movement [62]. Furthermore, the same M1 neurons that have relatively direct, even monosynaptic, connections to particular spinal motoneuron pools recently have been shown to have functional throughput contributing to muscle contraction during some motor behaviors, but no effective throughput during other behaviors [63]. So although M1 neurons make a necessary contribution to the control of dexterous hand movements, especially individuated finger movements, under some circumstances even some CM-cells can be dissociated rapidly from the muscle contractions that produce movements.

Such dissociation of M1 activity from movement also becomes apparent in closed-loop control of brain–machine interfaces (BMIs). When non-human primates practice switching from controlling movement of a cursor with movement of their native limb (e.g. holding a joystick) to controlling the cursor through a BMI by activating a recorded sub-population of M1 neurons, many of the recorded M1 neurons change their directional tuning over several days of practice [64]. Eventually, the monkey may stop making overt arm movements, and even stop contracting muscles. Meanwhile, the cortical neurons continue to be active to control the cursor in the absence of detectable movement or muscle contraction. Within a single session, and then across sessions, the directional activity of neurons adapts to improve direct, closed-loop control of the cursor, such that after several days of practice a small population of neurons can switch readily between control of the native limb and control of the BMI [65].

Dissociation of M1 neuron activity from movement of the native limb might seem to be a bizarre consequence of the unnatural paradigm provided by BMIs. Yet for many years it has been known that certain M1 neurons, including many that send axons to the spinal cord, discharge for hundreds of milliseconds during instructed-delay periods as the subject waits to execute an instructed movement without actually moving. Given that the spikes of CM-cells influence EMG activity with latencies in the range of 5–20 ms, such instructed-delay period activity constitutes a form of temporal dissociation of M1 neuron activity from movement. In behaviors that separate target direction from movement direction, other M1 neurons dissociate from the limb movement per se and discharge instead in relation to the target direction [66]. Furthermore, some corticospinal neurons recently have been shown to have mirror properties [67]. Such neurons discharge similarly whether the subject executes a particular hand movement or observes another monkey (or human) performing a similar hand movement. Mirror neurons thus appear to discharge during movements the subject performs with the native limb, and discharge as well when the subject visually observes a similar movement performed by another primate, during which the subject's own limb is quiescent. Dissociated activation of M1 while a subject withholds movement, including increased excitability of corticospinal neurons, presumably underlies the fMRI activation evident when human subjects imagine performing particular movements [68], something humans do quite naturally.

6 Conclusions

Cortical control of the hand thus is constrained by both peripheral and central factors. Many of the constraints present in the human hand—including passive coupling by soft tissues and tendons, incomplete compartmentalization of muscles, and short-term synchronization between motor units in different muscles—can be viewed as the remainder of an evolutionary process that has favored independence of the fingers, but has not yet achieved complete independence. To the extent that many fundamental uses of the hand for basic grasping (as can be observed in rodents and cats) do not actually require independent finger movements, cortical control may take advantage of representations of limited sets of movement primitives (involving the simultaneous motion of multiple joints in fixed proportion) and/or muscle synergies (involving the simultaneous contraction of multiple muscles in fixed proportion), which might be organized largely in the evolutionarily older centers including the spinal grey matter, the brainstem reticular formation and the red nucleus.

To achieve more dexterous control, however, the motor cortex in monkeys, and especially in humans, appears to have evolved the capacity to express even greater flexibility than that available from the numerous possible combinations of movement primitives and muscle synergies. The ability to influence spinal motoneurons directly in a great diversity of combinations may enable the motor cortex to sculpt basic synergies so as to individuate finger movements. This flexibility is achieved in part by enabling the discharge of a given M1 neuron to participate in some active movements but not in other similar movements, and to discharge in the absence of movements at other times. Understanding the factors that constrain such flexibility remains a challenge for future investigation.

This same flexibility also poses a challenge for the development of practical brain-machine interfaces. If the activity of cortical neurons were invariably related to particular kinematic or dynamic features of intended movement, whether the movement of the native limb or not, then the intended movement could be decoded from the activity of those neurons. But if the activity of M1 neurons changes depending on the movement context—native limb versus cursor; actual limb movement versus observed or imagined—then decoding M1 activity to control movement of a prosthetic limb through a BMI may be complicated by rapid changes in the discharge properties of M1 neurons. The subject may need to learn to produce the patterns of activity required to drive the BMI. Once again, understanding the factors that are responsible for the flexibility of M1 neuron activity, as well as the factors that constrain such flexibility, remains a challenge for future investigation.

Acknowledgments The author thanks Marsha Hayles for editorial comments. This work was supported in part by NINDS R01s NS065902 and NS079664, as well as NIBIB R01 EB010100.

References

1. J.N. Ingram et al., The statistics of natural hand movements. Exp. Brain Res. **188**, 223–236 (2008)
2. M.H. Schieber, M. Santello, Hand function: peripheral and central constraints on performance. J. Appl. Physiol. **96**, 2293–2300 (2004)
3. N.B. Macfarlane, M.S. Graziano, Diversity of grip in Macaca mulatta. Exp. Brain Res. **197**, 255–268 (2009)
4. J.R. Napier, The prehensile movements of the human hand. J. Bone Joint Surg. Br. **38**, 902–913 (1956)
5. M.H. Schieber, Individuated finger movements of rhesus monkeys: a means of quantifying the independence of the digits. J. Neurophysiol. **65**, 1381–1391 (1991)
6. C.E. Lang, M.H. Schieber, Human finger independence: limitations due to passive mechanical coupling versus active neuromuscular control. J. Neurophysiol. **92**, 2802–2810 (2004)
7. M.H. Schieber et al., Tension distribution to the five digits of the hand by neuromuscular compartments in the macaque flexor digitorum profundus. J. Neurosci. **21**, 2150–2158 (2001)
8. M.H. Schieber et al., Tension distribution of single motor units in multitendoned muscles: comparison of a homologous digit muscle in cats and monkeys. J. Neurosci. **17**, 1734–1747 (1997)
9. S.L. Kilbreath et al., Distribution of the forces produced by motor unit activity in the human flexor digitorum profundus. J. Physiol. **543**, 289–296 (2002)
10. D.A. Keen, A.J. Fuglevand, Role of intertendinous connections in distribution of force in the human extensor digitorum muscle. Muscle Nerve **28**, 614–622 (2003)
11. K.T. Reilly et al., Short-term synchronization between motor units in different functional subdivisions of the human flexor digitorum profundus muscle. J. Neurophysiol. **92**, 734–742 (2004)
12. D.A. Keen, A.J. Fuglevand, Common input to motor neurons innervating the same and different compartments of the human extensor digitorum muscle. J. Neurophysiol. **91**, 57–62 (2004)
13. A.K. Datta et al., Central nervous pathways underlying synchronization of human motor unit firing studied during voluntary contractions. J. Physiol. **432**, 401–425 (1991)
14. S.F. Farmer et al., Changes in motor unit synchronization following central nervous lesions in man. J. Physiol. **463**, 83–105 (1993)
15. M.H. Schieber, Muscular production of individuated finger movements: the roles of extrinsic finger muscles. J. Neurosci. **15**, 284–297 (1995)
16. K.T. Reilly, M.H. Schieber, Incomplete functional subdivision of the human multitendoned finger muscle flexor digitorum profundus: an electromyographic study. J. Neurophysiol. **90**, 2560–2570 (2003)
17. M.H. Schieber, Constraints on somatotopic organization in the primary motor cortex. J. Neurophysiol. **86**, 2125–2143 (2001)
18. P. Andersen et al., Mapping by microstimulation of overlapping projections from area 4 to motor units of the baboon's hand. Proc. R. Soc. Lond. B Biol. Sci. **188**, 31–36 (1975)
19. H.G. Kuypers, Some aspects of the organization of the output of the motor cortex. Ciba Found. Symp. **132**, 63–82 (1987)
20. J.A. Rathelot, P.L. Strick, Muscle representation in the macaque motor cortex: an anatomical perspective. Proc. Natl. Acad. Sci. USA **103**, 8257–8262 (2006)
21. Y. Shinoda et al., Divergent projection of individual corticospinal axons to motoneurons of multiple muscles in the monkey. Neurosci. Lett. **23**, 7–12 (1981)
22. E.E. Fetz, P.D. Cheney, Postspike facilitation of forelimb muscle activity by primate corticomotoneuronal cells. J. Neurophysiol. **44**, 751–772 (1980)
23. R.N. Lemon et al., Corticospinal facilitation of hand muscles during voluntary movement in the conscious monkey. J. Physiol. **381**, 497–527 (1986)

24. P.D. Cheney, E.E. Fetz, Comparable patterns of muscle facilitation evoked by individual corticomotoneuronal (CM) cells and by single intracortical microstimuli in primates: evidence for functional groups of CM cells. J. Neurophysiol. **53**, 786–804 (1985)
25. R.N. Holdefer, L.E. Miller, Primary motor cortical neurons encode functional muscle synergies. Exp. Brain Res. **146**, 233–243 (2002)
26. A.V. Poliakov, M.H. Schieber, Limited functional grouping of neurons in the motor cortex hand area during individuated finger movements: a cluster analysis. J. Neurophysiol. **82**, 3488–3505 (1999)
27. S.A. Overduin et al., Modulation of muscle synergy recruitment in primate grasping. J. Neurosci. **28**, 880–892 (2008)
28. C.B. Hart, S.F. Giszter, A neural basis for motor primitives in the spinal cord. J. Neurosci. **30**, 1322–1336 (2010)
29. A.G. Davidson et al., Bilateral spike-triggered average effects in arm and shoulder muscles from the monkey pontomedullary reticular formation. J. Neurosci. **27**, 8053–8058 (2007)
30. J. Roh et al., Modules in the brain stem and spinal cord underlying motor behaviors. J. Neurophysiol. **106**, 1363–1378 (2011)
31. M.H. Schieber, L.S. Hibbard, How somatotopic is the motor cortex hand area? Science **261**, 489–492 (1993)
32. G.W. Huntley, E.G. Jones, Relationship of intrinsic connections to forelimb movement representations in monkey motor cortex: a correlative anatomic and physiological study. J. Neurophysiol. **66**, 390–413 (1991)
33. J.N. Sanes et al., Shared neural substrates controlling hand movements in human motor cortex. Science **268**, 1775–1777 (1995)
34. P. Dechent, J. Frahm, Functional somatotopy of finger representations in human primary motor cortex. Hum. Brain Mapp. **18**, 272–283 (2003)
35. M.S. Graziano et al., Complex movements evoked by microstimulation of precentral cortex. Neuron **34**, 841–851 (2002)
36. C. Ethier et al., Linear summation of cat motor cortex outputs. J. Neurosci. **26**, 5574–5581 (2006)
37. E.V. Evarts, Relation of pyramidal tract activity to force exerted during voluntary movement. J. Neurophysiol. **31**, 14–27 (1968)
38. A.P. Georgopoulos et al., On the relations between the direction of two-dimensional arm movements and cell discharge in primate motor cortex. J. Neurosci. **2**, 1527–1537 (1982)
39. A.B. Schwartz et al., Primate motor cortex and free arm movements to visual targets in three-dimensional space. I. Relations between single cell discharge and direction of movement. J. Neurosci. **8**, 2913–2927 (1988)
40. M.H. Schieber, G. Rivlis, A spectrum from pure post-spike effects to synchrony effects in spike-triggered averages of electromyographic activity during skilled finger movements. J. Neurophysiol. **94**, 3325–3341 (2005)
41. M.H. Schieber, G. Rivlis, Partial reconstruction of muscle activity from a pruned network of diverse motor cortex neurons. J. Neurophysiol. **97**, 70–82 (2007)
42. M.C. Park et al., Consistent features in the forelimb representation of primary motor cortex in *Rhesus macaques*. J. Neurosci. **21**, 2784–2792 (2001)
43. H.C. Kwan et al., Spatial organization of precentral cortex in awake primates. II. Motor outputs. J. Neurophysiol. **41**, 1120–1131 (1978)
44. R.J. Nudo et al., Use-dependent alterations of movement representations in primary motor cortex of adult squirrel monkeys. J. Neurosci. **16**, 785–807 (1996)
45. J. Classen et al., Rapid plasticity of human cortical movement representation induced by practice. J. Neurophysiol. **79**, 1117–1123 (1998)
46. A. Jackson et al., Long-term motor cortex plasticity induced by an electronic neural implant. Nature **444**, 56–60 (2006)
47. C.S.R. Li et al., Neuronal correlates of motor performance and motor learning in the primary motor cortex of monkeys adapting to an external force field. Neuron **30**, 593–607 (2001)

48. U. Rokni et al., Motor learning with unstable neural representations. Neuron **54**, 653–666 (2007)
49. R.J. Nudo et al., Neural substrates for the effects of rehabilitative training on motor recovery after ischemic infarct. Science **272**, 1791–1794 (1996)
50. L.G. Cohen et al., Motor reorganization after upper limb amputation in man. A study with focal magnetic stimulation. Brain **114**, 615–627 (1991)
51. P. Giraux et al., Cortical reorganization in motor cortex after graft of both hands. Nat. Neurosci. **4**, 691–692 (2001)
52. H.X. Qi et al., Reorganization of primary motor cortex in adult macaque monkeys with long-standing amputations. J. Neurophysiol. **84**, 2133–2147 (2000)
53. P.A. McNulty et al., Cortically evoked neural volleys to the human hand are increased during ischaemic block of the forearm. J. Physiol. **538**, 279–288 (2002)
54. C. Mercier et al., Mapping phantom movement representations in the motor cortex of amputees. Brain **129**, 2202–2210 (2006)
55. K.T. Reilly et al., Persistent hand motor commands in the amputees' brain. Brain **129**, 2211–2223 (2006)
56. T.A. Kuiken et al., Targeted muscle reinnervation for real-time myoelectric control of multifunction artificial arms. JAMA **301**, 619–628 (2009)
57. P.D. Marasco et al., Sensory capacity of reinnervated skin after redirection of amputated upper limb nerves to the chest. Brain **132**, 1441–1448 (2009)
58. K.M. Mosier et al., Remapping hand movements in a novel geometrical environment. J. Neurophysiol. **94**, 4362–4372 (2005)
59. S.M. Radhakrishnan et al., Learning a novel myoelectric-controlled interface task. J. Neurophysiol. **100**, 2397–2408 (2008)
60. E.E. Fetz, D.V. Finocchio, Operant conditioning of isolated activity in specific muscles and precentral cells. Brain Res. **40**, 19–23 (1972)
61. E.E. Fetz, M.A. Baker, Operantly conditioned patterns on precentral unit activity and correlated responses in adjacent cells and contralateral muscles. J. Neurophysiol. **36**, 179–204 (1973)
62. C.T. Moritz et al., Direct control of paralysed muscles by cortical neurons. Nature **456**, 639–642 (2008)
63. A.G. Davidson et al., Rapid changes in throughput from single motor cortex neurons to muscle activity. Science **318**, 1934–1937 (2007)
64. D.M. Taylor et al., Direct cortical control of 3D neuroprosthetic devices. Science **296**, 1829–1832 (2002)
65. K. Ganguly, J.M. Carmena, Emergence of a stable cortical map for neuroprosthetic control. PLoS Biol. **7**, e1000153 (2009)
66. S. Kakei et al., Muscle and movement representations in the primary motor cortex. Science **285**, 2136–2139 (1999)
67. G. Vigneswaran et al., M1 corticospinal mirror neurons and their role in movement suppression during action observation. Current Biology **23**, 236–243 (2013)
68. J. Munzert et al., Cognitive motor processes: the role of motor imagery in the study of motor representations. Brain Res. Rev. **60**, 306–326 (2009)

Author Biography

Marc H. Schieber was born in St. Louis, MO, in 1953. He received his A.B. (in biology) in 1974, and M.D./Ph.D. (in Neural Sciences) in 1982, all from Washington University in St. Louis.

He currently is Professor of Neurology and of Neurobiology at the University of Rochester, and Attending Neurologist on the Brain Injury Rehabilitation Unit at Unity Health, Rochester, NY. His research focuses on how the nervous system controls muscles to perform dexterous finger movements.

Dr. Schieber is a member of the Society for Neural Control of Movement and the Society for Neuroscience. He has received an NINDS Javits Investigator Merit Award and has served as Chair of the NIH Sensorimotor Integration Study Section.

Chapter 2
Synergistic Control of Hand Muscles Through Common Neural Input

Marco Santello

Abstract Skilled grasping and manipulation rely on spatial and temporal coordination of multiple hand muscles. This chapter describes the phenomenon of common neural input to hand muscles as one of the mechanisms through which the Central Nervous system might coordinate the neural activation of groups of hand muscles acting on a single or multiple digits. The heterogeneous distribution of common input to intrinsic and extrinsic hand muscles is discussed in relation to its functional role for the coordination of hand muscles.

Keywords Synchrony · Coherence · Motor units · EMG

1 Introduction

The objective of this chapter is to provide an overview of how the Central Nervous System (CNS) controls hand muscles in tasks that require dexterous digit force control. The chapter focuses on studies that have characterized the neural control of hand muscles through the application of time and frequency domain analyses of electromyographical (EMG) signals. By determining and quantifying neural inputs that are common to concurrently active motor units of hand muscles, this work can provide significant insights into how the CNS coordinates the activity of multiple

This publication was made possible by grant number 2R01 AR47301 from the National Institute of Arthritis, Musculoskeletal and Skin Diseases (NIAMSD) at the National Institutes of Health (NIH). Its contents are solely the responsibility of the authors and do not necessarily represent the official views of NIAMSD or NIH.

M. Santello (✉)
School of Biological and Health Systems Engineering, Arizona State University, Tempe, AZ 85287, USA
e-mail: marco.santello@asu.edu

R. Balasubramanian and V. J. Santos (eds.), *The Human Hand as an Inspiration for Robot Hand Development*, Springer Tracts in Advanced Robotics 95,
DOI: 10.1007/978-3-319-03017-3_2, © Springer International Publishing Switzerland 2014

muscles during skilled grasping and manipulation. Therefore, the chapter discusses how experimental evidence from research on common neural input could be used to improve our understanding of muscle synergies and their implications for neural control of the hand.

The chapter starts by introducing the complex interactions between descending and ascending inputs to spinal motor nuclei of hand muscles (Sect. 2). A review of the most commonly used techniques to quantify common neural input follows, together with a brief review of the literature on their applications to characterize correlated inputs to motor unit pairs and populations (Sect. 3). The phenomenon of heterogeneous distribution of common neural input across intrinsic and extrinsic hand muscles is then discussed in relation to recent work on motor unit coherence (Sect. 4). Lastly, common neural input is discussed within the theoretical framework of synergies (Sect. 5), followed by a discussion of open questions and directions for future research (Sect. 6).

2 Neural Control of Hand Muscles

2.1 Inputs to Motor Units of Hand Muscles

The spatial and temporal convergence of several inputs onto alpha motor neurons dictates its final output to skeletal muscle fibers, hence motor neurons are referred to as the 'final common path' [1]. Descending inputs are mediated by highly divergent cortical inputs (corticospinal tract), as well as rubrospinal and reticulospinal tracts, whereas peripheral inputs are mediated by networks integrating signals from individual sensory modalities (muscle spindles, Golgi tendon organs, tactile afferents, joint receptors). Spinal interneurons contribute to the processing of descending and ascending inputs. The divergence of descending inputs to alpha motor neurons of hand muscles is the subject of ongoing investigation (for more details see Chap. 1). It should be emphasized that the divergence of inputs to several hand muscles does not seem to be a characteristic that is unique to descending inputs. Specifically, cutaneous reflex EMG responses elicited by stimulation of digital nerve branches of one digit occur not only at the muscles acting on the stimulated digit, but also at muscles innervating other digits (e.g., [2–4] unpublished observations). The following section discusses the approaches investigators have used to characterize and understand the functional significance of the organization of inputs onto the spinal motor nuclei of hand muscles.

2.2 Experimental Approaches

Inferences about neural control of hand muscles can be performed through a variety of experimental approaches, such as invasive and non-invasive recordings of cortical neuronal activity, transcranial magnetic stimulation, intraneural recording or stimulation of peripheral nerves, and recording of electrical activity of hand muscles. This chapter focuses on inferences that have been made through the analysis of the EMG signals of hand muscles, and in particular on the quantification of temporal relationships between the activity of individual units (single motor unit recordings) and populations of motor units (interference EMG).

3 Common Neural Input: Methodological Considerations

3.1 *Time and Frequency Domain Measures of Correlated Inputs*

This section describes two techniques that have been widely used to quantify correlated inputs to motor units. One technique, *motor-unit synchrony*, quantifies synchronous activity of motor unit in the time domain, whereas the second technique, *motor-unit coherence*, quantifies correlations of motor unit activity in the frequency domain. Common neural input to motor units has also been defined by correlations between the firing rates of individual motor units, i.e., *common drive* [5–9]. For a discussion on methodological aspects of the common drive technique, the reader is referred to a simulation study by Lowery and Erim [10].

3.2 *Motor-Unit Synchrony*

The correlation for continuous functions $x(t)$ and $y(t)$ is computed by the cross-correlation function

$$R_{YX}(\tau) = (1/T) \int y(t)x(t-\tau)dt \tag{1}$$

where τ is the time interval and T is the period of integration. The special case of cross-correlation between spike trains uses computations based on peri-spike histograms [11, 12]. Motor-unit synchrony is defined as greater than chance tendency for concurrently active motor units to discharge at short time intervals from each other, e.g., [13] and has been used as an indirect measure of common synaptic input across motor neurons. More specifically, motor-unit synchrony has been attributed to excitatory or inhibitory postsynaptic potentials that arise from branched axons of common presynaptic neurons ([13–15]; Fig. 1). Motor unit synchrony can be subdivided into short- and long-term synchrony based on the width of the cross-correlogram peak, the latter type of synchrony being defined by time lags between motor unit discharges of larger than a few milliseconds [16]. Broader cross-correlogram peaks are thought to reflect synchrony due to separate presynaptic inputs onto the motor neurons (top, Fig. 1).

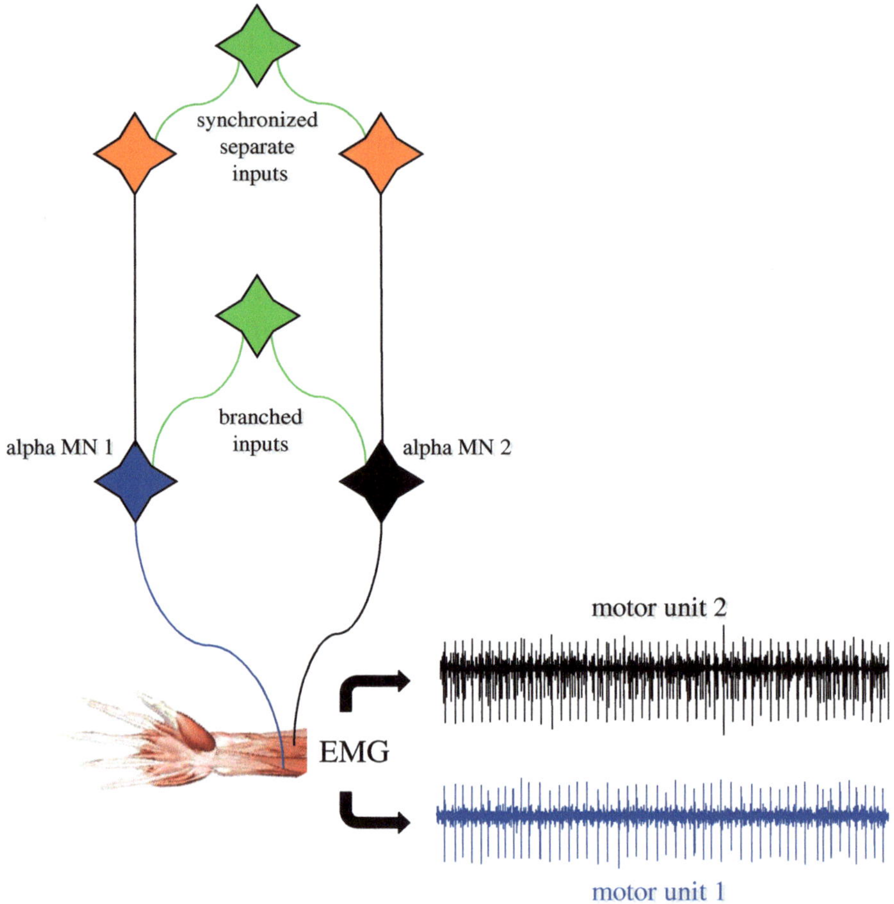

Fig. 1 Schematic description of common neural input through synchronized separate excitatory inputs and branched inputs to alpha motor neurons innervating two hand muscles

The strength of motor-unit synchrony is generally quantified by computing the ratio between the number of motor unit discharges within the central peak of the cross-correlogram histogram and that outside of the peak, e.g., time intervals between discharges of concurrently active motor units that are uniformly distributed and, therefore, not associated with synchronous discharges of the motor unit pair (Fig. 2a). The quantification of motor-unit synchrony focuses on well-defined temporal criteria. In contrast, frequency domain measures capture other features of relations that might exist in the discharges of two concurrently active motor units (see Sect. 3.3). Two examples of some of the most widely used techniques to quantify the strength of motor-unit synchrony are *common input strength* (CIS; [13]) and *k* [17]. The strength of motor-unit synchrony depends not only on

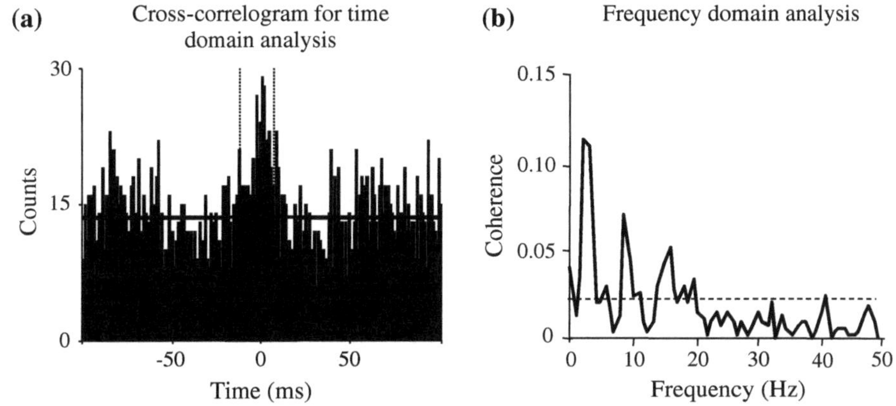

Fig. 2 **a** Cross-correlogram histogram obtained from cross-correlation of two motor unit spike trains. The *horizontal line* denotes mean counts outside of the cross-correlogram peak. **b** Coherence between spike trains of two motor units. The *horizontal line* denotes threshold for statistical significance

the number of shared inputs but also on complex interactions among several other factors such as background excitation and inhibition as well as the intrinsic properties of motor neurons [18–20].

3.3 Motor-Unit Coherence

Correlations in the discharge of a motor unit pair can occur at time intervals that might be longer than those defined by the quantification of short- or long-term motor-unit synchrony. These correlations are revealed by peaks and troughs occurring at consistent time intervals in the cross-correlogram and denote periodicities of correlated motor unit firing [21, 22]). Coherence can detect the existence of such periodicities by quantifying the linear relation between two motor unit spike trains in the frequency domain. Therefore, motor-unit coherence at a given frequency denotes correlated rhythmic activity between two motor units, which is thought to indicate the existence of a common periodic synaptic input ([23, 24]).

Coherence is computed as the modulus of cross-spectrum (f_{xy}) of two motor unit spike trains squared and normalized by the product of the autospectrum (f_{xx}, f_{yy}) of each spike train at each frequency (λ) in the frequency band of interest:

$$|R_{xy}(\lambda)|^2 = |f_{xy}(\lambda)|^2 / (f_{xx}(\lambda) f_{yy}(\lambda)) \tag{2}$$

After determining the frequency at which significant coherence occurs [25], the strength of coherence, bounded between 0 and 1, can be interpreted similarly to the strength of the coefficient of determination in linear statistics (Fig. 2b).

Most commonly, motor unit coherence studies focus on the peak value of coherence or the integral of coherence within specific frequency intervals (typically 0–55 Hz). Similar to motor-unit synchrony, motor-unit coherence reflects the efficacy with which a multitude of common inputs (see Sect. 2) of different strength and timing can elicit action potentials of the motor unit pair under investigation. Therefore, one may equate the strength of significant coherence and the frequency at which it occurs with the number of action potentials across two motor units that occur at similar times relative to the common input signal and the variability in the timing of discharges around the common input frequency. Coherence can also be computed from signals comprised of populations of motor units, e.g., interference EMG recorded through surface or intramuscular electrodes. In these circumstances, the strength of coherence will also depend on how many of the recorded motor units receive common neural input [26].

Besides assessing whether significant coherence may or may not occur between two motor units, often investigators are also interested in determining the frequency bands at which significant coherence occurs. This approach has been informative due to the fact that modulation of within- or across-muscle coherence to task conditions is often confined to specific frequency ranges (see Sect. 4.4 for more details). Furthermore, there is evidence to suggest that the sources of coherence differ across frequency bands. For example, healthy subjects exhibit significant coherence in both 1–12 and 16–32 Hz frequency bands. In contrast, coherence is markedly reduced in patients affected by cortical strokes, but only in the 16–32 Hz frequency band [22]. These authors suggested that the 1–12 Hz frequency may therefore reflect subcortical common inputs to hand muscles, as opposed to the 16–32 drive for which intact cortical inputs appear to be necessary.

3.4 Relation Between Motor-Unit Synchrony and Coherence

As indicated above, motor-unit synchrony and coherence capture different aspects of common neural input. Linear regression analysis has often been used to assess the extent of overlap in the information provided by motor-unit synchrony and coherence (e.g., [27]). The existence of significant linear correlations between these two measures of common neural input are interpreted as reflective of common periodic input to the motor neurons through branched presynaptic pathways [22, 28]. Overall, the literature suggests that linear correlations between across-muscle motor-unit synchrony and coherence are very weak and vary broadly across muscle pairs. This suggests that common neural inputs to hand muscles are delivered through different combinations of indirect and direct branched presynaptic pathways, as well as, independent pathways (weak motor-unit synchrony and coherence; [27]). The finding that motor units can exhibit significant motor-unit coherence and no motor-unit synchrony (e.g., [29, 30]) emphasizes the need to use both measures, when possible, to further delineate the nature of divergent inputs to motor units. For further discussion on the relation between time- and frequency-domain measures of common neural input to motor units, the reader is referred to Semmler et al. [31, 32] and computational modeling by Enoka and colleagues [19, 20, 33] and Lowery et al. [34]. Additional methodological issues, such as assessing EMG–EMG coherence through recording EMG activity of motor unit pairs versus motor unit populations, or the effect of EMG signal rectification on coherence estimation, are discussed in [26].

4 Common Neural Input to Hand Muscles: Experimental Evidence

4.1 Functional Considerations

The extent to which hand muscles in humans receive common input has been extensively studied by recording EMG from single motor units. These studies can be broadly classified based on whether they target correlated neural activity of motor units belonging to the same or different muscles, i.e., *within-* or *across-muscle* common neural input. Early studies of within-muscle motor unit synchrony of hand muscles were performed to quantify the strength of divergent inputs to motor units innervating a single hand muscle (often the first dorsal interosseus, FDI; [35]) or for comparing the strength of motor unit synchrony among hand muscles [36] as well as other upper or lower limb muscles [6, 37]. Studies of correlated neural activity across motor units belonging to different hand muscles often aimed at comparing the strength of synchrony of motor units belonging to the same versus different muscles. An important observation is that motor-unit synchrony across hand muscles tends to be weaker than within-muscle synchrony ([37–40]; but see Sect. 4.2). Similar considerations and experimental questions motivated studies of coherence between motor units within and across hand muscles (see Sect. 4.3).

More recently, however, the objective of quantifying correlated neural activity across hand muscles has been pursued to better understand constraints that might contribute to the coordination of digit movement or forces. Specifically, tasks involving individuated digit motion or forces, as well as reaching to grasp object with different shapes, have described the tendency for joint excursions and forces across digits to be correlated (this is discussed in more detail in Sect. 6). Therefore, correlated neural input across motor units of hand muscles has been studied as a potential mechanism that might contribute to constrain neural drive to hand muscles, hence contributing to the synergistic coupling of digit actions.

An important distinction, however, should be made between the contexts in which common neural input affects the hand behavior. Specifically, tasks that require independent action of one digit (e.g., index finger) would be potentially penalized by the existence of common neural input to muscles acting on the ‘instructed’ digit *as well as* the digits that are required to remain stationary (e.g., thumb, middle, ring, and little finger). Thus, it has been suggested that common neural input across finger muscles is one of the factors responsible for the limited extent to which fingers can move or generate forces independently [41, 42], therefore causing 'unwanted' coupling among the digits [43, 44]. Besides peripheral constraints such as passive linkages among muscles and tendons, central limitations contribute to limit the degree to which fingers can be controlled independently. However, superficial finger flexors appear to be under more independent volitional control than deep flexors [45] (for recent studies on the issue of independent finger control, the reader is referred to [46] and [47]; the interaction between central and peripheral factors in hand control is discussed in more detail in Chap. 1).

It has also been suggested that such neural constraints, limiting the independent action of the fingers, might be considered as a desirable feature in contexts where coupling of digit actions is required by the task, e.g., to prevent object slip while holding an object against gravity [48]. Specifically, a tendency for correlated inputs to motor units acting on different hand muscles or compartments of a multi-tendoned muscle, e.g. flexor digitorum profundus (FDP), would enhance the temporal coupling of digit forces [49, 50]. These considerations motivated studies of common neural input across hand muscles to address the following main questions: *What is the strength of correlated neural inputs across motor units of hand muscles? Does the strength of common neural input vary across hand muscles? Is the strength of correlated neural input task-dependent?*

4.2 Strength and Distribution of Across-Muscle Motor-Unit Synchrony

Motor unit studies of force production and object hold tasks have reported moderate to strong common neural input across thumb and extrinsic finger muscles and compartments, as well as across compartments of finger flexors and extensors [30, 40, 42, 48, 51]. Interestingly, the strength of common neural input is not uniformly distributed across hand muscles or muscle compartments. For example, Winges et al. [52] found moderate to strong synchrony across motor units innervating FDP compartments as well as between FDP compartments and flexor pollicis longus (FPL). However, synchrony across FPL and the FDP index finger compartment was significantly stronger than for FPL and any other FDP compartment. Similarly, significant differences have been found when comparing motor unit synchrony measured on extrinsic versus intrinsic muscle pairs. Specifically, intrinsic muscle pairs (FDI and palmar interossei, FPI) are characterized by weaker across-muscle motor unit synchrony than extrinsic muscles [48, 53]. This is a striking finding when considering that within-muscle synchrony to FDI and FPI is significantly stronger than across them [30, 48]. Therefore, there appears to be an important difference between within- and across-muscle common neural input, which has led to the speculation that these two forms of correlated inputs and their distribution among hand muscles may reflect differences in the muscles' functional roles (this is further discussed in Sect. 4). Another important observation is that motor units of adjacent finger flexor compartments tend to receive stronger common neural input to motor unit pairs than non-adjacent compartments [36, 42], thus lending support to the above notion that common input may contribute to the limited ability of moving adjacent fingers independently (see also a recent study on motor units of extensor digitorum profundus by van Duinen et al. [46]). The distribution of common neural input across motor units of thumb and finger flexor muscles during object hold does not follow the same gradient, however, this difference is likely due to methodological and task differences (see [48] for details).

4.3 Strength and Distribution of Across-Muscle Motor-Unit Coherence

Johnston et al. [27] reported significant across-muscle coherence across a thumb flexor muscle, FPL, and FDP compartments, as well as across FDP compartments during a five-digit object hold task. Interestingly and consistent with the above results on motor-unit synchrony, coherence across FPL and the index finger compartment of FDP was significantly stronger than across FPL and other FDP compartments. From a functional perspective, it is noteworthy that coherence between thumb and index finger forces in the low-frequency range (2–10 Hz) was significantly stronger than across thumb and other fingers [54]. Examination of across-muscle coherence to hand muscles has recently been extended to a larger number of muscles comprising intrinsic and extrinsic muscles of thumb, index, and middle fingers [55, 56]. This work has also revealed the existence of significant across-muscle coherence occurring primarily at low frequencies (<15 Hz), even though coherence strength varied significantly across muscle groups (the heterogeneous distribution of coherence across hand muscles is further discussed in Sect. 4).

4.4 Task-Dependency of Common Neural Input

Motor unit studies of within- and across-muscle synchrony have also addressed the question of whether and the extent to which correlated neural input is task-dependent. To date, a definite answer to this question is not available; therefore a general consensus has yet to be reached on whether the above described distribution of common input to motor neurons of hand muscles can be changed acutely or chronically. Lack of a clear answer is mostly due to differences in how task-dependency has been studied or defined, as well as contrasting results from studies of motor-unit synchrony versus coherence. It is also important to note that clear evidence for task-dependency would imply a functional role for common neural input in relation to a behavioral goal, e.g., coordinating multi-digit movements to grasp an object (this is further discussed in Sect. 5). However, the literature discussed below suggests that common neural input appears to be sensitive to some task parameters but not others.

Task-dependency has been described for motor-unit synchrony within and across hand muscles as a function of finger movement direction [57]. The strength of within-muscle motor-unit synchrony is also sensitive to the type of muscle contraction, e.g., motor unit synchronization is greater during lengthening than shortening contractions [28]. Task-dependency of motor-unit synchrony has also been studied in the context of grasping, in particular on the effects of grip type (power vs. precision). Winges et al. [30] hypothesized that synchrony between hand muscle motor units will be dependent on grip type. This expectation was based on evidence from a study of hand muscle interference EMG [58] indicating a higher synchronization for power than precision grips. The interpretation of this finding was that the control of power grip might be simplified by having common neural drive to all digits, whereas precision grips might require a higher degree of independence among digits. Winges et al. [30] asked subjects to hold the same grip device using either thumb and index or thumb and middle finger while measuring EMG from single motor units of FPL, FDP2 and FDP3. The experimental question was whether the weaker synchrony exhibited by thumb-middle finger flexors relative to that of thumb-index finger muscles when holding the object with five digits [48] would still be found when using two-digits. Such result would support the notion of muscle-pair specificity of common neural input despite the re-organization of force relations required to hold an object with two instead of five digits. If, however, thumb-middle finger flexors had exhibited a significantly stronger synchrony when holding the device with two versus five digits, the authors would have concluded that across-muscle motor-unit synchrony can be modulated based on grip type. It was found that across-muscle motor unit synchrony of thumb and index finger flexors was stronger than that from thumb and middle finger flexors regardless of the number of digits (five vs. two) and the digit pair (thumb-index finger or thumb-middle finger) used to hold the device. The stronger common motor-unit synchrony across motor nuclei of thumb and index finger muscles could reflect neuromuscular adaptations to their greater degree of involvement in many types of dexterous manipulation relative to other digit pairs. Specifically, thumb and index finger are engaged in nearly all hand-object interactions, whereas other thumb-fingers pairs are not. Winges et al. [30] therefore concluded that common neural input is distributed in a muscle-pair and/or specific fashion, a proposition that received further support through studies of EMG–EMG coherence (this is further discussed in Sect. 4).

It should also be noted that motor unit synchrony is not correlated with digit force output during five-digit object hold. Specifically, Winges et al. [48] reported that motor unit synchrony was strong (CIS: 0.48) across FDP muscle compartments (e.g., FDP3-5) innervating digit pairs (middle-ring fingers) that exerted significantly less normal forces than most digit pairs. However, the long flexors of the thumb and index finger were characterized by synchrony of similar strength (CIS: 0.49) and were also the digits that exerted the largest grip force. The lack of correlation between the strength of common neural input and forces exerted by most muscle pairs receiving such input indicates that motor unit synchrony is not directly associated with grip force. Later studies have extended this observation by showing that a significant increase in maximal index finger abduction force elicited by strength training (4–8 weeks) does not affect synchrony of motor units innervating an intrinsic muscle (FDI; [59]).

With regard to evidence from studies of coherence, two main frequency ranges, $\sim$1–12 Hz and $\sim$16–32 Hz, have been defined based on the extent to which the strength of coherence can be modulated to task requirements. Coherence in the low frequency band ($\sim$1–12 Hz) is stronger during position holding and lengthening muscle contractions when compared to shortening contractions [28]. It has been suggested that coherence in the higher frequency range reflects a "binding" mechanism [60] enabling efficient activation of task-related groups of neurons [61, 62]. In contrast to coherence occurring at higher frequencies, the low frequency range does not appear to be affected by the degree of object compliance during grasping [63], although a later study showed that displacement accounted for the coherence modulation [64]. Furthermore, Kakuda et al. [65] observed an increase in the magnitude of 6–12 Hz coherence between motor units from the Extensor Carpi Radialis during slow wrist movements versus position holding.

Evidence for task-dependent modulation of common neural input in the low frequency range has also been provided by studies of grasping. Winges et al. [30] found that coherence was significantly stronger during two- than five-digit object hold although, importantly, this observation was not associated with a re-distribution of common neural input across different muscle pairs for variations in task requirements (two vs. five-digit grip). Specifically, coherence across FPL and FDP2 was still stronger than across FPL and FDP3. The authors interpreted these findings as indicative of an invariant muscle-pair specific *distribution* of common neural input that, nevertheless, still allows for modulation of common neural input *strength*. It should also be noted that grip type affected across-muscle coherence but not motor-unit synchrony [30], further stressing the potential independence of these two mechanisms underlying correlated neural input and suggesting that motor-unit coherence might be more sensitive to modulation of neural drive as a function of task characteristics. This might be mostly due to the ability of coherence to detect correlations in motor-unit activity at longer time lags than those required for detection of short-term synchrony, e.g., outside of the central peak region of the cross-correlogram (Fig. 2a; see Sect. 3).

Probably one of the most convincing arguments for plasticity of motor-unit coherence has been provided by Semmler and colleagues [32]. This study reported that within-muscle coherence in the 16–32 Hz frequency range from motor units of the FDI was stronger in strength-trained than skill-trained and untrained subjects. The authors interpreted this finding by considering the potential functional consequences of coherence in relation to habitual muscle usage. Specifically, weaker across-muscle coherence might correlate with generating a more independent control of motor units relative to tasks that would benefit from a more global activation of motor units for the production of large forces (the relation between common neural input and muscle synergies is further discussed in Sect. 5).

To examine whether digit force affects the strength of across-muscle coherence, Poston et al. [56] studied brief submaximal isometric force production tasks using a three-digit grip (thumb, index, and middle fingers) while interference EMG activity was simultaneously recorded from six intrinsic and six extrinsic hand muscles. The total target forces (the sum of all digit normal forces) were 5, 20, 40, 60, and 80 % of the maximal voluntary force. To quantify changes in coherence, the integral of coherence was computed on pooled coherence from all muscle pairs ($n = 66$). Interestingly, force modulation over the voluntary range did not affect the strength of coherence [56]. This result suggests the existence of a force-independent distribution of excitatory drive to the hand muscles that were studied and is consistent with the results of the analysis of EMG amplitude from the same muscles. Specifically, the distribution of EMG amplitudes of each muscle remained consistent across target forces. Therefore, despite force-induced changes in motor unit recruitment and discharge rate to implement force modulation—and quantifiable as a modulation of EMG amplitude, no significant changes occurred in how neural drive was distributed to hand muscles. The finding of a invariant distribution of EMG amplitude *and* across-muscle coherence should not be considered obligatory, as a dissociation between these two variables has also been found, i.e., fatiguing contractions cause an increase in both EMG amplitude and EMG–EMG coherence [55]. To conclude our discussion of studies on the effect of force on common neural input, it has been found short term strength training does not elicit a modulation of across- or within-muscle coherence, respectively. These findings, together with the above observation that coherence is affected by long-term *skill* training, further support the notion of correlated neural input as a functional mechanism for coordinating the activity of motor units within a muscle and, possibly, across muscles.

The extent to which correlated neural input to hand muscles might be sensitive to sensory input from tactile receptors has been addressed by several studies. The question is particularly important when assigning a functional role to across-muscle motor-unit synchrony or coherence. For example, tonic input from the fingertip mechanoreceptors elicited by static grasping or holding of an object might potentially modulate the efficacy with which a diverging descending neural drive can temporally couple the activity of target motor neurons. Similarly, such a modulatory effect driven by tactile mechanoreceptors might also be elicited by different tactile stimuli, e.g., frictional properties of the contacts. It should be noted, however, that these questions assume a functional role of common neural input whereby its modulation might fulfill a given behavioral goal that might rely on tactile sensing, e.g., preventing an object from slipping when sensing a low friction object surface. Our preliminary work (unpublished) addressed these questions by having subjects hold against gravity objects with different frictional properties (sandpaper and rayon). We found that, although holding the object while contacting a more slippery surface elicited larger digit normal forces, across-muscle coherence was not significantly affected by contacting surfaces with different textures. These results further point to a dissociation between the modulation of total digit force magnitude and coherence (see also Sect. 5). Most importantly, however, they suggest that changes in tactile afferent activity associated with sensing the object's different frictional properties do not significantly modulate the strength of common neural input to hand muscles. These findings are consistent with other reports showing that the strength of across-muscle motor unit synchrony is not affected by tactile input during a simulated two-digit grip [66]. However, there appears to be other instances in which differences in tactile afferent activity can modulate common neural input. Following digital nerve anaesthesia, Fisher et al. [67] observed a reduction in the high frequency coherence across hand muscles during the static phase following a dynamic phase of a force production task. These authors suggested that tactile afferents might play a role in modulating across-muscle coherence when neural drive has to be changed during transitions from dynamic to static muscle contractions. Therefore, it would appear that tactile input might play a role in certain, but not all tasks.

To further investigate the issue of task-dependency of common neural input, we have compared the task of holding an object against gravity (unpublished data) with generating the grip forces of comparable magnitude on an object that was clamped to the table [56]. Our preliminary analyses indicate that object hold is associated with significantly weaker across-muscle coherence than generating forces on a fixed object. These two tasks are characterized by several differences. For example, the weighting of sensory modalities involved in monitoring ongoing task performance is mainly visually-driven for the force production task (visual display of the target force on a computer monitor), whereas tactile input is likely to dominate object hold against gravity to prevent the object from slipping. Another difference between the two tasks is that little or no load forces are generated or required during the force production task, whereas during object hold subjects modulate digit force vectors such that load forces match the gravitational force acting on the object. A more important factor, however, might be the mechanical requirement of generating net zero normal forces and moments associated with object hold, but not explicitly required by the force production task, e.g., exertion of non-zero torques on the object do not interfere with producing the target force as the object is clamped to the table. Therefore, holding an object might require a greater degree of independence in the neural drive to hand muscles to enable continuous force compensations across digits, such that changes in force at one digit are compensated by force modulation at other digits, thus minimizing net torques. This proposition, however, requires further testing across a wider variety of task conditions and is the subject of ongoing investigations.

5 Heterogeneous Distribution of Common Input to Hand Muscles

5.1 Motor-Unit Synchrony and Coherence: Studies of Single Motor Units

Bremner and colleagues [36, 37] first reported that motor-unit synchrony is heterogeneously distributed across motor units of hand muscles revealed through isometric contractions of the index and/or middle finger. Specifically, the occurrence and strength of motor-unit synchrony was higher for hand muscles with similar mechanical actions on different digits than muscles with different actions inserting on the same digit. Studies of object hold using a multi-digit grip have extended these observations by revealing significantly stronger motor unit synchrony and coherence across motor units of thumb and index finger flexors (FPL-FDP2) than thumb and all other FDP compartments [27, 52]. These findings demonstrated that the strength of across-muscle synchrony, traditionally considered to be weak [36, 37], can be significant. These authors speculated that the strong across-muscle synchrony exhibited by FPL-FDP2 might reflect the relatively more important role of thumb and index finger among all thumb-finger combinations for grasping and fine manipulation. Therefore, the stronger synchrony of thumb-index flexors relative to all other thumb-finger muscles suggested a distribution of common neural input that might be muscle-pair specific. Subsequent work [30, 48] further revealed the existence of heterogeneous distribution of common neural input to motor units innervating intrinsic (1DI-1PI) and intrinsic-extrinsic muscle pairs (FPL-1PI, FPL-1DI), the former muscle pair being characterized by weak synchrony. This was an unexpected result as, traditionally, synchrony across intrinsic muscles has been reported to be stronger than across extrinsic muscles [36, 37]. We also recorded motor unit synchrony within each intrinsic muscle and found that within-intrinsic muscle synchrony (1DI-1DI, 1PI-1PI) was three times stronger than across intrinsic muscle synchrony (1DI-1PI). These results are consistent with results reported by McIsaac and Fuglevand [53] about weak synchrony across motor units of two intrinsic muscles, adductor pollicis and FDI. The same group also reported that synchrony of motor units belonging to a given compartment of flexor digitorum superficialis (FDS) was significantly stronger than across adjacent FDS compartments, such difference being even larger when comparing within-compartment synchrony versus across non-adjacent compartments [68].

The above results support the notion that the distribution of across-muscle motor unit synchrony may reflect the functional role of specific digit pairs. Pairs of extrinsic flexors of the digits, capable of generating large forces and essential for coordination of grip forces receive stronger common neural input than pairs of intrinsic muscles that are important synergists for controlling force direction but not strong force producers. Similarly, stronger common neural input within than across finger flexor compartments might reflect the functional requirement of decoupling the action of individual fingers. Nevertheless, additional or alternative factors underlying the heterogeneity with which common neural input is distributed to hand muscle should be considered (see below).

5.2 Motor-Unit Synchrony and Coherence: Studies of Interference EMG

Coherence analysis has further confirmed and extended previous observations of synchrony between motor-unit pairs regarding the existence of a heterogeneous distribution of common neural input to muscle and muscle compartment pairs [27, 30, 48, 55, 56]. For example, the muscles innervating the thumb and index finger (FPL and FDP2) and thumb and little finger (FPL and FDP4) were characterized by the strongest coherence among all muscle pairs during a five-digit object hold [27]. Furthermore, during a static three-digit force production task, Poston et al. [56] recorded intramuscular EMG from 12 concurrently active hand muscles innervating the thumb, index, and middle fingers. The coordination of hand muscles was quantified in two ways, one based on EMG amplitude of each muscle (this is further discussed in Sect. 5) and one as across-muscle coherence from all muscle combinations ($n = 66$). A main finding of this study was that coherence was muscle-group dependent, the strongest and weakest coherence being found across extrinsic and intrinsic muscle pairs, respectively. Interestingly, this pattern was preserved across a wide range of digit forces (5–80 % of maximal voluntary force). Such force-independent distribution of common neural input suggests that increases in descending drive to the motor neuron pool, which result in an increase in motor unit recruitment and rate coding, do not significantly change how correlated inputs are distributed to multiple motor nuclei of hand muscles. These data, consistent with the above described results from motor-unit synchrony [48], point to a muscle-pair specific distribution of common neural input. Poston et al. [56] speculated that the stronger coherence exhibited by extrinsic versus intrinsic muscle groups could be associated with differences in their functional properties, e.g., how much force they can produce, and/or their innervations. As described above, extrinsic and intrinsic muscles can also be distinguished based on their functional role for grasping and manipulation. However, it should be noted that the muscle-pair specificity of common neural input should not be interpreted as a discrete and invariant categorization of how neural drive is distributed to hand muscles for a number of reasons. First, from a behavioral perspective both sets of muscles interact and contribute to modulate fingertip force vectors. Second, inputs to spinal motor neurons can undergo plastic changes. This plasticity might, in turn, alter how motor neurons respond to common neural input and, therefore, in how it is distributed across hand muscles. Last but not least, the observations of muscle-specific distribution of common neural input have been obtained through a fairly restricted set of experimental conditions. Therefore, the extent to which they generalize to other tasks involving coordinated actions among the digits require further experimental work.

The finding of stronger coherence across extrinsic than intrinsic muscles is particularly interesting because it provides further insight into the principles underlying how correlated neural input is distributed across hand muscles. As noted above for coherence measured from motor unit pairs (Sect. 5.2), weaker coherence across intrinsic than extrinsic muscles might be related to their different role in manipulation and force production capabilities. Ongoing work aims at determining the key factors responsible for the heterogeneous distribution of common neural input among hand muscles. Of particular interest are the questions of (a) whether the heterogeneity of common neural input is relatively fixed and (b) whether there are factors—other than the above-mentioned functional differences among hand muscles—that might account for the wide range of common neural input strengths.

The question of whether the distribution of correlated neural input is fixed or, conversely, sensitive to task demands or training is reminiscent of the question discussed in Sect. 3 (*E, F*), e.g., whether the *strength* of common neural input is task-dependent. However, it has been reported that modulation of common neural input strength can occur within an invariant distribution among hand muscles [27]. Therefore, the question arises as to whether and the extent to which the force-independent distribution of common neural input generalizes to other tasks. In a follow-up study of muscle fatigue, Danna-dos Santos and colleagues [55] observed a similar, but not identical distribution among the same 66 muscle pairs studied by Poston et al. [56]. Specifically, the comparison of across-muscle coherence between the first and last quarter of a fatiguing contraction (three-digit force production) revealed that fatigue caused an increase in coherence but not in how coherence was distributed among hand muscles. Therefore, significant changes in the mechanisms responsible for motor unit recruitment and rate coding underlying either voluntary force modulation [56] or maintaining a constant force while fatigue develops [55] do not seem to affect the distribution of neural drive to hand muscle motor nuclei. Another important consideration is that the distribution of common neural input remains invariant regardless of whether coherence magnitude is modulated [55] or not [56].

A difference between the results of these two studies, however, was that in the study by Danna-dos Santos et al. [55], no statistically significant difference was found in coherence from extrinsic and intrinsic muscle pairs. These authors speculated that methodological and task differences may underlie the different result. Specifically, Danna-dos Santos and colleagues [55] measured coherence over a significantly longer time period than Poston et al. [56]. Another major difference between the two studies is the examination of fatiguing versus non-fatiguing contractions, respectively. Therefore, it is possible that the distribution pattern of correlated neural input is sensitive to the duration of a voluntary contraction and/or to peripheral and central impairments associated with fatigue, e.g., increased descending drive from supraspinal sources and recruitment of larger motor units, transmitter depletion at IA afferent terminals, and increased presynaptic inhibition of IA afferents mediated by activation of group III and IV afferents by muscle metabolite accumulation (for review, see [69]). One may conclude that the distribution of correlated neural input to hand muscles is not fixed or insensitive to the characteristics of a task. Further investigations are warranted to determine the relation between coherence magnitude and distribution in relation to task demands. However, the evidence from these interference EMG studies as well as single motor unit studies seem to indicate that (a) correlated neural input to hand muscles is heterogeneously distributed and that (b) this distribution appears to reflect anatomical and/or functional differences among hand muscle groups.

With regard to the question of whether factors other than functional differences among hand muscles might contribute to the heterogeneous distribution of common neural input, another factor that should be considered is the nerve supply to given muscle pairs. Specifically, it has been suggested that hand muscles that share the same nerve innervation might be characterized by stronger common neural input than muscle that are innervated by different nerves [37]. For example, the results of Winges et al. [48] could be interpreted as due to the fact that each pair of muscles (extrinsic, FPL-FDP2; intrinsic, 1DI-1PI) shares the same innervation (median and ulnar nerve, respectively), whereas the muscle pair characterized by the weakest coherence (1DI-FDP2) are innervated by two different nerves. Lastly, an additional factor that might account for the heterogeneous distribution of common neural input to different hand muscle pairs is the digit they insert on. Therefore, a digit-specific distribution of common neural input might also be expected. Preliminary (unpublished) results from our laboratory further indicate that muscles inserting into digits involved in a three-digit grasping receive common neural input of different strength. This result was obtained by comparing a three-digit object hold with force production performed by individual digits, the thumb, index, and middle fingers muscles being characterized by the strongest across-muscle coherence.

6 Common Neural Input and Hand Muscle Synergies

6.1 *Functional Role of Motor-Unit Synchrony: Theoretical Considerations*

As indicated in Sect. 4, probably the most important question about common neural input is: *"What are the functional consequences of common neural input on motor control?"* This is a difficult question to answer experimentally because divergence is a defining characteristic of descending neural inputs to hand muscles in primates (see Chap. 1). One would need, for example, to 'turn off' the mechanisms responsible for constraining the correlations in motor unit firings to measure the consequences in how digit forces are coordinated during grasping or manipulation.

Alternatively, one may wish to modulate the strength with which the neural drive to muscle pairs is correlated, and measure what aspect, if any, of grasp behaviors is affected. These approaches, however, are not feasible as they would require selective *and* tonic activation, deactivation, or modulation of specific cortical and/or spinal networks. Nevertheless, insights have been provided by studies of task-dependency of common neural input as well as by computational models. Specifically, the studies reviewed above (Sect. 4.4) suggest that the strength of common neural input in time and frequency domains does not correlate with the magnitude of digit forces, but does correlate negatively with the requirement of fine motor control of hand muscles, e.g., the extent of digit movement fractionation. This observation points to a potential role of common neural input for the temporal coordination of neural drive to hand muscles.

Santello and Fuglevand [70] addressed the question of whether common neural input might be important for regulating the temporal relations among grip forces using a motor unit simulation. The focus on the temporal coordination of grip forces was motivated by previous work on five-digit grasping [49, 71, 72]. Specifically, this work revealed that each pair of digit normal forces during object hold exhibits in-phase relations. Importantly, this behavior was highly reproducible in a wide variety of task conditions, e.g., regardless of whether subjects could anticipate the object's mass distribution on a trial-to-trial basis or used the dominant versus non-dominant hand. These early observations raised the question of whether this 'default' pattern of digit force synchrony was a mere byproduct of biomechanical constraints (e.g., multi-tendoned finger flexors) or whether neural factors, e.g., common neural input, should also be considered. To address this question, Rearick et al. [50] compared multi-digit grasping of an object that remained stationary on the table versus holding it against gravity. By matching the total normal digit force required by the 'force production' task with that elicited by the object 'hold' task, these authors could dissociate the (common) force output requirement from the mechanical constraints that were unique to each task, e.g., time-to-time fluctuations in force at one digit must be compensated by force modulation at one or more digits to prevent object slip in the 'hold', but not in the 'force production' task. Therefore, the authors hypothesized that in-phase digit force relations are the byproduct of digit force production *per se* and thus, force synchronization should be found in both tasks. Conversely, a task-dependent modulation of digit force synchronization would indicate that different neural mechanisms are selectively involved in the two tasks, the expectation being that holding an object against gravity—challenging grasp stability to a greater extent than force production—would be associated with more consistent force synchronization among digit pairs. The results supported the latter prediction, between-digit force synchronization being consistently high when holding the object, but much weaker and less consistent in the 'force production' task. This finding, therefore, points to neural mechanisms that, by interacting with biomechanical constraints, can independently modulate the temporal relationships between digit forces while maintaining the same force output.

Santello and Fuglevand [70] identified and tested motor unit synchrony as a mechanism that could potentially account for the experimental results of Rearick et al. [50]. These authors used a motor unit model developed by Andrew Fuglevand and colleagues [73, 74]. In the model, the authors imposed varying levels of across-muscle synchrony to two simulated motor unit populations that activated two 'virtual' muscles and quantified the correlations among simulated forces (a) to establish whether across-muscle motor unit synchrony could reproduce the above experimental findings on the coupling of grip forces and (b) to determine values of across-muscle synchrony that are behaviorally important. Both of these questions were highly significant as the functional consequences of across-muscle motor unit synchrony for motor control had not been investigated or quantified. Santello and Fuglevand [70] reported that that moderate across-muscle synchrony, normally found in motor-unit synchrony studies (e.g., CIS: 0.3), could lead to force coupling that was quantitatively similar to that found during multi-digit grasping (see above). This result is consistent with the notion of across-muscle motor unit synchrony as a mechanism that can modulate the temporal coordination of digit forces.

The results of the motor unit simulation study appears to contradict the above-mentioned unpublished results from a study of EMG–EMG coherence where the same tasks (force production vs. object hold) were compared using a three-digit grip (Sect. 4.4). Specifically, across-muscle coherence was stronger in the force production than in the object hold task, the interpretation being that a higher degree of independent control of digit forces might be necessary. However, a direct comparison between the simulated and experimental results might be premature because it is limited by three main factors: [1] the simulation study examined motor-unit synchrony, defined as near-coincident firing of motor units, whereas the identification of coherence is not restricted to synchronous events (see Sect. 2); [2] motor-unit synchrony and coherence can operate independently, e.g., the task-dependent modulation of coherence might have not involved a modulation of motor-unit synchrony; and [3], to date no simulation study has quantified the affect of across-muscle coherence modulation on digit force coordination, and therefore the speculation that weaker coherence is associated with the requirement of higher digit force individuation during object hold remains to be validated.

6.2 Hand Muscle Synergies

The concept of common neural input as a mechanism that might play a functional, task-dependent role for the neural control of the hand is related to the broader concept of *muscle synergies*. Briefly, muscle synergies are defined as combinations of groups of muscles. The criteria underlying how certain combinations are selected by the CNS, the extent to which they are fixed or flexibly adapted to task conditions, and whether they serve a functional purpose for the control of movement are questions that are still debated and are subjects of ongoing investigations (for a more comprehensive discussion of muscle synergies and related issues, the reader is referred to a recent review by [75]; see also a recent study on forelimb muscle synergies in non-human primates associated with reach-to-grasp and manipulation by [76]). Note that the concept of 'muscle synergies' described here is different from the definition of 'synergies' used by clinicians in relation to neurological disorders, e.g., leg flexion synergy also known as 'Babinski sign'.

Within the framework of the above-reviewed studies of common neural input, constraints on the temporal relations in the activity of motor unit populations within and across hand muscles, and/or how common input is distributed across hand muscles, can be viewed as building blocks for muscle synergies. Specifically, it has been suggested that the correlated firing of motor units, as measured by coherence, might be a mechanism by which the central nervous system reduces the number of independent degrees of freedom to be controlled (e.g., motor units, forces) [32, 61, 77]. Although the coordination of motor units within a given muscle can be considered an example of 'muscle synergy', often this term is used in relation to the spatio-temporal coordination of multiple muscles. When examining neural control of the hand, this problem is equivalent to fingertip forces are modulated within a digit or, for grasping and manipulation tasks, across digits.

A more viable approach to study common neural input as a mechanisms contributing to muscle synergies is interference EMG, e.g., the recording of electrical activity of multiple motor units from concurrently active muscles. Studies of interference EMG of hand muscles have described a tendency for covariation in EMG amplitude of multiple hand muscles acting on one digit [78, 79] or multiple digits [55, 56]. Such covariation results from neural constraints through which the orderly recruitment of motor units and increase in motor unit discharge rate, indirectly quantified as EMG amplitude, is coordinated across multiple muscles in a fairly stereotypical fashion, e.g., as a scaling of the EMG vector length but not orientation, across the range of voluntary forces. As discussed above, the observation that both EMG–EMG coherence and the distribution of EMG amplitude across muscles does not change as a function of grip force might suggest a functional link between these two phenomena.

To further test the phenomenon of covariation of EMG amplitude across multiple muscles, a recent study of interference EMG examined the relation between EMG of intrinsic and extrinsic hand muscles during a two-digit grip as a function of wrist joint angle [80]. This study was designed to determine whether EMG activity of intrinsic muscles, whose length does not change with wrist angle, might be modulated in response to an expected EMG modulation of extrinsic hand muscles. Two alternative scenarios were envisioned: (a) both sets of muscles are modulated as a function of wrist angle, or (b) EMG amplitude modulation occurs only in those muscles in which changes in muscle length with wrist angle requires a concurrent modulation of neural drive to maintain the desired force output. Johnston et al. [80] reported that EMG amplitude was modulated in both intrinsic and extrinsic muscles, thus suggesting the existence of a muscle synergy as defined above. Common neural input across these two sets of muscles might contribute to their concurrent modulation in order to generate the desired force output.

Last but not least, and as mentioned in Sect. 2, divergence of inputs to motor units is not limited to descending inputs as indicated by one-to-many divergence of tactile stimuli to EMG reflex responses in muscles. One such recent observation shows that electrotactile stimuli to branches of the digital nerve elicit cutaneous reflexes across multiple hand muscles (unpublished observations). Most importantly, however, cutaneous reflexes occur in several hand muscles innervating different digits *regardless* of which digit is stimulated. This observation points to the existence of spinal circuitry that transmits sensory information from one stimulated digit to muscles that act on stimulated *and* non-stimulated digit(s). This concept is similar to that introduced above of a force- or fatigue-independent distribution of common neural input to hand muscles, both underscoring the usefulness of analyzing patterns through which correlated inputs might be revealed by quantifying motor-unit synchrony and/or coherence. The concept of hand muscle synergies can also inspire the design and control algorithms of robotic hands. Specifically, for anthropomorphic robotic hands, controls signals that are shared by task-specific groups of joint actuators—the artificial counterpart of biological common neural input—could potentially simplify the spatial and temporal coordination of digit forces for grasping and manipulation.

7 Conclusion

The analyses of common neural input, quantified through motor-unit synchrony and coherence, has provided significant insights into how the Central Nervous System controls hand muscles, including the task-dependency of common neural input and a functional gradient in its distribution.

Plasticity has been described for within-muscle motor-unit synchrony and coherence both at the acute and chronic level. Interestingly, however, short-term force training has no affect on the strength of common neural input, whereas long-term skill training does. These observations have been interpreted as indicative of a functional role of common neural input whereby constraints on motor unit recruitment can be modulated according to task requirements, e.g. movement direction and lengthening versus shortening contraction. However, task-dependency (grip type, frictional properties of the contacts) has been found for across-muscle coherence but not motor-unit synchrony measured during multi-digit tasks (force production or object hold). These and similar observations suggest separate mechanisms as well as differential sensitivity of motor-unit synchrony and coherence in detecting common neural inputs. Several important questions, however, remain, the most important one being what is the functional role of common neural input. This question has been addressed mostly using motor unit simulations indicating a potential role for motor-unit synchrony for the temporal coordination of digit forces, but further simulation work is needed to further characterize the functional consequences of common neural input. Although it is difficult to directly measure the effect of common neural input modulation on grasp or manipulative behaviors, experimental work is also needed to improve our understanding of the link between plasticity and function. If, as suggested by some authors, common neural input is associated with the requirement of coordinating the action of multiple muscles, further studies are needed to define the range of task conditions within which common neural input should or should not be modulated.

Even though the functional role of common neural input remains to be understood, analyses of the pattern of distribution of motor-unit synchrony have revealed important features of how neural drive is shared among hand muscle motor nuclei. Specifically, common neural input appears to be distributed in a muscle-specific fashion. The features of these patterns suggest the existence of a 'functional gradient' along which common neural input tends to be stronger across extrinsic than intrinsic hand muscles. Importantly, voluntary force modulation or fatiguing contractions do not affect the distribution of common neural inputs, therefore suggesting a fairly invariant neural network that might reflect functional differences across muscle groups. However, invariance of muscle-specific distribution of common neural input during short-term experimental manipulations does not rule out the possibility of plasticity in response to more prolonged practice or skill learning. A related question is whether the strength and distribution of common neural input is sensitive to sensory inputs. Although it appears that tactile feedback is not essential for across-muscle motor-unit synchrony or coherence, further work is needed to establish the role of sensory modalities that might be relevant to the performance of grasp or manipulation tasks.

Acknowledgments Marco Santello thanks Drs. Thomas Hamm, Brach Poston, and Mark Jesunathadas for their valuable comments on the chapter.

References

1. E. Pierrot-Deseilligny, D. Burke, *The Circuitry of the Human Spinal Cord: Its Role in Motor Control and Movement Disorders*, vol 129 (Cambridge University Press, Cambridge, 2005), pp. 551–554 (ISBN: 978-0-521-82581-8 Brain Feb. 2006)
2. M.R. Caccia, A.J. McComas, A.R. Upton, T. Blogg, Cutaneous reflexes in small muscles of the hand. J. Neurol. Neurosurg. Psychiatry **36**, 960–977 (1973)
3. R. Chen, P. Ashby, Reflex responses in upper limb muscles to cutaneous stimuli. Can. J. Neurol. Sci. **20**, 271–278 (1993)
4. R. Garnett, J.A. Stephens, The reflex responses of single motor units in human first dorsal interosseous muscle following cutaneous afferent stimulation. J. Physiol. **303**, 351–364 (1980)
5. C.J. De Luca, B. Mambrito, Voluntary control of motor units in human antagonist muscles: coactivation and reciprocal activation. J. Neurophysiol. **58**, 525–542 (1987)
6. C.J. De Luca, A.M. Roy, Z. Erim, Synchronization of motor-unit firings in several human muscles. J. Neurophysiol. **70**, 2010–2023 (1993)

7. C.J. De Luca, Z. Erim, Common drive of motor units in regulation of muscle force. Trends Neurosci. **17**, 299–305 (1994)
8. A. Adam, C.J. De Luca, Z. Erim, Hand dominance and motor unit firing behavior. J. Neurophysiol. **80**, 1373–1382 (1998)
9. C.J. De Luca, Z. Erim, Common drive in motor units of a synergistic muscle pair. J. Neurophysiol. **87**, 2200–2204 (2002)
10. M.M. Lowery, Z. Erim, A simulation study to examine the effect of common motoneuron inputs on correlated patterns of motor unit discharge. J. Comput. Neurosci. **19**, 107–124 (2005)
11. D.H. Perkel, G.L. Gerstein, G.P. Moore, Neuronal spike trains and stochastic point processes I. The single spike train. Biophys. J. **7**, 391–418 (1967)
12. D.H. Perkel, G.L. Gerstein, G.P. Moore, Neuronal spike trains and stochastic point processes II. Simultaneous spike trains. Biophys. J. **7**, 419–440 (1967)
13. M.A. Nordstrom, A.J. Fuglevand, R.M. Enoka, Estimating the strength of common input to human motoneurons from the cross-correlogram. J. Physiol. **453**, 547–574 (1992)
14. P.A. Kirkwood, T.A. Sears, The synaptic connexions to intercostal motoneurones as revealed by the average common excitation potential. J. Physiol. **275**, 103–134 (1978)
15. T.A. Sears, D. Stagg, Short-term synchronization of intercostal motoneurone activity. J. Physiol. **263**, 357–381 (1976)
16. P.A. Kirkwood, On the use and interpretation of cross-correlations measurements in the mammalian central nervous system. J. Neurosci. Methods **1**, 107–132 (1979)
17. P.H. Ellaway, K.S. Murthy, The source and distribution of short-term synchrony between gamma-motoneurones in the cat. Q. J. Exp. Physiol. **70**, 233–247 (1985)
18. M.D. Binder, R.K. Powers, Relationship between simulated common synaptic input and discharge synchrony in cat spinal motoneurons. J. Neurophysiol. **86**, 2266–2275 (2001)
19. A.M. Taylor, R.M. Enoka, Quantification of the factors that influence discharge correlation in model motor neurons. J. Neurophysiol. **91**, 796–814 (2004)
20. A.M. Taylor, R.M. Enoka, Optimization of input patterns and neuronal properties to evoke motor neuron synchronization. J. Comput. Neurosci. **16**, 139–157 (2004)
21. G.P. Moore, D.H. Perkel, J.P. Segundo, Statistical analysis and functional interpretation of neuronal spike data. Annu. Rev. Physiol. **28**, 493–522 (1966)
22. S.F. Farmer, F.D. Bremner, D.M. Halliday, J.R. Rosenberg, J.A. Stephens, The frequency content of common synaptic inputs to motoneurones studied during voluntary isometric contraction in man. J. Physiol. **470**, 127–155 (1993)
23. J.R. Rosenberg, D.M. Halliday, P. Breeze, B.A. Conway, Identification of patterns of neuronal connectivity–partial spectra, partial coherence, and neuronal interactions. J. Neurosci. Methods **83**, 57–72 (1998)
24. D.M. Halliday, Weak, stochastic temporal correlation of large scale synaptic input is A major determinant of neuronal bandwidth. Neural Comput. **12**, 693–707 (2000)
25. J.R. Rosenberg, A.M. Amjad, P. Breeze, D.R. Brillinger, D.M. Halliday, The Fourier approach to the identification of functional coupling between neuronal spike trains. Prog. Biophys. Mol. Biol. **53**, 1–31 (1989)
26. J.A. Johnston, G. Formicone, T.M. Hamm, M. Santello, Assessment of across-muscle coherence using multi-unit versus single-unit recordings. Exp. Brain Res. **207**, 269–282 (2010)
27. J.A. Johnston, S.A. Winges, M. Santello, Periodic modulation of motor-unit activity in extrinsic hand muscles during multidigit grasping. J. Neurophysiol. **94**, 206–218 (2005)
28. J.G. Semmler, K.W. Kornatz, D.V. Dinenno, S. Zhou, R.M. Enoka, Motor unit synchronisation is enhanced during slow lengthening contractions of a hand muscle. J. Physiol. **545**, 681–695 (2002)
29. J.G. Semmler, K.W. Kornatz, R.M. Enoka, Motor-unit coherence during isometric contractions is greater in a hand muscle of older adults. J. Neurophysiol. **90**, 1346–1349 (2003)

30. S.A. Winges, J.A. Johnston, M. Santello, Muscle-pair specific distribution and grip-type modulation of neural common input to extrinsic digit flexors. J. Neurophysiol. **96**, 1258–1266 (2006)
31. J.G. Semmler, M.A. Nordstrom, C.J. Wallace, Relationship between motor unit short-term synchronization and common drive in human first dorsal interosseous muscle. Brain Res. **767**, 314–320 (1997)
32. J.G. Semmler, M.V. Sale, F.G. Meyer, M.A. Nordstrom, Motor-unit coherence and its relation with synchrony are influenced by training. J. Neurophysiol. **92**, 3320–3331 (2004)
33. C.T. Moritz, E.A. Christou, F.G. Meyer, R.M. Enoka, Coherence at 16–32 Hz can be caused by short-term synchrony of motor units. J. Neurophysiol. **94**, 105–118 (2005)
34. M.M. Lowery, L.J. Myers, Z. Erim, Coherence between motor unit discharges in response to shared neural inputs. J. Neurosci. Methods **163**, 384–391 (2007)
35. A.K. Datta, J.A. Stephens, Synchronization of motor unit activity during voluntary contraction in man. J. Physiol. **422**, 397–419 (1990)
36. F.D. Bremner, J.R. Baker, J.A. Stephens, Variation in the degree of synchronization exhibited by motor units lying in different finger muscles in man. J. Physiol. **432**, 381–399 (1991)
37. F.D. Bremner, J.R. Baker, J.A. Stephens, Correlation between the discharges of motor units recorded from the same and from different finger muscles in man. J. Physiol. **432**, 355–380 (1991)
38. J. Gibbs, L.M. Harrison, J.A. Stephens, Organization of inputs to motoneurone pools in man. J. Physiol. **485**(Pt 1), 245–256 (1995)
39. E.J. Huesler, M.A. Maier, M.C. Hepp-Reymond, EMG activation patterns during force production in precision grip. III. Synchronisation of single motor units. Exp. Brain Res. **134**, 441–455 (2000)
40. D.A. Keen, A.J. Fuglevand, Common input to motor neurons innervating the same and different compartments of the human extensor digitorum muscle. J. Neurophysiol. **91**, 57–62 (2004)
41. K.T. Reilly, M.H. Schieber, Incomplete functional subdivision of the human multitendoned finger muscle flexor digitorum profundus: an electromyographic study. J. Neurophysiol. **90**, 2560–2570 (2003)
42. K.T. Reilly, M.A. Nordstrom, M.H. Schieber, Short-term synchronization between motor units in different functional subdivisions of the human flexor digitorum profundus muscle. J. Neurophysiol. **92**, 734–742 (2004)
43. S.L. Kilbreath, S.C. Gandevia, Limited independent flexion of the thumb and fingers in human subjects. J. Physiol. **479**(Pt 3), 487–497 (1994)
44. V.M. Zatsiorsky, Z.M. Li, M.L. Latash, Enslaving effects in multi-finger force production. Exp. Brain Res. **131**, 187–195 (2000)
45. T.J. Butler, S.L. Kilbreath, R.B. Gorman, S.C. Gandevia, Selective recruitment of single motor units in human flexor digitorum superficialis muscle during flexion of individual fingers. J. Physiol. **567**, 301–309 (2005)
46. H. van Duinen, W.S. Yu, S.C. Gandevia, Limited ability to extend the digits of the human hand independently with extensor digitorum. J. Physiol. **587**, 4799–4810 (2009)
47. W.S. Yu, H. van Duinen, S.C. Gandevia, Limits to the control of the human thumb and fingers in flexion and extension. J. Neurophysiol. **103**, 278–289 (2010)
48. S.A. Winges, K.W. Kornatz, M. Santello, Common input to motor units of intrinsic and extrinsic hand muscles during two-digit object hold. J. Neurophysiol. **99**, 1119–1126 (2008)
49. M. Santello, J.F. Soechting, Force synergies for multifingered grasping. Exp. Brain Res. **133**, 457–467 (2000)
50. M.P. Rearick, A. Casares, M. Santello, Task-dependent modulation of multi-digit force coordination patterns. J. Neurophysiol. **89**, 1317–1326 (2003)
51. G.B. Hockensmith, S.Y. Lowell, A.J. Fuglevand, Common input across motor nuclei mediating precision grip in humans. J. Neurosci. Off. J. Soc. Neurosci. **25**, 4560–4564 (2005)
52. S.A. Winges, M. Santello, Common input to motor units of digit flexors during multi-digit grasping. J. Neurophysiol. **92**, 3210–3220 (2004)

53. T.L. McIsaac, A.J. Fuglevand, Common synaptic input across motor nuclei supplying intrinsic muscles involved in the precision grip. Exp. Brain Res. **188**, 159–164 (2008)
54. W.S. Yu, S.L. Kilbreath, R.C. Fitzpatrick, S.C. Gandevia, Thumb and finger forces produced by motor units in the long flexor of the human thumb. J. Physiol. **583**, 1145–1154 (2007)
55. A. Danna-Dos Santos, B. Poston, M. Jesunathadas, L.R. Bobich, T.M. Hamm, M. Santello, Influence of fatigue on hand muscle coordination and EMG–EMG coherence during three-digit grasping. J. Neurophysiol. **104**, 3576–3587 (2010)
56. B. Poston, A. Danna-Dos Santos, M. Jesunathadas, T.M. Hamm, M. Santello, Force-independent distribution of correlated neural inputs to hand muscles during three-digit grasping. J. Neurophysiol. **104**, 1141–1154 (2010)
57. F.D. Bremner, J.R. Baker, J.A. Stephens, Effect of task on the degree of synchronization of intrinsic hand muscle motor units in man. J. Neurophysiol. **66**, 2072–2083 (1991)
58. E.J. Huesler, M.C. Hepp-Reymond, V. Dietz, Task dependence of muscle synchronization in human hand muscles. NeuroReport **9**, 2167–2170 (1998)
59. D.J. Kidgell, M.V. Sale, J.G. Semmler, Motor unit synchronization measured by cross-correlation is not influenced by short-term strength training of a hand muscle. Exp. Brain Res. **175**, 745–753 (2006)
60. C.M. Gray, Synchronous oscillations in neuronal systems: mechanisms and functions. J. Comput. Neurosci. **1**, 11–38 (1994)
61. S.F. Farmer, Rhythmicity, synchronization and binding in human and primate motor systems. J. Physiol. **509**(Pt 1), 3–14 (1998)
62. J.H. McAuley, C.D. Marsden, Physiological and pathological tremors and rhythmic central motor control. Brain J. Neurol. **123**(Pt 8), 1545–1567 (2000)
63. J.M. Kilner, M. Alonso–Alonso, R. Fisher, R.N. Lemon, Modulation of synchrony between single motor units during precision grip tasks in humans. J. Physiol. **541**, 937–948 (2002)
64. J.M. Kilner, S. Salenius, S.N. Baker, A. Jackson, R. Hari, R.N. Lemon, Task-dependent modulations of cortical oscillatory activity in human subjects during a bimanual precision grip task. NeuroImage **18**, 67–73 (2003)
65. N. Kakuda, M. Nagaoka, J. Wessberg, Common modulation of motor unit pairs during slow wrist movement in man. J. Physiol. **520**(Pt 3), 929–940 (1999)
66. T.L. McIsaac, A.J. Fuglevand, Influence of tactile afferents on the coordination of muscles during a simulated precision grip. Exp. Brain Res. **174**, 769–774 (2006)
67. R.J. Fisher, M.P. Galea, P. Brown, R.N. Lemon, Digital nerve anaesthesia decreases EMG–EMG coherence in a human precision grip task. Exp. Brain Res. **145**, 207–214 (2002)
68. T.L. McIsaac, A.J. Fuglevand, Motor-unit synchrony within and across compartments of the human flexor digitorum superficialis. J. Neurophysiol. **97**, 550–556 (2007)
69. J.L. Taylor, S.C. Gandevia, A comparison of central aspects of fatigue in submaximal and maximal voluntary contractions. J. Appl. Physiol. (Bethesda, Md.: 1985) **104**, 542–550 (2008)
70. M. Santello, A.J. Fuglevand, Role of across-muscle motor unit synchrony for the coordination of forces. Exp. Brain Res. **159**, 501–508 (2004)
71. M.P. Rearick, M. Santello, Force synergies for multifingered grasping: effect of predictability in object center of mass and handedness. Exp. Brain Res. **144**, 38–49 (2002)
72. M.P. Rearick, G.E. Stelmach, B. Leis, M. Santello, Coordination and control of forces during multifingered grasping in Parkinson's disease. Exp. Neurol. **177**, 428–442 (2002)
73. A.J. Fuglevand, D.A. Winter, A.E. Patla, D. Stashuk, Detection of motor unit action potentials with surface electrodes: influence of electrode size and spacing. Biol. Cybern. **67**, 143–153 (1992)
74. A.J. Fuglevand, D.A. Winter, A.E. Patla, Models of recruitment and rate coding organization in motor-unit pools. J. Neurophysiol. **70**, 2470–2488 (1993)
75. M.C. Tresch, A. Jarc, The case for and against muscle synergies. Curr. Opin. Neurobiol. **19**, 601–607 (2009)
76. S.A. Overduin, A. d'Avella, J. Roh, E. Bizzi, Modulation of muscle synergy recruitment in primate grasping. J. Neurosci. Off. J. Soc. Neurosci. **28**, 880–892 (2008)

77. J.G. Semmler, Motor unit synchronization and neuromuscular performance. Exerc. Sport Sci. Rev. **30**, 8–14 (2002)
78. F.J. Valero-Cuevas, Predictive modulation of muscle coordination pattern magnitude scales fingertip force magnitude over the voluntary range. J. Neurophysiol. **83**, 1469–1479 (2000)
79. F.J. Valero-Cuevas, F.E. Zajac, C.G. Burgar, Large index-fingertip forces are produced by subject-independent patterns of muscle excitation. J. Biomech. **31**, 693–703 (1998)
80. J.A. Johnston, L.R. Bobich, M. Santello, Coordination of intrinsic and extrinsic hand muscle activity as a function of wrist joint angle during two-digit grasping. Neurosci. Lett. **474**, 104–108 (2010)

Chapter 3
MRI-Based Skeletal Hand Movement Model

Georg Stillfried, Ulrich Hillenbrand, Marcus Settles and Patrick van der Smagt

Abstract The kinematics of the human hand is optimal with respect to force distribution during pinch as well as power grasp, reducing the tissue strain when exerting forces through opposing fingers and optimising contact faces. Quantifying this optimality is of key importance when constructing biomimetic robotic hands, but understanding the exact human finger motion is also an important asset in, e.g. tracking finger movement during manipulation. The goal of the method presented here is to determine the precise orientations and positions of the axes of rotation of the finger joints by using suitable magnetic resonance imaging (MRI) images of a hand in various postures. The bones are segmented from the images, and their poses are estimated with respect to a reference posture. The axis orientations and positions are fitted numerically to match the measured bone motions. Eight joint types with varying degrees of freedom are investigated for each joint, and the joint type is selected by setting a limit on the rotational and translational mean discrepancy. The method results in hand models with differing accuracy and complexity, of which three examples, ranging from 22 to 33 DoF, are presented. The ranges of motion of the joints show some consensus and some disagreement with data from literature. One of the models is published as an implementation for the free OpenSim simulation environment. The mean discrepancies from a hand model built from MRI data are compared against a hand model built from optical motion capture data.

G. Stillfried (✉) · U. Hillenbrand
Institute of Robotics and Mechatronics, German Aerospace Center (DLR), Wessling, Germany
e-mail: georg.stillfried@dlr.de

U. Hillenbrand
e-mail: ulrich.hillenbrand@dlr.de

M. Settles
Klinikum rechts der Isar, University hospital of TU München, Munich, Germany
e-mail: msettles@tum.de

P. van der Smagt
Faculty of Informatics, Technische Universität München, Munich, Germany

R. Balasubramanian and V. J. Santos (eds.), *The Human Hand as an Inspiration for Robot Hand Development*, Springer Tracts in Advanced Robotics 95,
DOI: 10.1007/978-3-319-03017-3_3, © Springer International Publishing Switzerland 2014

Keywords Human hand · Robot hand · Hand kinematics · MR imaging · 3D object localisation

Abbreviations

Bones

MC	Metacarpal bone
PP	Proximal phalanx
PM	Medial phalanx
PD	Distal phalanx

Joints

CMC	Carpometacarpal joint
IMC	Intermetacarpal joint
MCP	Metacarpophalangeal joint
PIP	Proximal interphalangeal joint
DIP	Distal interphalangeal joint
IP1	Thumb interphalangeal joint

Other

DoF	Degree(s) of freedom
LOOCV	Leave-one-out cross-validation
MRI	Magnetic resonance imaging
MoCap	(Optical) motion capture

The abbreviations for bones and joints are augmented by numbers from 1 for thumb to 5 for little finger. For the location of the joints and bones, see Fig. 2. Abbreviations for the joint *types* are found in Fig. 6.

1 Introduction

Many robot hands have been built after the human example, one of the latest being the DLR Hand Arm System [1] (Fig. 1). The design of its kinematics was guided by simple length measurements of a human hand, by functional considerations (e.g. how do the joint axes need to be inclined in order to achieve a robust opposition grasp, Fig. 2) and by intuitive appeal of different models to human subjects [2]. Each movement degree of freedom (DoF) of the hand is supplemented by a stiffness DoF, so that 21 movement DoF of the hand and wrist are driven by 42 antagonistically placed motors. By tensioning non-linear spring elements between the motors and the joints, the mechanical stiffness can be adjusted. This allows to mimic humans' stiffness variation in manipulation tasks, to store energy, e.g. for finger flicking, and to survive heavy collisions. With the motors placed in the

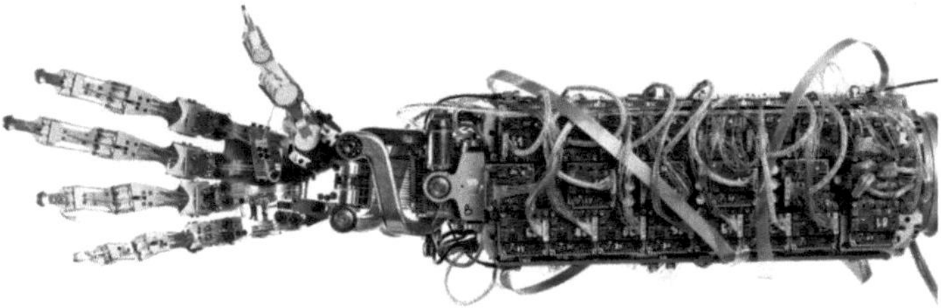

Fig. 1 Hand, wrist and forearm of the DLR Hand Arm System [1]

robotic "forearm", the hand is embedded in a Hand Arm System, which provides five additional movement DoF for shoulder, elbow and forearm rotation.

Here we present a method that aims at very precisely measuring and modelling movement of the human hand skeleton, in order to verify and further improve robot hand kinematics. The method is demonstrated by creating hand models that cover all joints of fingers and thumb as well as the palm arching movement of the metacarpus (Fig. 3). The models are based on magnetic resonance imaging (MRI) of an individual hand in 51 different postures, from which bone poses are extracted, localised, and used to optimise a parametrised general hand model.

Existing methods for modelling hand kinematics are mostly based on cadaver measurements [3–8] and optical surface tracking [9–16].

In our opinion, measurements at hand cadavers cannot be used to reconstruct the *active* kinematics of the hands, since the muscle synergies cannot be taken into account, plus the fixture by the ligaments is no longer realistic; consequently, such models will lead to artefacts in the kinematic representation. Models based on tracking the surface of the fingers, on the other hand, lead to unknown inaccuracies due to non-linear and varying movement of the soft tissue (e.g. the skin) with respect to the bones [17, 18]; indeed, such models often ignore phalanx rotation around their longitudinal axis. To overcome these disadvantages, we measure bone poses in vivo using 3D medical imaging. We conjecture that our method therefore leads to more accurate models of human hands.

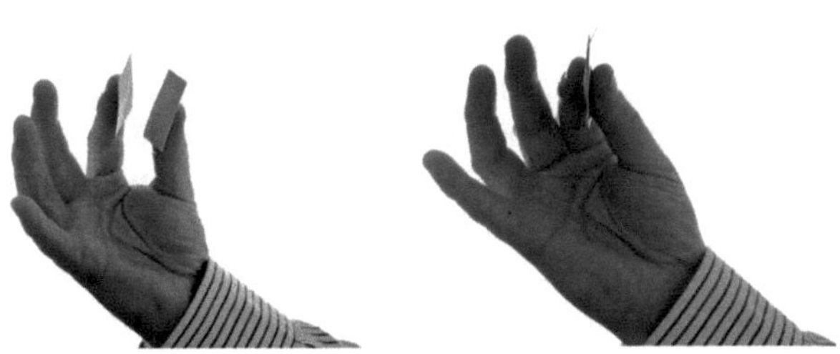

Fig. 2 Human hand kinematics enables optimal grasps by aptly orienting surfaces with respect to each other. Here the finger pads of thumb and index finger are brought face-to-face

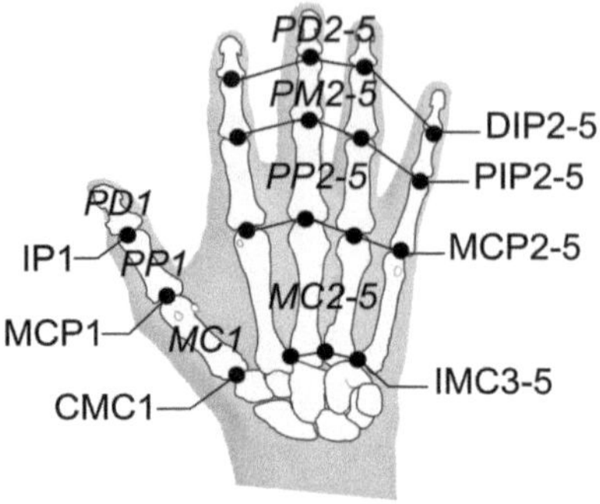

Fig. 3 The joints and bones of the fingers, the thumb and the metacarpus that are investigated here. Bone contours adapted from [32]

The first and probably only previous work that used MRI images for measuring finger kinematics is by Miyata et al. [19]. They recorded one reference posture and three other postures to determine helical movement axes for the flexion axes of the proximal and distal interphalangeal joints of the index finger (PIP2 and DIP2).

We aim to extend the MRI-based kinematic analysis in three ways:

(1) covering all joints of the hand, including multi-DoF joints;
(2) covering the range of motion of all the joints; and
(3) constructing a continuous representation of the hand kinematics from which any intermediate posture can be generated.

For solving these issues, we need to reproducibly measure the movement of fixed parts of the fingers, i.e., the bones.

We model the human hand as a kinematic chain with an arbitrary number of DoF per joint. The optimal number of DoF for each joint, as well as the static parameters of each of these joints, are optimised from the recorded data. Once these parameters are determined for each of the joints, a model is generated with a fixed number of DoF. Since the model is targeted towards the subsequent implementation in a robotic system [20], only rotary joints are considered—any more complex joint in a robotic system will probably create friction and control difficulties.

Data recording is focussed on single subjects rather than statistical averages. When reproducing the kinematics of the human hand, statistics are of no great help: that approach would average over a number of participants without taking principal components of the variation into account. Instead, the kinematics of one adult individual is measured and reproduced.

The resulting model is made available within the OpenSim simulation environment [21]. OpenSim is a software that is able to match motion capture data to skeletal models, and can be extended to include tendons and muscles. Much of the human skeleton is already covered: legs, torso, neck, arms and a cadaver-based thumb and index finger [22]. The hand model presented here will be a step towards the completion of the human skeletal model.

2 Methods

In order to determine the kinematics of the human hand, the rigid bones—forming the endoskeletal structure of the hand—are assumed as key reference points in the kinematics. The *active* kinematics is investigated, i.e., as induced by actively moving the joints through muscle activation, but *passive* kinematics, i.e., as induced by putting external forces on the fingers of the hand, is ignored. The presented reconstruction of the active kinematics of the human hand is based on the following scheme:

(1) first, 3-dimensional MRI images of the hand are recorded in a large number of predefined postures (Sect. 2.1). The postures must be chosen so as to represent the full kinematic range of the hand. Among a large number of other postures, the Kapandji Test [23] for the assessment of thumb motion is used;
(2) second, in each 3-D MRI image the bones are segmented, using automatic grey-level-based segmentation followed by manual correction;
(3) third, the position and orientation of each of the bones of the hand is determined. This is done automatically using novel image-processing algorithms, in which statistical methods are used to localise known objects in a 3-dimensional visual scene (Sect. 2.2);
(4) finally, from a range of possible joint models (Sect. 2.3), the optimal model for each joint is selected and the parameters of the joints are determined in order to minimise the errors in the model (Sect. 2.4). From that, kinematic chains are defined for each of the fingers, thus ending up with the full kinematic model.

The results of these steps are presented in Sect. 3.

2.1 MRI Images and Segmentation

The MRI images are taken on a Philips Achieva 1.5 T unit with a Philips SENSE eight-channel head coil to receive a more homogeneous signal and to improve the signal-to-noise ratio (SNR). Generally, SNR is proportional to the voxel[1] volume and to the square root of the net scan duration:

$$r \propto v\sqrt{t}, \tag{1}$$

where r is the SNR, v is the voxel volume and t is the net scan duration, i.e., the time actually spent for signal acquisition. Thus for every application an individual compromise has to be found optimally balancing the needs for a small v (high

[1] *voxel* "*vo*lume pi*xel*" = basic volume element of a 3-D image; analogous to *pixel* in 2-D images.

spatial resolution), small t (short scan times to minimise potential motion artefacts) and large r (image quality sufficient for either diagnosis or—as in this case—the segmentation of certain anatomic structures).

An optimal compromise is found with a total scan duration (which is always longer than the net scan duration) of between two and two and a half minutes and a spatial resolution of $(0.76\ \text{mm})^3$. Note that, from Eq. (1), a voxel volume of $(0.38\ \text{mm})^3$ would require 64 times the scan duration in order to achieve the same SNR. To further minimise motion artefacts the hand is stabilised using modelling clay. For post-processing, the spatial resolution is interpolated to $(0.38\ \text{mm})^3$ in order to achieve sub-voxel resolution in the segmentation process. In the processing step after the segmentation, the grey value information is discarded. The interpolation helps retain some of the information that is contained in the grey values.

For scanning, the sequence type *balancedFFE* is used (also known as trueFISP or balancedSSFP) with T_R/T_E/flip angle = 4.8 ms/2.4 ms/45°. The repetition time T_R is the time between two successive excitation pulses. The transverse component of the magnetization is read out at echo time T_E after each pulse.

The advantage of balancedFFE is that it yields a strong signal at short T_R. (In fact, the signal of the balancedFFE sequence becomes independent of T_R, which can be as low as 2.5 ms with the limiting factors being the readout time and the avoidance of peripheral and heart muscle stimulation.)

As a drawback, balancedFFE is prone to the so-called banding artefacts appearing as black stripes across the bone. This artefact can in principle be overcome by applying the balancedFFE offset averaging technique (also known as CISS or FIESTA-C), but requires twice the scan time.

Another artefact occurring in these sequences is opposed phase fat/water cancelling, where voxels containing both fat and water, i.e., at corresponding tissue boundaries, appear dark, because the magnetisation vectors of fat and water point in opposite directions.

Also a *cine*-sequence, i.e., a continuous-motion sequence with two to five images per second, is recorded. However, only one image layer for the whole hand can be recorded, which renders this method unusable for the purpose of exact bone localisation.

The images are taken of a 29 year old female subject with no history of hand problems who gave informed consent to the procedure. Fifty images are taken in different hand postures with the aim of reflecting each joint's range of motion.

From the MRI volume images, the bones are segmented. In fact, not the whole bone volume is segmented but the signal-intense volume inside the bone that corresponds to the cancellous bone. The tissue between the trabeculae of the cancellous bone is bone marrow consisting mainly of fat, which yields high signal intensity in the balancedFFE sequence.

The cortical bone, which forms the outer calcified layer of the bone, hardly contains any free fat or water protons and therefore stays dark in the MRI image. Near the bones there are other low-signal structures like tendons, which makes it

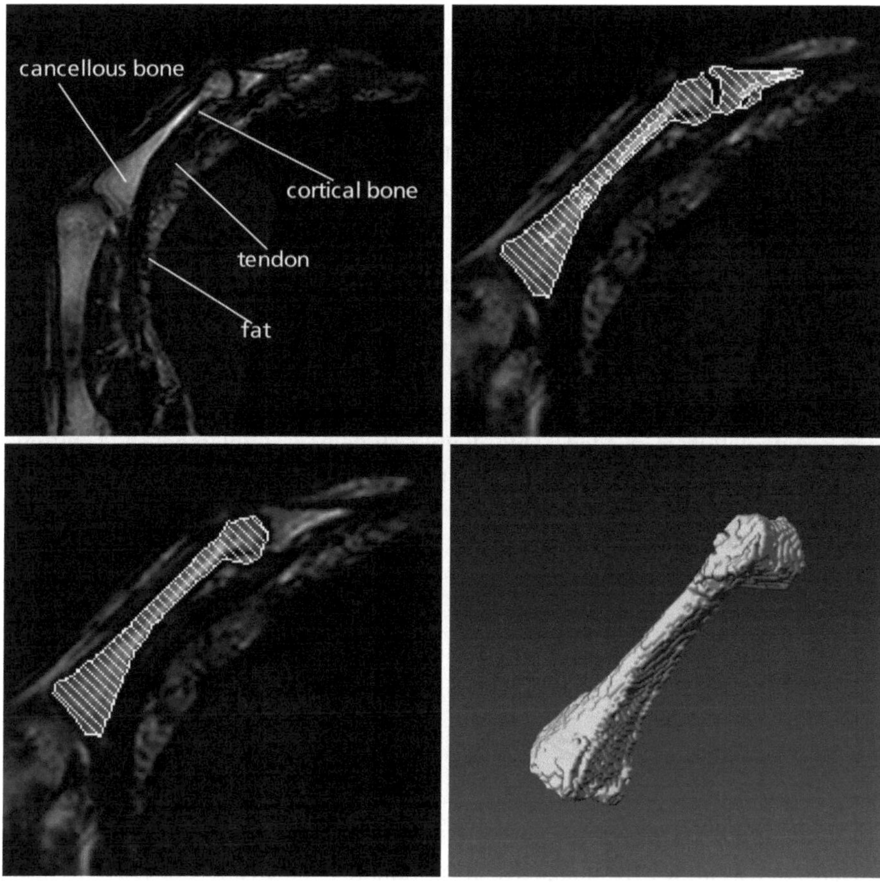

Fig. 4 Segmentation process. *Top left:* Slice of an MRI image, showing the middle finger metacarpal (MC3). Tissue types can be discriminated by the intensity of the signal that is emitted. Segmentation is done at the boundary between cancellous and cortical bone. *Top right:* Threshold-based preselection. *Bottom left:* Manually refined selection. *Bottom right:* Segmented volume consisting of the selected areas from all slices

difficult to determine the outer bone surface. Therefore, the boundary between cancellous and cortical bone is used for segmentation (Fig. 4).

The bones are segmented from the image by highlighting the cancellous bone area in each slice of the MRI image. In the medical imaging software Amira (Visage Imaging GmbH, Berlin, Germany), the area is preselected by adjusting a threshold and refined manually (Fig. 4).

2.2 Motion Estimation

For the purpose of estimating the rigid motion of bones between different hand postures, some geometric structure rigidly related to each bone has to be extracted from the MRI images that can be reliably recovered with little shape variation between images. Automatic reconstruction of the bone geometry is a challenge, as the image density of cancellous bone, cortical bone and surrounding tissue can vary greatly between and across images. Also manual segmentation, besides being tedious work, is prone to introducing shape variation.

Hence a double strategy is pursued. The border between cancellous and cortical bone often produces a marked contrast edge at reproducible locations. These border points can hence be detected by selection of high-contrast points. In the absence of such a marked density contrast, on the other hand, guidance by manual bone segmentation is needed. This double strategy is implemented as follows. First, the bone segments are padded with zero-density voxels to fit in a cuboid volume. Then a dipolarity score of the padded density within each $3 \times 3 \times 3$-voxel sub-volume is computed, as

$$Dipolarity(c_1, c_2, c_3) = \left\| \sum_{\substack{(i,j,k) \in \{c_1-1, c_1+1\} \\ \times \{c_2-1, c_2+1\} \\ \times \{c_3-1, c_3+1\}}} [I(i,j,k) - I(c_1, c_2, c_3)] \begin{pmatrix} i - c_1 \\ j - c_2 \\ k - c_3 \end{pmatrix} \right\| .$$

Here $I(i,j,k)$ is the MRI image density as function of the voxel indices (i,j,k), and (c_1, c_2, c_3) are the indices of the centre voxel within the considered $3 \times 3 \times 3$-voxel sub-volume. The sum computes the density-weighted centroid of voxels around the voxel at (c_1, c_2, c_3); its Euclidean norm quantifies the degree of dipolarity of the density at the centre voxel. It attains high values for centre voxels close to a strong density edge. Finally, the centre voxels with the top q percent of dipolarity are selected as representing bone-related points. The grey value information is discarded in the selected points, but the interpolation mentioned in Sect. 2.1 is used to refine the point set. The quantile q is chosen to produce a data set of between 2,000 and 20,000 points, depending on the size of the bone. This way, points on the manually determined bone border are selected in the absence of high-contrast edges in the image; while high-contrast image edges dominate the selected points where available.

The above procedure produces sets of points that are close to the surface of the bones. However, missing parts and shape variation cannot be avoided. Moreover, there is no correspondence of points across different data sets of the same bone. A robust estimator of motion between such data sets hence has to be employed. A correspondence-free alignment that is also robust to geometric deviations [24] is provided within the framework of parameter-density estimation and maximization, or parameter clustering. This is a robust estimation technique based on location statistics in a parameter space where parameter samples are computed from data samples [25, 26]. The estimator may be viewed as a continuous version of a generalised, randomised Hough transform. In the present variant, samples are drawn from the 3-D points selected through the high-dipolarity criterion above.

Let $X, Y \subset \mathbb{R}^3$ be the point sets extracted from two MRI images of the same bone. A motion hypothesis can be computed from a minimum subset of three points from X matched against a minimum subset of three points from Y. The sampling proceeds thus as follows:

1. Randomly draw a point triple $\boldsymbol{x}_1, \boldsymbol{x}_2, \boldsymbol{x}_3 \in X$.
2. Randomly draw a point triple $\boldsymbol{y}_1, \boldsymbol{y}_2, \boldsymbol{y}_3 \in Y$ that is approximately congruent to the triple $\boldsymbol{x}_1, \boldsymbol{x}_2, \boldsymbol{x}_3 \in X$.
3. Compute the rigid motion that aligns $(\boldsymbol{x}_1, \boldsymbol{x}_2, \boldsymbol{x}_3)$ with $(\boldsymbol{y}_1, \boldsymbol{y}_2, \boldsymbol{y}_3)$ in the least-squares sense.
4. Compute and store the six parameters of the hypothetical motion.

Random drawing of approximately congruent point triples in step 2 of the sampling procedure is efficiently implemented using a hash table of Y-point triples indexed with the three X-point distances $(\| \boldsymbol{x}_1 - \boldsymbol{x}_2 \|, \| \boldsymbol{x}_2 - \boldsymbol{x}_3 \|, \| \boldsymbol{x}_3 - \boldsymbol{x}_1 \|)$ as the key. Least-squares estimation of rigid motion in step 3 computes the rotation $R \in \mathrm{SO}(3)$ and translation $\boldsymbol{t} \in \mathbb{R}^3$ as

$$(R, \boldsymbol{t}) = \underset{(R', \boldsymbol{t}') \in \mathrm{SE}(3)}{\arg\min} \big[{}' \boldsymbol{x}_1 + \boldsymbol{t}' - \boldsymbol{y}_1 \|^2 + \| R' \boldsymbol{x}_2 + \boldsymbol{t}' - \boldsymbol{y}_2 \|^2 + \| R' \boldsymbol{x}_3 + \boldsymbol{t}' - \boldsymbol{y}_3 \|^2 \big].$$

The special three-point method of [27] is used to obtain a closed-form solution. The parametrisation of rigid motions chosen for sampling step 4 may have an influence on the result. In fact, the parameter density from which the samples are taken depends upon this choice. A parametrisation that is consistent for clustering is used here, in the sense of [25].

By repeatedly executing the sampling procedure 1–4 above (in the order of 10^6 times), samples are obtained from the parameter density for the rigid alignment problem. This parameter density is similar in spirit to a posterior density, but without assuming a probabilistic observation model.

The parameter samples can be stored in an array or a tree of bins. The sampling stops when a significant cluster of samples has formed, as judged from the bin counts. Then the location of maximum parameter density is searched by repeatedly starting a mean-shift procedure [28, 29] from the centre of the bins with high parameter counts. From all the local density maxima found through mean shift, the location in the 6D parameter space of the largest maximum is returned as the motion estimate of a bone, in the following denoted as R_e and t_e. Details of the implementation are presented elsewhere [26].

The main sources of error in the procedure for estimating bone motion are

- the variation in bone geometry erroneously represented in the point sets extracted from different images of the same bone, resulting from variation in manual segmentation or dipolarity values computed from the images;
- the approximate rotational symmetry about the longitudinal axis of a bone, especially in case of poor geometric representation lacking shape details.

To get rid of grossly wrong motion estimates, an interactive cluster analysis is performed on the estimated rotations. Making use of the stochastic nature of the estimation algorithm, each motion estimate is repeated 100 times with different subsets of the data being sampled, resulting in motion estimates $\{R_{e1}, t_{e1}\} \dots \{R_{e100}, t_{e100}\}$. If the angular distance between any two of the 100 motion estimates exceeds a threshold, clusters of rotation parameters are identified and the correct cluster $\mathcal{C} \subset \{1, \dots, 100\}$ is selected through visual inspection (Fig. 5).

The angular distance between two rotations is defined as the angle of a third rotation that would have to be appended to the the first rotation in order to make it identical to the second rotation. It is calculated as follows:

$$\text{angdist}(R_1, R_2) := \arccos\left(\frac{1}{2}\left(\text{trace}\left(R_2 R_1^{-1}\right) - 1\right)\right), \qquad (2)$$

where R_1 and R_2 are the rotation matrices of the first and second rotation.

The final rotation estimate $\overline{R}$ is determined as the rotation that minimises the sum of squared angular distances to all rotations in the cluster, i.e., the mean rotation in the difference-rotation-angle metric,

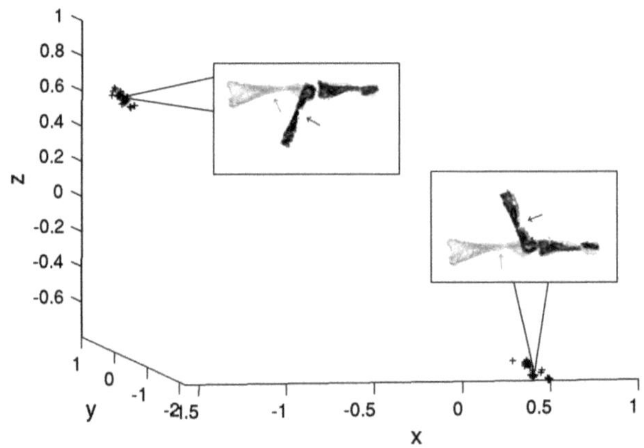

Fig. 5 Visual inspection of pose estimates. The rotational part of 100 randomly repeated pose estimates is plotted in three dimensions as the product of rotation axis and angle. In this example there are two distinct clusters. One element in each cluster is inspected by regarding the more strongly curved side of the neighbouring bone (*arrows*). The motion of the *bottom right* cluster element implies a large, anatomically impossible, longitudinal rotation of the bones. Therefore the *top left* cluster is taken as the correct cluster $\mathcal{C}$

$$\overline{R} = \underset{R' \in \mathrm{SO}(3)}{\arg\min} \left[\sum_{i \in \mathcal{C}} \mathrm{angdist}\left(R', R_{\mathrm{ei}}\right)^2 \right], \tag{3}$$

Likewise, the final translation estimate $\bar{\boldsymbol{t}}$ is determined as the translation that minimises the sum of squared Euclidean distances to all translations in the cluster, i.e., the ordinary mean value of valid translations,

$$\bar{\boldsymbol{t}} = \frac{1}{n} \sum_{i \in \mathcal{C}} \boldsymbol{t}_{\mathrm{cei}}, \tag{4}$$

where n is the number of elements in the correct cluster $\mathcal{C}$, and $\boldsymbol{t}_{\mathrm{cei}}$ is the i-th translation estimate of the bone centroid. The translation estimate of the bone centroid is calculated as follows:

$$\boldsymbol{t}_{\mathrm{cei}} = \boldsymbol{R}_{\mathrm{ei}} \boldsymbol{c} + \boldsymbol{t}_{\mathrm{ei}} - \boldsymbol{c},$$

where $\boldsymbol{c}$ is the bone centroid, i.e., the mean of all points in $\boldsymbol{X}$. If the correct cluster contains less than ten elements, the respective bone pose is discarded from the modelling process. Furthermore, all pose estimates are checked optically and obviously wrong estimates are discarded.

A natural confidence weight of the final rotation estimates is obtained from the variance of the sample mean values, i.e.,

$$\sigma_r^2 = \frac{1}{n(n-1)} \sum_{i=1}^{n} angdist(\overline{R}, R_{\mathrm{ei}})^2. \tag{5}$$

This confidence weight enters in the estimation of orientation of rotational axes for the kinematic hand model below. Likewise, a confidence weight of the final translation estimates is given by

$$\sigma_t^2 = \frac{1}{n(n-1)} \sum_{i=1}^{n} \| \bar{\boldsymbol{t}} - \boldsymbol{t}_{\mathrm{cei}} \|^2, \tag{6}$$

and used in the estimation of the position of rotational axes for the kinematic hand model below.

2.3 Determining Joint Models

In the fingers of the human hand contain different types of joints. The 1-DoF joints all are hinge joints (Fig. 6); 2-DoF joints can be divided into two types. The metacarpal joint of the thumb is a saddle joint. In contrast, the metacarpal joints of the fingers are condyloid. The main difference between saddle and condyloid joints is that condyloid joints have (roughly) intersecting axes, which saddle joints do not have.

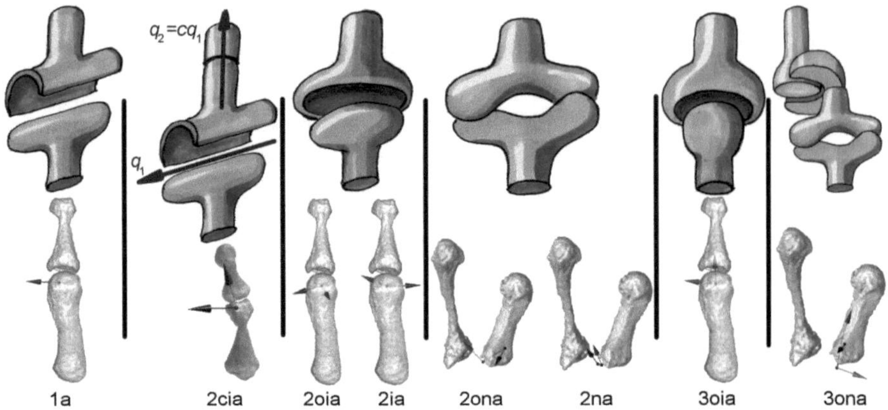

Fig. 6 Joint types used in the presented method. *From left to right:* Hinge joint (one axis, "1a"), hinge joint with combined longitudinal rotation (two coupled intersecting axes, "2cia"), condyloid joint (two orthogonal/oblique intersecting axes, "2oia"/"2ia"), saddle joint (two orthogonal/oblique non-intersecting axes, "2ona"/"2na"), ball joint (three orthogonal intersecting axes, "3oia") and 3-DoF joint with orthogonal non-intersecting axes ("3ona", combination of a saddle and a pivot joint). Upper row images from [33]

The condyloid and saddle joint types are further divided into joints whose rotation axes are orthogonal and joints whose rotation axes are at any arbitrary angle to each other. Additionally, a hinge joint with a coupled longitudinal rotation, a ball joint and a 3-DoF joint with non-intersecting axes are defined (Fig. 6).

As mentioned above, one of the goals is to compute an optimal kinematic model for a robotic system. For that reason, but also for reasons of computational efficiency and easy representation, the joints are rotational joints with axes fixed to the proximal bone or the preceding axis in multi-DoF joints. This certainly has an effect on the accuracy of the model, but this accuracy remains within the accuracy of the recording and reconstruction method.

Typically there is a trade-off between the complexity and the accuracy of a joint type. For each joint, the joint type is selected by setting a limit on the mean deviation between the measured and modelled bone poses, and by selecting the simplest joint type that fulfils it. Joints that have fewer axes are considered simpler than joints with more axes, intersecting axes simpler than non-intersecting axes and orthogonal axes simpler than freely oriented ones. The mean deviation is an outcome of the identification of the joint parameters (Sect. 2.4).

2.4 Identification of Joint Parameters

The joint parameters (positions and orientations of the rotation axes) are identified on a joint-by-joint basis by numerically minimising the discrepancy between the measured and modelled relative motion of the joint's distal bone with respect to the proximal bone. To calculate the relative motion, the absolute motion of the proximal bone is inversely applied to the absolute motion of the distal bone:

$$R_\mathrm{r} = \overline{R}_\mathrm{p}^{-1} \overline{R}_\mathrm{d} \tag{7}$$

and

$$\boldsymbol{t}_\mathrm{r} = \overline{R}_\mathrm{p}^{-1} \left(\boldsymbol{c}_\mathrm{d} + \bar{\boldsymbol{t}}_\mathrm{d} - \bar{\boldsymbol{t}}_\mathrm{p} - \boldsymbol{c}_\mathrm{p} \right) + \boldsymbol{c}_\mathrm{p} - \boldsymbol{c}_\mathrm{d}, \tag{8}$$

where $\{R_\mathrm{r}, \boldsymbol{t}_\mathrm{r}\}$ is the relative motion of the distal bone with respect to the proximal bone, $\{\overline{R}_\mathrm{p}, \bar{\boldsymbol{t}}_\mathrm{p}\}$ and $\{\overline{R}_\mathrm{d}, \bar{\boldsymbol{t}}_\mathrm{d}\}$ are the absolute motions of the proximal and distal bone according to Eqs. (3) and (4), and $\boldsymbol{c}_\mathrm{p}$ and $\boldsymbol{c}_\mathrm{d}$ are the vectors of Cartesian coordinates of the centroids of the proximal and distal bone.

In order to reduce the dimensionality of the search space, the identification of the axis orientations and positions is split up into two steps. In the first step, the axis orientations are identified by minimising the angular difference between the measured orientations and the modelled orientations.

The modelled orientation R_m of the bone is calculated as follows:

$$R_\mathrm{m} = \prod_{k=1}^{n_\mathrm{a}} \mathrm{rot}(\boldsymbol{a}_k, q_k) \tag{9}$$

where $n_\mathrm{a} \in \{1, 2, 3\}$ is the number of rotation axes of the joint, $\boldsymbol{a}_k$ is the orientation of the kth axis and q_k is the rotation angle around the kth axis. The operator $\mathrm{rot}(\boldsymbol{a}, q)$ yields the rotation matrix of a rotation around an axis $\boldsymbol{a}$ by an angle q:

$$\mathrm{rot}(\boldsymbol{a}, q) := \begin{pmatrix} c + c' a_x^2 & c' a_x a_y - a_z s & c' a_x a_z + a_y s \\ c' a_x a_y + a_z s & c + c' a_y^2 & c' a_y a_z - a_x s \\ c' a_x a_z - a_y s & c' a_y a_z + a_x s & c + c' a_z^2 \end{pmatrix}, \tag{10}$$

with

$$\begin{aligned} c &= \cos q, \\ c' &= 1 - \text{ and} \\ s &= \sin q, \end{aligned}$$

where a_x, a_y and a_z are the Cartesian elements of the unit orientation vector $\boldsymbol{a}$. The position and orientation vectors of the rotation axes are given in the coordinate system of the MRI system, and with respect to the bones in the reference posture.

The orientations of the rotation axes and the rotation angles are identified by numerically minimising the weighted mean square angular difference over all postures:

$$\left(\boldsymbol{a}_1, \ldots, \boldsymbol{a}_{n_\mathrm{a}}, \boldsymbol{q}_1, \ldots, \boldsymbol{q}_{n_\mathrm{p}}\right) = \underset{\left(\boldsymbol{a}'_1, \ldots, \boldsymbol{a}'_{n_\mathrm{a}}, \boldsymbol{q}'_1, \ldots, \boldsymbol{q}'_{n_\mathrm{p}}\right)}{\arg\min} \left[\sum_{j=1}^{n_\mathrm{p}} w_{\mathrm{r}j} \operatorname{angdist}\left(R_{\mathrm{r}j}, R_{\mathrm{m}j}\left(\boldsymbol{a}'_1, \ldots, \boldsymbol{a}'_{n_\mathrm{a}}, \boldsymbol{q}'_j\right)\right)^2\right], \tag{11}$$

with

$$w_{\mathrm{r}j} = \frac{1}{\sigma^2_{\mathrm{rp}j} + \sigma^2_{\mathrm{rd}j}} \tag{12}$$

where n_p is the number of postures, $\boldsymbol{a}_1, \ldots, \boldsymbol{a}_{n_\mathrm{a}}$ are the orientation vectors of the rotation axes, $\boldsymbol{q}_1, \ldots, \boldsymbol{q}_{n_\mathrm{p}}$ are the vectors of joint angles for each posture $j \in \{1, \ldots, n_\mathrm{p}\}$, where $\boldsymbol{q}_j = \left(q_{1j}, \ldots, q_{n_\mathrm{a}j}\right)^\mathrm{T}$ contains the joint angles for each rotation axis, $w_{\mathrm{r}j}$ is the confidence weight due to the variances $\sigma^2_{\mathrm{rp}j}$ and $\sigma^2_{\mathrm{rd}j}$ of the rotation estimates of the proximal and distal bone in posture j as calculated in Eq. (5), angdist is the angular distance operator according to Eq. (2), $R_{\mathrm{r}j}$ is the measured relative orientation of the bone in posture j according to Eq. (7) and $R_{\mathrm{m}j}$ is the modelled relative orientation of the bone according to Eq. (9).

The positions of the rotation axes are identified by minimising the mean squared distance between the measured and modelled position of the bone centroid:

$$\left(\boldsymbol{p}_1, \ldots, \boldsymbol{p}_{n_\mathrm{a}}\right) = \underset{\left(\boldsymbol{p}'_1, \ldots, \boldsymbol{p}'_{n_\mathrm{a}}\right)}{\arg\min} \left[\sum_{j=1}^{n_\mathrm{p}} w_{\mathrm{t}j} \left\| \boldsymbol{t}_{\mathrm{m}j}\left(\boldsymbol{p}'_1, \ldots, \boldsymbol{p}'_{n_\mathrm{a}}\right) - \boldsymbol{t}_{\mathrm{r}j} \right\|^2\right], \tag{13}$$

with

$$\vec{t}_{\mathrm{m}j}\left(\vec{p}'_1\right) = \left(\prod_{k=1}^{n_\mathrm{a}} \operatorname{rot}(\vec{a}_k, q_{kj})\right)\left(\vec{c}_\mathrm{d} - \vec{p}'_1\right) + \vec{p}'_1 - \vec{c}_\mathrm{d} \tag{14}$$

for joints with intersecting axes,

$$\begin{aligned} t_{\mathrm{m}j}\left(\vec{p}'_1, \ldots, \vec{p}'_{n_\mathrm{a}}\right) = &\left(\prod_{k=1}^{n_\mathrm{a}} \operatorname{rot}(\vec{a}_k, q_{kj})\right)\vec{c}_\mathrm{d} \\ &+ \left(\sum_{k=1}^{n_\mathrm{a}} \left(\prod_{l=1}^{k-1} \operatorname{rot}(\vec{a}_l, q_{lj}) - \prod_{l=1}^{k} \operatorname{rot}(\vec{a}_l, q_{lj})\right)\vec{p}'_k\right) - \vec{c}_\mathrm{d} \end{aligned} \tag{15}$$

for joints with non-intersecting axes and

$$w_{\mathrm{t}j} = \frac{1}{\sigma^2_{\mathrm{tp}j} + \sigma^2_{\mathrm{td}j}}, \tag{16}$$

where $\boldsymbol{p}_1,\ldots,\boldsymbol{p}_{n_\mathrm{a}}$ are the position vectors of the rotation axes, $\boldsymbol{t}_{\mathrm{m}j}$ are the modelled translations of the bone centroid, $\boldsymbol{t}_{\mathrm{r}j}$ are the measured translations of the bone centroid, $\boldsymbol{a}_k$ and q_{kj} are the rotation axes and angles as derived from Eq. (11) and $\boldsymbol{c}_\mathrm{d}$ is the position vector of the distal bone centroid.

In order to perform the optimisations described in Eqs. (11) and (13), the *fminsearch* function of the *Matlab* computation software is used, which implements the Nelder-Mead simplex algorithm [30]. The algorithm is called with broadly different starting points to increase the chance of finding the global optimum, and not only a local optimum. For Eq. (11), a nested optimisation is conducted, with an outer optimisation for the axis orientations $\boldsymbol{a}_1,\ldots,\boldsymbol{a}_{n_\mathrm{a}}$. Within each iteration step of the outer optimisation, a number of n_p inner optimisations are carried out to find the optimum joint angles $\boldsymbol{q}_1,\ldots,\boldsymbol{q}_{n_\mathrm{p}}$. For the outer optimisation, the axis orientations are parametrised by two spherical coordinates (azimuth and elevation), in order to reduce the search space by one dimension as compared to Cartesian coordinates for axis orientation.

2.5 Leave-one-out Cross-Validation of Joint Parameters

In order to check to what extent the results apply to the investigated hand in general as opposed to being overfit to the investigated postures, a leave-one-out cross-validation is performed. For this, the parameters of the joints are identified n_p times, with n_p being the number of measured bone poses, where in each round one of the poses is left out. The joint parameters (axis orientations and positions) resulting from each identification are used to move the bone as close as possible to the omitted pose. The rotational and translational discrepancy between the modelled and measured bone pose is calculated, and the weighted mean of rotational and translational discrepancies between the modelled and measured bone poses is calculated.

2.6 Comparison with Optical Motion Capture

Kinematic hand models are built based on MRI data and based on optical motion capture (MoCap) data. The residual rotational and translational discrepancies of both models are compared.

Another subject is recruited for MRI and MoCap measurements, because the previous subject was not available anymore. Due to time constraints, only one reference posture and 19 other postures are recorded with MRI, using a *turboFFE* sequence and a spatial resolution of (1 mm)^3. For MoCap, a Vicon system (OMG plc, Oxford, UK) with seven 0.3-megapixels cameras is used. One finger is recorded at a time, with three retroreflective markers per finger segment. One reference time sample and nineteen representative other samples are selected from the capture data. In the reference frame, bone coordinate systems are attached manually to the marker triples of each segment. The motion of the finger segment from the reference posture to the other postures is determined using the least-squares method by [27].

One joint instead of three joints is used to model the palm, because the motion of the single metacarpal bones is difficult to discriminate with MoCap. The same fifteen joints for fingers and thumb as described above are used. The thumb CMC joint is modelled with two non-orthogonal, non-intersecting axes (2na), the MCP joints are modelled with two orthogonal, intersecting axes (2oia) and the remaining joints are modelled with single axes (1a). The axis parameters and residual rotational and translational discrepancies are modelled as described above.

Additionally, whole finger postures are matched with both methods. For this, the joints are concatenated to form kinematic chains. The global pose and the joint angles are optimised to minimise the mean rotational and translational discrepancies between the modelled and measured bone poses. For this, a weighting between the rotational and translational discrepancy is decided. One millimetre of translational discrepancy is treated with the same weight as one degree of rotational discrepancy.

The means of the residual discrepancies are tested with a two-tailed Student's t test for unpaired data, with a significance threshold $\alpha = 5\%$, to find out if they are statistically significantly different. We conjecture that the MRI method will lead to lower residuals than MoCap, because the measurements are not disturbed by soft tissue artefacts. The null hypothesis is that the mean residuals are equal.

3 Results

The calculation steps described in the previous Methods section lead to optimised joint parameters. By setting a limit on the modelling error, the joint types for each joint are found (Sect. 3.1). The modelling error is computed for each joint and checked by a leave-one-out cross-validation (Sect. 3.2). The measurement error is assessed by a repeatability test (Sect. 3.3). As far as available, results are compared to data from literature (Sect. 3.4). The software implementation of the hand model is introduced (Sect. 3.5).

3.1 Joint Types

The main results of the presented method are movement models of the analysed human hand. Depending on the desired accuracy in terms of discrepancy between modelled and measured bone poses, hand models with different complexity are generated. In Fig. 7, different hand models from simple (*top*) to complex (*bottom*) are presented.

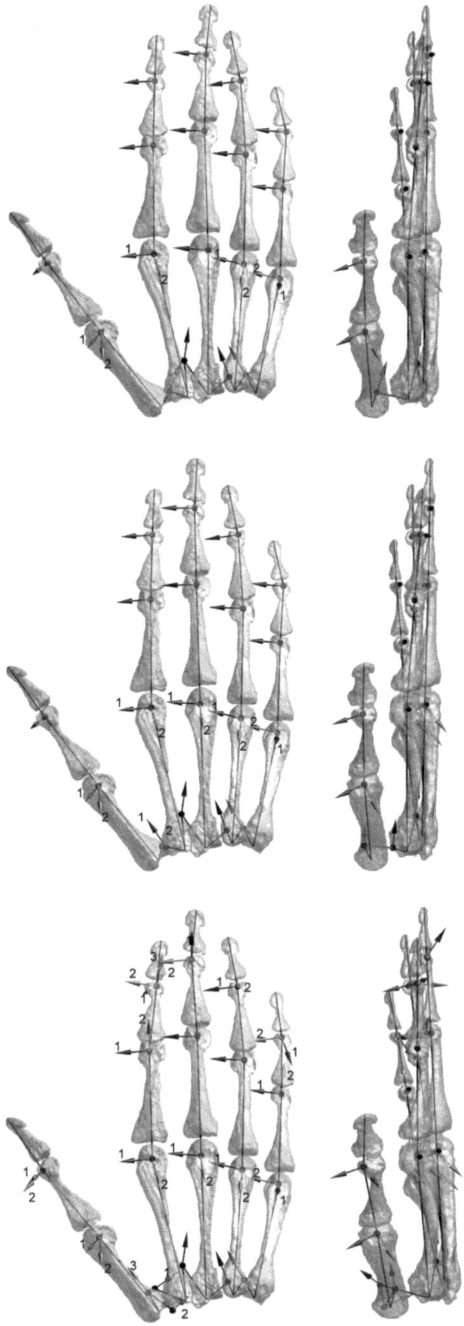

Fig. 7 Variants of the kinematic model at different accuracy constraints, dorsal view (*left*) and radial view (*right*). *Top:* 22 DoF, rotational deviation $<9°$, translational deviation <6 mm. *Middle:* 24 DoF, rotational deviation $<6°$, translational deviation <3 mm. *Bottom:* 33 DoF, rotational deviation $<3°$, translational deviation <2 mm. In joints with more than one axis, the first one is marked "1", the second one "2", and, if existing, the third one "3"

In the simple model, four joints are modelled as 2-DoF universal joints: thumb, index, ring and little finger MCP. The other joints are modelled as 1-DoF hinge joints.

The intermediately complex hand model (*middle*) differs from the simple one by providing two DoF each to MCP3 and CMC1. The joint axes of MCP3 intersect, while the ones of CMC1 do not.

The most complex model (*bottom*) models CMC1 with three non-intersecting axes, with the third one allowing a longitudinal rotation (pro-/supination) of MC1. A longitudinal rotation is also enabled in DIP2 and PIP5, while PIP2 allows a combined longitudinal rotation and sidewards movement. The little finger DIP joint allows a longitudinal rotation only in an extended position. Additional DoF for sidewards movement are found in DIP2, DIP3, DIP4 and IP1.

The weighted-mean rotational deviation per joint ranges from 1.6° in IMC3 to 5.5° in IP1. The maximum rotational deviation in a single hand posture is 17.2° in CMC1. Weighted-mean translational deviation ranges from 0.9 mm (PIP4) to 2.6 mm (CMC1), and the maximum translational deviation in a single hand posture is 7.2 mm, and also occurs in CMC1. The examples in Fig. 8 are supposed to give a feel of these values.

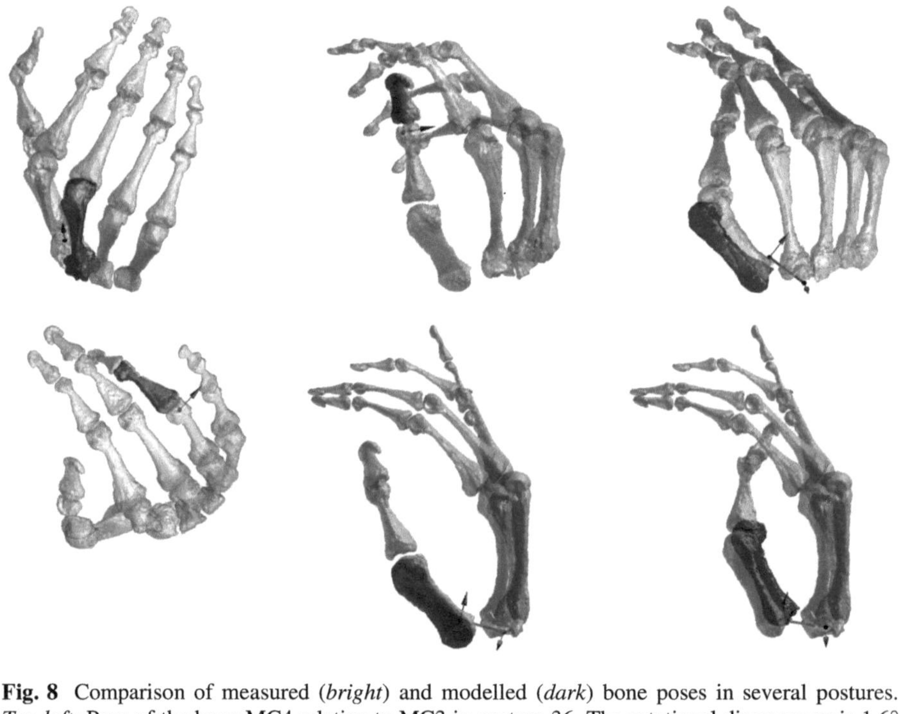

Fig. 8 Comparison of measured (*bright*) and modelled (*dark*) bone poses in several postures. *Top left:* Pose of the bone MC4 relative to MC3 in posture 36. The rotational discrepancy is 1.6° and the translational discrepancy is 1.0 mm. The arrow is the rotation axis of the modelled IMC4 joint that connects MC3 and MC4. *Top middle:* DP1 relative to PP1 in posture 1. Discrepancy: 5.5°, 1.4 mm. IP1 joint. *Top right:* MC1 relative to MC2 in posture 29. Discrepancy: 17.2°, 6.4 mm. CMC1 joint. *Bottom left:* MP4 relative to PP4. Discrepancy: 2.6°, 0.9 mm. PIP4 joint. *Bottom middle:* MC1 relative to MC2 in posture 24. Discrepancy: 5.5°, 2.6 mm. *Bottom right:* MC1 relative to MC2 in posture 35. Discrepancy: 5.1°, 7.2 mm

3.2 Cross-Validation of the Results

For most joints, there is only a slight increase of the rotational and translational modelling error from the whole data mean error (columns 3 and 5) to the LOOCV mean error. For example, in the thumb MCP joint, the mean angular when using all poses is 2.5°, and the mean angular error of the LOOCV is 2.9°. In the same joint, the mean translational error is 1.2 mm when taking into account all poses and 1.3 mm in the LOOCV analysis. This means that the results are generally valid for the investigated individual hand and do not depend on certain postures.

All differences for the translational error are within 0.2 mm and all differences for the rotational error are within 1.0° except for the thumb interphalangeal joint, where the difference is 1.2° and the little finger metacarpophalangeal joint, where it is 3.0°. In these exceptional cases the joint parameters depend strongly on the selection of the subset of bone poses. This means that there are single extreme poses is the data that are not adequately represented by the other poses.

3.3 Motion Estimation Repeatability

The repeatability of the motion estimations is examined by repeating it 100 times with randomly permuted point sets. The standard deviation of the rotation and translation estimate is given in Table 1 as the square root of the variance described in Eqs. (5) and (6).

3.4 Comparison with Results from Literature

Comparing the range of motion (RoM) of the MCP joints with the values provided by [31], in most points the results agree, but some are different [31] states that

(1) the flexion RoM "falls short of 90° for the index but increases progressively with the other fingers",
(2) "active extension [...] can reach up to 30° or 40°" and
(3) "the index finger has the greatest side-to-side movement".

In the hand we used in our measurements, a smaller flexion RoM for the index finger (80°) is measured, while there is an increase towards in the little finger: middle and ring finger are similar with 86° and 84°, respectively, and the little finger has a flexion RoM of 95°. The active extension RoM is 30° (index), 33° (middle), 43° (ring) and 52° (little finger) and therefore higher than the ones in [31] for the ring and little finger. We agree that, for the hand we investigated, index finger side-to-side movement is greater than that of the other fingers with 59° (index), 43° (middle), 44° (ring) and 54° (little finger).

Table 1 Standard deviation of the motion estimation for the rotational (σ_r) and translational (σ_t) part. The minimum, maximum and mean over all images are given

Bone	σ_r (°)			σ_t (mm)		
	Min	Max	Mean	Min	Max	Mean
MC1	1.0	5.3	3.0	0.1	0.2	0.1
PP1	1.6	5.7	3.2	0.1	0.3	0.1
PD1	1.2	5.4	1.8	0.1	0.3	0.1
MC2	1.7	8.0	3.2	0.1	0.4	0.2
PP2	1.0	5.9	2.8	0.1	0.2	0.1
PM2	1.2	3.7	2.1	0.0	0.1	0.1
PD2	3.2	4.9	3.9	0.0	0.1	0.1
MC3	1.1	23.7	2.6	0.1	0.5	0.2
PP3	1.3	5.8	3.1	0.1	0.2	0.1
PM3	1.0	2.9	1.7	0.0	0.1	0.1
PD3	2.9	5.1	3.3	0.1	0.4	0.2
MC4	1.4	7.6	3.3	0.1	0.5	0.1
PP4	0.9	8.9	3.4	0.1	0.2	0.1
PM4	1.2	8.5	2.5	0.0	0.4	0.1
PD4	2.1	4.3	2.9	0.1	0.2	0.1
MC5	1.4	11.2	3.6	0.1	0.4	0.1
PP5	1.5	7.1	3.3	0.0	0.1	0.1
PM5	1.0	4.3	2.8	0.0	0.1	0.1
PD5	0.1	4.4	3.1	0.0	0.1	0.0
All	0.1	23.7	2.9	0.0	0.5	0.1

For the PIP and DIP joint, Kapandji states that

(1) the "range of flexion in the PIP joints is greater than 90°" and
(2) in the DIP joints "slightly less than 90°",
(3) the range "increases from the second to fifth finger" to 135° (PIP5) and
(4) to "a maximum of 90°" (DIP5);
(5) the range for PIP extension is zero and
(6) for active DIP extension zero or trivial.

Our measurements agree with points 1 and 2. An increase from second to fifth finger (points 3 and 4) is not observed. Small PIP and DIP extension angles (up to 22.5°) are measured.

3.5 OpenSim

The OpenSim model of the hand contains simplified bone geometries and the full set of joints. The joints can be moved by sliders (Fig. 9). The OpenSim framework also provides for the extension of the model with tendons and muscles [21]. The model is available for download as a zip-compressed file at http://www.robotic.dlr.de/human_hand.

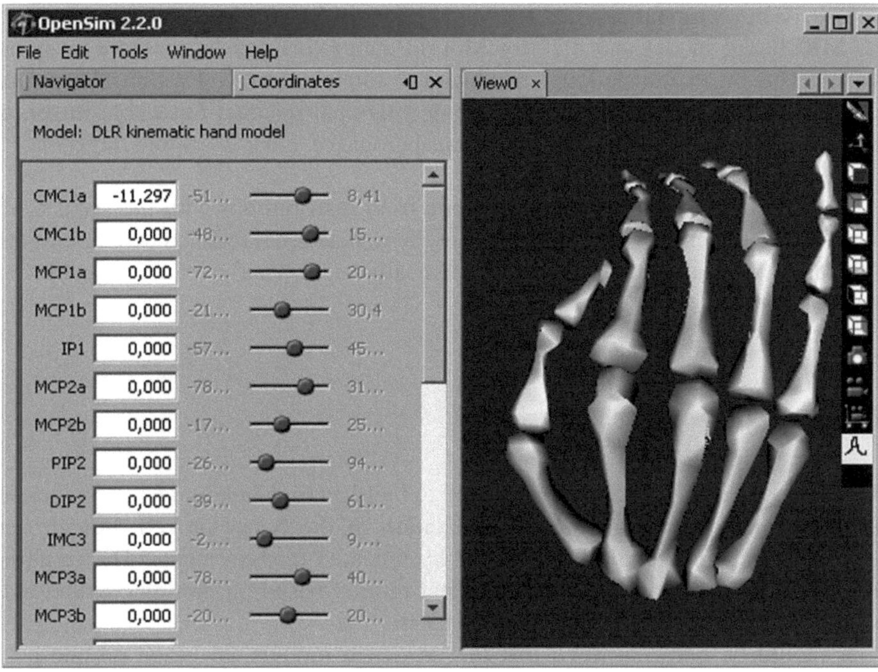

Fig. 9 Screenshot of the OpenSim application. Each DoF of the skeletal model can be moved by a slider

3.6 Dependence of Results on Starting Points

The result of the parameter identification is in some joints sensitive to the optimisation starting point and in others not. For example, the parameters of the CMC1 joint were optimised with three different starting points for each of the two axis orientations and three different starting points for the axis offset. The results are slightly sensitive to the axis orientation starting points, with rotational error ranging from 2.95° to 3.18°. The variation of the axis offset starting point has no effect on the results. In other joints, for example the IP1 joint, the optimisation starting points have no effect on the result.

3.7 MRI versus MoCap

MRI and optical motion capture (MoCap) were compared with respect to the residual discrepancies, as described in Sect. 2.6. The results of this comparison are shown in Fig. 12.

The mean rotational residual for single joints is 4.4° for MRI data explained by the MRI-based model and 4.7° for MoCap data explained by the MoCap-based model. The mean translational residual for single joints is 1.4 mm (MRI) and 1.2 mm (MoCap) respectively. The mean rotational residual for whole fingers is 3.5° and 3.6°, respectively. The mean translational residual for whole fingers is 1.2 and 1.5 mm, respectively.

The null hypothesis that the data come from distributions with the same mean could not be rejected except in the case of the translational residual of single joints. The mean translational residual is statistically significantly smaller for MoCap than for MRI.

4 Discussion

We showed that MRI images can be used to build a movement model of the human hand. Estimating the longitudinal orientation of the bones proved difficult, but reasonable accuracy could be achieved by repeated measurements. The resulting model can be used for visualising skeletal motion based on motion capture data, using the freely available software OpenSim.

It should be borne in mind that the presented models are based on one individual hand. Since different people have different hands in terms of e.g. size, flexibility and possibly kinematics, the models should be used with caution. In OpenSim there is, however, a functionality to scale skeletal models according to measured marker movements. This might compensate part of the inter-individual differences.

4.1 Consideration of Errors

There are six sources for errors in the kinematic modelling process:

(1) selection of postures,
(2) MRI acquisition,
(3) segmentation,
(4) motion estimation,
(5) joint definition, and
(6) joint parameter identification.

It is impossible to consider all possible postures of each joint as they are infinite. Ideal, therefore, would be a very dense sampling of postures during a large number of different movements. This is not possible in MRI due to cost and time constraints. Hand postures for this work are selected so that for each joint, the extremes and some intermediate positions are covered. The selection of postures influences the resulting model in the way that multiple recordings of similar joint postures assign them a greater weight compared to postures that occur only once.

In MRI acquisition, same tissue can be represented by different grey values. Artefacts such as missing parts, motion artefacts, artefacts due to the surrounding tissue and possibly distortions can occur. A discretisation error occurs due to the spacial resolution of $(0.76\ \text{mm})^3$.

In the segmentation process, the segmented shape depends on the way the operator defines the grey value thresholds and manually refines the selection. The combined error of MRI acquisition and segmentation is illustrated by the distributions of grey value and segmented volume (Figs. 10 and 11).

An attempt was made at measuring the error of the MRI image acquisition by taking images of an animal bone without surrounding tissue, in order to discard the need for segmentation. However, the images showed hardly any signal, which might be due to the missing surrounding tissue or to a lack of humidity within the bone.

The motion estimation error depends on the quality of the segmented point clouds and the robustness of the algorithm with respect to differences in shape and

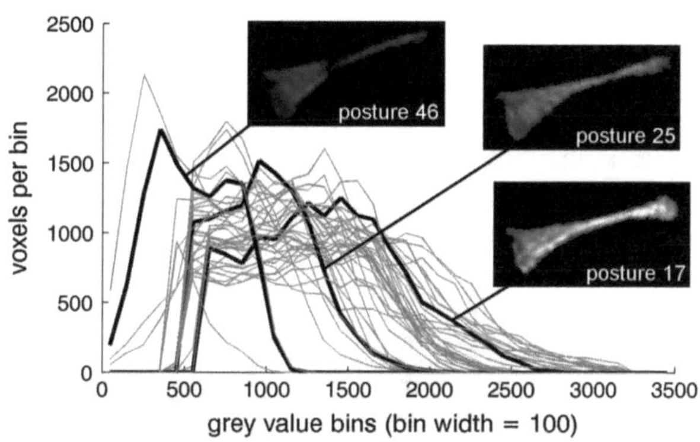

Fig. 10 Histograms of the grey value distributions of middle finger medial phalanx in different segmented MRI images. Three clearly different examples are highlighted. Of these the central sagittal slice of the MRI image is shown

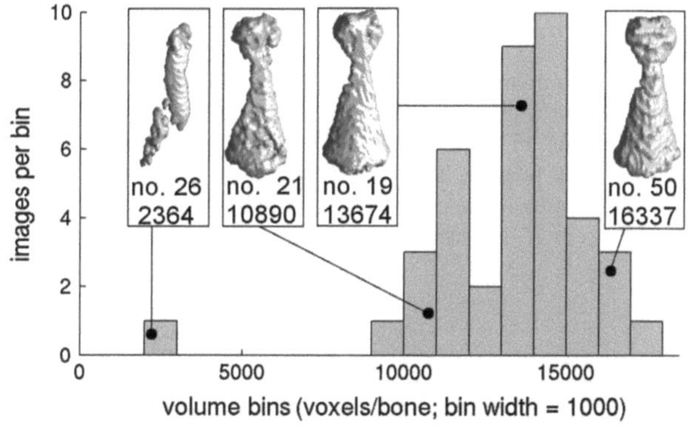

Fig. 11 Histogram of the segmented volumes of the middle finger medial phalanx in different MRI images. Surface renderings of four examples are shown, with the image numbers and volumes in number of voxels

grey value distribution. The combined error of steps 2 to 4 is partly expressed by the repeatability values in Table 1, which however do not reflect a potential bias.

In this work, joints are modelled as rotational joints with constant parameters. In the case of a 1-DoF joint, this corresponds to rigid joint surfaces with perfectly circular cross-sections orthogonal to the joint axis. The 3-DoF joint with intersecting axes would be ideally represented by spherical joint surfaces. These are simplifications of the human joints with elastic cartilage and more complex surfaces.

The parameters of the defined joints are identified by way of numerical optimisations. These may introduce error by finding local optima. This error is assessed by the robustness of the result in the face of varying starting points (Sect. 3.6).

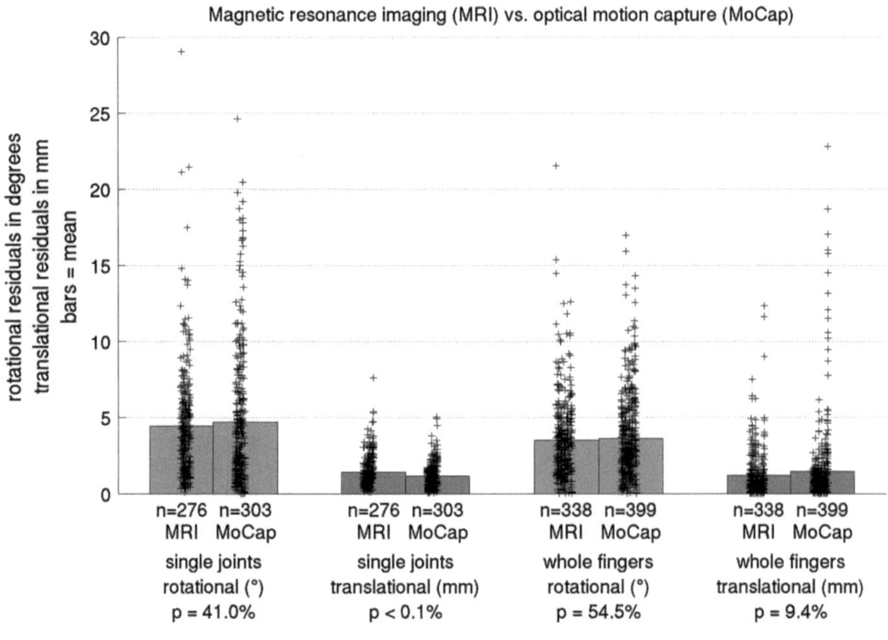

Fig. 12 Comparison of MRI and MoCap. On the left, single bone poses are measured, and on the right, whole finger postures. The points marked by "plus" signs show the residual rotational discrepancies in degrees and the translational discrepancies in mm, respectively, of each bone pose in each posture. The bars show the mean values. The value n is the number of bone poses, and the p-value is the probability that a difference of means equal or larger than the observed one would occur if data were sampled from normal distributions with an identical mean

4.2 Joint Types

The model in the middle of Fig. 7 seems to be the most natural one with two DoF for the thumb CMC and the MCP joints and one DoF for all other joints. The simpler model lacks a second DoF in the thumb CMC joint and in the middle finger MCP joint. On the other hand, the more complex model at the bottom has many additional DoF: in the thumb CMC and IP joint, in the PIP joints of index and little finger and in all the DIP joints. The tendon structure of the hand makes it seem unlikely or impossible that these axes represent independently actuated DoF. They probably compensate bone pose measurement errors or deviations of the bones from a perfectly circular path that really occur in the hand.

4.3 Thumb CMC Joint

At the thumb CMC joint, the largest translational error occurs. This might be due to the fact that the thumb metacarpal poses are determined with respect to the index finger metacarpal. However, the bone that the thumb metacarpal articulates with is the trapezium bone, which is one of the carpal bones. Another possible explanation is that the motion of this joint is more complex, so that simple rotation axes are not sufficient to fully model it.

4.4 Comparison Between MRI and Optical Motion Capture

Fitting a model with equal number of DoF to either MRI or optical motion capture (MoCap) data yielded no statistically significant differences in the mean residuals in three of four comparisons, and one statistically significant difference in favour of MoCap (Fig. 12). This contradicts our initial hypothesis that MRI data can be fitted with significantly smaller residuals. The effect of the soft tissue artefact on MoCap data seems not to be as strong as initially postulated.

Acknowledgments The authors would like to thank Karolina Stonawska for the tedious work of segmenting the bones. This project was partly funded by the EU project The Hand Embodied (FP7-ICT-248587).

References

1. M. Grebenstein, A. Albu-Schäffer, T. Bahls, M. Chalon, O. Eiberger, W. Friedl, R. Gruber, U. Hagn, R. Haslinger, H. Höppner, S. Jörg, M. Nickl, A. Nothhelfer, F. Petit, J. Reill, N. Seitz, T. Wimböck, S. Wolf, T. Wüsthoff, G. Hirzinger, The DLR hand arm system, in *2011 IEEE International Conference on Robotics and Automation* (2011)

2. M. Grebenstein, M. Chalon, G. Hirzinger, R. Siegwart, A method for hand kinematics designers—7 billion perfect hands, in *Proceedings of 1st International Conference on Applied Bionics and Biomechanics* (2010)
3. Y. Youm, T.E. Gillespie, A.E. Flatt, B.L. Sprague, Kinematic investigation of normal MCP joint. J. Biomech. **11**, 109–118 (1978)
4. K.N. An, E.Y. Chao, I.W.P. Cooney, R.L. Linscheid, Normative model of human hand for biomechanical analysis. J. Biomech. **12**, 775–788 (1979)
5. B. Buchholz, T.J. Armstrong, S.A. Goldstein, Anthropometric data for describing the kinematics of the human hands. Ergonomics **35**(3), 261–273 (1992)
6. A. Hollister, W.L. Buford, L.M. Myers, D.J. Giurintano, A. Novick, The axes of rotation of the thumb carpometacarpal joint. J. Orthop. Res. **10**, 454–460 (1992)
7. A. Hollister, D.J. Giurintano, W.L. Buford, L.M. Myers, A. Novick, The axes of rotation of the thumb interphalangeal and metacarpophalangeal joints. Clin. Orthop. Relat. Res. **320**, 188–193 (1995)
8. J.L. Sancho-Bru, A. Pérez-González, M. Vergara-Monedero, D. Giurintano, A 3-D dynamic model of human finger for studying free movements. J. Biomech. **34**, 1491–1500 (2001)
9. G.S. Rash, P. Belliappa, M.P. Wachowiak, N.N. Somia, A. Gupta, A demonstration of the validity of a 3-D video motion analysis method for measuring finger flexion and extension. J. Biomech. **32**(12), 1337–1341 (1999)
10. L.-C. Kuo, F.-C. Su, H.-Y. Chiu, C.-Y. Yu, Feasibility of using a video-based motion analysis system for measuring thumb kinematics. J. Biomech. **35**, 1499–1506 (2002)
11. X. Zhang, L. Sang-Wook, P. Braido, Determining finger segmental centers of rotation in flexion-extension based on surface marker measurement. J. Biomech. **36**, 1097–1102 (2003)
12. P. Cerveri, N. Lopomo, A. Pedotti, G. Ferrigno, Derivation of centers of rotation for wrist and fingers in a hand kinematic model: Methods and reliability results. Ann. Biomed. Eng. **33**, 402–412 (2005)
13. L.Y. Chang, N.S. Pollard, Constrained least-squares optimization for robust estimation of center of rotation. J. Biomech. **40**(6), 1392–1400 (2007)
14. L.Y. Chang, N.S. Pollard, Robust estimation of dominant axis of rotation. J. Biomech. **40**(12), 2707–2715 (2007)
15. L.Y. Chang, N.S. Pollard, Method for determining kinematic parameters of the in vivo thumb carpometaracpal joint. IEEE Trans. Biomed. Eng. **55**(7), 1879ff (2008)
16. Dexmart, Deliverable D1.1 kinematic model of the human hand, Dexmart, Technical Report, 2009
17. J.H. Ryu, N. Miyata, M. Kouchi, M. Mochimaru, K.H. Lee, Analysis of skin movement with respect to flexional bone motion using mr images of a hand. J. Biomech. **39**, 844–852 (2006)
18. K. Oberhofer, K. Mithraratne, N. Stott, I. Anderson, Error propagation from kinematic data to modeled muscle-tendon lengths during walking. J. Biomech. **42**, 77–81 (2009)
19. N. Miyata, M. Kouchi, M. Mochimaru, T. Kurihaya, Finger joint kinematics from MR images, in *IEEE/RSJ International Conference on Intelligent Robots and Systems* (2005)
20. M. Grebenstein, P. van der Smagt, Antagonism for a highly anthropomorphic hand-arm system. Adv. Robot. **22**(1), 39–55 (2008)
21. S.L. Delp, F. Anderson, S. Arnold, P.J. Loan, A. Habib, C. John, E. Guendelman, D. Thelen, OpenSim: open-source software to create and analyze dynamic simulations of movement. IEEE Trans. Biomed. Eng. **54**, 1940–1950 (2007)
22. R.V. Gonzales, T.S. Buchanan, S.L. Delp, How muscle architecture and moment arms affect wrist flexion-extension moments. J. Biomech. **30**, 705–712 (1997)
23. A. Kapandji, Cotation clinique de l'opposition et de la contre-opposition du pouce [clinical test of opposition and counter-opposition of the thumb]. Ann. Chir. Main. **5**(1), 67–73 (1986)
24. U. Hillenbrand, Non-parametric 3D shape warping, in *Proceedings International Conference on Pattern Recognition (ICPR)* (2010)
25. U. Hillenbrand, Consistent parameter clustering: definition and analysis. Pattern Recogn. Lett. **28**, 1112–1122 (2007)

26. U. Hillenbrand, A. Fuchs, An experimental study of four variants of pose clustering from dense range data. Comput. Vis. Image Underst. **115**(10), 1427–1448 (2011). http://www.sciencedirect.com/science/article/pii/S1077314211001445
27. B.K. Horn, Closed-form solution of absolute orientation using unit quaternions. J. Opt. Soc. Am. A **4**(4), 629–642 (1987)
28. K. Fukunaga, L.D. Hostetler, The estimation of a gradient of a density function, with applications in pattern recognition. IEEE Trans. Inf. Theory **21**, 32–40 (1975)
29. D. Comaniciu, P. Meer, Mean shift: a robust approach toward feature space analysis. IEEE Trans. Pattern Anal. Mach. Intell. **24**, 603–619 (2002)
30. J.A. Nelder, R. Mead, A simplex method for function minimization. Comput. J. **7**(4), 308–313 (1965)
31. A. Kapandji, *The Physiology of the Joints* (Churchill Livingstone, Edinburgh, 1998)
32. Wikipedia, Hand (2009) http://en.wikipedia.org/wiki/Hand
33. Wikipedia, Hinge joint (2006) http://en.wikipedia.org/wiki/Hinge_joint

Chapter 4
Transmission of Musculotendon Forces to the Index Finger

Sang Wook Lee and Derek G. Kamper

Abstract This chapter reviews work completed by the authors and others to examine the mechanisms of musculotendon force transmission to the index finger and, ultimately, to generation of desired force at or movement of the fingertip. Specifically we examined the roles of finger posture, passive joint impedance, anatomical pulleys, and the extensor hood in mapping muscle forces to finger dynamics. Results from in vivo and in vitro experiments, as well as from biomechanical modeling of the musculotendon structure of the index finger, are provided. These findings can inform both the study of motor control of the hand and the potential design of robotic end-effectors.

Keywords Index finger · Musculotendon · Moment arm · Posture · Passive joint impedance · Pulley · Extensor hood

S. W. Lee (✉)
Department of Biomedical Engineering, The Catholic University of America, Washington, DC 20064, USA
e-mail: leeb@cua.edu

S. W. Lee
Center for Applied Biomechanics and Rehabilitation Research, MedStar National Rehabilitation Hospital, Washington, DC 20010, USA

D. G. Kamper
Department of Biomedical Engineering, Illinois Institute of Technology, Chicago, IL 60616, USA
e-mail: kamper@iit.edu

D. G. Kamper
Sensory Motor Performance Program, Rehabilitation Institute of Chicago, Chicago, IL 60611, USA

R. Balasubramanian and V. J. Santos (eds.), *The Human Hand as an Inspiration for Robot Hand Development*, Springer Tracts in Advanced Robotics 95,
DOI: 10.1007/978-3-319-03017-3_4, © Springer International Publishing Switzerland 2014

1 Introduction

1.1 Complexity of the Human Hand

The human hand consists of a remarkably complex musculotendon structure designed to maximize power and agility while minimizing mass and bulk of the hand. A total of 39 extrinsic and intrinsic musculotendon units control the movements of the 27 bones in each hand. Over 25 degrees-of-freedom (DOF) are available to generate the sophisticated finger and wrist movements possible with the human hand [1].

1.2 Index Finger Musculotendon Structure and Tendon Force Transmission

1.2.1 Musculoskeletal Structure of the Index Finger

The index finger alone is comprised of three joints: metacarpophalangeal (MCP), proximal interphalangeal (PIP), and distal interphalangeal (DIP) joints with a total of four DOF (one each for DIP and PIP, and two for MCP). These joints connect the four bones of the index finger: metacarpal, proximal phalanx, middle phalanx, and distal phalanx. Motion about the joints is controlled by seven different muscles: four extrinsic muscles which are located in the forearm and three intrinsic muscles contained within the hand. The extrinsic muscles consist of two long extensor muscles, extensor digitorum communis (EDC) and extensor indicis proprius (EIP) and two long flexor muscles, flexor digitorum superficialis (FDS) and flexor digitorum profundus (FDP). As the relatively large muscle bellies for these muscles reside in the forearm, only their tendons run through the hand, thereby reducing the necessary weight and size of the hand. The intrinsic muscles consist of the first dorsal interosseous (FDI), first palmar interosseous (FPI), and lumbrical (LUM) muscles (Fig. 1). The intrinsic muscles are generally smaller in size (aside from FDI) than the extrinsic muscles and are thought to have a greater

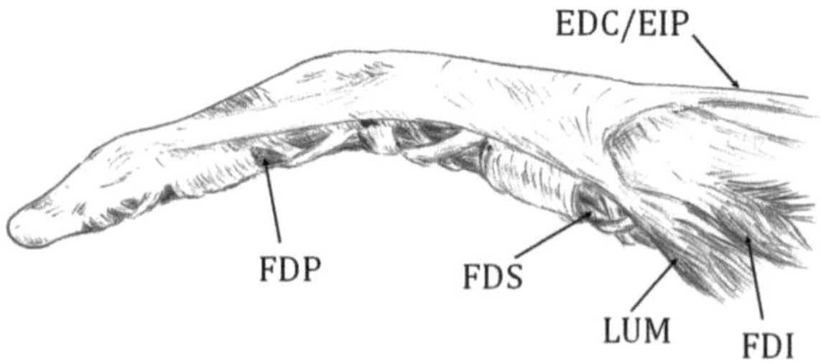

Fig. 1 Index finger musculotendon structure (*Right finger, radial view*)

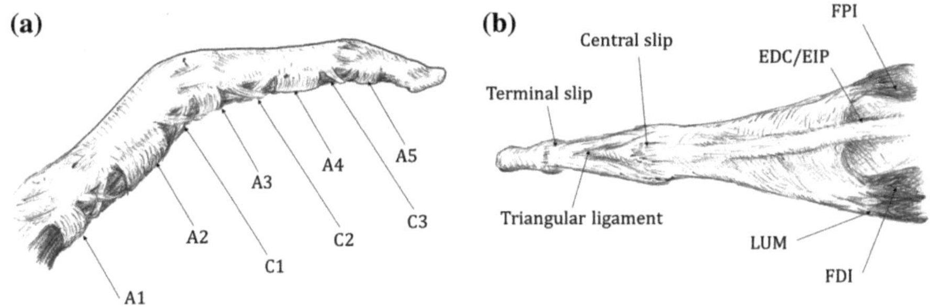

Fig. 2 Biomechanical structures within the index finger: **a** *Lateral view* pulley structures constraining tendon path, **b** *Dorsal view* extensor hood

impact on fingertip force direction than magnitude [2]. Each of these index finger muscles impacts multiple DOF.

1.2.2 Musculotendon Connections on Palmar Side

On the palmar side of the fingers, both FDP and FDS travel through a series of anatomical pulleys that shape the tendon path (Fig. 2a). The pulley system is comprised of five annular pulleys and three cruciform pulleys, which arise either from the phalanges or from the volar plates of the three joints [3]. The direct tendon pulling forces are applied to only more distal segments, i.e., the distal phalanx for FDP and the middle phalanx for FDS. The pulleys restrict bowstringing of the flexor tendons as the finger joints flex. This has the effect of transmitting some of the force within the FDP and FDS tendons to other segments of the finger. The magnitude and direction of the force applied to the pulley is dependent upon the joint posture. Thus, the anatomical pulleys impact fingertip motion or fingertip force production resulting from a specific muscle activation pattern.

1.2.3 Musculotendon Connections on the Dorsal Side

On the dorsal side of each finger, tendons from multiple muscles insert into the extensor hood, an aponeurotic sheet, also referred to as the "extensor apparatus" [4], "dorsal aponeurosis" [5], "tendinous rhombus" [6], and "extensor mechanism" [7] (Fig. 2b). In the index finger, the hood is comprised of the tendons for EDC, EIP, FPI, and LUM, and sometimes FDI. Transverse and oblique fibers connect the tendons to each other (i.e., EDC to the lateral bands from LUM and FPI) or to the palmar plate near the MCP joint (sagittal bands, [8]) (Fig. 2b). The structure is also connected by transverse and oblique retinacular ligaments to the palmar plate at the PIP joint and into the pulleys of the flexor tendons. The extensor hood has two primary insertions into the phalanges: the central slip from the EDC and EIP tendons inserts into the base of the middle phalanx just distal to the PIP joint and the terminal slip, arising from two lateral bands with contributions from EDC, EIP, FPI, and LUM, inserts into the dorsal base of the distal phalanx just beyond the DIP joint.

Due to its complex structure, tendon force transmission within the extensor hood and its contribution to the finger kinetics is still not well understood. It has been studied by using optimization techniques, which usually minimize the weighted sum of muscle forces or maximize the endurance [9–12]. Strains and geometric changes in selected regions of the extensor mechanism have also been directly measured by varying tendon loading conditions and finger postures [7, 13, 14].

1.2.4 Tendon Force Transmission Within the Hand

Ultimately, the different structures of the finger work together to position the fingertip or to create desired forces between the fingertip and an object. Fundamental to this control is the translation of musculotendon force into fingertip movement/force. Both in vivo and in vitro studies have examined relationships between muscle activations, or tendon forces, and fingertip forces/moments [2, 15–17]. The contribution of individual tendons to finger kinetics, however, has not been evaluated explicitly. The direct measurement techniques that correlate the measured joint moment to the applied tendon force via in vitro experiments have been implemented to evaluate the contribution of tendon forces to joint moments [18–20], but the measured joint moments usually include passive joint moments produced by extrinsic (inactive muscle–tendon units) and intrinsic (cartilage, ligament, joint capsules) tissues of the hand. Several dynamic models of the finger have also been developed to elucidate the kinetic functions of the finger flexor muscles [9, 10, 17, 21]; however, most of these dynamic models employed inverse dynamics, for which the mechanism involved in the transformation of muscle forces into joint torques cannot be examined in detail.

As noted, almost all of the tendons in the finger cross more than one joint. Thus, forces are transmitted from the insertion sites to more proximal segments via joint reaction forces and resistance torques generated by passive joint impedance (i.e., stiffness and damping). This passive impedance seems especially important for torque transmission to proximal joints [10, 22, 23]. Local stiffness is highly dependent upon joint angle [22]. Similarly, joint posture affects muscle length [24], as well as contact and pulley forces, thereby impacting the mapping of tendon to fingertip forces. Thus, knowledge of finger posture is vital for control of the hand. Proprioception provides key information regarding the kinematic state of the joints. Proprioceptive perception is derived from a variety of sensory inputs such as muscle spindle afferents, cutaneous mechanoreceptors, joint receptors, and Golgi tendon organs (for details, see the chapter '*Proprioceptive mechanisms and the human hand*' by Walsh, Taylor, and Gandevia).

1.3 Outline of the Chapter

The goal of this chapter is to elucidate the means by which musculotendon forces are transmitted to the index finger to yield desired force production at or movement of the fingertip. This mapping in the fingers is complicated by four key factors: passive joint impedance, multi-articular musculotendons, anatomical pulleys, and the extensor hood. Results of in vivo and in vitro experiments, as well as of biomechanical modeling of the musculotendon structure of the index finger, are provided.

2 Joint Impedance

The mass and inertia of the finger segments are quite small relative to other segments. Accordingly, gravitational and inertial forces have a smaller impact on finger dynamics. Rather, the resistance torques generated by passive joint impedance (e.g., stiffness and damping) play a much larger role in determining finger movement. This impedance arises from the mechanical properties of surrounding passive tissues such as ligaments, skin, and inactive muscles and tendons. A variety of measurement techniques have been implemented to estimate these passive properties and the resulting impedance.

2.1 Parameter Estimation: Different Models

2.1.1 Mathematical Model of Nonlinear Viscoelastic Joint Passive Moment [25]

In this study, the authors aimed to identify the viscoelastic property of the passive moment, i.e., the dissipative part of the passive moment of the finger MCP joints. They utilized two different tests. First, the relaxation of the moment was measured over a range of MCP joint angles in the index finger. Then, in a constant-velocity test, the joint moments were measured while the joint was cycled through its full range of motion. Measurements were made at five different angular speeds to determine the effect of rate of joint rotation on passive joint moment. A customized device, consisting of a vertical cantilever beam and a strain gauge, was used to measure the MCP joint moment. During measurements, the index finger was strapped to a brace so that the DIP and PIP joint angles remained fixed throughout the MCP motion.

Based on the experimental data obtained from three subjects, empirical models of moment relaxation and instantaneous elastic response of the finger joints were determined. While the proposed mathematical model was capable of describing the behavior of the passive finger joint moments, a large degree of inter-subject variability in the identified model parameters was also observed. More importantly, the behavior of the dissipative part of the passive moment required modeling by joint speed raised to a power less than one. This indicates that the dissipative part of the passive moment cannot be described by traditional linearly viscous models or a Coulomb friction model. Values were not obtained for the PIP or DIP joints.

2.1.2 Intrinsic and Extrinsic Contribution to MCP Passive Moment [26]

The particular objective of this study was to delineate the relative contributions of the intrinsic tissues crossing a single joint, such as articular cartilage, ligaments and inactive intrinsic muscles, and of the extrinsic tissues crossing multiple joints, i.e., inactive multi-articular musculotendons (e.g., FDP and FDS tendons), on the passive impedance of the MCP joint. The passive moment about the MCP joint was measured by the same custom-design device used in [25] during a complete cycle of joint extension and flexion at various wrist positions. Therefore, the intrinsic component was modeled as a function of the MCP joint angle, and the extrinsic component as a function of the MCP joint angle and wrist angle.

The results of this study indicated that the contribution of extrinsic components becomes significant near the flexion and/or extension limits of the MCP joint angle, while the contribution of intrinsic tissues is dominant when the wrist is flexed or extended sufficiently to slacken the extrinsic tissues.

2.1.3 System Identification Approach [22, 27]

In this study, the finger passive joint impedance was estimated in an attempt to accurately describe the role of the long finger flexors (i.e., FDP and FDS) on index finger movement generation. The authors employed a system identification technique to estimate the dynamic characteristics of the passive joint impedance (i.e., stiffness and damping). Pseudo-random binary sequences of ±2° in amplitude were imposed at different operating points (i.e., flexion angles) of each joint, and the dynamic characteristics of the joint passive impedance were computed by fitting a linear 2nd-order inertia-damping-stiffness dynamic system model at each operating point (Eq. 1). Local stiffness was found to vary substantially with joint angle for all three joints, but the damping coefficients remained relatively constant across postures.

$$\tau = I\ddot{\theta} + B\dot{\theta} + K(\theta - \theta_0) \tag{1}$$

Due to limitations with the system, however, only one joint could be rotated at any one time; the other two joints were kept in the neutral (0°) flexion/extension posture while the third joint was manipulated. Thus, exploration of the finger workspace was limited. Recently, this group developed a novel actuated exoskeleton for the index finger, the FingerBot, which can independently actuate each of the three joints [27]. A system identification technique similar to the one used in the previous study [22] is currently being employed to measure the passive joint impedance of the index finger throughout the sagittal plane workspace.

2.1.4 Theoretical Model of Joint Passive Moment [10]

The biomechanical model of the finger developed by Sancho-Bru et al. [10] attempted to theoretically model the passive MCP joint moment produced by ulnar and radial ligaments over the MCP joint using a quadratic relationship between the ligament force and its elongation. This model utilized in vitro experimental data from literature to determine the insertion points of the ligaments [28], the rotation and orientation of the joint rotation axes [29], and the ligament stiffness values [30].

2.2 *Impact of Joint Impedance*

Kamper et al. [22] included their experimentally determined values of passive joint impedance in a planar model of index finger dynamics. The authors modeled the FDS and FDP tendons as cables running through pulleys attached to the different finger segments. The inclusion of the passive joint impedance in the finger model was crucial to accurate replication of the movement pattern (i.e., spatial joint coordination pattern) of the in vivo data obtained by the electrical stimulation of the FDS and FDP muscles. The concurrent flexion of all three finger joints that was seen experimentally could only be achieved with simulations including the passive joint impedance.

Similarly, Lee and Kamper [23] assessed the importance of passive joint impedance in tendon force transmission to proximal joints by examining the effects of the removal of passive components using a biomechanical model of the index finger. While the original biomechanical model was capable of accurately predicting the movement pattern generated by the stimulation of FDP muscle (Fig. 3a vs. b), removal of passive components from the model resulted in an abnormal movement pattern of rapid DIP flexion followed by PIP flexion, while the MCP joint was slightly extended (Fig. 3c). This confirms the importance of passive impedance in the tendon force transmission to the proximal joints; without the passive impedance, the proximal joint flexion can only occur after the distal joint reaches its physical flexion boundary, at which point it acts as a rigid rather than rotational connection.

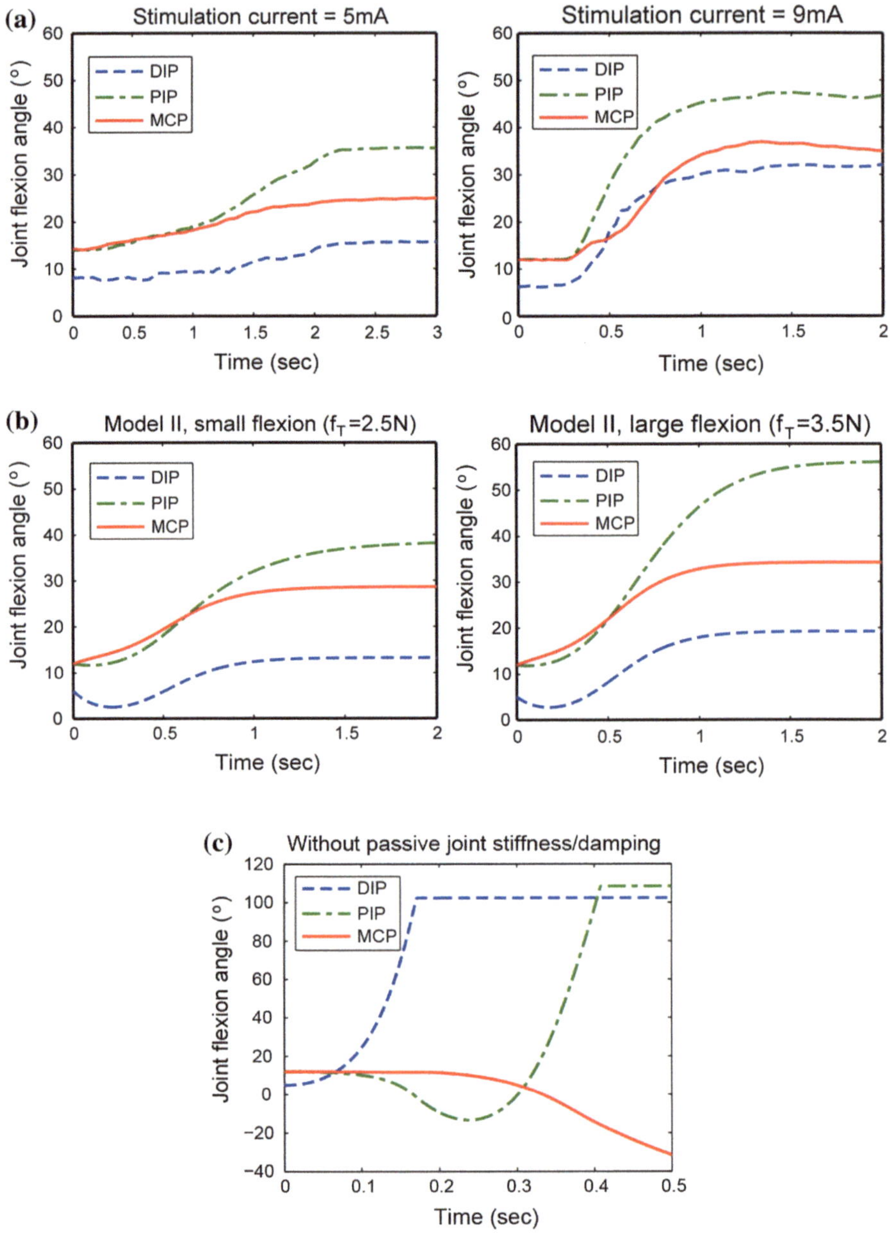

Fig. 3 Finger flexion movements obtained from the experiment (**a**) and the simulation (**b**). Spatial coordination of the joint flexion angles of the model-generated movements (**a**) in response to two levels of tendon force (2.5 vs. 3.5 N) were compared with that of representative flexion movements generated in response to electrical stimulation of FDP(I) (Subject 1) (**a**). **c** Simulation without passive stiffness and damping. FDP tendon force $=$ 0.01 N. Removal of passive joint properties resulted in rapid DIP flexion, followed by PIP flexion; MCP joint was extended. Here, both DIP and PIP joint flexion reached their upper bound. Figure adapted from [23]. Used with permission from IEEE

3 Multi-Articular Musculotendons

Finger musculotendons are multi-articular, thereby impacting multiple joints simultaneously. This makes the mapping between tendon force and joint rotation or moment much more complex than for uniarticular muscles. For example, shortening of FDS could result in rotation of PIP, rotation of MCP, or some combination of the two depending on the states of the joints. The traditional use of geometric moment arms [31] may not fully capture this mapping.

The translation of muscle activation into fingertip movement/force generation is highly dependent upon posture of the entire finger. Joint angle affects internal joint contact forces [32], passive joint impedance, and the extensor hood force distribution [33], in addition to musculotendon length (and therefore force generating capacity [34]). Thus, muscle activation patterns may have to change substantially to create the same result (e.g., fingertip force generation in a desired direction) when the finger posture changes [35, 36].

3.1 Effective Static Moment Arms [32]

To address possible limitations in the use of geometric moment arms, Lee et al. [32] developed a novel parameter, the effective static moment arm (ESMA), to compactly represent the net effects of the tendon force on joint moment production during static force generation. They measured this parameter through cadaveric experiments (Fig. 4). The authors specifically aimed to elucidate the postural effect on the transmission of tendon forces through the extensor hood to joint moments. Thus, the ESMAs for the five tendons contributing to the finger extensor apparatus were estimated by directly correlating the applied tendon force to the measured resultant joint moments in nine different finger postures.

Substantial postural effects on tendon force transmission were confirmed, as repeated measures analysis of variance revealed that the finger posture, specifically interphalangeal (IP) joint angles, had significant impact on the measured ESMA values in seven out of 20 conditions (four DOFs $\times$ five muscle-tendons) (Table 1). Strikingly, MCP ESMA values of the EDC and EIP tendons increased along with the IP joint flexion. ESMA values, therefore, were dependent upon the posture of joints other than the one for which the effective moment arm was being computed; with geometric moment arms only the posture of the joint being examined is considered. Additionally, the abduction ESMA values of all tendons except EDC were mainly affected by MCP flexion.

The ESMA values were generally smaller than the geometric moment arm values obtained from cadaveric experiments [31] or imaging studies [37], which implies that tendon forces are not fully converted to generate joint moments. This reduction may result from a change in tendon force distribution into different

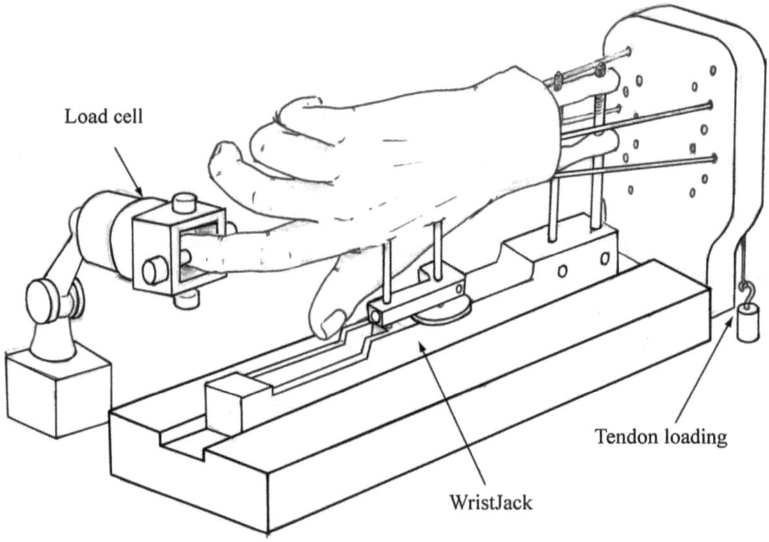

Fig. 4 Experimental setup for the measurement of fingertip force/moment in the cadaveric specimen. This setup was adapted from the work of Valero-Cuevas, et al. [53]. Each specimen was mounted on a fixation device (Agee-Wristjack, Hand Biomechanics Lab, Sacramento, CA), and the index fingertip was secured to a 6 degrees-of-freedom load cell (JR3, Inc., Woodland, CA). Cables connected to the tendons were directed through a metal plate which served as a pulley. Loads were applied to the tendons by hanging weights from the ends of the cables. Figure adapted from [32]. Used with permission from Pergamon Press Ltd

tendon insertions within the extensor apparatus [17]. Note that the MCP moment arm estimated from the geometric excursion method [31] corresponds to the moment-generating capability of the tendon if it were to insert directly into the phalanx on the distal side of the joint. In actuality, the MCP joint extension moment is generated by the tendon forces at two distal insertion sites, as no tendon has an apparent insertion into the proximal phalanx. Another possible mechanism that explains the moment reduction is force dissipation into the surrounding ligaments or joint capsules. Previous studies have reported the insertion/adhesion of part of the extensor apparatus to the MCP joint capsule [38, 39] and have identified components such as oblique retinacular ligaments that connect the extensor hood and the proximal phalanx [40]. Our own dissections of selected specimens confirmed the existence of these connections in most cases, which may serve as pathways for the potential tendon force dissipation. The loss of force would decrease joint moments, and thus fingertip force. Reduction in the ESMA values relative to the geometric moment arms may represent the amount of force dissipation into the surrounding structure.

Table 1 Estimated ESMA mean (SD) values across different postures (unit: mm) (a) affected by IP postures (b) affected by MCP postures (a: abduction, e: extension)

Posture	(a) ESMA				Posture	(b) ESMA			
	EDC MCPe	EIP MCPe	FPI DIPe	LUM DIPe		EIP MCPa	FDI MCPa	FPI MCPa	LUM MCPa
IP1	4.29 (3.50)	4.29 (3.50)	4.29 (3.50)	4.29 (3.50)	IP1	−2.64 (1.64)	−2.64 (1.64)	−2.64 (1.64)	−2.64 (1.64)
IP2	7.88 (1.46)	7.88 (1.46)	7.88 (1.46)	7.88 (1.46)	IP2	−2.06 (1.05)	−2.06 (1.05)	−2.06 (1.05)	−2.06 (1.05)
IP3	8.92 (2.33)	8.92 (2.33)	8.92 (2.33)	8.92 (2.33)	IP3	−0.81 (0.49)	−0.81 (0.49)	−0.81 (0.49)	−0.81 (0.49)

Table adapted from [32]. Used with permission from Pergamon Press Ltd

3.2 Changes in Activation Patterns [36]

As the kinetic function of the finger musculotendons is significantly affected by the joint angles, such postural dependency may directly affect the motor control of the finger musculature, thereby requiring a change in the muscle activation patterns. Accordingly, this study attempted to elucidate the effects of finger posture on the muscle activation patterns of the index finger. The activities of six index finger muscles were recorded with intramuscular electrodes when subjects produced submaximal isometric fingertip forces in six specific orthogonal directions at each of nine different joint postures.

A post hoc statistical analysis showed that muscle activation patterns vary significantly with IP joint posture ($p < 0.001$). Such IP postural dependence did not consist of a uniform scaling of EMG amplitude but rather impacted different muscles in different ways, as the interaction terms between the muscle and IP posture were significant in most fingertip force directions, aside from the dorsal direction (see Sect. 5). The results of this study suggest that joint posture can have a profound impact on muscle activation patterns of the index finger. Changes were seen in activation of both intrinsic and extrinsic muscles. Interestingly, MCP posture did not significantly affect activation patterns despite the fact that MCP rotation has a greater impact on muscle fiber length than PIP or DIP rotation. Thus, alterations in musculotendon length did not appear to contribute to the changes in activation patterns; rather, changes in ESMAs and joint contact forces had the greatest impact.

4 Anatomical Flexor Pulleys

As the two extrinsic flexor tendons, FDP and FDS, run along the volar aspect of each finger, their paths are restrained by a number of ligamentous or retinacular structures called pulleys. The pulley system consists of three major components: transverse carpal ligament, palmar aponeurosis pulley, and the digital flexor pulley system composed of five annular pulleys and three cruciform pulleys. These pulleys constrain the tendon paths and keep the tendons close to the finger joint centers of rotation, thereby preventing bowstringing and promoting mechanical efficiency in finger flexion [41]. Therefore, damage to the pulley system would have a fundamental effect on the dynamics of finger movement generation. The pulleys also impact the translation of tendon force into joint contact forces and the mechanical stress applied to articular cartilage. These stresses can lead to detrimental changes within the joint structure of the hand, such as osteoarthritis.

4.1 Effect of Pulley Damage

4.1.1 Experimental [42–46]

Due to the aforementioned importance of the anatomical pulleys, researchers examined the effects of pulley damage on the kinematic and/or dynamic aspects of the finger movements via in vitro experiments. The loss of function due to the pulley excision was generally assessed by the change in tendon excursion, i.e., range of motion, and/or in the work of flexion, which represents the kinematic and dynamic effects of the pulley excision. A majority of these studies concluded that the effects of pulley excision on finger kinematics and/or dynamics are substantial, as a significant amount of change in both the tendon excursion and the work of flexion results from the loss of integrity of a pulley and/or multiple pulleys. After the pulley excision, the amount of tendon excursion required to produce a given amount of finger flexion significantly increased [42, 43], or a substantial loss of motion was observed when the same tendon excursion was applied [44]. Furthermore, the detrimental effects of pulley damage on the finger dynamics, assessed by the change in the work of flexion, were found to be even more significant than the change in the kinematics [43].

Most of these studies, however, examined the detrimental effects of the complete removal of the major annular pulleys. Conversely, a partial excision (up to 25 %) of these annular pulleys (i.e., A2 and A4) was found to have no significant effects on either excursion or work during flexion [45]. In addition, damage to the minor pulleys (i.e., A1 and/or A5) did not significantly affect the tendon excursion or work efficiency [46].

4.1.2 Modeling [11, 12]

Vigouroux et al. [11] employed a biomechanical model of the finger [47] in order to estimate the forces acting on finger flexor tendons and pulleys during sport-climbing grip techniques. The results of this study indicate that the type of grip may have a significant effect on the magnitude of the forces acting on the pulleys, as well as on the coordination of FDP and FDS muscle forces. The outcome of this study also emphasizes the importance of joint posture in the dynamics of the finger.

Lee and Kamper [23] utilized a biomechanical model of the index finger in order to examine the detrimental effects of pulley removal on finger movement generation. Two types of pulley damage were simulated in this study: complete excision of the A1 pulley and a partial excision of the A2 pulley. The excision of the A1 pulley, simulated by adjusting the geometrical parameters of the model, resulted in a roughly 10° increase in MCP flexion angle when the FDP force was applied, while the partial excision of the distal A2 pulley increased the PIP flexion angle by 5° (Fig. 5). Note that the simulation results of this study are generally in

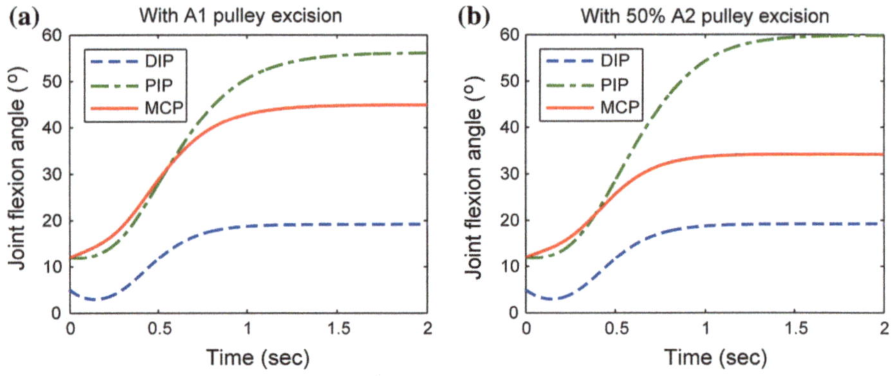

Fig. 5 Simulation of pulley excisions. Compared to the intact condition (Fig. 3b), **a** A1 pulley excision resulted in an increase in MCP flexion angle by ~10° **b** partial A2 distal pulley excision (50 %) increased the PIP flexion angle by ~4°. Figure adapted from [23]. Used with permission from IEEE

agreement with the experimental results from the in vitro studies mentioned in the previous section [45, 46]. Other simulations of full A2 excision and removal of more pulleys, however, have shown substantial kinematic changes [22].

4.2 *Joint Contact Forces*

4.2.1 Possible Involvement in Osteoarthritis [48]

As emphasized by the impact of the pulley mechanisms, the tendons convey force, not moments. These forces must be countermanded by structures within the joints to maintain joint integrity. These joint reaction forces, in addition to being important to propagation of moments to more proximal joints, can also lead to pathological changes. The internal joint force produced by muscle contraction has been found to be associated with the development and progression of knee and hip osteoarthritis (OA) [49], for example. The neuromechanical pathways through which OA develops and/or progresses remain largely unexplored in hand OA. Lee and Kamper [48] hypothesized that potential neuromechanical change in the muscle coordination of the hand is responsible for the hand OA development/progression. In this pilot study, the authors compared coordination patterns of FDS and EDC muscles of two individuals with hand OA and two control subjects during submaximal grip force generation tasks. OA patients generally demonstrated much higher levels of co-contraction of the antagonist muscle (EDC) across all grip force levels. Accordingly, the estimated internal force at the PIP joint of OA patients was much greater than that of control subjects, even though their grip force was generally smaller than control subjects. Overall, higher antagonist co-contraction in OA subjects led to an inefficient strategy that created higher detrimental internal force while generating lower end-point (grip) force. The outcome of this study suggests that the internal force generated by the extrinsic flexor (FDP and FDS) and extensor (EDC) muscle contractions may be responsible for the detrimental change in the internal mechanics of the hand, resulting in the hand OA development and/or progression.

5 Extensor Hood

The four (or five) tendons on the dorsal aspect of the index finger, EDC, EIP, FPI, LUM, (and sometimes FDI), enter into a complex tendon network, commonly referred to as the 'extensor hood' (see Fig. 1a). This structure interconnects the tendons, attaches to the soft tissue on the palmar aspect (e.g., palmar plates and/or pulleys), and forms insertions into the distal phalanx (i.e., distal slip) and the proximal phalanx (i.e., central slip). Force distribution between the two tendon slips has a critical impact on the dynamics of the finger movement and on static fingertip force generation. Tendon force dissipation to the soft tissue attachment sites also affects the efficiency of the tendon force conversion into fingertip force.

Force distribution patterns within the extensor hood can be deduced from its mechanical response to forces applied to the tendons. Accordingly, strains and/or geometric changes in different regions of the extensor hood, which are the indicators of the magnitude of the force transmitted to the corresponding part of the apparatus, have been measured under varying tendon loading conditions (see [7, 13, 14]) in attempts to elucidate the force transmission characteristics within the extensor hood. Generally, the tendon force-strain relationship within the extensor hood is thought to be non-linear due to its heterogeneous tissue properties (i.e., stiffness) within the hood [7, 13]. Indeed, we have observed significant variation in the mechanical stiffness, as quantified by the Young's modulus, across different regions of the extensor hood [50]. Additionally, the force transmission pattern within the extensor hood was observed to be dependent upon joint posture [32, 33].

5.1 Distribution of Force Between Central and Terminal Slips

5.1.1 Indirect Estimation [33]

Lee et al. [33] estimated the magnitude of the tendon forces transmitted to the terminal and central slips of the extensor hood by combining the force/strain measurement data from cadaveric experiments with the biomechanical model of the index finger musculoskeletal structure [23]. Strain resulting from loading of a given tendon inserting into the extensor hood was measured with a 3D camera system for different regions within the extensor hood. Fingertip force/moment and regional strains of the extensor apparatus were measured in cadaver specimens for nine different finger postures. The amount of tendon force transmitted into the terminal and central slips of the extensor hood across nine postures were estimated from the experimental strain data.

While finger posture had a substantial impact on absolute force transmission, it did not affect the relative force distribution between the two slips. Repeated measures analysis of variance revealed that the total amount of tendon force transmitted to each of the two tendon slips was significantly affected by finger posture, specifically by the IP joint angles ($p < 0.01$). Absolute tendon force transmitted to both the central and terminal slips was found to decrease with IP flexion. The main effect of the MCP joint angle was not as consistent as the IP angle, but there was a strong MCP × IP interaction effect for which MCP flexion led to large decreases in the slip forces (>30 %) when the IP joints were extended. The estimated force distribution ratio of terminal slip: central slip remained relatively constant across postures at approximately 1.7:1. The outcome of this study suggests that the impact of finger posture should be carefully considered when studying finger motor control or examining injury mechanisms in the extensor apparatus.

5.1.2 Direct Measurement [51]

Valero-Cuevas, et al. [51] hypothesized that the peripheral structure (i.e., tendon network of the extensor hood) itself plays a critical role in the motor control of the joint torque coordination. In order to address this research question, the authors systematically changed the proportion of tendon forces (EDC versus interosseous tendons) and directly measured the tension delivered to the terminal and central slips in cadaver specimens. The results of this study indicate that the distribution of input tensions in the tendon network, i.e., ratio of interosseous: extensor tensions, regulates how forces propagate to finger joints. They showed that the tendon network acts as a switch, or logic gate, enabling different torque production capabilities and permitting a wide range of joint actuation patterns that would not have been possible to achieve with a simpler tendon structure. The authors concluded that the structural complexity of the tendon network may be critical to understanding brain-body co-evolution and neuromuscular control.

It should be noted that, in this study, the force distribution ratio between terminal and central slips identified from the experiment (around 1:2) was different from the ratio obtained from simulation (around 2:1), which was similar to the value obtained from our indirect estimation [33]. Potential errors in the biomechanical models in both studies may be responsible for the dissimilarity between the direct measurement and indirect estimation.

6 Discussions

The dexterity of the hand is greatly enhanced by the large number of DOF and the capability of the digits to work in close proximity. Not only can fingers make contact with adjacent fingers along their lengths, but the digits can even overlap each other. This is possible due to the actuation mechanism of the digits. The majority of the strongest actuators of the digits, as determined by the physiological cross-sectional area, reside in the forearm. This helps to minimize the bulk of the digits and, as importantly, the mass of the hand. By reducing the hand mass, the torques required at the shoulder and elbow to support the hand are reduced. In fact, all of the hand muscles that actuate the digits are located proximal to the MCP joints; only tendinous structures run along the digit phalanges. This provides tremendous advantage in terms of maximizing the attainable workspace of the tip of each digit. For example, in some individuals, the distal finger segment can actually contact the proximal segment, which is quite remarkable. The digits easily curl into the palm to create a fist. This contrasts with the situations at the elbow and knee, where flexion is limited by the bulk of the biceps brachii and by the hamstrings and gastrocnemius, respectively. Additionally, torques needed to move the digits are minimized in the fingers as no muscle mass has to be accelerated or decelerated.

In order to provide the large passive range of motion, a more complex control scheme is required. Generation of even an isometric force in a specified direction often requires the activation of all seven muscles [15–17]. The nervous system in humans, with disproportionally large cortical representations in motor and somatosensory cortices [52], has obviously adapted to provide this control. This neural organization, however, is supported by a number of specialized biomechanical structures that serve to transmit the muscle forces across the multiple joints of the digits. The extensor hood, the dorsal aponeurosis spanning the length of each finger, is a prime example. Tendons from four or five independent muscles insert into this structure, which in turn inserts into the middle and distal phalanges, with additional connections to the proximal phalanx. The extensor hood thus affects how forces from the different muscles are mapped into joint torques. Experimental studies by Lee et al. [32, 33] and Valero-Cuevas et al. [51] showed how the nature of this mapping can be affected by joint posture and the relative force levels of the different muscles. The adaptations in mapping can reduce the requirements on the neural controller. Intriguingly, we found that, unlike for other force directions, dorsally directed isometric forces could be generated for a variety of finger postures without altering the basic muscle activation pattern [36]. Production of this force was dominated by contraction of EDC and EIP, two of the primary contributors to the extensor hood. This suggests that the biomechanical structure of the hood may help to accommodate changes in task requirements without necessitating a change in control strategy.

Similarly, the anatomical pulleys, which keep the long flexor tendons close to the phalanges, improve control. Less tendon excursion (produced physiologically by muscle activation) is needed to produce a desired amount of finger flexion when the pulleys are intact. Thus, less muscle contraction and less energy are needed to achieve the same amount of finger movement.

The passive properties of the muscles and joints themselves help to produce coordinated movement among the joints of a given digit. Electrical stimulation of the long flexor FDP, for example, generates concurrent flexion of all three joints (MCP, PIP, and DIP). Removal of this passive joint impedance in computer simulations led to disruption of this coordination [22], with the DIP and MCP joints actually extending rather than flexing. These results suggest that, due to the passive joint impedance, a very functional movement in which all three joints are flexed concurrently can be achieved through the activation of a single muscle. Without this impedance, a more complex muscle activation pattern would be required.

Ultimately, muscles produce force. This force is applied through tendinous structures to different parts of the digits. These forces will create joint moments, but they will also create joint reaction forces which will be transmitted to more proximal segments. Force-based models may thus be preferable to moment-arm based models for examining certain multi-articular characteristics of the digits, such as the impact of joint contact forces on the development of osteoarthritis.

Acknowledgments This work was supported in part by Coleman Foundation, and in part by the National Institute of Health under Grant 1R01NS052369-01A1 (NINDS).

References

1. R. Tubiana, *The Hand* (W.B. Saunders, Philadelphia, 1981)
2. T.E. Milner, S.S. Dhaliwal, Activation of intrinsic and extrinsic finger muscles in relation to the fingertip force vector. Exp. Brain Res. **146**, 197–204 (2002)
3. J.R. Doyle, Anatomy of the finger flexor tendon sheath and pulley system. J. Hand Surg. **25A**, 489–498 (1988)
4. R. Wheeler Haines, The extensor apparatus of the finger. J. Anat. **85**, 251–259 (1951)
5. J.M.F. Landsmeer, The anatomy of the dorsal aponeurosis of the human finger and its functional significance. Anat. Rec. **104**, 31–44 (1949)
6. E. Zancoli, *Structural and Dynamic Bases of Hand Surgery*, 2nd edn. (Lippincott, Philadelphia, 1979)
7. M. Garcia-Elias, K.N. An, L.J. Berglund, R.L. Linscheid, W.P. Cooney, E.Y. Chao, Extensor mechanism of the finger. I. A quantitative geometric study. J. Hand Surg. **16A**, 1130–1136 (1991)
8. J.A. Clavero, P. Golano, O. Farinas, X. Alomar, J.M. Monill, M. Esplugas, Extensor mechanism of the fingers: MR imaging-anatomic correlation. Radiographics **23**, 593–611 (2003)
9. N. Brook, J. Mizrahi, M. Shoham, J. Dayan, A biomechanical model of index finger dynamics. Med. Eng. Phys. **17**, 54–63 (1995)
10. J.L. Sancho-Bru, A. Perez-Gonzalez, M. Vergara-Monedero, D. Giurintano, A 3-D dynamic model of human finger for studying free movements. J. Biomech. **34**, 1491–1500 (2001)
11. L. Vigouroux, F. Quaine, A. Labarre-Vila, F. Moutet, Estimationof finger muscle–tendon tensions and pulley forces during specificsport-climbing grip techniques. J. Biomech. **39**, 2583–2592 (2006)

12. L. Vigouroux, F. Quaine, A. Labarre-Vila, D. Amarantini, F. Moutet, Using EMG data to constrain optimization procedure improves finger tendon tension estimations during static fingertip force production. J. Biomech. **40**, 2846–2856 (2007)
13. S.K. Sarrafian, L.E. Kazarian, L.K. Topouzian, V.K. Sarrafian, A. Siegelman, Strain variation in the components of the extensor apparatus of the finger during flexion and extension: a biomechanical study. J. Bone Joint Surg. **52**, 980–990 (1970)
14. P.T. Hurlbut, B.D. Adams, Analysis of finger extensor mechanism strains. J. Hand Surg. **20A**, 832–840 (1995)
15. M.A. Maier, M.C. Hepp-Reymond, EMG activation patterns during force production in precision grip, I. Contribution of 15 finger muscles to isometric force. Exp. Brain Res. **103**, 108–122 (1995)
16. M.A. Maier, M.C. Hepp-Reymond, EMG activation patterns during force production in precision grip, II. Muscular synergies in the spatial and temporal domain. Exp. Brain Res. **103**, 123–136 (1995)
17. F. J. Valero-Cuevas, F. E. Zajac, C. G. Burgar, Large index finger tip forces are produced by subject-independent patterns of muscle excitation. J. Biomech. **31**, 693–703 (1998)
18. R.T. Lauer, K.L. Kilgore, P.H. Peckham, N. Bhadra, M.W. Keith, The function of the finger intrinsic muscles in response to electrical stimulation. IEEE Trans. Rehabil. Eng. **7**, 19–26 (1999)
19. W.L. Buford, S. Koh, C.R. Andersen, S.F. Viegas, Analysis of intrinsic-extrinsic muscle function through interactive three-dimensional kinematic simulation and cadaver studies. J. Hand Surg. **30A**, 1267–1275 (2005)
20. S. Koh, W.L. Buford, C.R. Andersen, S.F. Viegas, Intrinsic muscle contribution to the metacarpophalangeal joint flexion moment of the middle, ring, and small fingers. J. Hand Surg. **31A**, 1111–1117 (2006)
21. J.Z. Wu, K.N. An, R.G. Cutlip, K. Krajnak, D. Welcome, R.G. Dong, Analysis of musculoskeletal loading in an index finger during tapping. J. Biomech. **41**, 668–676 (2008)
22. D.G. Kamper, T.G. Hornby, W.Z. Rymer, Extrinsic flexor muscles generate concurrent flexion of all three finger joints. J. Biomech. **35**, 1581–1589 (2002)
23. S.W. Lee, D.G. Kamper, Modeling of multiarticular muscles: Importance of inclusion of tendon–pulley interactions in the finger. IEEE Trans. Biomed. Eng. **56**, 2253–2262 (2009)
24. F.E. Zajac, Muscle and tendon: properties, models, scaling, and application to biomechanics and motor control. Crit. Rev. Biomed. Eng. **17**, 359–411 (1989)
25. A. Esteki, J.M. Mansour, An experimentally based nonlinear viscoelastic model of joint passive moment. J. Biomech. **29**, 443–450 (1996)
26. J.S. Knutson, K.L. Kilgore, J.M. Mansour, P.E. Crago, Intrinsic and extrinsic contributions to the passive moment at the metacarpophalangeal joint. J. Biomech. **33**, 1675–1681 (2000)
27. E.G. Cruz, D.G. Kamper, Use of a novel robotic interface to study finger motor control. Ann. Biomed. Eng. **38**, 259–268 (2010)
28. Y. Youm, T.T. Gillespie, A.E. Flatt, B.L. Sprague, Kinematic investigation of normal MCP joint. J. Biomech. **11**, 109–118 (1978)
29. K. N. An, W. P. Cooney, in *Biomechanics of the hand,* ed. B.F. Morrey. Joint Replacement Arthroplasty (Churchill Livingstone, New York, 1991)
30. A. Minami, K.N. An, W.P. Cooney, R.L. Linscheid, Ligament stability of the MCP joint: a biomechanical study. J. Hand Surg. **10A**(2), 255–260 (1985)
31. K.N. An, Y. Ueba, E.Y. Chao, W.P. Cooney, R.L. Linscheid, Tendon excursion and moment arm of index finger muscle. J. Biomech. **16**, 419–425 (1983)
32. S.W. Lee, H. Chen, J.D. Towles, D.G. Kamper, Estimation of the effective static moment arms of the tendons in the index finger extensor mechanism. J. Biomech. **41**, 1567–1573 (2008)
33. S.W. Lee, H. Chen, J.D. Towles, D.G. Kamper, Effect of finger posture on the tendon force distribution within the finger extensor mechanism. J. Biomech. Eng. T. ASME **130**, 051014 (2008)

34. A.M. Gordon, A.F. Huxley, F.J. Julian, The variation in isometric tension with sarcomere length in vertebrate muscle fibres. J. Physiol. **184**, 170–192 (1966)
35. D.G. Kamper, H.C. Fischer, E.G. Cruz, Impact of finger posture on mapping from muscle activation to joint torque. Clin. Biomech. **21**, 361–369 (2006)
36. D. Qiu, H. C. Fischer, D. G. Kamper, Muscle activation patterns during force generation of the index finger, in *Proceedings of IEEE Engineering in Medicine and Biology Society Conference (EMBC 2009)*, pp. 3987–3990, 2009
37. N. K. Fowler, A. C. Nicol, B. Condon, D. Hadley, Method of determination of three dimensional index finger moment arms and tendon lines of action using high resolution MRI scans. J. Biomech. **34**, 791–797 (2001)
38. R. Tubiana, P. Valentin, The anatomy of the extensor apparatus of the finger. Surg. Clin. North Am. **44**, 897–906 (1964)
39. R.A. Beekman, A.E. Abbot, N.L. Taylor, M.P. Rosenwasser, R.J. Strauch, Extensor mechanism slide for the treatment of proximal interphalangeal joint extension lag: an anatomic study. J. Hand Surg. **29A**, 1063–1068 (2004)
40. T.A. El-Gammal, C.M. Steyers, W.F. Blair, J.A. Maynard, Anatomy of the oblique retinacular ligament of the index finger. J. Hand Surg. **18A**, 717–721 (1993)
41. J.R. Doyle, Palmar and digital flexor tendon pulleys. Clin. Orthop. Relat. Res. **383**, 84–96 (2001)
42. P.W. Brand, K.C. Cranor, J.C. Ellis, Tendon and pulleys at the metacarpophalangeal joint of a finger. J. Bone Joint Surg. **57A**, 779–784 (1975)
43. W.W. Peterson, P.R. Manske, B.A. Bollinger, P.A. Lesker, J.A. McCarthy, Effect of pulley excision of flexor tendon biomechanics. J. Orthop. Res. **4**, 96–101 (1986)
44. G.T. Lin, P.C. Amadio, K.N. An, W.P. Cooney, Functional anatomy of the human digital flexor pulley system. J. Hand Surg. **14**, 949–956 (1989)
45. G. Mitsionis, J.A. Bastidas, R. Grewal, H.J. Pfaeffle, K.J. Fischer, M.M. Tomaino, Feasibility of partial A2 and A4 pulley excision: effect on finger flexor tendon biomechanics. J. Hand Surg. **24A**, 310–314 (1999)
46. D. Rispler, D. Greenwald, S. Shumway, C. Allan, D. Mass, Efficiency of the flexor tendon pulley system in human cadaver hands. J. Hand Surg. **21A**, 444–450 (1996)
47. I. Roloff, V.R. Schöffl, L. Vigouroux, F. Quaine, Biomechanical model for the determination of the forces acting on the finger pulley system. J. Biomech. **39**, 915–923 (2006)
48. S. W. Lee, D. G. Kamper, Higher antagonist co-contraction in hand osteoarthritis leads to detrimental joint mechanics, in *Proceedings of the 34th Annual Meeting* of *American Society of Biomechanics*, Providence, 2010
49. M.D. Lewek, D.K. Ramsey, L. Snyder-Mackler, K.S. Rudolph, Knee stabilization in patients with medial compartment knee osteoarthritis. Arthritis Rheum. **52**, 2845–2853 (2005)
50. K. Traylor, S. W. Lee, B. Ellis, J. Weiss, D. Kamper, Tensile properties of the extensor hood in support of a finite element model, in *Proceedings of Biomedical Engineering Society Annual Meeting*, Hartford, 2011
51. F.J. Valero-Cuevas, J.W. Yi, D. Brown, R.V. McNamara III, C. Paul, H. Lipson, The tendon network of the fingers performs anatomical computation at a macroscopic scale. IEEE Trans. Biomed. Eng. **54**, 1161–1166 (2007)
52. W. Penfield, T. Rasmussen, *The Cerebral Cortex of Man. A Clinical Study of Localization of Function* (Macmillan, New York, 1950)
53. F.J. Valero-Cuevas, J.D. Towles, V.R. Hentz, Quantification of fingertip force reduction in the forefinger following simulated paralysis of extensor and intrinsic muscles. J. Biomech. **33**, 1601–1609 (2000)

Author Biographies

Sang Wook Lee (M'10) received the B.S. and M.S. degrees in mechanical design and production engineering from Seoul National University, Seoul, Korea, in 1997 and 1999, respectively, and the Ph.D. degree in mechanical engineering from University of Illinois at Urbana-Champaign, IL in 2006.

He currently holds a joint appointment as an Assistant Professor in the Department of Biomedical Engineering at the Catholic University of America and as a Research Scientist at the Center for Applied Biomechanics and Rehabilitation Research in the National Rehabilitation Hospital. His current research interests include biomechanics of hand and upper extremity, and rehabilitation of neurologically impaired patients.

Derek G. Kamper (M'97) received the B.E. degree in electrical engineering from Dartmouth College, Hanover, NH, in 1989, and the M.S. and Ph.D. degrees in biomedical engineering from the Ohio State University, Columbus, OH, in 1992 and 1997, respectively.

He currently holds a joint appointment as an Associate Professor in the Department of Biomedical Engineering at the Illinois Institute of Technology and as a Research Scientist at the Rehabilitation Institute of Chicago. His research interests include neurorehabilitation, mechatronics, and upper extremity neuromechanics.

Chapter 5
The Control and Perception of Finger Forces

Lynette A. Jones

Abstract The human hand demonstrates remarkable dexterity in its capacity to control precisely the forces involved in manipulating objects and the timing of movements during the execution of skilled motor tasks. In all of these endeavors, mechanoreceptors in the skin play a critical role in encoding the timing, magnitude, direction and spatial distribution of fingertip forces. When cutaneous inputs are absent or deficient, the hand is unable to compensate rapidly when an object begins to slip between the digits, and misdirected finger movements are not recognized and corrected. The control and perception of forces generated by the hand therefore relies on a close interplay between the sensory and motor systems. When sensory information changes, the capacity to control and modulate force can be disrupted and this in turn influences the perceived magnitude of the forces being produced. Cutaneous mechanoreceptors provide crucial information about the forces produced by the fingers and these inputs together with centrally generated corollary discharges are fundamental to the human perception of force.

Keywords End effectors · Force feedback · Grasping · Tactile sensors

1 Introduction

The ability to control the forces generated by the hand is a requirement for most skilled activities from grasping a small object between the index finger and thumb to turning a key in a lock. In these activities finger forces are exquisitely adapted to the contact conditions between the hand and the object and vary as the weight, surface texture or shape of the object changes [1, 2]. The skillful manipulation

L. A. Jones (✉)
Department of Mechanical Engineering, Massachusetts Institute of Technology,
77 Massachusetts Ave, Cambridge, MA 02139, USA
e-mail: LJones@MIT.edu

R. Balasubramanian and V. J. Santos (eds.), *The Human Hand as an Inspiration for Robot Hand Development*, Springer Tracts in Advanced Robotics 95,
DOI: 10.1007/978-3-319-03017-3_5,

entailed in these tasks requires that the central nervous system (CNS) has accurate information about the magnitude and direction of load forces acting tangential to the skin surface, and normal forces orthogonal to the skin surface. The results from numerous experiments have demonstrated that normal (grip) forces are precisely coordinated in space and time and are of sufficient, but not excessive, magnitude to prevent the object from slipping between the fingers [2]. When an object begins to slip, cutaneous mechanoreceptors sense the displacement of the object on the skin with the result that there is an automatic and rapid ($\sim$70 ms) increase in grip force that results in a more stable grasp [3]. Afferent feedback from mechanoreceptors in the skin is therefore essential for the control and modulation of force during skillful manipulation. When it is impaired due to damage in the peripheral nerves or cortical lesions, precise control of prehensile force is lost and the hand becomes clumsy and inefficient, even though the muscles controlling finger movements are not affected.

Non-prehensile activities such as typing on a keyboard or playing a piano also require that the forces produced by the fingers are controlled either to avoid musculoskeletal strain injuries (e.g. when typing) or to vary the sound intensity of the music produced (piano playing). The performance of experts engaged in these activities has been studied and compared to that of novices to provide insight into which aspects of these motor skills are explicitly controlled and which elements of performance differentiate the responses of novices and experts. Most computer keyboards used for typing have activation forces of less than 1 N and the break-away force at which tactile feedback indicates that the key has registered ranges from 0.04 to 0.25 N [4]. For skilled typists the mean keystroke force is around 0.9 N and the force does not vary significantly across the fingers [4]. Peak forces are usually higher, in the order of 2–3 N, and are correlated with the velocity of the fingertip prior to key contact [5]. The forces produced during typing are quite variable which suggests that force is a rather loosely controlled variable that is simply programmed to exceed the key activation force [6]. The most obvious factor that differentiates the performance of expert and novice typists is that experts type faster (i.e. have shorter inter-stroke intervals) with much lower error rates. The faster performance of experts is achieved by overlapping their finger movements so that two and occasionally three keystrokes are occurring concurrently as successive letters are processed in parallel. In contrast, novices typically adopt a serial hunt and peak method when typing [2].

The forces produced when playing an acoustic piano determine the sound intensity of the notes and contribute to the dynamics of the performance and so in contrast to typing are explicitly controlled by experts. In their extensive study of the performance of expert and amateur pianists, Krampe and Ericsson [7] found that expert pianists typically applied more force than amateurs and were more variable in the forces applied. It has also been noted that expert performance is characterized by a superior ability to reproduce reliably the timing and loudness variations in a musical score in consecutive performances [8]. As was noted for prehensile activities such as grasping and manipulation, feedback from mechanoreceptors in the skin is important for non-prehensile skills such as typing and playing the piano. When the fingers are anesthetized, which eliminates tactile feedback from the skin but does not affect the ability to move the fingers, there is an increase in misdirected movements during typing with the majority of errors involving striking a key adjacent to the correct key. In contrast to normal performance when the interval before the next keystroke increases after mistyping a letter, inter-stroke intervals do not usually increase after errors when the fingers are anesthetized. This suggests that the errors are not recognized and that tactile feedback contributes to the accuracy of finger movements [9].

The control of finger forces is not only essential for dexterous activities but is also required when the hand is used for tactile and haptic exploration. In these activities the normal and tangential forces are optimized to recruit tactile receptors in order to achieve a sensory goal, such as detecting changes in the surface friction between the fingers and an object or the presence of a lump in soft tissue [10, 11]. Mechanical deformation of the fingertip surface during tactile or haptic exploration is signaled by cutaneous mechanoreceptors that convey the critical spatial and temporal information that permits the human observer to detect features such as a texture composed of periodically ordered bars only 0.06 μm high [12]. The latter study and many others have highlighted the importance of tangential forces on the fingertips to haptic perception (see Sect. 5).

In all of these activities, from perceiving texture to grasping an object, the hand often serves dual roles as an active sensory organ and as the medium for motor activity. In many situations both functions must be performed concurrently, for example, when we reach and grasp an object using force appropriate for its weight, surface texture and compliance, and so the central nervous system has to optimize the performance of the task to satisfy both objectives.

2 Force Sensing: Peripheral and Central Mechanisms

2.1 Cutaneous Mechanoreceptors

Forces applied to the fingers are sensed by different populations of mechanoreceptors in the skin that respond to various features of the stress–strain field and motion during contact. Slowly adapting type 1 (SA I) units are particularly responsive to fine spatial features such as edges, bars and curvature rather than the overall indentation, although they do respond to the application of normal force and have sustained responses to lateral stretch of the finger pad [13]. Their receptive fields are relatively small, in the order of 2–3 mm in diameter, with highly sensitive local areas within the receptive field. Psychophysical studies have suggested that SA I units are maximally sensitive to very low frequencies of vibration in the 0.4–3 Hz range, which are associated with sensations of pressure consistent with the neurophysiological studies described above [14]. Slow adapting type II (SA II) units are much more sensitive to skin stretch than SA I units and are thought to contribute to the perception of object motion and force when the skin is stretched as a result of object motion. Fast adapting type I (FA I) units are poorer at discriminating fine spatial details than SA I units but respond well to transient deformation of the finger pad and to the low frequency vibration that occurs during the initial contact with an object. Similar to SA I units, their receptive fields are small, about 3–5 mm in diameter, but they respond uniformly across the entire receptive field and so are not sensitive to the fine spatial features of a stimulus. They discharge when there is relative motion between the skin and an object such as when an object beings to slip between the fingers. Fast adapting type II (FA II) units are less densely distributed in the hand than either FA I and SA I units, and have extremely large receptive fields, rendering them poor at discriminating the spatial details of a stimulus. They are, however, extremely sensitive to transient stimulation, particularly vibratory stimuli in the 40–500 Hz range. Collectively, cutaneous mechanoreceptors provide the requisite cues that enable people to perceive variations in texture, the tangential forces associated with movement of an object across the skin and the normal forces generated during contact. (See chapter by M. Tavakoli).

2.2 Muscle Receptors

When the hand actively explores the environment, forces are generated by muscles and these are sensed peripherally via the afferent discharges arising from mechanoreceptors in muscles and can be detected centrally from changes in the voluntarily generated motor command. In muscle, forces are sensed by Golgi tendon organs which are encapsulated receptors usually found at the junction between the collagen strands that comprise the muscle tendon and a small group of extrafusal muscle fibers, the main contractile element. In human muscles, each receptor is in series with 10–20 muscle fibers, and discharges in response to the forces developed only by these fibers. Although individual Golgi tendon organs respond non-linearly to the total force generated by the muscle, the ensemble response of a population of tendon organs is monotonically related to muscle force and so provides the CNS with information regarding active muscle tension [15, 16]. There have been very few recordings from the afferent fibers arising from Golgi tendon organs in human subjects, in part because tendon organ receptors are less numerous and more variable in number than muscle spindles. In one of the rare descriptions of a perceptual response to micro-stimulation of a Golgi tendon organ afferent fiber, Macefield et al. [17] reported an illusion of muscle lengthening. This is consistent with the observation that Golgi tendon organs respond to changes in muscle length in addition to force during contractions and that the perception of force is primarily derived from descending efferent commands, as described below.

2.3 Corollary Discharges

The CNS monitors the output of the motor areas of the brain that produce contractions in muscles. These correlates or copies of the motor command were historically referred to as sensations of innervation and more recently have been termed efference copies or corollary discharges. Information about the forces generated by muscles can be sensed via these copies of the motor commands that are transmitted to the sensory structures in the brain and reflect the magnitude of the efferent command generated [18]. The results from a number of experiments indicate that whenever there is an increase in the efferent signal sent to the motoneurons controlling a muscle, there is a corresponding increase in the perceived magnitude of the force of contraction. This overestimation of muscle force occurs even when the force produced by the muscle remains constant. The increase in the efferent signal may result from changes in the excitability of the muscle due to fatigue, interference with transmission at the neuromuscular junction, decreased excitability of the spinal cord because of cerebellar damage, or of the motor cortex following damage to the corticofugal pathways. In each of these situations the forces generated by the affected muscle are overestimated in magnitude [18, 19]. The finding that the perceived amplitude of muscle forces is derived from neural correlates of the descending efferent command should not be interpreted as dismissing a contribution from Golgi tendon organs to the awareness of muscle force. Afferent input is required to provide a signal that the force generated by the muscle is adequate for the task being performed and these inputs are critically important in peripheral feedback pathways [16]. Reflex inputs from joint, muscle and skin receptors can inhibit or facilitate motoneurons in the spinal cord and in so doing influence the magnitude of the centrally generated motor command.

3 Force Control

The hand is capable of performing a remarkable variety of actions which have provided the basis for numerous taxonomies developed to characterize its diverse range of function (e.g. [20, 21]). Most of these classificatory schemes make a distinction between the two dominant prehensile postures, the precision grip and the power grip, each of which may take several forms. Three types of prehensile grips are often differentiated: the tip pinch in which a small object is held between the tips of the index finger and thumb; the lateral or key pinch that involves contact between the pad of the thumb and the lateral surface of the index finger, and the palmar pinch in which the pad of the thumb opposes the finger pads of the index and middle fingers (see Fig. 1). The most commonly used prehensile grips are the tip and lateral pinch. The maximum forces produced with these prehensile grips range from 50 to 120 N, depending on the particular grasp, and age and sex of the person [22]. The peak forces produced with the palmar and lateral pinch grips are approximately 40 % higher than those generated with the tip pinch. When force is the primary objective the power grip is used (see Fig. 1); this grip typically involves an extensive area of contact between the hand and the object (e.g. when holding a hammer) and movement of the object held in the hand is accomplished by moving the whole hand using the wrist and arm. The maximum forces produced with the power grip are large (up to 600 N) and vary substantially as a function of the age and sex of the individual [22].

For many activities in which prehensile grips are used the forces generated at the fingertips are much lower than the peak forces and are typically around 10 N or less. At these low forces, the contact area on the fingertip increases rapidly as the normal force increases, with 70 % of the overall change in area with force occurring at forces below 1 N [23]. This means that if a normal force of 1–2 N at the fingertips is adequate to perform a task, little additional tactile information is available at higher forces because the increase in contact area is so small.

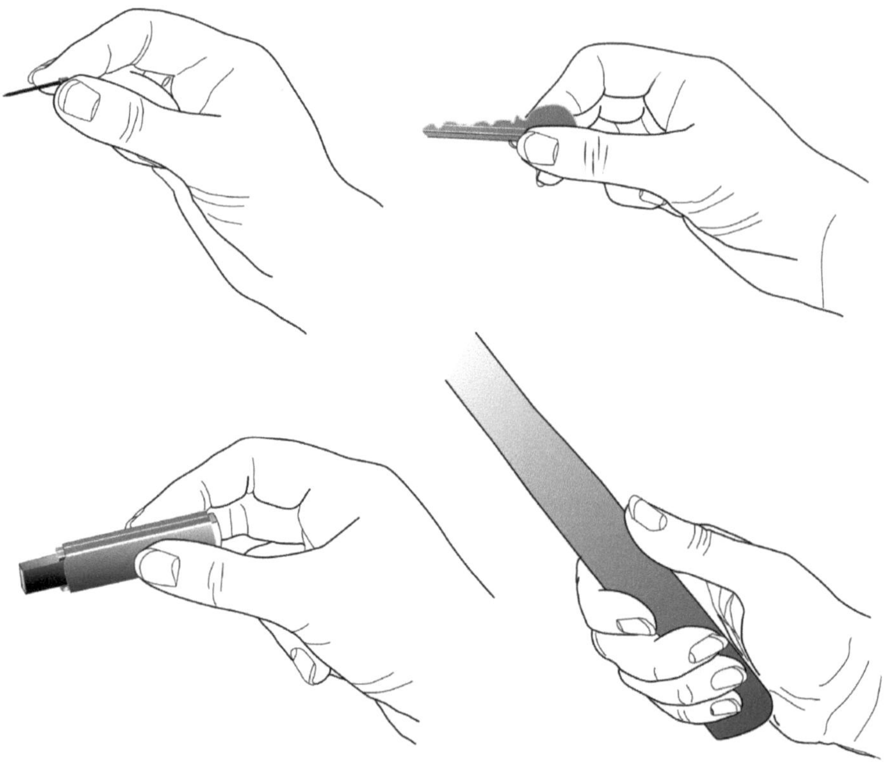

Fig. 1 Common prehensile patterns. *Upper right*: tip pinch between the pulps of the opposed thumb and index finger. *Upper left*: the lateral pinch in which the thumb pulp is in contact with the lateral surface of the middle phalanx of the index finger. *Lower left*: the palmar pinch in which the pulp surface of the thumb opposes the finger pads of the index and middle finger. *Lower right*: the power grip is shown on the *lower right*. From [2] with permission of Oxford University Press

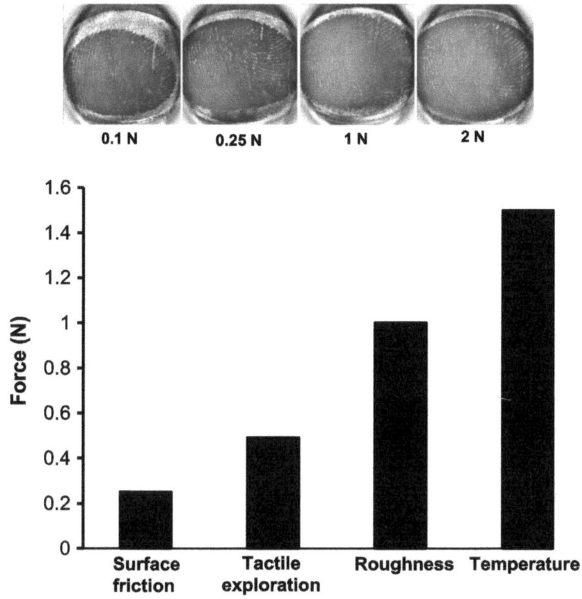

Fig. 2 Average contact forces and contact areas (finger pad images) at these forces when people are asked to make judgments about surface friction [10], the presence of an asperity in a smooth surface [11], the roughness of raised-dot surfaces [24] and the temperature of different materials [25]. Adapted from [2] and [23] with permission of Oxford University Press and ASME

Measurements of the contact forces used during tactile and haptic exploration of objects reflect this finding, in that the forces are typically less than 1.5 N, and for some tasks such as locating a small feature on a smooth surface or perceiving the friction between the finger and a surface are less than 0.5 N [10, 11]. The average contact forces and contact areas measured at the fingertip when people make judgments about various material properties of objects are shown in Fig. 2. These data illustrate that people adopt an optimal strategy to perceive surface features that minimizes the normal forces between the finger and the surface while maximizing contact area.

3.1 Sensory Feedback and Force Control

There has been an extensive number of experiments on the control of force during grasping many of which have used variations of an apparatus and procedure first described by Johansson and his colleagues in the 1980s [3]. The experimental protocol requires that the subject grasp an object between the thumb and index finger, lift it a fixed distance from a support surface and then lower it to a resting position. The temporal sequence of the forces produced is analyzed in terms of a loading phase in which there is a parallel increase in the grip force normal to the object surface and the load force tangential to the surface, until the load force overcomes gravity and the object is lifted to the target position. In these studies, the object usually has displacement or acceleration sensors attached to its surface and force transducers are mounted on the object to record the normal and tangential forces. The object may be passive, for example a box whose weight, shape and surface texture may vary, or active such as a servo-controlled motor, which can be perturbed as it is manipulated. The results from numerous experiments have demonstrated that when an object is grasped and lifted from a supporting surface the rate at which the normal and tangential forces increase is maintained at an approximately constant ratio, indicating a coordinated pattern of muscle activation in the hand and forearm [1]. The grip forces are modulated as a function of the weight of the object and the friction between the object and the skin. If friction is varied independently at the contact surfaces on the thumb and index finger, the fingertip forces are automatically adjusted to the surface conditions at each digit [26]. Afferent input from cutaneous mechanoreceptors is essential to the regulation of grip forces when manipulating objects. The mechanoreceptors' capacity to detect mechanical events at the digit-object interface permits the hand to maintain stable and efficient grasps during manipulation.

Various experiments have examined how accurately finger forces can be controlled under different feedback conditions. In these studies, subjects are typically provided with visual feedback to reach a target force and are then asked to maintain the force at a constant amplitude for a specified period of time using either visual and haptic feedback or only haptic feedback [27, 28]. When only haptic feedback is available subjects are able to maintain finger forces ranging from 2 to 6 N to within 1 N over a 120 s time period [27]. In this situation the only cues available regarding the forces being generated come from mechanoreceptors in the skin and muscles and from central feedback pathways that monitor the output from motor cortex. When visual feedback of the forces being produced is also provided the accuracy with which forces are maintained improves considerably and the mean absolute error is now 0.22 N. The precision with which finger forces can be controlled is reflected in the coefficient of variation (i.e. standard deviation/mean) which averages 13 % when only haptic feedback is available and decreases to 4 % when both visual and haptic feedback is provided as shown in Fig. 3 [27]. The latter number represents the optimal performance of voluntary force control over relatively long periods. At much shorter intervals (5 s), the coefficient of variation has been reported to be considerably lower (1–2 %) [29].

It is interesting to note in this context that the accuracy (measured in terms of absolute and constant errors) and precision (evaluated by the coefficient of variation) with which forces are controlled has not been found to differ significantly for more proximal muscles such as the elbow flexors in the upper arm as compared to the distal muscles in the forearm controlling movements of the fingers. With visual and haptic feedback, the coefficient of variation associated with maintaining a constant force averages 4 % across both muscle groups when the forces are scaled to the operating range of the muscle groups. However, even for the same

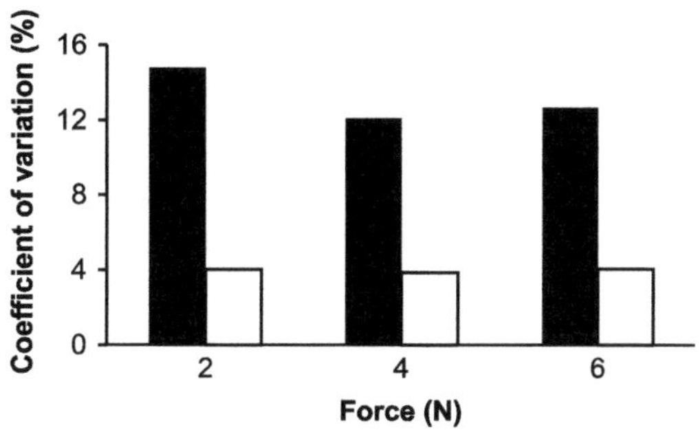

Fig. 3 Group (N = 10) mean coefficients of variation of force when maintaining a constant force with the index finger flexors using either haptic feedback (*black bars*) or both visual and haptic feedback (*white bars*). Data are from [27]

force (4 N), Jones [27] found no significant difference between the index finger and elbow flexor muscles in the precision with which the force was controlled. This suggests that there is not a proximal to distal gradient in force control as has been reported for limb position and movement and that forces are controlled with same relative degree of precision across different muscle groups.

3.2 Multi-digit Force Control

Many of the muscles that control the fingers and wrist act over many joints and so the forces and movements produced by these muscles are not controlled independently and coactivation occurs frequently [30]. Three of the extrinsic hand muscles in the forearm (the flexor digitorum superficialis, the flexor digitorum profundus and the extensor digitorum communis) connect to each of the four fingers and within these muscles are motor units that can activate several fingers concurrently [31]. The extent to which each of these muscles can produce a force in an isolated finger depends on the extent of compartmentalization within the muscle. This means that there are both neural and mechanical limitations in the degree to which individual fingers can produce forces when they are acting synergistically. The effect of concurrent activation of the fingers on the maximum forces produced is considerable as demonstrated by the 37 % increase in the peak forces produced by the fingers individually as compared to the maximum force produced when all four fingers flex simultaneously [32]. The decrease in the maximum force that can be produced by a finger as more fingers are involved in force production is assumed to reflect a limit in the central neural drive sent to the muscles. As more muscles are involved in the task, there is a reduction in the neural input each muscle receives with a concomitant deficit in force production. Mechanical and neural coupling also influences the production of submaximal forces (i.e. 40–80 % of the maximum force) and the extent of coactivation varies across the fingers. Reilly and Hammond [33] showed that coactivation was least in the thumb and then increased progressively when the index, middle, little and ring fingers generated a target force. In their study, forces were measured across the digits while subjects were asked to produce a target force with a single finger. Additional evidence that indicates the limitations in controlling individual finger forces comes from force tracking experiments in which subjects have to generate different combinations of forces using two or three fingers. The force combinations that subjects have most difficulty producing are those that involve large differences in the amplitudes of forces produced by adjacent fingers (e.g. index finger 40 % maximum force, middle finger 10 % maximum force and little finger 40 % maximum force) [34].

The ability to coordinate and control forces across the fingers concurrently has also been evaluated during prehensile grasping with three, four or five digits. The results from these experiments indicate that the largest change in finger force as the weight of the object increases or the surface friction decreases occurs at the index finger [35]. Interestingly, the relative contribution of each finger to the overall force remains relatively constant independent of variations in surface texture or weight [36]. This means that the control of individual forces during grasping involves a fairly simple scaling strategy that creates fixed relations between the forces produced at the fingertips.

4 Force Perception

As described in Sect. 2 (Force Sensing), perceptual information about muscle force is primarily derived from centrally generated sensations arising from the innervation of the efferent pathways. Numerous experiments have detailed the underlying mechanisms involved in sensing muscle force and how the perception of voluntarily generated force changes when the state of the muscle or the peripheral sensory apparatus is altered [37, 38]. In contrast to active force perception, forces applied passively to an immobilized finger are sensed via the responses of cutaneous mechanoreceptors. Wheat et al. [39] demonstrated that these forces can be scaled consistently, with the perceived magnitude of force increasing monotonically as normal force increases.

4.1 Contralateral Limb-Matching Paradigm

Many of the experiments on the perception of voluntarily generated forces have used the contralateral limb matching procedure to study perception. In this paradigm the forces exerted by a muscle group in one limb, designated the reference limb, are matched in perceived magnitude by contractions of the corresponding muscle group in the other limb, the matching limb. The subject usually has feedback of the force produced by the reference limb and attempts to match the force sensations in both limbs during the matching interval. In some studies the limbs are loaded with weights and subjects request adjustments in the loading of the matching limb until the weights are perceived to be the same [37]. The matching forces or weights indicate the perceived force or heaviness on the reference side. Provided that there is a control condition in which the accuracy of force matching under normal conditions is measured, this paradigm can be used to analyze the effect of a range of perturbations (e.g. fatigue, electrical stimulation) on force perception. One advantage of the matching technique is that some of the response biases associated with numerical estimation of stimulus magnitude are avoided.

Under normal conditions the average error associated with matching forces in two corresponding muscle groups is around 3.5 %, when the forces are expressed in terms of the maximum voluntary contraction (MVC) of each muscle group [40]. The accuracy with which forces are matched varies somewhat across different muscle groups, but does not follow any proximal to distal gradient in sensitivity. There is no difference in the precision with which forces can be matched by the first dorsal interosseous (FDI) and the adductor pollicis (AP) muscles, both of which are intrinsic muscles in the hand, the flexor pollicis longus (FPL) and the flexor digitorum profundus (FDP) muscles, extrinsic hand muscles in the forearm, and the biceps brachii (BB) an elbow flexor muscle in the upper arm [40, 41]. These findings are illustrated in Fig. 4 where it can be seen that the coefficient of variation does not vary significantly across the different muscles (range: 13.5–17.6 %) and is surprisingly high at an average of 15.6 %. The absence of any gradient in sensitivity indicates that the recruitment of many small motor units in a distal muscle is not necessarily associated with a superior accuracy in perceiving the forces produced.

The matching paradigm has been used to explore whether the perception of finger force changes as a function of how forces are produced. Although the thresholds for discriminating changes in force are similar when measured isometrically (constant muscle length) and anisometrically (changing muscle length) as described below [40, 42], the perceived magnitude of forces does vary. If both

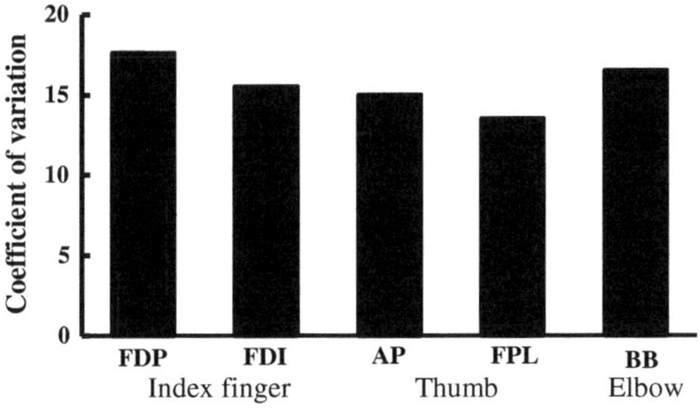

Fig. 4 Coefficient of variation for matching forces produced by muscles controlling the index finger: flexor digitorum profundus (FDP) and first dorsal interosseus (FDI); the thumb: adductor pollicis (AP), and flexor policis longus (FPL); and the elbow: biceps brachii (BB). Data are from [37, 40] and [41]. From [34] with permission of Mac Keith Press

the reference and matching muscle group are producing forces under the same conditions (i.e. either both isometric or both anisometric), forces are reasonably accurately matched, although matching is more accurate with anisometric force production. However, if the reference force is produced anisometrically and matched isometrically then the forces are overestimated in magnitude, and in the reverse condition forces are underestimated [43]. This means that the same force is judged to be much greater if the muscle changes length during production, a result that probably reflects the effects of sensory inputs from the spindle receptors in muscles being stretched on the motoneurons innervating the muscle.

4.2 Force Constancy

Perceptual constancy is a feature of most sensory systems and refers to the temporal and spatial continuity of the perceived world in the presence of changing sensory inputs. In some situations the human haptic system appears to adopt a principle of constancy as demonstrated by the observation that the weight of an object is not perceived to change when it is supported first by the fingers and then the whole hand. Interactions with the environment would be much more difficult if this were not the case, since the perceived properties of objects or tools would vary as a function of the limb segments and muscles used to interact with them. Extensive training would be required before the neuromuscular system could determine what torques should be produced at which joints to perform a specific task. Associated with this perceptual constancy in the haptic system is the poor localization of action internally, where the focus again is on the end result of the action rather than on the muscles involved in achieving that objective.

Although the perception of force appears to remain relatively invariant with changes in the muscle groups contributing to force production, the perceived magnitude of a force does change when other forces are produced concurrently by adjacent muscle groups. Kilbreath and Gandevia [44] showed that the perceived magnitude of forces produced by flexing the thumb increased when the index finger was also generating a force, but was not affected by forces produced concurrently by muscles in the leg. Moreover, the magnitude of the change in the perceived force generated by a single finger increased progressively with the amplitude of the forces generated concurrently by an adjacent digit. These findings have been interpreted in terms of a perceptual inability to segregate the motor output sent to muscles commonly used together, such as the long flexors of the thumb and index finger used for grasping.

The properties of objects may be perceived to be invariant, however, in the absence of an external referent does the haptic system still demonstrate perceptual constancy? This question has been addressed by asking subjects to match the force produced in one muscle group by generating the same force with another non-homologous muscle group. In that experiment forces ranging from 2 to 10 N were produced and matched by the elbow flexor muscles, the muscles involved in the palmar prehensile grasp and the index finger flexors [45]. The results showed that the perceived magnitude of a force varied as a function of the muscle group generating the force. When reference forces produced by the elbow flexor muscles were matched in perceived magnitude by flexing the index finger, they were consistently underestimated in magnitude: an elbow flexion force of 8 N was perceptually equivalent to a finger force of 4.5 N. In contrast, a reference finger force of 2 N was equivalent to an elbow flexion force of 7 N. These findings are illustrated in Fig. 5. When these matching forces are scaled with respect to the maximum forces that each muscle group can produce, they span a remarkably similar range in amplitude. It appears that forces are perceived relatively and scaled with respect to the operating range of the muscles. The relative force amplitudes produced by the reference and matching muscle groups are not precisely matched however, with forces being underestimated for the finger (21 % MVC matched by 13 % MVC) and overestimated for the elbow (8 % MVC matched by 15 % MVC), which suggests that the scaling takes into account the absolute force produced by the muscles. These results clearly demonstrate that the perceived magnitude of a force changes as a function of the muscle group producing the force and that there is little evidence for perceptual constancy in force perception [45].

4.3 Thresholds

Most experiments that have used the contralateral limb matching procedure have examined the perception of supra-threshold forces and have focused on understanding the sensory mechanisms involved in force perception. It is also of interest to determine how well people can discriminate between forces. Discrimination is usually measured in terms of the difference threshold which is the amount of change in a stimulus required to produce a just noticeable difference (jnd) in sensation. For many sensory modalities the change in stimulus intensity that can be discriminated is a constant fraction of the intensity of the stimulus and this is known as the Weber fraction for that modality. The results from a number of experiments have shown that the Weber fraction (as a percentage) for force is approximately 7 % [40, 42]. This is within the range of Weber fractions reported for haptic weight discrimination in which subjects discriminate between weights lifted either simultaneously or successively. In the latter studies, thresholds are typically around 6–12 % depending on the range of weights presented and the experimental procedure [46]. When only tactile cues are available, for example when weights are placed on a hand that rests passively on a support surface, differential thresholds increase (i.e. sensitivity diminishes) and the Weber fraction is approximately 50 % higher than that obtained during active lifting [46]. When weight is sensed passively, the properties of the object in contact with the skin such as its volume, shape and temperature can influence its perceived heaviness. Various illusions such as the size-weight, shape-weight and temperature-weight

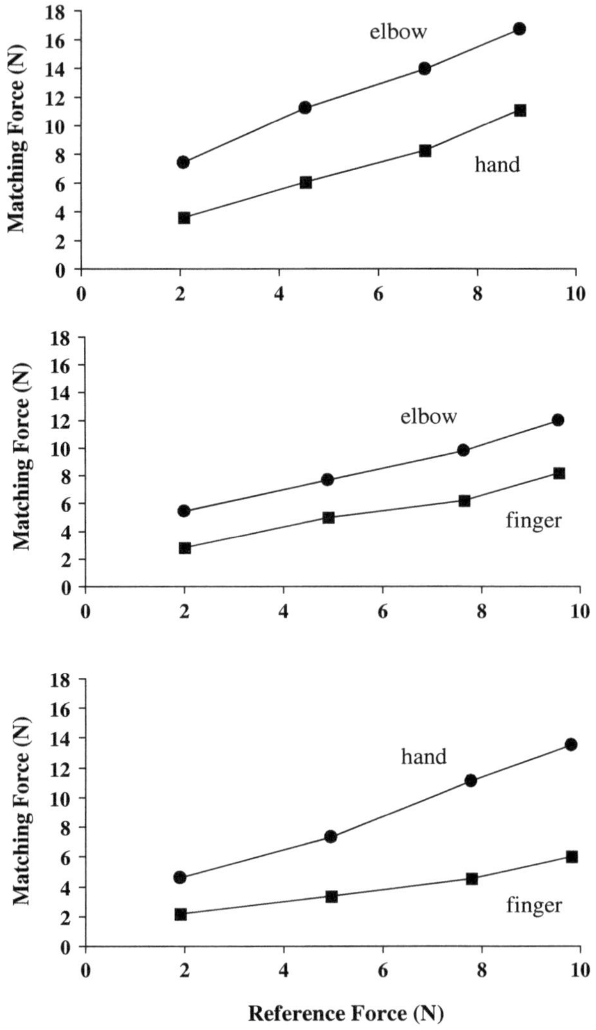

Fig. 5 In all three graphs the data are the group (N = 10) means. In the *top figure* the reference forces produced by the index finger are matched using the palmar pinch and elbow flexor muscles. In the *middle figure* the reference forces produced by the palmar pinch muscles are matched by the index finger and elbow flexors. In the *bottom figure* the reference forces produced by the elbow flexors are matched by the index finger and the palmar pinch muscles

illusions attest to the profound effect that the geometric and material properties of an object can have on its perceived heaviness.

Force discrimination thresholds do not appear to vary significantly as a function of the muscle group or range of forces produced, although at very low forces the Weber fraction increases, consistent with the results from other sensory modalities at low stimulus intensities. The invariance of force thresholds across different muscle groups is demonstrated by the findings from Jones [40] and Pang et al. [42]. The latter measured force thresholds using an electromechanical device that was grasped between the thumb and index finger and was designed to display a constant resistive force as it was squeezed to a predetermined location [42]. They reported thresholds of around 7 % for forces ranging from 2.5 to 10 N, and noted that thresholds did not change as a function of the method of termination of the finger displacement (i.e. a "cliff" or a "wall"). Jones [40] used an isometric force task involving flexion of the elbow joint to measure thresholds over a force range of 25–410 N and also reported a threshold of 7 %. The absence of any proximal to distal gradient in differential thresholds is consistent with the findings reported earlier (Sect. 3.1), where it was also noted that there was no evidence for such a gradient in the ability to control force.

4.4 Effects of Tactile Feedback

The perceived magnitude of forces produced by the fingers can be influenced by feedback from cutaneous mechanoreceptors. These effects vary across the hand and as a function of the type of tactile feedback. When the skin on the thumb is anesthetized, the perceived heaviness of weights lifted by thumb flexion increases by over 40 % as compared to control conditions, whereas anesthesia of the index finger results in only a 13 % increase in perceived heaviness. Surprisingly, anesthesia of the ring and little finger results in a decrease in the perceived heaviness of weights lifted by these digits [47]. The differential effects of anesthesia are presumed to reflect the net loss of cutaneous reflex facilitation (thumb and index finger) or inhibition (ring and little fingers) on the motoneuron pool activated by the descending motor command. These results also suggest that cutaneous inputs from the thumb play a more critical role in modulating forces generated by the thumb as compared to tactile inputs from the skin on the index finger and that finger's force production.

The contribution of tactile feedback to the perception of force has also been compared at different sites on the hand and arm. Jones and Piateski [38] constrained tactile feedback from the contact surfaces during force production with the index finger, hand and forearm using rigid splints to eliminate spatial cues about the forces being produced. They found that when spatial information was attenuated, forces were underestimated as compared to control conditions and that the effect did not vary across the three muscle groups tested. This suggests that tactile cues from the contact surface on the hand convey spatial cues about the forces produced that influence the perceived magnitude of forces, even when those forces are produced by muscles in the upper arm. In contrast to the effects of skin anesthesia which eliminates all sensory feedback from the skin with the result that forces are overestimated in magnitude [47], attenuating sensory feedback from the skin causes forces to be underestimated. This suggests that the contribution of cutaneous mechanoreceptor feedback to the perception and control of force is task and finger dependent, and that subtle changes in tactile feedback can modulate the forces being produced.

4.5 Effects of Thermal Feedback

When the hand makes contact with an object at room temperature there is a fairly rapid decrease in skin temperature in the initial 2–3 s of contact and then the temperature declines more slowly at a rate that depends on the thermal properties of the skin and object [23, 48]. Due to their thermal properties, objects made from materials such as stainless steel and copper result in greater heat flux being conducted out of the skin during contact than objects made from plastic or foam. In addition to the thermal properties of the objects, the change in skin temperature is influenced by the contact force exerted. As the force increases from 0.1 to 6 N there is a concomitant decrease in skin temperature as illustrated in Fig. 6, where the temperature has decreased by an average of 3.2 °C after 20 s. This decrease presumably results from the increase in contact area with force which provides a larger surface for heat transfer between the hand and the object. At higher forces it is also likely that the increased compression of the finger pad restricts blood flow in the capillary network beneath the skin, further contributing to a reduction in temperature. The changes in skin temperature as a function of force are well above threshold and so are perceptible; this suggests that thermal cues could be used to discriminate between different contact forces if other sources of sensory information were unreliable or absent.

Although the contact force influences the change in skin temperature, the perceived magnitude of voluntarily generated forces is not affected by the temperature of the contact surface. Galie and Jones [48] used the contralateral limb matching procedure to evaluate the perception of forces exerted by the index finger and found that the perceived magnitude of forces (1–8 N) did not change as the reference contact surface varied from 22 to 38 °C. These results are in marked contrast to those obtained when weights or forces are estimated using only tactile (pressure) cues and the temperature of the object in contact with the skin surface varies. Under these conditions, cold objects are perceived to be much heavier than objects of the same weight at neutral temperatures (30–36 °C), and warm objects are also perceived to be heavier than neutral objects but the effect is smaller, a phenomenon known as the thermal-weight illusion [49]. It appears that corollaries from the centrally generated motor command provide information about voluntarily generated forces that is not influenced by peripheral thermal inputs; in contrast thermal-tactile interactions reflect either the influence of thermoreceptors on the perception of pressure or thermally induced changes in the responses of cutaneous mechanoreceptors.

4.6 Tangential Forces

In comparison to the fairly extensive body of research on the perception and control of normal forces, there are relatively few studies on the perception of tangential forces applied to the hand. These forces are important during object

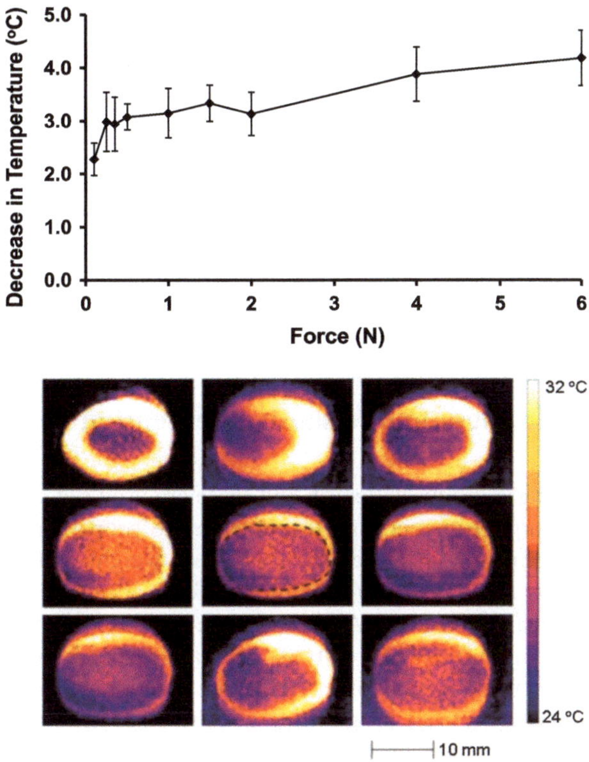

Fig. 6 *Upper*: Mean (N = 10) decrease in skin temperature as a function of force after 20 s of contact. The error bars represent the standard errors of the means. *Lower*: Thermal images from one subject of change in temperature as a function of force after 20 s of contact. First row forces: 0.1, 0.25, 0.35 N. Second row: 0.5, 1.0, 1.5 N. Third row: 2.0, 4.0, 6.0 N. From [48] with permission of Springer

manipulation and palpation and are optimized during certain activities to maximize the extraction of tactile information [11]. For example, when subjects are asked to stroke featureless smooth surfaces and rate them on a scale from "most sticky" to "most slippery" they typically move their fingers across the surface by varying the tangential force while keeping the normal force relatively constant [10].

Neurophysiological studies have demonstrated that all classes of cutaneous mechanoreceptors (SA I, SA II, FA I and FA II) are activated with the application of normal and tangential forces to the finger pad [13]. However, only the SA I afferents respond with a sustained discharge as the skin is stretched and they have a strong directional bias. It is challenging to study the perception of tangential forces experimentally as the device presenting stimuli to the finger pad must also either apply a normal force to prevent it from slipping on the skin surface, or be glued to the finger pad which affects skin mechanics and results in an unnatural stimulus. Wheat et al. [39] used the procedure of first applying a constant normal force to the finger pad and then imposing a tangential force on the finger and found that subjects could scale and discriminate tangential forces ranging from 0.25 to 1.75 N consistently. The magnitude estimates of the forces increased monotonically with increasing tangential force, and when the normal force was held constant at either

2.5 or 4 N it did not influence the perceived magnitude of the tangential forces. However, when both the tangential and normal forces varied randomly from trial to trial there was a small but consistent effect of normal force on the perception of tangential force, with larger normal forces reducing the perceived amplitude of the tangential forces applied. The threshold (Weber fraction) for discriminating tangential force was 16 % when calculated using standard tangential forces of 1 and 1.6 N [39]. Collectively, these results indicate that cutaneous mechanoreceptors provide precise information about both the magnitude and direction of forces applied to the skin and that information about these two components of force can be extracted independently from the ensemble response of cutaneous afferents.

5 Stiffness and Viscosity Perception

Changes in muscle length and force are used not only to sense the internal state of the organism, but also to perceive the properties of objects encountered in the environment. Features such as mass, weight, stiffness (force/displacement), viscosity (force/velocity) and inertia (force/acceleration) are typically sensed through manual exploration. All of these are derived variables in the sense that there are no known stiffness, viscosity or inertial sensors in muscle or skin. The human proprioceptive system must therefore integrate sensory information regarding the forces generated by muscles and the associated movements of limbs to perceive these variables. By understanding how people derive information about these properties and which cues are the most salient, effective haptic displays can be designed that present realistic cues to human users.

5.1 Perception of Stiffness and Compliance

The sensitivity of people to changes in stiffness or compliance has been studied in a number of experiments using a variety of servo-controlled electromechanical systems and physical stimuli. A major finding from this research has been the loss in perceptual resolution when limb movement and force cues are integrated so as to perceive stiffness. Jones and Hunter [50] examined the perception of stiffness using the contralateral limb-matching procedure, in which subjects adjusted the stiffness of a computer servo-controlled motor connected to one arm until it was perceived to be the same as that of the motor connected to the other arm. They found that over a stiffness range extending from 0 to 6260 N/m, subjects could adjust and match the stiffness of the two motors with an accuracy that paralleled the matching of forces and movements by the same muscle groups. The ability to discriminate the stiffness of the two motors varied across the wide range of stiffness amplitudes studied, but was essentially stable from 670 to 6260 N/m at a mean value of 23 %. These findings were confirmed in another experiment in which subjects discriminated the compliance (the inverse of stiffness) of a mechanical system compressed between the thumb and index finger. Tan et al. [51] reported that when work and terminal force cues were eliminated (as was the case in [50]), compliance resolution was poor relative to force and length discrimination, with a Weber fraction of 22 %. This number stands in contrast to the 7 and 9 % reported for force and displacement, respectively, and suggests that the human proprioceptive system is much less consistent when perceiving changes in stiffness as compared to force and displacement.

The thresholds for stiffness just described were obtained using rigid electromechanical systems in which there was no deformation of the contact surface as subjects interacted with the device. The perception of stiffness of compliant materials such as silicone rubber has also been studied, often with the objective of determining the relative contribution of cutaneous and kinesthetic (force and displacement) cues to the perceptual judgments. With compliant materials the Weber fraction is around 15 % when both cutaneous and kinesthetic cues are available; when surface deformation cues are eliminated performance deteriorates and thresholds treble [52]. These results and other similar findings point to the importance of cutaneous cues in discriminating the compliance of deformable materials and indicate that such judgments are primarily based on the haptic perception of surface deformation. For rigid interfaces, however, cutaneous mechanoreceptors do not appear to be a major source of information regarding stiffness, as the capacity to discriminate the stiffness of hand-held springs is not affected by cutaneous anesthesia.

5.2 Perception of Viscosity

The perception of viscosity has also been studied using servo-controlled motors and physical stimuli, generally with the objective of measuring the sensitivity of human observers to changes in viscosity. Consistent with the findings reported for stiffness, Jones and Hunter [53] found that for viscosities ranging from 2 to 1024 N.s/m subjects were accurate at setting the viscosity of a matching motor to equal that of the reference motor connected to the other arm. The matching function was linear with a slope of 0.88, which is comparable to the slopes measured when matching forces (0.84) and positions (0.95) with the same muscle group. Differential thresholds for viscosity were calculated and found to be even larger than those reported for stiffness. They varied from 15 to 50 % across the amplitude range studied, but were relatively stable at approximately 34 % for viscosities ranging from 32 to 1024 N.s/m. Differential thresholds for viscosity have also been measured using adaptive psychophysical techniques in which subjects did not have direct control over the viscosity of the motors (as in the matching method), but simply indicated on each trial which of two motors had the greater viscosity. Using this method, Jones et al. [54] determined that the Weber fraction for viscosity was 19 %. Consistent with the results from studies of stiffness and compliance, subjects were less sensitive to changes in viscosity as compared to force and displacement, which suggests that there is a loss in resolution when force and velocity cues are integrated to perceive viscosity.

Analyses of the forces and movements produced by subjects as they sensed the viscosity of the electromechanical systems revealed a remarkable consistency in the amplitudes of the movements generated. In contrast, both the forces and the velocities of the movements changed significantly with different reference viscosities which indicates that subjects did not simply apply a constant velocity or force to perceive the viscosity of the system. It is interesting to note that these thresholds are very similar to those reported using physical stimuli (e.g. silicone oils) that are manually explored [55].

6 Conclusions

The human hand demonstrates remarkable dexterity in its capacity to control precisely the forces involved in manipulating objects and the timing of movements during the execution of skilled tasks such as typing or playing a piano. In all of these endeavors, cutaneous mechanoreceptors play a crucial role in encoding the timing, magnitude, direction and spatial distribution of fingertip forces. When these inputs are absent or deficient, the hand is unable to compensate rapidly as an object begins to slip between the digits, and misdirected finger movements are not recognized. It is also important to bear in mind as Napier [56] noted that the qualities of the human hand that "elevate it to the peerage lie in the degree of differentiation of its musculature, the intricacy of its nerve supply, and its generous representation in the higher centers of the brain." It is the latter and in particular the capacity of the cortical maps of the body to be modified as a function of experience that contributes to the remarkable capabilities of the human hand.

Kinematic analyses of complex keyboard tasks such as typing and playing the piano have offered insight into the internal representation of motor activities and have shown that the muscles controlling movements of the fingers are not controlled independently and that coactivation is common. Although keyboard tasks are often thought of as serial activities, analyses of the sequencing of finger movements in skilled practitioners reveal that there is an anticipatory component to performance particularly when keystrokes are executed by different hands. In contrast to the precise control of the trajectory of the fingers' movements when typing, force appears to be a rather loosely controlled variable that is simply programmed to exceed the key activation force. When playing the piano, however, expert pianists do modulate force to vary the sound intensity of the notes and reliably reproduce the same forces in consecutive performances of the same musical score.

The ability to control the forces produced by the hand using only haptic feedback is not particularly precise and has not been found to vary significantly across the broad range of muscle groups involved in controlling movements of the hand and arm. There is no evidence of a proximal to distal gradient in sensitivity in force control or perception as has been reported for the control and perception of limb movements. This suggests that the recruitment of smaller motor units in a distal muscle in the hand does not impart a superior accuracy in controlling or perceiving the forces produced. Studies of the perception of forces produced by the hand and arm have revealed that people are able to discriminate relatively small changes in force, as reflected in the Weber fraction of 7 %, and that their performance in discriminating forces is comparable to their capacity to discriminate changes in limb position or movement. The thresholds for force are relatively invariant across different muscle groups and are similar to those measured for haptic weight discrimination. However, when force and displacement cues have to be combined so as to perceive the stiffness and viscosity of an object, there is a substantial increase in the Weber fractions which are larger by a factor of three to four than those measured for force and limb movement. The process of integrating two sensory channels each with their associated noise appears to result in a loss in perceptual resolution.

The perceived magnitude of supra-threshold forces does vary across muscle groups and is perceived to change when other forces are produced concurrently by adjacent muscle groups. The latter finding has been interpreted as indicating that there is a perceptual inability to segregate the motor output sent to muscles commonly used together. Forces produced by muscle groups not normally used together (e.g. the flexors in the leg and hand) have no effect on the perceived magnitude of the forces produced. When people are asked to match the forces produced by two non-homologous muscle groups, they scale the forces with reference to the operating range of the muscle, indicating that forces are sensed relatively and not absolutely. The perceived magnitude of these forces does, however, change when tactile cues are eliminated or attenuated.

In summary, the control and perception of forces generated by the hand relies on a close interplay between the sensory and motor systems. When sensory information changes, the capacity to control and modulate force can be disturbed and this in turn influences the perceived magnitude of the forces being produced. Cutaneous mechanoreceptors provide crucial information about the magnitude and direction of forces on the fingers and these inputs together with centrally generated corollary discharges are fundamental to the human perception of force.

Acknowledgments This work was supported in part by the U.S. National Science Foundation.

References

1. R.S. Johansson, J.R. Flanagan, Coding and use of tactile signals from the fingertips in object manipulation tasks. Nat. Rev. Neurosci. **10**, 345–359 (2009)
2. L.A. Jones, S.J. Lederman, *Human Hand Function* (Oxford University Press, New York, 2006)
3. R.S. Johansson, G. Westling, Signals in tactile afferents from the fingers eliciting adaptive motor responses during precision grip. Exp. Brain Res. **66**, 141–154 (1987)

4. D. Rempel, E. Serina, E. Klinenberg, B.J. Martin, T.J. Armstrong, J.A. Foulke, S. Natarajan, The effect of keyboard keyswitch make force on applied force and finger flexor muscle activity. Ergonomics **40**, 800–808 (1997)
5. J.T. Dennerlein, C.D. Mote, D.M. Rempel, Control strategies for finger movement during touch typing: the role of the extrinsic muscles during a keystroke. Exp. Brain Res. **121**, 1–6 (1998)
6. B.J. Martin, T.J. Armstrong, J.A. Foulke, S. Natarajan, E. Klinenberg, E. Serina, D. Rempel, Keyboard reaction force and finger flexor electromyograms during computer keyboard work. Hum. Factors **38**, 654–664 (1996)
7. R.T. Krampe, K.A. Ericsson, Maintaining excellence: deliberate practice and elite performance in young and older pianists. J. Exp. Psych. Gen. **125**, 331–359 (1996)
8. C. Palmer, Mapping musical thought to musical performance, J. Exp. Psychol. Human Percept. Perform. **15,** 331–346 (1989)
9. A.M. Gordon, J.F. Soechting, Use of tactile afferent information in sequential finger movements. Exp. Brain Res. **107**, 281–292 (1995)
10. A.M. Smith, S.H. Scott, Subjective scaling of smooth surface friction. J. Neurophysiol. **75**, 1957–1962 (1996)
11. A.M. Smith, C.E. Chapman, M. Deslandes, J.-S. Langlais, M.-P. Thibodeau, Role of friction and tangential force variation in the subjective scaling of tactile roughness. Exp. Brain Res. **144,** 211–223 (2002)
12. R.H. LaMotte, M.A. Srinivasan, Surface microgeometry: tactile perception and neural encoding, in *Information Processing in the Somatosensory System*, ed. by O. Franzen, J. Westman (Macmillan, London, 1991), pp. 49–58
13. M.A. Srinivasan, J.M. Whitehouse, R.H. LaMotte, Tactile detection of slip: surface microgeometry and peripheral neural codes. J. Neurophys. **63**, 1323–1332 (1990)
14. S.J. Bolanowski, G.A. Gescheider, R.T. Verrillo, C.M. Checkosky, Four channels mediate the mechanical aspects of touch. JASA **84**, 1680–1694 (1988)
15. L. Jami, Golgi tendon organs in mammalian skeletal muscle: functional properties and central action. Physiol. Rev. **72**, 623–661 (1992)
16. M.P. Mileusnic, G.E. Loeb, Mathematical models of proprioceptors. II. Structure and function of the Golgi tendon organ. J. Neurophysiol. **96**, 1789–1802 (2006)
17. G. Macefield, S.C. Gandevia, D. Burke, Perceptual responses to microstimulation of single afferents innervating joints, muscles and skin of the human hand. J. Physiol. **429**, 113–129 (1990)
18. S.C. Gandevia, Kinesthesia: roles for afferent signals and motor commands, in *Handbook of physiology. Section 12. Exercise: Regulation and integration of multiple systems*, ed. by L. Rowell, J.T. Shepherd (Oxford University Press, New York, 1996), pp. 128–172
19. L.A. Jones, The senses of effort and force during fatiguing contractions, in *Fatigue: Neural and Muscular Mechanisms*, ed. by S.C. Gandevia, R.M. Enoka, A.J. McComas, D.G. Stuart, C.K. Thomas (Plenum, New York, 1995), pp. 305–313
20. M.R. Cutkosky, P.K. Wright, Modeling manufacturing grips and correlations with design of robotic hands. Proc. IEEE ICRA **3**, 1533–1539 (1986)
21. C.L. MacKenzie, T. Iberall, *The Grasping Hand* (North-Holland, Amsterdam, 1994)
22. V. Mathiowetz, N. Kashman, G. Volland, K. Weber, M. Dowe, S. Rogers, Grip and pinch strength: normative data for adults. Arch. Phys. Med. Rehab. **66**, 69–72 (1985)
23. H.-N. Ho, L.A. Jones, Modeling the thermal responses of the skin surface during hand-object interactions. J. Biomech. Eng. **130**, 21005-1–21005-8 (2008)
24. E.M. Meftah, L. Belingard, C.E. Chapman, Relative effects of the spatial and temporal characteristics of scanned surfaces on human perception of tactile roughness using passive touch. Exp. Brain Res. **132**, 351–361 (2000)
25. L.A. Jones, M. Berris, Material discrimination and thermal perception, in *IEEE Proceedings 11th* on *Haptic Interfaces* for *Virtual Environment* and *Teleoperator Systems,* pp. 171–178 (2003)

26. B.B. Edin, G. Westling, R.S. Johansson, Independent control of human finger-tip forces at individual digits during precision lifting. J. Physiol. **450**, 547–564 (1992)
27. L.A. Jones, Visual and haptic feedback in the control of force. Exp. Brain Res. **130**, 269–272 (2000)
28. N. Mai, M. Avarello, P. Bolsinger, Maintenance of low isometric forces during prehensile grasping. Neuropsychol. **23**, 805–812 (1985)
29. H.Z. Tan, B. Eberman, M.A. Srinivasan, B. Cheng, Human factors for the design of force-reflecting haptic interfaces. Proc. ASME Dyn. Sys. Cont. Div. DSC. **55–1**, 353–359 (1994)
30. M. Schieber, M. Santello, Hand function: Peripheral and central constraints in performance. J. Appl. Physiol. **96**, 2293–2300 (2004)
31. K.T. Reilly, M.H. Schieber, Incomplete functional subdivision of the human multitendoned finger muscle flexor digitorum profundus: An electromyographic study. J. Neurophysiol. **90**, 2560–2570 (2003)
32. T. Ohtsuki, Decrease in grip strength induced by simultaneous bilateral exertion with reference to finger strength. Ergonomics **24**, 37–48 (1981)
33. K.T. Reilly, G.R. Hammond, Independence of force production by digits of the human hand. Neurosci. Lett. **290**, 53–56 (2000)
34. L.A. Jones, Manual dexterity, in *The Psychobiology of the Hand*, ed. by K.J. Connolly (Mac Keith Press, London, 1998), pp. 47–62
35. H. Kinoshita, S. Kawai, K. Ikuta, Contributions and co-ordination of individual fingers in multiple finger prehension. Ergonomics **38**, 1212–1230 (1995)
36. R. Reilmann, A.M. Gordon, H. Henningsen, Initiation and development of fingertip forces during whole-hand grasping. Exp. Brain Res. **140**, 443–452 (2001)
37. S.L. Kilbreath, S.C. Gandevia, Neural and biomechanical specialization of human thumb muscles revealed by matching weights and grasping objects. J. Physiol. **472**, 537–556 (1993)
38. L.A. Jones, E. Piateski, Contribution of tactile feedback from the hand to the perception of force. Exp. Brain Res. **168**, 298–302 (2006)
39. H.E. Wheat, L.M. Salo, A.W. Goodwin, Human ability to scale and discriminate forces typical of those occurring during grasp and manipulation. J. Neurosci. **24**, 3394–3401 (2004)
40. L.A. Jones, Matching forces: constant errors and differential thresholds. Perception **18**, 681–687 (1989)
41. S.C. Gandevia, S.L. Kilbreath, Accuracy of weight estimation for weights lifted by proximal and distal muscles of the human upper limb. J. Physiol. **423**, 299–310 (1990)
42. X.D. Pang, H.Z. Tan, N.I. Durlach, Manual discrimination of force using active finger motion. Percept. Psychophys. **49**, 531–540 (1991)
43. N. Mai, P. Schreiber, J. Hermsdörfer, Changes in perceived finger force produced by muscular contractions under isometric and anisometric conditions. Exp. Brain Res. **84**, 453–460 (1991)
44. S.L. Kilbreath, S.C. Gandevia, Independent digit control: Failure to partition perceive heaviness of weights lifted by digits of the human hand. J. Physiol. **442**, 585–599 (1991)
45. L.A. Jones, Perceptual constancy and the perceived magnitude of muscle forces. Exp. Brain Res. **151**, 197–203 (2003)
46. E.E. Brodie, H.E. Ross, Sensorimotor mechanisms in weight discrimination. Percept. Psychophys. **36**, 477–481 (1984)
47. S.L. Kilbreath, K. Refshauge, S.C. Gandevia, Differential control of the digits of the human hand: evidence from digital anaesthesia and weight matching. Exp. Brain Res. **117**, 507–511 (1997)
48. J. Galie, L.A. Jones, Thermal cues and the perception of force. Exp. Brain Res. **200**, 81–90 (2010)
49. J.C. Stevens, J.E. Hooper, How skin and object temperature influence touch sensation. Percept. Psychophys. **32**, 282–284 (1982)
50. L.A. Jones, I.W. Hunter, A perceptual analysis of stiffness. Exp. Brain Res. **79**, 150–156 (1990)

51. H.Z. Tan, N.I. Durlach, G.L. Beauregard, M.A. Srinivasan, Manual discrimination of compliance using active pinch grasp: the roles of force and work cues. Percept. Psychophys. **57**, 495–510 (1995)
52. W.M. Bergmann Tiest, A.M.L. Kappers, Cues for haptic perception of compliance. IEEE Trans. Haptics **2**, 189–199 (2009)
53. L.A. Jones, I.W. Hunter, A perceptual analysis of viscosity. Exp. Brain Res. **94**, 343–351 (1993)
54. L. Jones, I. Hunter, S. Lafontaine, Viscosity discrimination: a comparison of an adaptive two-alternative forced-choice and an adjustment procedure. Perception **26**, 1571–1578 (1997)
55. W.M. Bergmann Tiest, A.C.L. Vrijling, A.M.L. Kappers, Haptic perception of viscosity. EuroHaptics 2010, part 1, LNCS 6191, pp. 29–34 (2010)
56. J.R. Napier, *The Human Hand* (Carolina Biological Supply, Burlington, 1976)

Chapter 6
Proprioceptive Mechanisms and the Human Hand

Lee D. Walsh, Janet L. Taylor and Simon C. Gandevia

Abstract The hand is complex and used in many functions, including eating and communication. To control the hand accurately the brain needs information about the position, velocities and forces around each joint, which is provided by proprioception. Despite being studied for over a century, there is much to learn about this enigmatic sense. The first part of this chapter summarises the key historical debates and the evidence that shaped the current view of proprioception. The main part then highlights recent evidence that has profoundly changed the understanding of proprioception. One recent development is the discovery that the firing rates of muscles spindles depend upon the contraction history of the muscle and that it is possible for muscle spindles to become insensitive to movement of the joint, remaining quiet during small joint movements. This alters the perceived position of joints. Other experiments show that illusions of joint movement can be induced by stretching the skin, providing evidence that slowly adapting cutaneous receptors contribute to movement sense. However, at least at finger joints, it seems that rapidly adapting cutaneous receptors interfere with the detection of the direction of movement. Recent evidence reveals illusions of altered joint position and movement with voluntary efforts during paralysis and anaesthesia. Thus command signals generated by the brain provide direct information about joint position and movement. Other experiments using anaesthesia have shown key roles of muscle receptors in generating the body maps stored by the brain. Together this recent work shows that the textbook view of proprioception needs revision.

L. D. Walsh · J. L. Taylor · S. C. Gandevia (✉)
Neuroscience Research, Australia and the University of New South Wales, Sydney, NSW, Australia
e-mail: s.gandevia@neura.edu.au

L. D. Walsh
e-mail: l.walsh@neura.edu.au

J. L. Taylor
e-mail: j.taylor@neura.edu.au

R. Balasubramanian and V. J. Santos (eds.), *The Human Hand as an Inspiration for Robot Hand Development*, Springer Tracts in Advanced Robotics 95,
DOI: 10.1007/978-3-319-03017-3_6, © Springer International Publishing Switzerland 2014

Keywords Proprioception · Muscle spindles · Cutaneous receptors · Joint receptors · Body representation · Body ownership

1 Introduction

The hand has a remarkable structure with more than 15 joints and more than 20 muscles which can move them. It is positioned with seeming ease to perform a wide range of essential tasks from manipulation and communication to grooming and eating. The sensory and motor mechanisms which permit this have long interested scientists and clinicians. One overarching contribution on the sensory side is proprioception. The present chapter reviews some background to this topic and recent developments. It is not coincidental that many of the mechanisms subserving proprioception have been exposed in psychophysical studies of the human hand.

The brain generates movement by activation of skeletal muscles. Activation of a skeletal muscle causes it to pull on the bones, causing them to move relative to each other. However, the brain can only make skeletal muscles shorten, it cannot lengthen them directly. Thus each muscle can only move a joint in one direction. To move the joint back to where it started a second opposing, or antagonist, muscle is required. Furthermore many joints contain multiple degrees of freedom and therefore require more than two muscles to control them. Some joints are controlled by remote muscles. Muscles located in the forearm move the fingers, and their tendons cross several joints to reach the fingertip. Even simple movements require the control of multiple joints, so clearly controlling movement of the skeleton is a complex task.

To perform movements accurately the brain requires information about the position, velocity and forces around each joint. Perception of this information is known as *proprioception*. This term is used now to refer to the sensation of any movements and forces that occur within or are imposed on the body. Proprioception is provided via two broad mechanisms. The first is a feedback mechanism and involves information being transduced in the peripheral parts of the body and sent to the brain. This is *afferent* information and it is generated by sensory receptors located in the muscles and skin as well as the joints themselves. Sensory receptors in the muscles, called muscle spindles, transduce information about the length and rate of change of length of the muscles while Golgi tendon organs, located mostly in the musculo-tendinous junction, signal the forces generated by muscles. Some slowly adapting receptors in the skin signal stretch of the skin around a joint as the joint moves. Similarly, joint receptors signal stretch of the joint capsule. The second mechanism involves the brain using a motor command signal and a stored model to generate information about how the body responds to the *efferent* command signals sent to the muscles. Where the term afferent refers to information moving *towards* the brain, the term efferent refers to information associated with generation of motor output, often termed corollary discharge or efference copy (see Sect. 3.4). The term efferent is used here because the proprioceptive information is derived from an efferent signal, although once derived, the information may stay in the brain.

How proprioceptive information is generated and used by the brain has been studied for more than a century and has often been a source of controversy. However, despite the long history, there is still a lot that is unknown, and there have been important recent developments in the understanding of proprioceptive mechanisms. Over the history of research into proprioception many of the key experiments have been performed in the hand [e.g. 1–3] and the hand continues to play an important role in the study of proprioception.

2 History

2.1 Muscle Spindles Versus Joint Receptors

The idea of having a sense that signals the position and movements of the limbs goes back at least as far as Bell in the 1830s [4] and even early on there was controversy about whether this sense was based on afferent signals [e.g. 5–7] or centrally-generated signals [e.g. 8]. Sherrington's [7] 'muscle sense' favoured afferent information, specifically from muscle receptors, such as muscle spindles. Muscle spindles do not detect joint angle as such, but the length and rate of change of length of the skeletal muscles. If the length of all the muscles around a joint is known then the position of the joint and velocity of joint movements can be determined. Sherrington's view became dominant in the early 1900s [9–11] but muscle receptors subsequently fell out of favour as contributors to sensation. By the 1950s joint receptors were seen as the dominant source of information about joint position and movement. This view was supported by the extensive work of Skoglund [12]. Further support came from the discovery of neurons in the sensory cortex that responded to joint movement and joint probing [13], combined with evidence that muscle receptors did not project into the cortex [14]. The latter result was eventually proved wrong and was due to the class of anaesthetics used in animal experiments at the time [15].

Muscle spindles again became the favoured source for the sense of joint position and movement after experiments by Goodwin, McCloskey and Matthews [16] showed that vibration of the muscle's tendon induced illusions of altered joint position and joint movement. Muscle spindles are known to be sensitive to low-amplitude vibration (frequency ~80 Hz, amplitude ~1 mm) [e.g. 17, 18]. Furthermore illusions of movement are produced when the muscle is stretched by pulling on an exposed tendon without any movement of the joint [19, 20, cf. 21] or when the muscle afferents are stimulated electrically [22]. This strong evidence of a role for muscle spindles was accompanied by evidence that joint receptors were not ideal as transducers of joint position and movement. It was shown that joint receptors were usually not good at encoding joint angle in cats [23] or in the human hand [24]. Furthermore, they produced ambiguous signals because their output was often similar at both ends of the joint range [23]. In addition, after total hip replacement surgery, which removes the joint capsule and presumably all joint receptors, position sense at the hip is still intact [25, 26]. Similar results have been found for the knee [27] and the shoulder [28]. Currently it is thought that muscle spindles are the dominant afferent source of information about the position and movement of the joints, and that joint receptors are important for signalling the extremes of joint range.

2.2 Cutaneous Receptors

When joint movement occurs, the skin around the joint is stretched and thus cutaneous receptors, usually thought of as being simply used for the sense of touch, are able to provide proprioceptive information. This was first noted in 1929 by Adrian and Umrath [29] but not until 1979 was it shown that cutaneous receptors in the human hand respond to the movement of nearby joints [30]. Hulliger and colleagues found that rapidly adapting type two (RA-II) receptors and slowly adapting type two (SA-II) receptors were the most responsive to joint movement. In addition SA-II afferent nerve fibres encoded the static position of a joint. This fitted with previous evidence that SA-II afferents responded to skin stretch and were directionally sensitive [31]. These studies demonstrated that signals from SA-II receptors were capable of contributing to proprioception because they produced stable firing rates that depended upon the amount of skin stretch around a joint. However a problem with cutaneous afferent signals is that the same afferents also respond to skin stretch that is not related to movement of position of the joint. This means that cutaneous signals must undergo some processing if they are to be used for proprioception.

While it was clear that cutaneous receptors *could* contribute to proprioception, it was not straightforward to determine if they actually had an important role. One problem was that the experimental techniques that remove cutaneous signals, for example blocking the digital nerves of the finger with local anaesthetic, usually also removed the signals from joint receptors. Some authors assumed that skin receptors did not contribute directly to proprioception when they blocked the digital nerves to study the contribution of joint receptors [1, 32, see also 2]. This separation is less difficult in the knee and thus Clark et al. [33] believed they were able to independently block skin and joint afferents using local anaesthetic. They suggested that skin receptors did not contribute to static position sense, but they noted that that the ability to match limb position degraded with the removal of cutaneous input. They suggested that signals from the skin may be more important for movement sense in the hand than for proximal joints. Further studies suggested that cutaneous receptors contributed to limb position and movement sense [21, 34], but opposition continued [35]. One problem at the time was that it was not clear if cutaneous signals provided a specific signal of limb position and/or movement or whether they provided central facilitation to aid in the decoding of muscle and joint receptor input [2, 36, 37]. This issue has been addressed in recent studies which are discussed in Sect. 3.2.

2.3 Tendon Organs

Proprioception also includes the ability to perceive force and heaviness, the history of which has been less controversial than the senses of limb position and movement [38]. The sense of force refers to the ability to perceive the force that is generated by the muscles and its primary receptor is the Golgi tendon organ. Tendon organs are located mostly near the musculo-tendinous junction [39] and their firing rates are closely related to muscle force [18, 40]. However they are more sensitive to force produced by an actively contracting muscle than passive forces [41]. In addition to tendon organs, information about force can be obtained from cutaneous receptors [42] because when a limb moves and applies force to something the skin is compressed or sheared in the region of contact. A third source of information about force is the centrally generated motor command which is directly related to the signal driving the muscle to contract. This signal may increase when a muscle is effectively weakened [43]. However, this information is not ideal because it is corrupted by factors, such as muscle fatigue, that change the relationship between motor command output and muscle force output. This is where the sense of effort is distinguished from the sense of force. An example of the divergence of the sense of force and effort can be seen when carrying a heavy suitcase some distance. Over time the suitcase feels heavier, even though its weight has not changed. This is because the loaded muscles are fatiguing and therefore have to be driven harder to maintain the force required to lift the suitcase. The increase in motor command required to drive the muscles is perceived as increased effort and interpreted as increased heaviness of the suitcase. Central signals of motor command are thought to be a major source of the sense of effort, however it is possible that muscle spindles provide input as well [44].

It is common to break proprioception into 'sub-senses' to simplify investigation. In the history presented above it was broken into senses of limb position, limb movement, force and effort. All proprioceptive signals provide information simultaneously when a movement is made. Some receptors, such as muscle spindles and tendon organs, are specialised to signal one type of information. This does not mean that they only provide that information. As an example, tendon organs signal force, but could also detect lengthening during muscle contraction because if the muscle shortens they will be stretched. Likewise, muscle spindles *could* also signal muscle force. Because they are 'co-activated' by fusimotor drive during voluntary contractions, they also fire more when the muscle contracts (see Sect. 3.1). Although it is possible to focus and perceive the angle, velocity or force at one joint, this is not how we generally perceive our body. For example, we are consciously aware of the position of our hands, but we do not perceive the information from individual joint and muscles, we simply perceive the position of the hand. This emphasises that proprioceptive information is combined into a synthesised representation of the body, and we perceive that representation which is updated continuously by sensory information.

3 Recent Developments

3.1 Muscle Spindles

When skeletal muscle contracts and shortens it might be expected that the muscle spindles would also shorten, and therefore fall slack. However this is overcome by the structure of the muscle spindles which comprise small muscle fibres inside a spindle-shaped capsule. These intrafusal muscle fibres have their own motor supply, the fusimotor system. Fusimotor activation acts to shorten the muscle spindles and thus, ideally to ensure that they are at an appropriate length to signal changes in muscle length. However the presence of the fusimotor system also creates its own complications for the muscle spindle signal. The contractile parts of the intrafusal muscle fibres are predominantly located at both ends of the spindles. This means that when the fusimotor system is activated both ends of the intrafusal muscle fibres shorten and the centre of the fibres, where the afferent nerve terminals are located, is stretched. During an isometric contraction there is little change in total muscle length (i.e. the length of the muscle–tendon unit), although muscle fibres would shorten slightly. However, the fusimotor system is usually activated along with the rest of the muscle [45–49, cf. 50, 51] and this will stretch the muscle spindle endings and increase their firing rate. Hence despite the actual muscle length remaining relatively constant, the spindle signal during an isometric contraction should indicate that the muscle is lengthening, which is consistent with a movement or change in position opposite to the direction of the contraction. In the 1970s, it was suggested that we do not perceive such a change because the part of the muscle spindle signal that is due to the fusimotor activation is subtracted out using a corollary of the motor command [16, 20, 37]. However there is no evidence that this is the case. In fact recent studies have shown that we do perceive a change in joint position during an isometric contraction [52, 53], although the change is in the opposite direction to that expected from the fusimotor activation of muscle spindles. So while fusimotor activation can assist in putting muscle spindles at an appropriate length to indicate muscle length accurately, it also corrupts the signals and makes them potentially ambiguous. In situations in which the muscle spindle signal from the agonist muscles is ambiguous, the signal from muscle spindles in the passive antagonist muscle could be very useful [54].

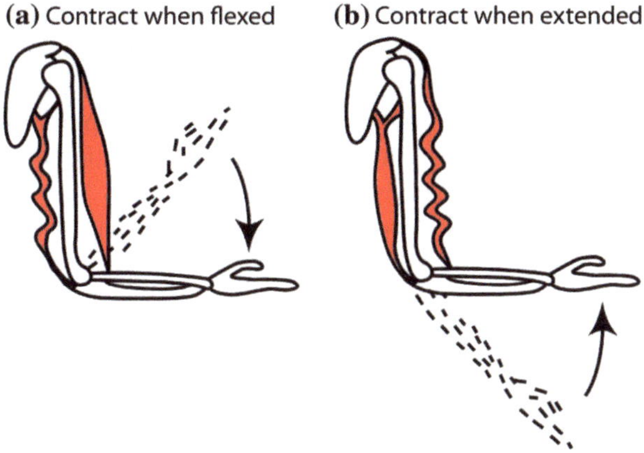

Fig. 1 The effect of thixotropy on passive muscle. The *dotted arms* show the position at which an isometric contraction was made. In (**a**) the elbow was extended after the contraction while biceps and triceps were relaxed. This action stretched biceps and shortened triceps, but because of the splinting caused by stable crossbridges the triceps falls slack. In this situation any passive lengthening or contraction of triceps will be taken up by the slack. A similar situation is depicted in **b**, the elbow has been passively flexed after an isometric contraction at a more extended position. This causes the biceps to fall slack. The intrafusal muscle fibres of the muscle spindles are affected by the same thixotropic property and spindles that fall slack will lower their firing rate and may go silent. Muscle spindles that are stretched (biceps in **a** and triceps in **b**) will increase their firing rates. The situation depicted in **a** is known as *flexion conditioning* and **b** is known as *extension conditioning*. The figure is adapted from [83] with permission

Another problem with the muscle spindle signal is that it depends on the history of contraction of the muscle. Skeletal muscle has thixotropic properties in that after contraction, or even after a period of rest at a set length, the behaviour of the crossbridges means that the muscle fibres have a resistance to length changes [55]. Hence, length changes imposed on passive muscle are met with some stiffness. The result is that a muscle contracted at a long length and then passively shortened may fall slack (Fig. 1) and a muscle contracted at a short length and then passively stretched may initially resist the increase in length. These thixotropic properties extend to the intrafusal muscle fibres of the muscle spindles [56–58] so that the state of the spindle endings and their firing rate in a passive muscle depend on the previous muscle contraction and any subsequent passive movement. The effects of muscle spindle thixotropy on the sense of joint position are large enough to produce illusions of altered joint position [e.g. 57, 59]. While the role of muscle spindles in the sense of joint position and the sense of joint movement is well established, they are not ideal detectors of muscle length or rate of change of muscle length. Activation of the fusimotor system can make the spindle signal ambiguous in active muscle and the thixotropic properties of the intrafusal fibres can make the signal ambiguous in passive muscle. Furthermore the initial response of spindle afferents to stretch is reset a short time after the last change in length [60]. Despite being accepted as the primary detector of joint position and movement, muscle spindles seem poorly suited to the task. However this is typical of proprioceptive signals, none are ideal and all have shortcomings, but they all combine to provide an accurate sense of what the body is doing.

Muscle spindles have generally been thought of as receptors for muscle length and rate of change of muscle length. Recent work investigating the firing of muscle afferents in the hand during 'natural' grasping movements has shown that muscle spindles may also encode the second derivative of muscle length, that is acceleration [61, 62]. Dimitriou and Edin also suggested that muscle length and velocity information is not simple to extract from muscle spindle signals in active muscle and that information about fusimotor activation and the ongoing load properties may be necessary. These authors have gone on to show that muscle spindles are also capable of acting as forward sensory models and predicting future kinematic states of their parent muscle [63]. Forward models are important in motor control because the ability to predict motor outcomes allows faster adaptation and correction 'on-the-fly'.

3.2 Cutaneous Receptors

Studies that blocked the nerves containing cutaneous afferents impaired the sense of limb position in the hand. However it was not known if signals from cutaneous receptors signalled position and movement directly or whether they simply provided facilitation to the signal from other receptors. More recently it has been shown that a facilitatory role is not likely because removing the input from the digital nerves of the middle finger did not impair movement detection of the proximal interphalangeal joint of the adjacent index finger, and adding input did not improve it [64]. However this study also showed that an electrical or natural cutaneous stimulus applied to the adjacent finger did impair movement detection, suggesting that there is an interaction between cutaneous input and proprioception. Edin and Johansson [65] suggested that afferents from skin receptors contributed to the perception of movement. They demonstrated that illusions of finger movement could be induced by manually stretching the skin around the proximal interphalangeal joint in a way that mimicked how it would normally be stretched during a movement of that joint. Similar illusions were shown for the metacarpophalangeal joint by Collins and Prochazka [3], who also produced these illusions by electrically stimulating the skin on the back of the hand. More recently, skin stretch was shown to produce illusions of movement of the finger, elbow and knee (Fig. 2) [66], although illusions in the hand occurred in more subjects than illusions at proximal joints. This suggests that perhaps skin receptors are most important for movement detection in the hand. Cutaneous receptors have a high density in the hand [67] which gives them the potential to provide greater, more accurate input. Furthermore the accuracy of muscle spindles is compromised when a muscle spans multiple joints [68], whereas cutaneous receptors in the fingers may provide more localised information about which joint moved.

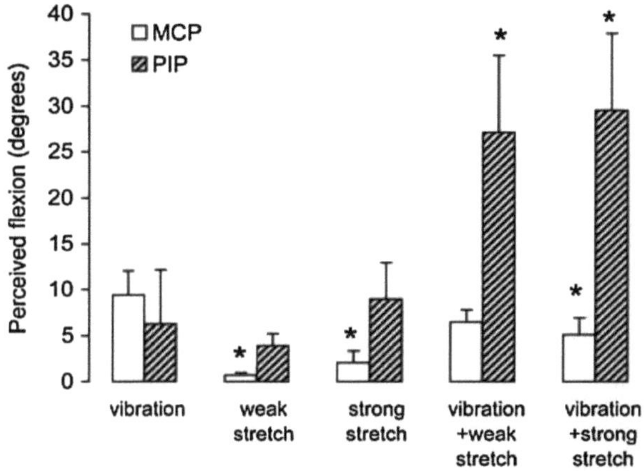

Fig. 2 Illusions of altered joint position induced by tendon vibration and stretch of the skin around joints of the index finger. Vibration results in errors in the perception of joint angle at the metacarpophalangeal (*MCP*) and proximal interphalangeal (*PIP*) joints of the index finger. Skin stretch can also induce these illusions. The effect of skin stretch is enhanced by simultaneous vibration and is greater at the *PIP* joint which is distal compared to the *MCP* joint. * indicates a result that is significantly different from the vibration only result ($p < 0.05$). This figure is adapted from [66] with permission

Another recent development in proprioception concerns whether some input from the cutaneous receptors can interfere with proprioception. Very low-amplitude (20–50 μm peak-to-peak) high-frequency (300 Hz) vibration favours input from Pacinian corpuscles, while lower frequency (30 Hz) favours input from Meissner corpuscles. Low-amplitude, high-frequency stimulation applied to either the middle finger or the thenar eminence reduces the ability to detect movements of the proximal interphalangeal joint in the index finger [69]. This reduction in the ability to detect movement is not seen with a lower frequency stimulus, which suggests that input from Pacinian corpuscles can interfere with movement detection in the hand. If the digital nerves of the adjacent finger are blocked, removing input from skin and joint receptors, there is no impairment produced by Pacinian corpuscle input (Fig. 3) [70]. Furthermore, removal of the muscle receptor contribution makes no difference to the effect of the high frequency vibration. The evidence from these two studies shows that cutaneous receptors can interfere with proprioception in the hand.

3.3 Pain

Pain is another factor that has been shown to interfere with proprioception. Studies done on the elbow flexor muscles have shown that muscle pain induced by eccentric muscle damage or injection of hypertonic saline causes subjects to make errors

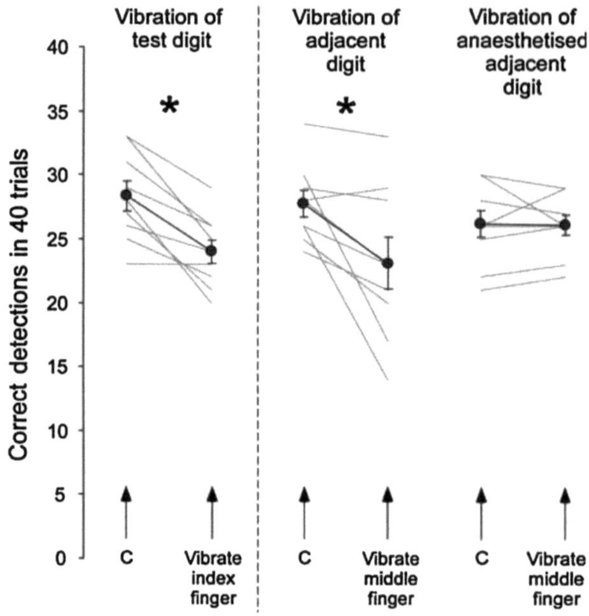

Fig. 3 The effect of high-frequency cutaneous vibration (designed to activate Pacinian corpuscles) on movement detection at the proximal interphalangeal joint (PIP) of the index finger (*test digit*). Movement detection was measured as the number of correct detections of the direction of movement at the PIP joint out of 40 trials, during three conditions: (1) vibration of the index finger, (2) vibration of the middle finger and (3) vibration of the anaesthetised middle finger. The *dark lines* and *symbols* show the group mean ± SEM. The *grey lines* show each of the nine subjects from each condition. There was a significant (*asterisk*, $p < 0.05$) decrease in the incidence of correct detection of movement direction during vibration of the index finger and vibration of the adjacent middle finger compared to control trials (*C*). No change occurred when vibration was applied to an anaesthetised finger. Reproduced from [70] with permission

when matching the force produced by the painful muscle with a force generated by the other arm [71–73]. Pain in the skin over the muscle also resulted in subjects making force matching errors [72]. Another study of the hand investigated the effect of muscle and skin pain on detection of movement of the thumb. Experimentally induced pain in the flexor pollicis longus muscle which moves the thumb, or in the skin of the thumb, impaired movement detection at the distal joint of the thumb [74]. However pain in the skin over the flexor pollicis longus muscle, or in the flexor carpi radialis muscle, which does not act on the thumb, did not impair movement detection in the thumb. One concept emerging from the work is that while pain may have some generalised effects on motor and sensory performance, nociceptive inputs may also exert specific effects on aspects of proprioception.

3.4 Motor Commands

If the brain is responsible for causing a movement, it can theoretically monitor the commands that were issued to enact that movement. Using that information and knowledge gained from experience about how the body responds, the brain could determine the outcome of its motor commands. For example, during a voluntary flexion of the index finger, the brain could know that the index finger is flexed because the muscle, skin and joint receptors are signalling finger flexion, or it could know that the finger is flexed because it commanded the finger to flex. The concept that we are consciously aware of the drive to our muscle may be as old as the 1500s [75], but it became more widely known in the late 1800s [8, 10]. In the 1950s it was proposed that corollary discharges [76] or efference copies [77] of the motor command interacted with afferent information to compensate for, or subtract, the corrupting influences of self-generated action on afferent information. From the 1970s until recently it was speculated that this was the only role of central motor commands in the sense of limb position and movement. That is, corollary discharges allowed subtraction of the increased firing of muscle spindles caused by fusimotor activation, but did not provide a direct signal of position or velocity [16, 36–38].

Recently it was shown that when subjects have one set of elbow flexors weakened by fatigue or eccentric muscle damage, they make errors in matching the angle of one arm, placed by the experimenter, with the other arm in the absence of vision [78–80]. It was proposed that subjects used the amount of effort required to hold their arm against gravity as a cue for the angle of their elbow, rather than afferent signals. Further experiments in the hand have shown that during anaesthesia and paralysis induced by ischaemic block, if subjects are asked to make a voluntary effort with their wrist muscles they perceive the hand to become displaced in the same direction as the effort [81]. For example, if a subject makes an effort into flexion, they report their wrist to be more flexed. Furthermore the size of the perceived displacement grades with the level and duration of the voluntary effort [82]. This result is consistent with a motor command signal directly influencing the sense of limb position. Similar, although smaller, illusions of displacement also occur when subjects who have all proprioceptive signals intact perform isometric contractions with their wrist muscles [52, 53].

The work described above is focused on the role of central motor commands in signalling joint position. The sense of joint movement is distinctly different from the sense of joint position. Joint position sense is concerned with signalling the static position of the limbs, whereas movement sense is more concerned with signalling velocity. A recent study, again performed in the hand, has shown that in addition to contributing to limb position sense, central motor command signals also contribute to the sense of limb movement and velocity (Fig. 4) [82].

In contrast to these studies, it has been shown that when illusions of altered joint position are produced by voluntary efforts about the elbow, they can be removed by manipulating the firing rates of muscle spindles using conditioning contractions of the muscle [83]. This suggests that the effects of voluntary efforts on position sense

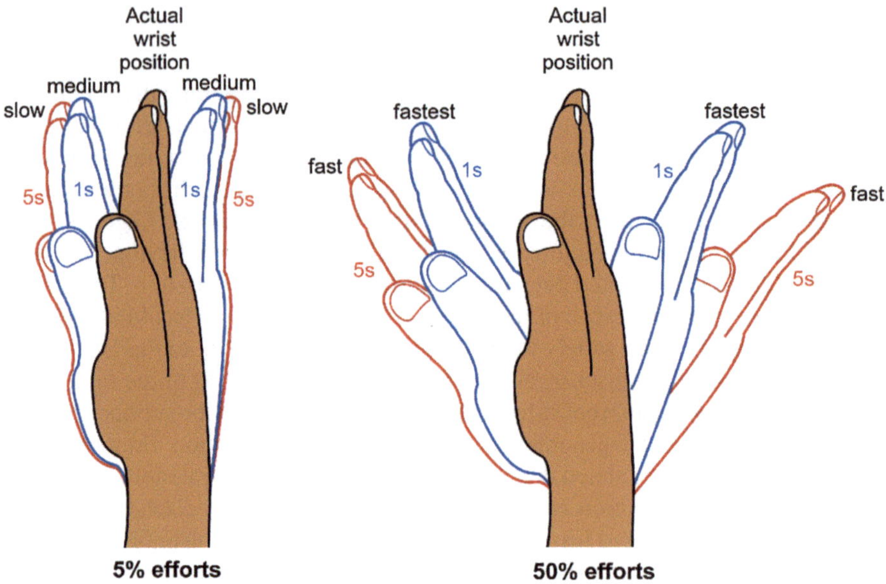

Fig. 4 Movements of a phantom hand during voluntary efforts. In our recent study [96] subjects had their forearm wrist and hand anaesthetised and paralysed by a cuff around the upper arm. When the block was complete subjects were asked to make voluntary efforts with their paralysed and anaesthetised wrist and indicate with a pointer any movements that they perceived. The filled hands (index and middle finger are shown) represent the actual position that the subjects' hand was in. The *blue outlined hands* show the mean size of the phantom movements indicated by subjects during a voluntary effort that was 1 s long. The *red outlined hand* shows the mean size of the phantom movements reported by subjects during a 5-second voluntary effort. The speeds of the movements produced by the four voluntary efforts shown are ranked by the terms slow, medium, fast and fastest and these varied from $11s^{-1}$ to $35s^{-1}$. Subjects indicated that they perceived movements of their phantom hand which were bigger when they made stronger efforts or longer efforts and faster if they made stronger efforts. Adapted from [96] with permission

may be due to changes in the firing rates of muscle spindles that occur during the muscle contractions that result from the efforts. However, at the wrist, an effect of the voluntary efforts is present even when the muscle spindle firing rates are controlled by careful conditioning of the muscles [53]. It is not yet clear why the effect of motor commands on position sense is apparently different at the elbow and the wrist. There may be a difference in position sense between the two joints, but it may also be due to a methodological difference. The elbow experiments involved a bilateral matching task in which one arm was used to indicate the position of the other. However, in the wrist experiments the subject indicated the position of their wrist with a pointer. In the two-arm task there may be an interaction between perception of the two arms that does not exist between one wrist and a pointer. Alternatively the wrist experiments may involve an effect related to the process of matching the position of an object in the visual space with the position of a joint. This conflict of results needs to be resolved in order to fully understand how central motor command signals contribute to the sense of limb position.

3.5 Body Representation

We can focus on the position and velocity of a single joint, but do not generally perceive the position of the body this way. The proprioceptive information about the position, velocity and forces at each joint are combined into a 'representation' of the body. For example, if the arm is held out in front of the body, rather than perceiving a shoulder angle, elbow angle, wrist angle and finger angles we tend to perceive the arm, as a whole, in a posture. This view is supported by evidence that subjects are better at judging the orientation of their limbs [84] and limb segments [85] rather than joint angles. Body representations must be updated continuously to remain accurate as the body is moved around. However the body representation does not only store information about movements, forces and joint positions.

To know the position of the fingertips, the brain must know the angles of all the joints in the hand and arm, but must also know the distances between joints. There are no receptors that provide information about the length of the body segments, so presumably this information is provided by vision and experience. Information about body segments can be updated less frequently because while the length of body segments is not constant, it changes very slowly. Even adolescent boys only grow a maximum of 80 to 100 mm per annum [86, 87].

Body representations can also be manipulated. Anaesthesia of a body part blocks its sensory information from reaching the brain, but perception of the body part does not cease. Instead perception of the body part continues as a 'phantom' limb. Phantom limbs can occur after amputation [88], but can also be induced experimentally using local anaesthetic, or by inflating a cuff around the limb to above arterial pressure to produce an ischaemic block. When the hand or finger is anaesthetised, subjects report that the anaesthetised body part feels larger [89–91]. This change in perceived size has been suggested to be due to the loss of small-diameter afferent nerve fibres [90], but must also involve large-diameter afferents because changes in perceived size are reported when large fibres are becoming blocked but small fibres are still mostly intact [91]. Changes in perceived position of joints in the hand also occur if the block is applied using a cuff around the upper arm which slowly paralyses and anaesthetises the hand, wrist and forearm. As the large-diameter afferent nerves, which carry proprioceptive and touch information, cease to function, subjects report that the fingers become more flexed or more extended [91]. The direction of these changes depends upon the actual position of the subject's hand before the block. An initially flexed hand leads to perceived extension and an extended hand to perceived flexion. The final position is not the same, suggesting that there is no 'default' position for a phantom and perhaps the body representation. Phantom limbs can also be moved. If subjects are asked to make voluntary efforts with their wrist muscles after they have been paralysed and anaesthetised, subjects report movement (see above) of the phantom in the same direction as the effort [82]. The amplitude and velocity of the movement grades with the level of effort. The amplitude also grades with the duration of the effort.

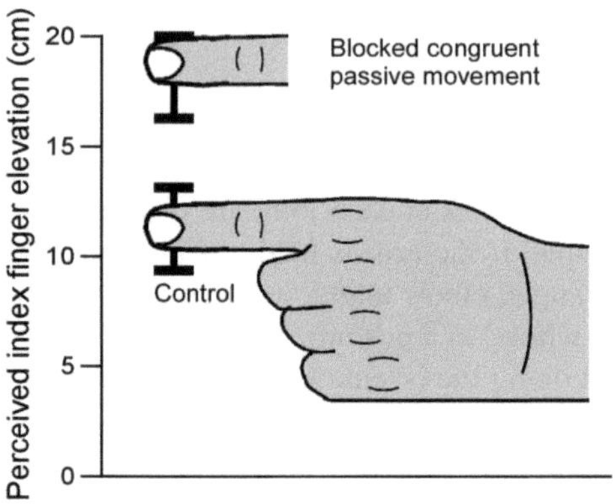

Fig. 5 Perceived height of the subject's index finger following simultaneous passive movement of the subject's finger and a plastic finger. The diagram shows the perceived elevation of the index finger (median and interquartile range) above the table on which the subject's hand was resting. Zero represents the level of the table. During the control condition (shown by the whole hand) the passive movement of the subject's finger and the plastic finger were unrelated to each other (incongruent). For the 'blocked congruent passive movement' condition (shown by the disembodied finger) the digital nerves of the subject's index finger were blocked with local anaesthetic. Also the plastic finger was coupled to the subject's index finger so that movement of one was reproduced by the other. That is movement was congruent between the subject's finger and the plastic finger Subjects perceived their finger to be at significantly different elevations in the two conditions ($p < 0.005$). Adapted from [97] with permission

Another body representation that can be manipulated is the map of which 'parts' belong to the body. We know that our body belongs to us without needing to 'interrogate' the body part in question. Presumably this sense of body ownership is generated from sensory information, and senses such as touch and proprioception must play an important role because they do not signal remote events but exclusively those events occurring in or on the body. With careful manipulation of sensory information subjects will perceive an artificial rubber hand as belonging to their body [92]. One method of doing this is to cover the subject's real hand so that they cannot see it and place the rubber hand where they can see it. The subject then watches the rubber hand while an experimenter synchronously strokes both the rubber hand and the subject's real hand in the same location. Once the illusion of ownership of the rubber hand is induced the subject will report that they feel the stroking on the rubber hand rather than their own hand. They will have physiological responses to threats made against the rubber hand [93] and the temperature of their real hand will drop [94] suggesting that the real hand is being neglected in favour of the adopted rubber hand. The rubber hand illusion does not require vision to be established [95]. It is also possible to establish a similar illusion in the finger with proprioceptive movement signals in the absence of tactile signals [97]. Congruent movements are applied to the proximal interphalangeal joint of the subject's index finger and a plastic finger. The plastic finger is visible to the subject and is set up to appear visually as though it is part of a false arm that could be the subject's arm. The subject's finger is not visible and both its digital nerves have been blocked with local anaesthetic, which excludes skin and joint receptors, but leaves intact the muscle receptors in the forearm muscles that control the finger. When an illusion of ownership over the plastic finger is established, and subjects are asked to judge the position of their finger in space, they report the position as close to the plastic finger, rather than close to their actual finger (see Fig. 5), suggesting that the illusion of ownership has biased their proprioception.

4 Conclusions

The direction of research into proprioception has changed. In the past, the focus was on determining which *afferent* signals contribute and the information they provide. Therefore, a lot is known about the peripheral receptors and their individual contributions to proprioception. However little is known about how multiple proprioceptive signals, afferent and efferent, interact with each other and combine to produce a coherent perception of the body and its actions. It is also unknown how stored information is generated and combined with the dynamic signals to produce continuously updated body representations through which we perceive our body. Recent research has begun to focus on these key points and exposing these mechanisms will be important for understanding proprioception. The human hand has been an important body part for studies on proprioception in the past and it continues to be important in current research.

Acknowldegments The authors and their labs are supported by funding from the National Health and Medical Research Council (of Australia).

References

1. K.A. Provins, The effect of peripheral nerve block on the appreciation and execution of finger movements. J. Physiol. **143**, 55–67 (1958)
2. S.C. Gandevia, D.I. McCloskey, Joint sense, muscle sense, and their combination as position sense, measured at the distal interphalangeal joint of the middle finger. J. Physiol. **260**, 387–407 (1976)
3. D.F. Collins, A. Prochazka, Movement illusions evoked by ensemble cutaneous input from the dorsum of the human hand. J. Physiol. **496**, 857–871 (1996)
4. C. Bell, *The Hand. Its Mechanism and Vital Endowments as Evincing Design* (William Pickering, London, 1833)
5. D. Ferrier, *The Functions of the Brain* (Smith Elder & Co., London, 1876)
6. W. James, *The Principles of Psychology*, vol. 1 (Dover Publications, Inc., New York, 1890)
7. C. Sherrington, The muscular sense, in *Text Book of Physiology*, vol. 2, ed. by E. Schafer (Pentland, Edinburgh, 1900), pp. 1002–1025
8. H. von Helmholtz, *Helmholtz's Treatise on Physiological Optics*, vol. 3 (Optical Society of America, Menasha, 1867)

9. W.B. Pillsbury, Does the sensation of movement originate in the muscle? Am. J. Psychol. **12**, 346–353 (1901)
10. J.H. Jackson, L. Paton, On some abnormalities of ocular movements. Lancet **173**, 900–905 (1909)
11. J.E. Winter, The sensation of movement. Psychol. Rev. **19**, 374–385 (1912)
12. S. Skoglund, Anatomical and physiological studies of knee joint innervation in the cat. Acta. Physiol. Scand. **36**(Supp 124), 1–101 (1956)
13. V. Mountcastle, T. Powell, Central nervous mechanisms subserving position sense and kinesthesis. B. Johns Hopkins Hosp. **105**, 173–200 (1959)
14. G. Brindley, P. Merton, The absence of position sense in the human eye. J. Physiol. **153**, 127–130 (1960)
15. A.K. McIntyre, U. Proske, J.A. Rawson, Cortical projection of afferent information from tendon organs in the cat. J. Physiol. **354**, 395–406 (1984)
16. G.M. Goodwin, D.I. McCloskey, P.B.C. Matthews, The contribution of muscle afferents to kinæsthesia shown by vibration induced illusion of movement and by the effects of paralysing joint afferents. Brain **95**, 705–748 (1972)
17. M.C. Brown, I. Engberg, P.B.C. Matthews, The relative sensitivity to vibration of muscle receptors of the cat. J. Physiol. **192**, 773–800 (1967)
18. D. Burke, K.E. Hagbarth, L. Lofstedt, B.G. Wallin, The responses of human muscle spindle endings to vibration of non-contracting muscles. J. Physiol. **261**, 673–693 (1976)
19. P. B. C. Matthews, A. Simmonds, Sensations of finger movement elicited by pulling upon flexor tendons in man, in *Proceedings of the Physiological Society*, London, 1974, pp. 27–28
20. D.I. McCloskey, S. Gandevia, E.K. Potter, J.G. Colebatch, Muscle sense and effort: motor commands and judgments about muscular contractions. Adv. Neurol. **39**, 151–167 (1983)
21. E. Moberg, The role of cutaneous afferents in position sense, kinaesthesia, and motor function of the hand. Brain **106**, 1–19 (1983)
22. S.C. Gandevia, Illusory movements produced by electrical stimulation of low-threshold muscle afferents from the hand. Brain **108**, 965–981 (1985)
23. P. Burgess, F. Clark, Characteristics of knee joint receptors in the cat. J. Physiol. **203**, 317–335 (1969)
24. D. Burke, S. Gandevia, G. Macefield, Responses to passive movement of receptors in joint, skin and muscle of the human hand. J. Physiol. **402**, 347–361 (1988)
25. P. Grigg, G. Finerman, L. Riley, Joint-position sense after total hip replacement. J. Bone Joint Surg. **55-A**, 1016–1025 (1973)
26. Y. Ishii, T. Tojo, K. Terajima, S. Terashima, J.E. Bechtold, Intracapsular components do not change hip proprioception. J. Bone Joint Surg. **81-B**, 345–348 (1999)
27. Y. Ishii, K. Terajima, S. Terashima, J.E. Bechtold, R.S. Laskin, Comparison of joint position sense after total knee arthroplasty. J. Arthroplasty **12**, 541–545 (1997)
28. F. Cuomo, M.G. Birdzell, J.D. Zuckerman, The effect of degenerative arthritis and prosthetic arthroplasty on shoulder proprioception. J. Shoulder Elb. Surg. **14**, 345–348 (2005)
29. E.D. Adrian, K. Umrath, The impulse discharge from the Pacinian corpuscle. J. Physiol. **68**, 139–154 (1929)
30. M. Hulliger, E. Nordh, A.E. Thelin, A.B. Vallbo, The responses of afferent fibres from the glabrous skin of the hand during voluntary finger movements in man. J. Physiol. **291**, 233–249 (1979)
31. M. Knibestöl, Stimulus-response functions of slowly adapting mechanoreceptors in the human glabrous skin area. J. Physiol. **245**, 63–80 (1975)
32. K. Browne, J. Lee, P.A. Ring, The sensation of passive movement at the metatarso-phalangeal joint of the great toe in man. J. Physiol. **126**, 448–458 (1954)
33. F.J. Clark, K.W. Horch, S.M. Bach, G.F. Larson, Contributions of cutaneous and joint receptors to static knee-position sense in man. J. Neurophysiol. **42**, 877–888 (1979)
34. S.C. Gandevia, L.A. Hall, D.I. McCloskey, E.K. Potter, Proprioceptive sensation at the terminal joint of the middle finger. J. Physiol. **335**, 507–517 (1983)

35. P.R. Burgess, J.Y. Wei, F.J. Clark, J. Simon, Signaling of kinesthetic information by peripheral sensory receptors. Annu. Rev. Neurosci. **5**, 171–187 (1982)
36. D.I. McCloskey, Kinaesthetic sensibility. Physiol. Rev. **58**, 763–820 (1978)
37. P.B.C. Matthews, Where does Sherrington's "muscular sense" originate? Muscles, joints, corollary discharges? Annu. Rev. Neurosci. **5**, 189–218 (1982)
38. S.C. Gandevia, Kinesthesia: roles for afferent signals and motor command, in *Handbook on Integration of Motor, Circulatory, Respiratory and Metabolic Control during Exercise*, ed. by L. Rowell, J. Shepherd (American Physiological Society, Bethesda, 1996), pp. 128–172
39. D. Barker, The innervation of mammalian skeletal muscle, in *Myotatic, Kinesthetic and Vestibular Mechanisms*, ed. by A.V.S. de Reuck, J. Knight (J. & A. Churchill Ltd., London, 1967)
40. Å.B. Vallbo, Discharge patterns in human muscle spindles afferents during isometric voluntary conrtactions. Acta. Physiol. Scand. **80**, 552–566 (1970)
41. J. Houk, E. Henneman, Responses of Golgi tendon organs to active contractions of the soleus muscle of the cat. J. Neurophysiol. **30**, 466–481 (1967)
42. D.I. McCloskey, Muscular and cutaneous mechanisms in the estimation of the weights of grasped objects. Neuropsychologia **12**, 513–520 (1974)
43. G. Holmes, The symptoms of acute cerebellar injuries due to gunshot injuries. Brain **40**, 461–535 (1917)
44. B.L. Luu, B.L. Day, J.D. Cole, R.C. Fitzpatrick, The fusimotor and reafferent origin of the sense of force and weight. J. Physiol. **589**, 3135–3147 (2011)
45. R. Granit, The functional role of the muscle spindle's primary end organs. P. Roy. Soc. Med. **61**, 69–78 (1968)
46. K.-E. Hagbarth, Å.B. Vallbo, Discharge characteristics of human muscle afferents during muscle stretch and contraction. Exp. Neurol. **22**, 674–694 (1968)
47. Å.B. Vallbo, Human muscle spindle discharge during isometric voluntary contractions. Amplitude relations between spindle frequency and torque. Acta. Physiol. Scand. **90**, 319–336 (1974)
48. D. Burke, K.E. Hagbarth, L. Lofstedt, B.G. Wallin, The responses of human muscle spindle endings to vibration during isometric contraction. J. Physiol. **261**, 695–711 (1976)
49. A. Prochazka, M. Gorassini, Ensemble firing of muscle afferents recorded during normal locomotion in cats. J. Physiol. **507**, 293–304 (1998)
50. S.C. Gandevia, L. Wilson, P.J. Cordo, D. Burke, Fusimotor reflexes in relaxed forearm muscles produced by cutaneous afferents from the human hand. J. Physiol. **479**, 499–508 (1994)
51. E. Ribot-Ciscar, V. Hospod, J.-P. Roll, J.-M. Aimonetti, Fusimotor drive may adjust muscle spindle feedback to task requirements in humans. J. Neurophysiol. **101**, 633–640 (2009)
52. J.L. Smith, M. Crawford, U. Proske, J.L. Taylor, S.C. Gandevia, Signals of motor command bias joint position sense in the presence of feedback from proprioceptors. J. Appl. Physiol. **106**, 950–958 (2009)
53. L.D. Walsh, J.L. Smith, S.C. Gandevia, J.L. Taylor, The combined effect of muscle contraction history and motor commands on human position sense. Exp. Brain Res. **195**, 603–610 (2009)
54. I. Di Giulio, C.N. Maganaris, V. Baltzopoulos, I.D. Loram, The proprioceptive and agonist roles of gastrocnemius, soleus and tibialis anterior muscles in maintaining human upright posture. J. Physiol. **587**, 2399–2416 (2009)
55. U. Proske, D.L. Morgan, J.E. Gregory, Thixotropy in skeletal muscle and in muscle spindles: a review. Prog. Neurobiol. **41**, 705–721 (1993)
56. D.L. Morgan, A. Prochazka, U. Proske, The after-effects of stretch and fusimotor stimulation on the responses of primary endings of cat muscle spindles. J. Physiol. **356**, 465–477 (1984)
57. J.E. Gregory, D.L. Morgan, U. Proske, Aftereffects in the responses of cat muscle spindles and errors of limb position sense in man. J. Neurophysiol. **59**, 1220–1230 (1988)
58. L.R. Wilson, S.C. Gandevia, D. Burke, Increased resting discharge of human spindle afferents following voluntary contractions. J. Physiol. **488**, 833–840 (1995)

59. A. Wise, J. Gregory, U. Proske, The effects of muscle conditioning on movement detection thresholds at the human forearm. Brain Res. **735**, 125–130 (1996)
60. P.B.C. Matthews, *Mammalian Muscle Receptors and their Central Actions* (Edward Arnold Ltd., London, 1972)
61. M. Dimitriou, B.B. Edin, Discharges in human muscle spindle afferents during a key-pressing task. J. Physiol. **586**, 5455–5470 (2008)
62. M. Dimitriou, B.B. Edin, Discharges in human muscle receptor afferents during block grasping. J. Neurosci. **28**, 12632–12642 (2008)
63. M. Dimitriou, B.B. Edin, Human muscle spindles act as forward sensory models. Curr. Biol. **20**, 1763–1767 (2010)
64. K.M. Refshauge, D.F. Collins, S.C. Gandevia, The detection of human finger movement is not facilitated by input from receptors in adjacent digits. J. Physiol. **551**, 371–377 (2003)
65. B.B. Edin, N. Johansson, Skin strain patterns provide kinesthetic information to the human central nervous system. J. Physiol. **487**, 243–251 (1995)
66. D.F. Collins, K.M. Refshauge, G. Todd, S.C. Gandevia, Cutaneous receptors contribute to kinesthesia at the index finger, elbow, and knee. J. Neurophysiol. **94**, 1699–1706 (2005)
67. R.S. Johansson, A.B. Vallbo, Tactile sensibility in the human hand: relative and absolute densities of four types of mechanoreceptive units in glabrous skin. J. Physiol. **286**, 283–300 (1979)
68. D.L. Sturnieks, J.R. Wright, R.C. Fitzpatrick, Detection of simultaneous movement at two human arm joints. J. Physiol. **585**, 833–842 (2007)
69. N.S. Weerakkody, D.A. Mahns, J.L. Taylor, S.C. Gandevia, Impairment of human proprioception by high-frequency cutaneous vibration. J. Physiol. **581**, 971–980 (2007)
70. N.S. Weerakkody, J. Taylor, S. Gandevia, The effect of high-frequency cutaneous vibration on different inputs subserving detection of joint movement. Exp. Brain Res. **197**, 347–355 (2009)
71. U. Proske et al., Force-matching errors after eccentric exercise attributed to muscle soreness. Clin. Exp. Pharmacol. P **30**, 576–579 (2003)
72. N. Weerakkody, P. Percival, B. Canny, D. Morgan, U. Proske, Force matching at the elbow joint is disturbed by muscle soreness. Somatosens. Mot. Res. **20**, 27–32 (2003)
73. U. Proske et al., Force matching errors following eccentric exercise. Hum. Movement Sci. **23**, 365–378 (2004)
74. N.S. Weerakkody, J.S. Blouin, J.L. Taylor, S.C. Gandevia, Local subcutaneous and muscle pain impairs detection of passive movements at the human thumb. J. Physiol. **586**, 3183–3193 (2008)
75. H.C. Bastian, The "muscular sense"; its nature and cortical localisation. Brain **10**, 5–137 (1887)
76. R.W. Sperry, Neural basis of the spontaneous optokinetic response produced by visual inversion. J. Comp. Physiol. Psych. **43**, 482–489 (1950)
77. E. von Holst, Relations between the central nervous system and the peripheral organs. B. J. Anim. Behav. **2**, 89–94 (1954)
78. L.D. Walsh, C.W. Hesse, D.L. Morgan, U. Proske, Human forearm position sense after fatigue of elbow flexor muscles. J. Physiol. **558**, 705–715 (2004)
79. T.J. Allen, U. Proske, Effect of muscle fatigue on the sense of limb position and movement. Exp. Brain Res. **170**, 30–38 (2006)
80. L.D. Walsh, T.J. Allen, S.C. Gandevia, U. Proske, The effect of eccentric exercise on position sense at the human forearm in different postures. J. Appl. Physiol. **100**, 1109–1116 (2006)
81. S.C. Gandevia, J.L. Smith, M. Crawford, U. Proske, J.L. Taylor, Motor commands contribute to human position sense. J. Physiol. **571**, 703–710 (2006)
82. L.D. Walsh, S.C. Gandevia, J.L. Taylor, Illusory movements of a phantom hand grade with the duration and magnitude of motor commands. J. Physiol. **588**, 1269–1280 (2010)
83. G.E. Ansems, T.J. Allen, U. Proske, Position sense at the human forearm in the horizontal plane during loading and vibration of elbow muscles. J. Physiol. **576**, 445–455 (2006)

84. J.F. Soechting, Does position sense at the elbow reflect a sense of elbow joint angle or one of limb orientation? Brain Res. **248**, 392–395 (1982)
85. C.J. Worringham, G.E. Stelmach, Z.E. Martin, Limb segment inclination sense in proprioception. Exp. Brain Res. **66**, 653–658 (1987)
86. J.M. Tanner, *Growth at Adolescence*, 2nd edn. (Blackwell Scientific Publications, Oxford, 1962)
87. J. Visser, R.H. Geuze, A.F. Kalverboer, The relationship between physical growth, the level of activity and the development of motor skills in adolescence: differences between children with DCD and controls. Hum. Movement Sci. **17**, 573–608 (1998)
88. V.S. Ramachandran, D. Rogers-Ramachandran, Phantom limbs and neural plasticity. Arch. Neurol. **57**, 317–320 (2000)
89. S.C. Gandevia, C.M.L. Phegan, Perceptual distortions of the human body image produced by local anaesthesia, pain and cutaneous stimulation. J. Physiol. **514**, 609–616 (1999)
90. X. Paqueron, M.E. Gentili, J.C. Willer, P. Coriat, B. Riou, Time sequence of sensory changes after upper extremity block: swelling sensation is an early and accurate predictor of success. Anesthesiology **101**, 162–168 (2004)
91. N. Inui, L.D. Walsh, J.L. Taylor, S.C. Gandevia, Dynamic changes in the perceived posture of the hand during ischaemic anaesthesia of the arm. J Physiol **589**, 5775–5784 (2011)
92. M. Botvinick, J. Cohen, Rubber hands 'feel' touch that eyes see. Nature **391**, 756 (1998)
93. H.H. Ehrsson, K. Wiech, N. Weiskopf, R.J. Dolan, R.E. Passingham, Threatening a rubber hand that you feel is yours elicits a cortical anxiety response. Proc. Nat. Acad. Sci. USA **104**, 9828–9833 (2007)
94. G.L. Moseley, N. Olaf, A. Venema, S. Don, M. Wijers, A. Gallace, C. Spence, Psychologically induced cooling of a specific body part caused by the illusory ownership of an artificial counterpart. Proc. Nat. Acad. Sci. USA **105**, 13169–13173 (2008)
95. H.H. Ehrsson, N.P. Holmes, R.E. Passingham, Touching a rubber hand: feeling of body ownership is associated with activity in multisensory brain areas. J. Neurosci. **25**, 10564–10573 (2005)
96. L.D. Walsh, S.C. Gandevia, J.T. Taylor, Phantom hands: a window into how we perceive limb position and movement. Physiol. News **81**, 34–36 (2010)
97. L.D. Walsh, G.L. Moseley, J.L. Taylor, S.C. Gandevia, Proprioceptive signals contribute to the sense of body ownership. J. Physiol. **589**, 3009–3021 (2011)

Chapter 7
Tactile Feedback from the Hand

Sliman Bensmaia and Stephen I. Helms Tillery

Abstract The last decade has seen remarkable advances in upper-limb neuroprosthetics. Several groups have developed algorithms that enable control of devices by decoding the neural signals recorded over an array of electrodes. These advances have encouraged researchers to move onto control of neuroprosthetic hands, which faces two complications: (1) hands are geometrically far more complex than arms, (2) hands are also sensitive and sophisticated sensory systems. In this chapter, we review the role of tactile and proprioceptive sensation in hand function, with a focus on the integration of multiple inputs to extract information about our haptic interactions with objects. We argue here that creating a seamless somatosensory prosthetic system will require both a detailed understanding of how individual deformations of the skin result in modulation of neurons in primary somatosensory cortex, but also how those signals are combined to create a somatosensory image.

Keywords Neuroprosthetics · Somatosensory · Tactile · Neurofeedback

1 Introduction

In the 1890s, Frederick Mott and Charles Sherrington [1] reported on a series of experiments on non-human primates in which they severed the nerves that provided for sensation from the upper limb. They reported profound changes in how

---

S. Bensmaia
Department of Organismal Biology and Anatomy, University of Chicago, Chicago, USA

S. I. Helms Tillery (✉)
School of Biological and Health Systems Engineering, Arizona State University, Tempe, USA
e-mail: Steve.HelmsTillery@asu.edu

R. Balasubramanian and V. J. Santos (eds.), *The Human Hand as an Inspiration for Robot Hand Development*, Springer Tracts in Advanced Robotics 95,
DOI: 10.1007/978-3-319-03017-3_7,

the animals used their arms. Indeed, the animals did not use the impaired limb to walk or to cling to the cage bars. They did not use the limbs to reach and grasp offered fruits, and they even failed to use the limb normally during struggles when "held awkwardly". These observations were offered as evidence of the role of sensation in movement, and are still cited to suggest the role of proprioception in voluntary movement. Often unnoticed, however, is the observation made later in the papers, that partial denervation of the arm can have the same effect, so long as all sensation is removed from the hand, whereas a partial denervation which spares sensation to the hand has only a minor impact upon movement.

These observations, repeated several times since, show how crucial sensation is for use of the hand. We tend to think of the hand as a manipulative tool, but like the appendages on a star-nosed mole, our hand is also used continually to tell us about the objects with which we are interacting: the texture, weight, material, and even stability of grasped objects all can be discerned from the activity of sensors in the hand. This is to say that the hand is not just a motor apparatus, but it is also a sensory one. More appropriately, since sensation arising from peripheral receptive systems is used continually in routine manipulative activities, it is reasonable to describe the hand as a model system for sensorimotor integration. Likewise, if we are to build prosthetic hands that can mimic the function of the human hand and be seamlessly integrated into a user's body image, we will need a deep understanding of how information from those sensors is processed and used in routine hand movements.

In this review, we discuss some of the elements of tactile and proprioceptive sensation in the context of manipulative movements. Our main goal is to provide an overview of manual somatosensation with a view towards the development of the next generation of neuroprostheses, in which neural stimulation will be used to provide somatosensation. In that context, we discuss the projections of epicritic sensors in the nervous systems, i.e., sensors which provide signals related to self-movement and to contact with objects in the external environment. We also discuss the organization of information transmission in specific channels within this system, and the eventual coding of that information in the firing of neurons in primary somatosensory cortex (S1). Finally, we present recent data showing how multiple disparate pieces of information might be encoded in the firing of individual neurons in cortex.

We start the discussion at the periphery, with a description of the most important sensors that trigger tactile percepts.

2 Sensory Innervation of the Hand

Three types of mechanoreceptors innervate the glabrous skin of the primate hand, each of which conveys information about different aspects of skin deformation [2] (Fig. 1). Slowly adapting type 1 (SA1) afferents have small receptive fields and produce a tonic response to steady skin indentation. Electrical stimulation of

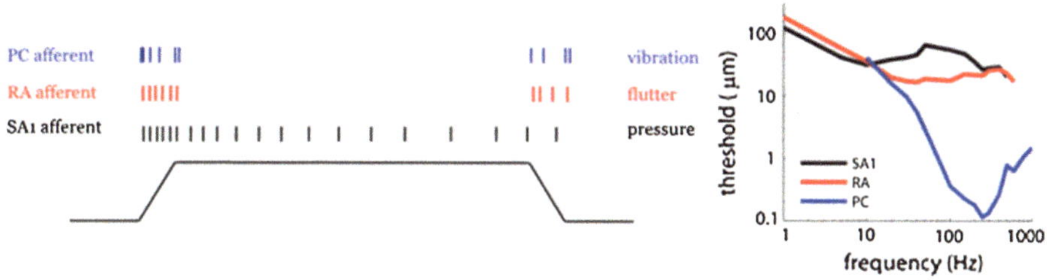

Fig. 1 Responses of the three types of afferents to a step indentation (*left*); sensations evoked when individual afferents of each type are stimulated (*middle*); threshold frequency function of the three afferent types that innervate the glabrous skin of primates (*right*) (adapted from [5])

individual SA1 afferents elicits sensations of pressure [3]. Like SA1 afferents, rapidly adapting (RA, often referred to as RA1) afferents have small receptive fields, but these fibers produce a phasic response at the onset and offset of a skin indentation (thus their designation as rapidly adapting). Electrical stimulation of individual RA afferents elicits sensations of tactile flutter. Pacinian (PC, often referred to as RA2) afferents have large receptive fields, are rapidly adapting, and are exquisitely sensitive to high-frequency vibrations. Electrical stimulation of PC afferents elicits sensations of vibration. The three types of mechanoreceptive afferents can also be distinguished by their sensitivities to sinusoidal vibrations delivered to the skin [4, 5]: SA1 afferents peak in sensitivity around 5–10 Hz, RA afferents around 40–50 Hz, and, PC afferents around 200–300 Hz. Remarkably, PC afferents can respond to vibrations at amplitudes on the order of 100 nm and respond robustly when puffs of air are delivered to the skin. An additional afferent type, slowly adapting type 2 (SA2) afferents occur only infrequently in the glabrous skin, but are found largely in the hairy skin on the dorsum of the hand [6]. These afferents are particularly responsive to skin shear, which can induce a sense of motion about a joint [7] and are thought to contribute to proprioception. We should note that the description of somatosensory afferents as responding selectively to a single category of stimulus is an oversimplification. For example, SA2 afferents respond both to the shear forces that occur during joint movements and to skin indentations [7, 8].

As mentioned above, the different types of afferents are thought to play different albeit overlapping sensory roles. SA1 afferents have been shown to play an important role in mediating our ability to perceive the shape and texture of objects [2]. Furthermore, the strength of the response of these afferents increases as the pressure exerted at the skin's surface increases over a wide range of pressures, so they likely play a crucial role in conveying information about how much pressure is exerted on a grasped object [9]. RA fibers respond strongly to stimuli brushed against the skin and have been shown to play a critical role in detecting slip between skin and a grasped object to trigger an adjustment in grip force [10]. PC afferents are thought to mediate the tactile perception of distal events (through vibrations transmitted through a tool, e.g. [11]) and the perception of surface microtexture [12].

A remarkable property of mechanoreceptive afferents is the repeatability of their responses. Indeed, repeated presentations of a stimulus evoke virtually identical responses in individual afferents, even when stimulation consists of complex spatio-temporal patterns of deformation. In fact, the stimulus–response transformation can be captured using relatively simple models that take into consideration both the filtering properties of the skin and the biophysical properties of the receptors and associated neurites themselves (see below).

First, because mechanoreceptors are embedded in the skin, any stimulus applied to the skin's surface must propagate through the tissue to activate them. Accordingly, the receptors do not experience the stimulus itself, but rather a version of the stimulus that has been filtered through the skin. This filtering leads to the enhancement of certain stimulus features (such as corners) and the obscuration of others (e.g. the internal features of a spatially complex stimulus). The spatial filtering of the skin has been effectively captured using continuum mechanical models [13] and finite element models [14].

Second, the three types of mechanoreceptors comprise different capsules and different mechanosensory elements and are coupled to the tissue in different ways. These differences confer different stimulus sensitivities to the different afferents, for example to stimulus frequency (as summarized above). The best understood example of how the surrounding structures affect the responses of the neurite is the Pacinian corpuscle: The onion-like capsule that surrounds the neurite confers to the receptor-afferent complex a sensitivity to the second derivative of indentation depth and, as a result, its rapidly adapting response to a steady indentation [15]. The various response properties of mechanoreceptive afferents can be captured using a relatively simple modeling framework, wherein the stimulus (expressed as indentation depth or pressure) and its first and second derivatives constitute the input to a linear filter, the summed output of which constitutes the input current to an integrate-and-fire neuron. With this relatively simple model, afferent responses to an arbitrary time-varying stimulus can be predicted with millisecond accuracy [16]. The combination of skin mechanics and transduction models can thus reproduce the responses of mechanoreceptive afferents to arbitrary spatiotemporal stimuli.

One approach to conveying sensory feedback in neuroprostheses for amputees, then, would consist in converting, using models of mechanotransduction, the output of pressure sensors on the limb into the patterns of neuronal activity that would be evoked in the native limb, and effecting these activity patterns through electrical stimulation of the residual nerve in an amputee [17, 18]. Indeed, as mentioned above, electrical stimulation of individual afferents produces percepts with a specific quality (pressure, flutter, vibration) localized to a specific skin location. Accordingly, a stimulating array could be implanted in the residual nerve of an amputee, and the type and projected field of the afferent at each electrode on the array could be mapped by delivering low-intensity pulse trains and having the patient report both the location on the phantom limb and the quality of the elicited percept. The appropriate neuronal model (SA1, RA or PC) would then convert the output from the corresponding sensor (at the location of the projected field on the prosthetic limb) to patterns of electrical stimulation to that afferent. For instance, if stimulation through an electrode elicits a pressure percept on the index fingertip, then the sensor on the prosthetic index fingertip could be linked to an SA1 model to produce patterns of electrical stimulation through that electrode. Given that integrate-and-fire models can easily be implemented in hardware and that chronically implantable electrode arrays for peripheral nerves are currently in development, this approach to conveying sensory feedback, strongly analogous to the cochlear implant, seems within reach.

3 Central Projections and Convergence/Divergence

Signals generated in primary afferents in the periphery access the spinal cord via the dorsal roots and enter the dorsal columns where they project onto neurons in the dorsal column nuclei (DCN) in the medulla. Here the inputs are structured somatotopically, such that the body surface is represented in an organized map within the nuclei. Individual second-order neurons (neurons removed by one synapse from the periphery) in the dorsal column nuclei receive inputs from around 300 peripheral afferents, generally from a relatively localized area of the skin [19]. The responses of these second-order neurons are typically dominated by a single modality to an extent that allows for a classification similar to that of peripheral afferents. However, it is unlikely that the synapses on any given second-order neuron all originate from the same type of peripheral receptor. Thus, not only is the input from a given receptor type often ambiguous as to what information it may be conveying (as described above for an SA1 and SA2 afferents), but second-order neurons receive input from multiple types of receptors, so the signals conveyed by individual DCN neurons are already highly multiplexed.

Axons from neurons in the dorsal column nuclei then project via the medial lemniscus into the ventral posterior lateral nuclei of the thalamus, where the body representation remains somatotopically organized and roughly segregated by modality. Cutaneous neurons with receptive fields on the fingers respond not only to light touch but also to certain movements of the fingers. Likewise, proprioceptive neurons often fire briskly to a light touch if it causes a change in the posture of the finger.

While submodality convergence is not extensive in the DCN and thalamus, it is extensive in S1, specifically in areas 3b, 1, and 2. Indeed, electrophysiological recordings suggest a convergence of signals from multiple afferent types onto individual cortical neurons [20]. Thus, while it is feasible to classify neurons, at a first level of approximation, as having responses dominated by input from one peripheral sensor type, this classification does not comprehensively capture the information conveyed by these neurons during the performance of naturalistic tasks. If one were to build a somatosensory neuroprosthesis that involves direct stimulation of structures within the central nervous system, it may be important to achieve a description of neural coding that captures subtler elements of signals generated during reach-to-grasp and object manipulation. For example, neurons might encode higher-level object features such as roughness or curvature instead of more basic features such as stimulus amplitude or frequency.

4 Neural Coding in Somatosensory Cortex

S1 is organized hierarchically, with receptive fields becoming larger and more complex as one progresses from area 3b to area 1 to area 2. Neurons in area 3b typically have small, well-defined receptive fields (which tend nonetheless to be larger than those of afferents). The responses of neurons in this area can partly be accounted for using simple linear spatio-temporal filters [21]. Neurons in areas 1 and 2 have progressively larger receptive fields, often including multiple finger-pads and even multiple digits, and their responses are more non-linear with respect to the afferent input than are those of their counterparts in area 3b. Neurons in areas 3b and 1 typically receive convergent input from multiple tactile submodalities (SA1, RA, and PC) [20] and neurons in area 2 exhibit cutaneous as well as proprioceptive responses; these neurons are sensitive not just to stimuli applied to the skin, but also to movements of the joints. Information about where an object contacts the skin and the pressure it exerts on the skin is represented in the responses of S1 neurons. Furthermore, different neuronal populations are sensitive to different stimulus features: Some populations respond to edges indented to the skin at specific orientations, others respond to stimuli brushed against the skin in specific directions, still others respond best to vibratory stimulation or to textured surfaces.

4.1 Location and Pressure

As mentioned above, neurons in area 3b typically have small, spatially restricted receptive fields, with nearby neurons having adjacent and partially overlapping receptive fields (that is, they are somatotopically organized). Accordingly, stimulation of a spatially restricted patch of skin (using a punctuate probe, e.g.) evokes activity in a spatially restricted population of neurons in area 3b. These two results combined suggest that information about where a stimulus is contacting the skin is conveyed as a place code, that is, in the location of the activated neuronal population within the body map in S1. In fact, the ability to discriminate two points as being distinct on the skin can be predicted from the overlap in the activation these produce in area 3b [22], which further bolsters this hypothesis. The somatotopy is more blurred in area 2 than it is in areas 3b and 1, which may suggest poorer place coding of contact location in area 2.

As a punctuate probe is indented deeper into the skin, the size of the activated population remains relatively constant, but the response of neurons in the hotspot of activity increases [23]. Information about the pressure a stimulus exerts on the skin is thus conveyed in the strength of the evoked neuronal activation. In area 1, the relationship between indentation pressure and neuronal response is less systematically monotonic than it is in area 3b [24]. Thus, area 3b seems to comprise an explicit and relatively acute representation of both location and pressure.

4.2 Stimulus Orientation

A large subpopulation of neurons in S1 (around 50 %) responds preferentially to an edge indented into the skin, and that response is selective for the orientation of that edge [25] (Fig. 2, left panel). Orientation selective neurons are found in areas 3b, 1 and 2, and the incidence of orientation selectivity is relatively constant across these three areas. This preponderance of orientation selectivity suggests that object shape is initially parsed as a series of oriented contours, as is the case in the early stages of visual form processing (in primary visual cortex or V1). Neurons in area 2 exhibit more complex receptive fields, such as a selectivity for curvature [26], which implies a hierarchical processing scheme wherein input from orientation detectors in areas 3b and 1 with progressively shifting preferred orientations is combined to determine the responses of these neurons. Neurons in secondary somatosensory

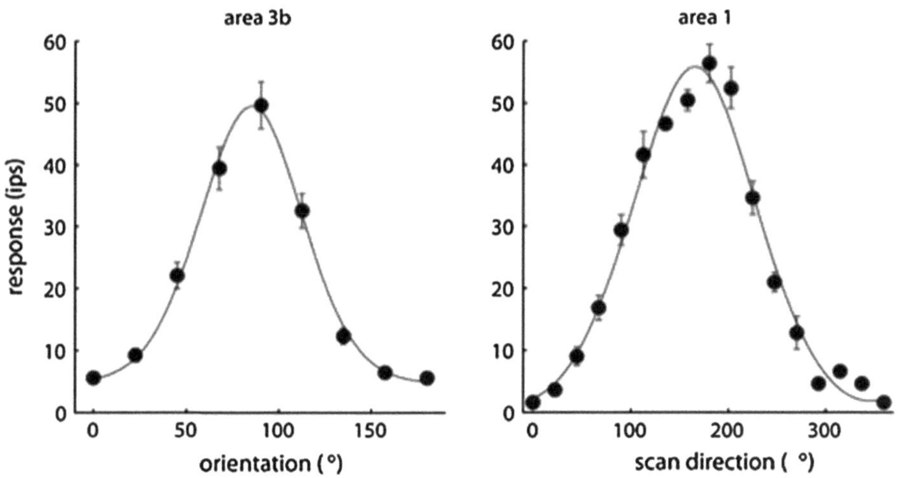

Fig. 2 *Left* Responses of a neurons in area 3b to bars indented at 8 orientations; this neuron responds most strongly when a bar oriented along the long axis of the finger is indented into the skin (data reproduced from [25]). *Right* Responses of a neuron in area 1 to *dot* patterns scanned across the skin. This neuron responds most strongly when the pattern is scanned at 180° (from *right* to *left*) (data reproduced from [28])

cortex exhibit even more elaborate feature selectivity, for both orientation and curvature, analogous to those observed in extrastriate visual area 4 [27].

4.3 Stimulus Motion

A subpopulation of neurons in S1 is sensitive to motion across the skin [28]. These neurons respond preferentially when a stimulus is scanned across the skin in a particular direction and their response decreases as the direction of stimulus motion diverges from their preferred direction (Fig. 2, right panel), analogously to motion-selective neurons in primary visual cortex. A subset of these neurons is also orientation selective, and their preferred orientation is orthogonal to their preferred direction of motion, as is the case with complex cells in V1. Another population of neurons, found primarily in area 1, encodes the direction of motion regardless of the shape of the object. These neurons have been shown to combine input from simple motion detectors using a vector-average mechanism, analogously to neurons in the middle temporal area (area MT), an area specialized for visual motion processing [29, 30].

4.4 Vibration and Texture

Vibratory stimuli applied to the skin excite populations of S1 neurons that receive RA and PC input. When a sinusoidal stimulus is delivered to the skin, a subpopulation of neurons in area 3b exhibits entrained responses, with the modal interspike interval matching the period of the stimulus for frequencies up to 800 Hz [31, 24]. Little is known about how texture is encoded in the brain, particularly the textural information related to surface microgeometry. Ablation of area 1 has been shown to produce selective impairments in texture discrimination [32], so this area likely plays an important role in texture processing. Furthermore, the perceived roughness of embossed dot patterns and gratings has been shown to be determined by the spatial variation in the SA1 response these evoke [33]. A subpopulation of neurons in cortex, whose receptive fields comprise an excitatory region flanked by inhibitory patches, seems to implement the algorithm necessary for this roughness computation [34].

5 Multiplexing of Somatic Signals During Manipulation

Ultimately, we use our hands to interact with the environment, and there are many overlapping goals for interpreting the signals which arise from skin deformation. Characterizing the responses of neurons during any natural task, however, is a complex problem. During manipulative tasks, the skin and motor apparatus (muscles, tendons, joints) are subject to forces and deformations that are unrelated to the actual interface between the hand and the manipulated object. During the simple case of a precision grip between the thumb and index finger, neurons with receptive fields on the glabrous skin of the thumb and index finger have characteristic responses. For example, SA1 type cells are sensitive to grip force, whereas RA cells are very responsive to object movement or slip. SA2 cells appear to be most responsive to forces oriented tangentially to the skin, as these induce the largest shear forces [10].

When extended to the domain of more complex tasks such as reach-to-grasp, neurons in S1 fire during all the phases of the task. Furthermore, over 60 % of S1 neurons produce their peak firing rates during the grasp and contact phases, whereas fewer than 10 % of these neurons show peak rates prior to contact with the object [35]. Similarly, fewer than 10 % of neurons in the hand area of primary motor cortex exhibit their main firing peak prior to object contact. Clearly, in this particular task, sensorimotor cortex is far more engaged during the actual manipulation of the object. However, this paradigm does not provide a clear distinction between activity that is driven by exteroceptive sensory input and activity that is related to the movement itself.

In an effort to clarify the relationship between movement- and contact-related neural firing, animals were trained to perform a reach-to-grasp task in a robot-enhanced virtual environment [36]. The experimental design provided a separation between activity related to object contact and activity related to hand movement by having animals perform similar movements, with and without contact with a physical object. S1 neurons with cutaneous receptive fields on the volar or dorsal surface of the palm or digits were identified using a combination of mechanical stimuli and passive joint manipulation. A unit response was considered task-related if the mean firing rate during any single task phase was significantly different from the mean rate during any other task phase. A neuron was attributed a contact type (C-type) response if it responded when the hand actually touched an object in the environment, and a movement type (M-type) response if it responded during hand movement in the absence of contact. Simple responses exhibited only one response trait, either contact or movement, while Mixed responses exhibited both types of responses.

Figure 3 shows example activity from a Simple contact-driven neuron (Fig. 3a) and a Simple movement-driven neuron (Fig. 3b). Fewer than half of the neurons recorded in S1 had Simple response types. Despite carefully and specifically selecting neurons with cutaneous receptive fields on the hands, less than a third of S1 neurons were modulated by the reach-to-grasp task in the manner shown in Fig. 3a, in which only information about object contact was encoded (C).

In fact, a slight majority of neural responses were *Mixed*, encoding both contact and movement. Examples of these unit responses are shown in Fig. 4a, b. These cells typically responded most strongly to either contact or movement, but were also modulated during all phases of the task, as indicated in the histograms in

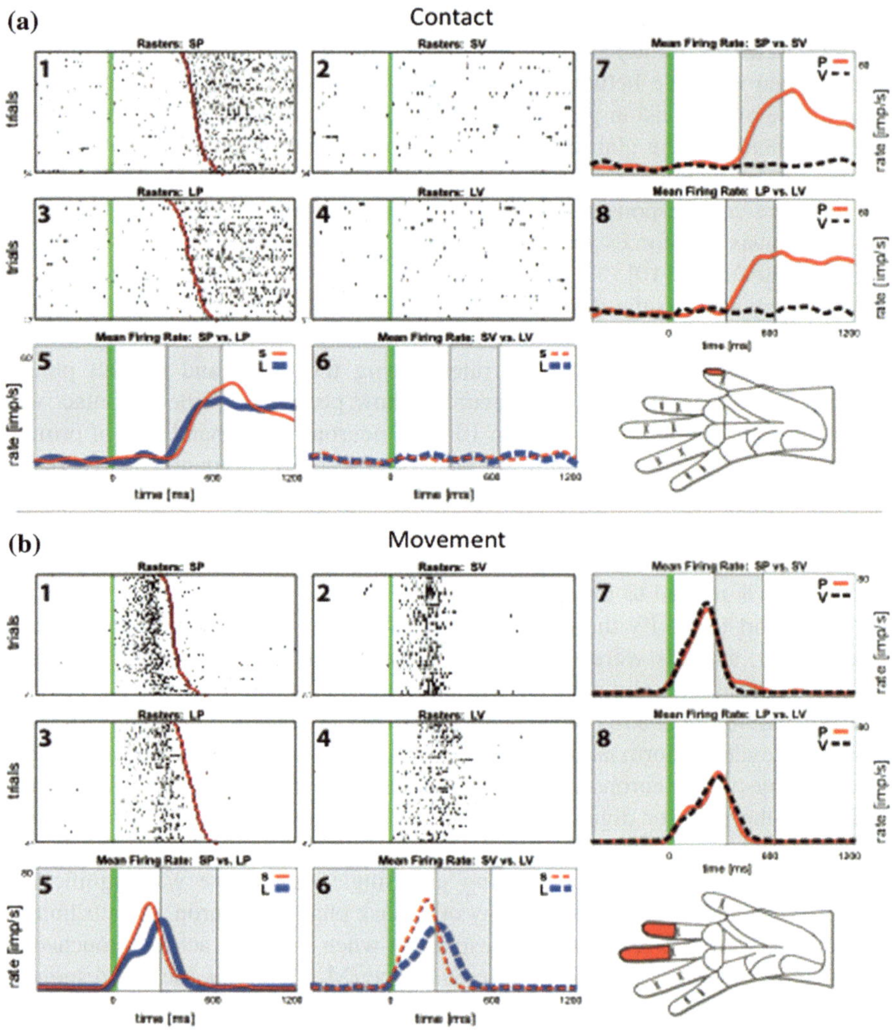

Fig. 3 Responses of two S1 neurons during reach to grasp of a physical object (*panels* 1, 3, and 5) or grasp of a virtual object (*panels* 2, 4, and 6). In both tasks, the animal grasped a small (*panels* 1, 2, and 7) and large (*panels* 3, 4, and 8) object. *Green* mark indicates movement start, *red* mark indicates object contact. *Panels* 5–8 show direct comparisons of neural firing in adjoining rasters. **a** A neuron which responded only during object contact. **b** A neuron which responded only during movement (Figure adapted from [36])

column iii. In total, the responses of 63 % of the sampled neural population were modulated during hand movements.

While a small majority of cortical units were contact-driven (C), nearly half of that population simultaneously exhibited significant pre-contact activity during the Reach phase of the task. Primarily movement-driven (M) units constituted a 30 %

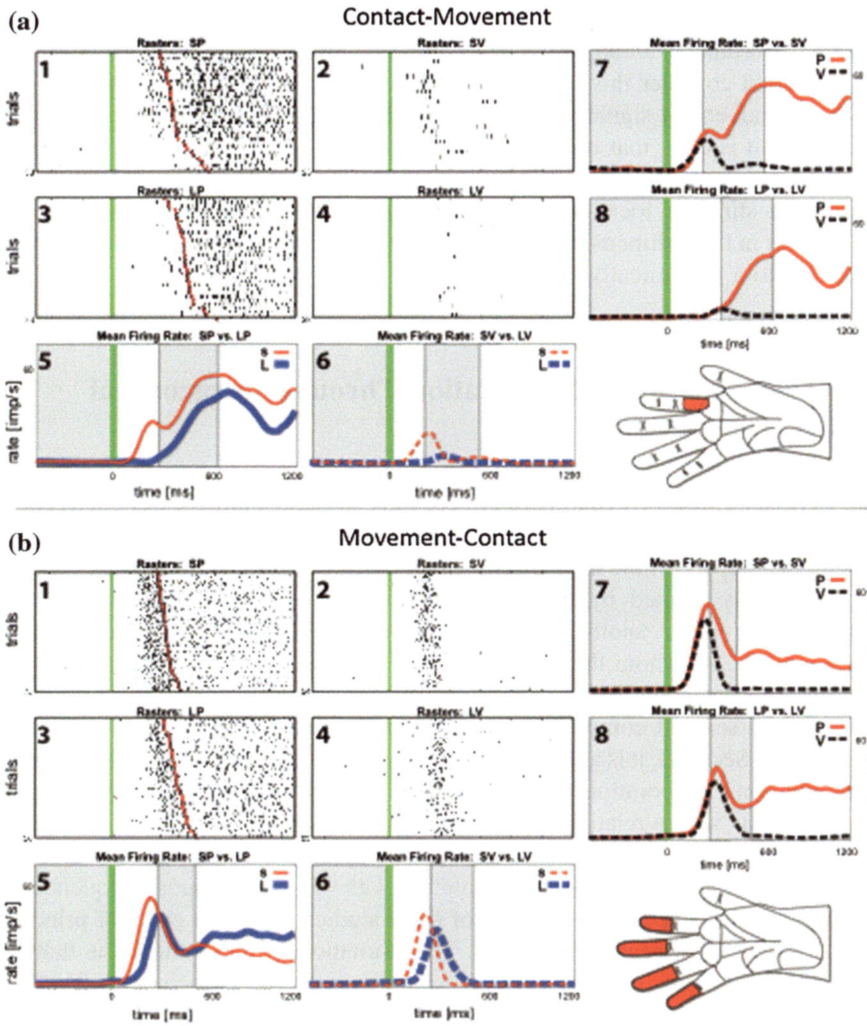

Fig. 4 Responses of two S1 neurons with responses which encoded both movement (firing between movement start and object contact) and contact (firing after object contact, only in the physical trials). **a** A neuron which responded only during object contact. **b** A neuron which responded only during movement (Figure adapted from [36])

of the overall sampled population, yet 70 % of these neurons also exhibited significant responses during the contact phase. Thus the majority of task-related neurons recorded in S1 encode multiple sensory phenomena.

This variety of individual unit responses suggests a broad distribution of response types in S1 with varying combinations of contact-driven and movement-driven response traits. Importantly, these differences in response properties do not appear to be associated with differences in receptive field structure, depth, or chamber coordinates amongst the cell types.

One might consider this multiplexing of functional signals as a matter of self-generated movement signals contaminating the tactile code. However, for haptic perception, it is clear that both types of signals are used and integrated. A typical somatotopic map, as described by a homunculus, is a two-dimensional representation of a stimulus location on the skin. Assigning somatic stimuli to spatial coordinates in three dimensional space requires transforming that somatotopic map according to a dynamically changing underlying proprioceptive map.

6 Conveying Sensory Information Through Intracortical Microstimulation

One advantage of conveying sensory feedback through intracortical microstimulation (ICMS) is that it exploits the native processing effected by the central nervous system. Indeed, one of the main disadvantages of stimulating at the periphery is that it is ultimately limited by the number of channels that can be independently stimulated. Of course, another limitation is that this method cannot be used in spinal injury patients for whom the communication between peripheral and central nervous systems has been severed. Because the representation of the environment in cortex is sparser and comprises explicit representations of behaviorally relevant features (see Sect. 2.), it is at least theoretically possible to make use of this inherent representational elaboration in attempting to convey information about complex sensory events with a relatively small number of channels.

Animals have been shown to be able to distinguish ICMS delivered to different neural populations or to the same population with different temporal sequences of electrical pulses [37, 38]. The results of these studies provide a proof of principle that ICMS elicits repeatable percepts. The limitation in these studies is that the animals were taught to make arbitrary stimulus–response associations. The percepts themselves may have consisted of poorly localized tingling sensations, for example. In a dexterous upper-limb prosthesis, however, creating arbitrary associations between a stimulus impinging on the prosthesis and a pattern of neuronal activation may be impractical given the limb's large number of degrees of freedom, and thus the enormous number of states it can adopt. Indeed, this approach is functionally analogous to sensory substitution and may involve a burdensome memory load for the patient.

Instead, it may be preferable to exploit native neuronal representations in conveying sensory feedback through ICMS. The most successful attempt at systematically conveying somatosensory information through ICMS was achieved by Ranulfo Romo and colleagues, who were able to elicit sensations of flutter at specific perceived frequencies by delivering trains of electrical pulses to cortical neurons at the corresponding frequencies [39]. The key point is that these investigators were able to produce percepts that had a specific sensory property, namely temporal frequency. While information about flutter frequency does not play a major role in object manipulation, this study showed that it is possible to create a systematic mapping between parameters of sensory stimulation (in this case pulse frequency) and properties of the elicited sensation (in this case flutter frequency). This general approach can then be used to convey specific types of stimulus information: Stimulation of orientation-selective neurons can be used to convey information about the orientation of a stimulus impinging upon the prosthesis, stimulation of motion-selective neurons to convey information about moving stimuli, and so on.

One problem with attempting to recruit neuronal populations with predetermined properties is that this approach is based on the assumption that nearby neurons have similar response properties (i.e., prefer the same orientation, the same direction of motion, etc.). Indeed, ICMS is not limited to a single neuron but leads to the activation of a population of neurons. Theoretically, the evoked percept can only be coherent to the extent that the responses of activated neurons are consistent with one another. This principle has been confirmed with ICMS of neurons in area MT: With small to moderate currents, the perceived motion direction can be systematically biased in the preferred direction of the activated neuronal population [40]. At high currents, which recruit a large pool of neurons whose preferred direction is no longer approximately homogeneous, this systematic biasing effect disappears and is replaced with a decrement in the animal's sensitivity to motion direction. Such organization has been observed in the spatial distribution of direction selectivity of proprioceptive neurons in S1, a very promising development for upper-limb neuroprostheses [41].

Acknowledgements The authors gratefully acknowledge the editing help of Hannes Saal. This work was supported in part by R01-NS050256 (SHT).

References

1. F.W. Mott, C.S. Sherrington, Experiments upon the influence of sensory nerves upon movement and nutrition of the limbs. Preliminary communication. Proc. R. Soc. Lond. **57**, 481–488 (1895)
2. K.O. Johnson, The roles and functions of cutaneous mechanoreceptors. Curr. Opin. Neurobiol. **11**, 455–461 (2001)
3. J.L. Ochoa, H.E. Torebjörk, Sensations evoked by intraneural microstimulation of single mechanoreceptor units innervating the human hand. J. Physiol. **342**, 633–654 (1983)
4. A.W. Freeman, K.O. Johnson, A model accounting for effects of vibratory amplitude on responses of cutaneous mechanoreceptors in macaque monkey. J. Physiol. **323**, 43–64 (1982)
5. M.A. Muniak, S. Ray, S.S. Hsiao, J.F. Dammann, S.J. Bensmaia, The neural coding of stimulus intensity: linking the population response of mechanoreceptive afferents with psychophysical behavior. J. Neurosci. **27**, 11687–11699 (2007)
6. M. Paré, C. Behets, O. Cornu, Paucity of presumptive ruffini corpuscles in the index finger pad of humans. J. Comp. Neurol. **456**(3), 260–266 (2003). doi:10.1002/cne.10519

7. B.B. Edin, Quantitative analyses of dynamic strain sensitivity in human skin mechanoreceptors. J. Neurophysiol. **92**(6), 3233–3243 (2004)
8. B.B. Edin, Quantitative analysis of static strain sensitivity in human mechanoreceptors from hairy skin. J. Neurophysiol. **67**, 1105–1113 (1992)
9. F. Vega-Bermudez, K.O. Johnson, SA1 and RA receptive fields, response variability, and population responses mapped with a probe array. J. Neurophysiol. **81**, 2701–2710 (1999)
10. R.S. Johansson, G. Westling, Signals in tactile afferents from the fingers eliciting adaptive motor responses during precision grip. Exp. Brain Res. **66**, 141–154 (1987)
11. T. Yoshioka, S.J. Bensmaia, J.C. Craig, S.S. Hsiao, Texture perception through direct and indirect touch: an analysis of perceptual space for tactile textures in two modes of exploration. Somatosens. Mot. Res. **24**, 53–70 (2007)
12. M. Hollins, S.J. Bensmaia, The coding of roughness. Can. J. Exp. Psychol. **61**, 184–195 (2007)
13. A.P. Sripati, S.J. Bensmaia, K.O. Johnson, A continuum mechanical model of mechanoreceptive afferent responses to indented spatial patterns. J. Neurophysiol. **95**, 3852–3864 (2006)
14. K. Dandekar, M.A. Srinivasan, A 3 dimensional finite element model of the monkey fingertip for predicting responses of slowly adapting mechanoreceptors. ASME Bioeng. Conf. BED-**29**, 257–258 (1995)
15. W.R. Loewenstein, M. Mendelson, Components of receptor adaptation in a Pacinian corpuscle. J. Physiol. **177**, 337–397 (1965)
16. S.S. Kim, A.P. Sripati, S.J. Bensmaia, Predicting the timing of spikes evoked by tactile stimulation of the hand. J. Neurophysiol. **104**(3), 1484–1496 (2010)
17. S.S. Kim, S. Mihalas, A. Russell, Y. Dong, S.J. Bensmaïa, Does afferent heterogeneity matter in conveying tactile feedback through peripheral nerve stimulation. IEEE Trans. Neural Syst. Rehabil. Eng. Publ. IEEE Eng. Med. Biol. Soc. **19**(5), 514–520 (2011). doi:10.1109/TNSRE.2011.2160560
18. S.S. Kim, A.P. Sripati, R.J. Vogelstein, R.S. Armiger, A.F. Russel, S.J. Bensmaia, Conveying tactile feedback in sensorized hand neuroprostheses using a biofidelic model of mechanotransduction. IEEE TBIOCAS **1**, 1–6 (2009)
19. E.G. Jones, Cortical and subcortical contributions to activity-dependent plasticity in primate somatosensory cortex. Annu. Rev. Neurosci. **23**, 1–37 (2000)
20. Y.C. Pei, P.V. Denchev, S.S. Hsiao, J.C. Craig, S.J. Bensmaia, Convergence of submodality specific input onto neurons in primary somatosensory cortex. J. Neurophysiol. **102**(3), 1843–1853 (2009)
21. A.P. Sripati, T. Yoshioka, P. Denchev, S.S. Hsiao, K.O. Johnson, Spatiotemporal receptive fields of peripheral afferents and cortical area 3b and 1 neurons in the primate somatosensory system. J. Neurosci. **26**, 2101–2114 (2006)
22. V. Tannan, B.L. Whitsel, M. Tommerdahl, Vibrotactile adaptation enhances spatial localization. Soc. Neurosci. Presented at the Society for Neuroscience. (2012)
23. S.B. Simons, J. Chiu, O.V. Favorov, B.L. Whitsel, M. Tommerdahl, Duration-dependent response of SI to vibrotactile stimulation in squirrel monkey. J. Neurophysiol. **97**(3), 2121–2129 (2007). doi:10.1152/jn.00513.2006
24. M.A. Harvey, H.P. Saal, J.F. Dammann, S.J. Bensmaia, Multiplexing stimulus information through rate and temporal codes in primate somatosensory cortex. PLoS Biol. **11**(5), e1001558 (2013)
25. S.J. Bensmaia, P.V. Denchev, J.F. Dammann III, J.C. Craig, S.S. Hsiao, The representation of stimulus orientation in the early stages of somatosensory processing. J. Neurosci. **28**, 776–786 (2008)
26. J.M. Yau, P.J. Fitzgerald, C.E. Connor, S.S. Hsiao, Early and intermediate representation of edge shape in the somatosensory system. Soc. Neurosci. [Soc. Neurosci. Abs. **37**, 620–626] (2007)
27. J.M. Yau, A. Pasupathy, P.J. Fitzgerald, S.S. Hsiao, C.E. Connor, Analogous intermediate shape coding in vision and touch. PNAS **106**(38), 16457–16462 (2009). doi:10.1073/pnas.0904186106

28. Y.C. Pei, S.S. Hsiao, J.C. Craig, S.J. Bensmaia, Shape invariant representations of motion direction in somatosensory cortex. PLoS Biol. **8**(2), e1000305 (2010)
29. Y.C. Pei, S.S. Hsiao, J.C. Craig, S.J. Bensmaia, Neural mechanisms of tactile motion integration in somatosensory cortex. Neuron **69**, 1–12 (2011)
30. C.D. Salzman, K.H. Britten, W.T. Newsome, Cortical microstimulation influences perceptual judgements of motion direction. Nature **346**, 174–177 (1990)
31. E. Salinas, A. Hernandez, A. Zainos, R. Romo, Periodicity and firing rate as candidate neural codes for the frequency of vibrotactile stimuli. J. Neurosci. **20**, 5503–5515 (2000)
32. M. Randolph, J. Semmes, Behavioral consequences of selective ablations in the postcentral gyrus of Macaca mulatta. Brain Res. **70**, 55–70 (1974)
33. C.E. Connor, K.O. Johnson, Neural coding of tactile texture: comparison of spatial and temporal mechanisms for roughness perception. J. Neurosci. **12**, 3414–3426 (1992)
34. J.J. DiCarlo, K.O. Johnson, S.S. Hsiao, Structure of receptive fields in area 3b of primary somatosensory cortex in the alert monkey. J. Neurosci. **18**, 2626–2645 (1998)
35. E.P. Gardner, J.Y. Ro, K.S. Babu, S. Ghosh, Neurophysiology of prehension. II. Response diversity in primary somatosensory (S-I) and motor (M-I) cortices. J. Neurophysiol. **97**(2), 1656–1670 (2007). doi:10.1152/jn.0. 1031.2006
36. L. Rincon-Gonzalez, J.P. Warren, D.M. Meller, S.H. Tillery, Haptic interaction of touch and proprioception: implications for neuroprosthetics. IEEE Trans. Neural Syst. Rehabil. Eng. Publ. IEEE Eng. Med. Biol. Soc. **19**(5), 490–500 (2011). doi:10.1109/TNSRE.2011.2166808
37. N.A. Fitzsimmons, W. Drake, T.L. Hanson, M.A. Lebedev, M.A. Nicolelis, Primate reaching cued by multichannel spatiotemporal cortical microstimulation. J. Neurosci. **27**, 5593–5602 (2007)
38. B.M. London, L.R. Jordan, C.R. Jackson, L.E. Miller, Electrical stimulation of the proprioceptive cortex (area 3a) used to instruct a behaving monkey. IEEE Trans. Neural Syst. Rehabil. Eng. **16**, 32–36 (2008)
39. R. Romo, A. Hernández, A. Zainos, E. Salinas, Somatosensory discrimination based on cortical microstimulation. Nature **392**, 387–390 (1998)
40. M.J. Nichols, W.T. Newsome, Middle temporal visual area microstimulation influences veridical judgments of motion direction. J. Neurosci. **22**, 9530–9540 (2002)
41. D.J. Weber, B.M. London, J.A. Hokanson, C.A. Ayers, R.A. Gaunt, R.R. Torres, B. Zaaimi et al., Limb-state information encoded by peripheral and central somatosensory neurons: implications for an afferent interface. IEEE Trans. Neural Syst. Rehabil. Eng. Publ. IEEE Eng. Med. Biol. Soc. **19**(5), 501–513 (2011). doi:10.1109/TNSRE.2011.2163145

Chapter 8
Human Grip Responses to Perturbations of Objects During Precision Grip

Michael De Gregorio and Veronica J. Santos

Abstract Grasp stability of a precision grip requires fine control of three-dimensional fingertip forces. This chapter begins with a review of the literature on how precision grip forces are affected by intrinsic object properties, anticipation, load direction, and sensory feedback. Previous studies have established that reactive, initial increases in grip forces (pulse-like "catch-up responses" in grip force rates) are elicited by unexpected translational perturbations and that response latency and strength scale with the direction of linear slip relative to the hand as well as gravity. To determine if catch-up responses are elicited by unexpected rotational perturbations and are strength-, axis-, and/or direction-dependent, we imposed step torque loads about each of two axes which were defined relative to the hand: the distal-proximal axis away from and towards the palm, and the grip axis which connects the two fingertips. First dorsal interosseous activity, marking the start of the catch-up response, began 71–89 ms after the onset of perturbation. Onset latency, shape, and duration (217–231 ms) of the catch-up response were not affected by axis, direction, or magnitude of the rotational perturbation, while strength scaled with axis of rotation and slip conditions. Rotations about the grip axis induced rotational slip at the fingerpads and elicited stronger catch-up responses than rotations about the distal-proximal axis. The chapter concludes with a discussion of this study that, to our knowledge, is the first to investigate grip responses to unexpected torque loads and to show characteristic, yet axis-dependent, catch-up responses for conditions other than pure linear slip.

Keywords Catch-up response · Fingertip forces · Precision grip · Rotational perturbation · Rotational slip · Torque load · Unexpected perturbation

---

M. De Gregorio · V. J. Santos (✉)
Mechanical and Aerospace Engineering, Arizona State University, Tempe, AZ, USA
e-mail: veronica.santos@asu.edu

R. Balasubramanian and V. J. Santos (eds.), *The Human Hand as an Inspiration for Robot Hand Development*, Springer Tracts in Advanced Robotics 95,
DOI: 10.1007/978-3-319-03017-3_8,

1 Introduction

With the advent of anthropomorphic robotic and prosthetic hands comes both the promise of increased functionality of artificial hands and the non-trivial challenges of controlling a large number of degrees of freedom. Engineers can reduce the cognitive burden on the operators of artificial hands by designing automated grip responses into the control of the hand itself. Of course, care must be taken so as not to undermine the independence of the human operator. If implemented well, automated grip responses (artificial reflexes) could intervene in a manner that is transparent to the user and buys time for processing and two-way communication between the operator and the manipulator.

Building intelligence into artificial systems has long been an approach in the field of robotics. In the 1980s, roboticists proposed a reflexive control architecture that directly linked sensing and actuation in artificial systems [1]. "Grasp pre-shaping" is a good example of how visual feedback on object shape has been used to implement pre-programmed grasp postures [2, 3]. While automated pre-shaping could aid in the initial grasp of an object, on-line grasp adjustments such as changes in digit placement or fingertip force vectors may still be necessary to account for erroneous digit placement or perturbations of the grasped object. Much effort has been put into developing dynamic tactile sensors (the reader is referred to [4–6]) and control algorithms that will enable on-line adjustments of fingertip forces using real-time feedback from tactile sensors.

We are particularly interested in on-line grasp adjustments during precision grip, which is fundamental to the human ability to grasp and manipulate objects. In precision grip, or opposition pinch, the grasped object is "pinched between the flexor aspects of the fingers and the opposing thumb" [7]. Grasp stability using a two-fingered precision grip is achieved through force closure, which only requires two soft finger contacts for an object in 3D having six degrees of freedom [8]. Assuming minimal changes in digit center of pressure location, grasp stability will depend critically on each digit's 3D force vector and its relationship to the friction cone determined by the grasped object and the fingerpad skin. Most of this chapter will, therefore, focus on the spatial and temporal coordination of fingertip forces during precision grip.

While biomimicry is not a necessity for engineered solutions to artificial grasp, it is difficult to ignore the ease with which the human hand can manipulate objects. Much inspiration for grasp control algorithms can be taken from neurophysiology studies on human grasp. This chapter begins with a brief overview of key findings from a subset of prior precision grip studies. For more comprehensive reviews, the reader is referred to [9–11]. The chapter will conclude with results from a study that, to our knowledge, is the first to investigate grip responses to unexpected torque loads and to show characteristic, yet axis-dependent, "catch-up responses" for conditions other than pure linear slip [12]. The catch-up response will be described in further detail in Sect. 1.2.2.

1.1 Influence of Intrinsic Object Properties on Fingertip Forces

An object's physical properties (extrinsic and intrinsic) can influence fingertip forces on the object during grasp and manipulation [11]. Examples of extrinsic object properties include the orientation and motion of the object. Examples of intrinsic object properties include weight, mass distribution, size, texture, frictional condition, shape, surface curvature, and fragility [11]. A subset of intrinsic object properties and their parametric influences on fingertip force control is addressed here.

1.1.1 Weight

In 1984, Westling and Johansson conducted an experiment in which each subject held an object in midair using a precision grip. Experimenters were able to pseudorandomly alter the weight of an object having symmetric grip surfaces without changing its visual appearance [13]. Subjects performed tasks in which they lifted the object from a table, held the object stationary in midair, and then replaced the object to the table (lift-hold-replace) or lifted the object, held the object in midair, and then allowed the object to slip from their fingertips (lift-hold-slip).

Fingertip forces were reported in normal and tangential components relative to the grip surface. Grip force was defined as the compressive force normal to the grip surface and load force referred to the lifting force tangential to the grip surface for the vertically oriented object. Parallel coordination of grip forces and load forces on the object is believed to be a general control strategy for achieving grasp stability, regardless of grasp configuration or task [11]. For instance, when subjects were tasked with holding objects of various weights, the static grip forces employed during the static grasp of the object in midair changed in proportion to the weight of the object [13, 14].

1.1.2 Frictional Condition Between Object and Skin

The preceding discussion on the influence of object weight on fingertip forces would not be complete without an analysis of the frictional condition between the object and the fingerpad skin. In the same set of experiments previously described, Westling and Johansson also varied frictional conditions by exchanging pairs of grip surfaces (sandpaper, suede, or silk) [13]. Static grip force changed in proportion to the inverse of the coefficient of static friction at the finger-object interface. That is, static grip force increased as the surface became more slippery. While the relationship between static grip force and weight remained approximately linear, the slope of this linear relationship increased as the surface became more slippery [13].

Slip force was defined as the minimum grip force necessary to prevent slip, as determined via trials during which subjects purposefully reduced grip forces until slip occurred [13]. The relationship between the tangential load force and the normal slip force was determined by the coefficient of friction for the object-skin contact condition. Thus, slip forces would increase as slipperiness increased. Subject-specific safety margins for prevention of slip were defined as the difference between actual grip forces and the slip forces. The safety margin increased with slip force and was, therefore, higher for more slippery surfaces [13]. Safety margins were relatively low such that excessive grip forces were avoided [13, 15]. It has been postulated that "the manipulative task is apparently underwritten by a program to prevent slips" [16].

1.1.3 Shape

Object shape was varied in a 1997 study by Jenmalm and Johansson [17]. The test object had symmetric grip surfaces and was designed such that the angle of the grip surface relative to the vertical plane could be varied from −40° (tapered downward) to 30° (tapered upward) in 10° increments. The grip surfaces were vertical and parallel for the 0° case.

Fingertip forces were reported in horizontal and vertical components. Horizontal forces were not always normal to the grip surface (except for the 0° case), but they were always pointed towards the object's midline. Horizontal forces increased as surface angle increased and the object tapered upward [17]. This relationship held prior to object lift-off and during the static grasp of the object in midair. Vertical forces were relatively independent of surface angle. Interestingly, the components of force normal and tangential to the grip surface maintained an approximately constant ratio regardless of surface angle. As a result, the safety margin for the prevention of linear slip was independent of surface angle [17].

1.1.4 Surface Curvature

In 1998, Goodwin et al. conducted an experiment in which curvature of the grip surface was varied for a task requiring object tilt [18]. The test object had two symmetric grip surfaces and was designed such that surface curvature could be varied (concave, flat, or convex). Subjects were instructed to grasp the test object and tilt the base of the object up and away from the body using elbow flexion and radial flexion of the wrist. Unlike the prior studies discussed here which addressed linear slip caused by forces tangential to the grip surface, this study addressed *rotational slip* [14] caused by tangential torques at the fingertips about an axis perpendicular to the grip surface. Such torques resulted from rotation of the grasped object about the *grip axis* (line connecting the two fingertips).

Rotary slip tests were performed in order to determine the slip force, or grip force below which rotational slip would occur [18]. Interestingly, the experimenters differentiated the importance of each digit by defining a critical digit as the digit that employed the greater of the two slip forces. Any decrease in grip force by this critical digit would lead to undesired rotational slip of the object. Safety margins for prevention of rotational slip were defined as the difference between actual grip forces and the slip force of the critical digit for that trial. It was found that safety margin increased as surface curvature increased and the surface became more convex [18]. However, the relative safety margin, the safety margin expressed as a fraction of grip force, was actually less dependent upon surface curvature [18, 19] and more dependent upon torque load [18]. Interestingly, the magnitude of the safety margin for rotational slips [18] was comparable to that for linear slips [13].

Grip force for the tilting task had an approximately linear relationship with surface curvature (higher grip forces for more convex surfaces) [18]. Regardless of surface curvature, grip force increased linearly with torsional load throughout the tilting movement [18]. As the object was tilted further upward, the tangential torque at the fingertips increased, and higher grip forces were applied. A similar experiment that varied surface curvature for a lifting task, as opposed to a tilting task, concluded that surface curvature moderately affected slip force but not grip force for linear slips [20]. It is important to note that the effects of surface curvature on grip forces were greater for tasks in which rotational slip was imminent (tangential torque loads dominated) [18, 19] as compared to those when linear slip was imminent (linear force loads dominated) [20].

1.2 Influence of Anticipation on Fingertip Forces

The discussion of anticipation effects on fingertip forces is limited here. For lengthier discussions on the importance of prediction of motor commands and sensory events with respect to control strategies during purposeful object manipulation, the reader is referred to [9, 21, 22].

1.2.1 Anticipated, Self-Imposed Perturbations in Bimanual Tasks

Anticipatory or preparatory motor responses are particularly useful when the subject has an expectation or prediction about the grasping task. This is the case for voluntary movements and when perturbations are imposed by subjects themselves, as with a bimanual task. In 1988, Johansson and Westling conducted a study in which subjects grasped a pair of vertical grip plates that were attached to a cup into which balls of various weights could be dropped from different heights [16]. In some trials, subjects dropped the ball themselves using the contralateral, non-dominant hand. In other trials, the experimenter dropped the ball such that the load perturbation was unexpected [16]. Subjects wore ear phones and had minimal audio feedback. Except for control trials in which subjects had visual feedback, subjects closed their eyes for all trials. Another set of control trials was conducted during which the experimenter interfered with the task by catching the ball dropped by subjects before the ball landed in the cup.

When subjects dropped the ball, grip forces increased after the ball was released in anticipation of the impending impact of the ball with the cup. Grip forces began to increase 150 ms before impact and continued 100 ms after load force peaked [16]. This preparatory motor response occurred whether or not the experimenter interfered with the impact of the ball and whether or not visual cues were available. The preparatory grip force magnitude increased with more slippery frictional conditions. The preparatory grip force rate scaled with the weight of the ball and the weight of the grip apparatus. The duration of the grip force increase scaled with the height through which the ball dropped [16].

When the experimenter dropped the ball, there were no preparatory motor responses by subjects. Rather, a reactive grip force response was triggered by the impact of the ball with the cup and began 70–80 ms after the onset of impact [16]. As with the preparatory grip force response, the triggered grip force magnitude scaled with the weight of the ball. However, the magnitude of the grip force increase was larger when the experimenter dropped the ball than when subjects dropped the ball. Regardless of who dropped the ball, both grip and load forces increased in parallel [16].

1.2.2 "Active" Objects with Unpredictable Properties

Thus far, the fingertip force adaptations described were observed during experiments in which subjects grasped passive objects whose intrinsic physical properties were constant and predictable. For such passive objects, sensorimotor memories could be used to make predictions about the objects for feedforward adjustments of fingertip force. In the early 1990s, Johansson and colleagues conducted a series of experiments [19–21] with active objects that exerted unpredictable pulling loads having different force amplitudes [23] and rates [24]. The manipulandum consisted of a pair of grip plates that could be pulled in the distal direction by a motor to impose a loading phase (ramped increase in force load), plateau phase (constant force load), and unloading phase (ramped decrease in force load) [23]. Subjects were instructed to restrain the object upon detection of the pulling load without visual feedback. For the static load phase, the safety margins for the prevention of linear slip appeared similar for passive and active objects [23].

As with passive objects, grip force changed with load force during restraint of the active objects. With passive objects, grip and load forces were initiated and scaled in parallel and the grip force rate had a bell-shaped profile [16]. With active objects, however, the grip force changes were delayed due to the unpredictable nature of the force loading and unloading [23]. The grip force rate had a biphasic profile and featured a rapid initial increase in grip force (a pulse-like *"catch-up response"* in grip force rate) that was followed by a period of steadily increasing grip force (a secondary "tracking response" in grip force rate) if the distal load continued to increase after the catch-up response [23]. The catch-up response allowed the delayed grip force to quickly catch up to the unexpected load force and maintain a sufficient safety margin for the prevention of linear slip. For a constant load force rate, features of the catch-up response such as latency (approximately 140 ms after onset of loading), shape, size, and duration (200–250 ms) were independent of load force amplitude [23]. When load force rate was varied, the amplitude of the catch-up response increased with load force rate and the grip response latency decreased with load force rate [24]. Interestingly, the duration of the the catch-up response was 200 ms, regardless of the load force rate [24]. The stereotypical features of the catch-up response, in particular its fixed duration, suggested that the catch-up response was a centrally programmed, default grip response that was released as a unit in response to unexpected increases in load [23, 24]. These findings supported prior observations of stereotypical evoked grip force responses (consistent response latencies of 60–90 ms and durations of 100–200 ms) that appeared to be automatic and not consciously mediated [25]. When finger sensibility was impaired via digital anesthesia, the tracking response disappeared, suggesting that the tracking response is a grip response mechanism that relies critically on cutaneous afferent input to make online grip force adjustments [26].

In 1993, Cole and Johansson used the same active object as in the prior series of experiments [23, 24, 26], but also varied frictional condition at the finger-object interface [27]. The loading profile featured a ramped increase, plateau phase, and ramped decrease in force applied in the distal direction [27]. The intertrial or preload grip force was defined as the grip force at the onset of the load perturbation. The preload grip force scaled with the inverse of the coefficient of friction [27]. The more slippery the grip surface, the greater the preload grip force. As in prior studies [23, 24], the grip force profile featured a catch-up and secondary tracking response [27]. Features of the catch-up response such as onset latency, shape, and duration were independent of frictional condition [27]. It was found that the preload grip force greatly affected the scaling of the catch-up and tracking responses, possibly to avoid unnecessarily large or dangerously small safety margins for linear slip at the peak load force [27]. Since the preload grip force contributes to the overall grip response, increases in this grip force baseline value resulted in decreases in the catch-up and/or tracking responses. Thus, it was postulated that initial state information such as frictional condition could "globally" affect the grip response to unexpected pulling loads by adjusting a "central scaling factor" or "gain" for the load-to-grip sensorimotor transformation [27]. Larger preload grip forces and, thus, higher safety margins at the start of the trial were employed after slip [20, 28] and when digital anesthesia was used [29].

1.3 Influence of Load Direction on Fingertip Forces

Multiple studies have reported that fingertip force responses evoked by unanticipated loads are affected by the direction of the loading [29–31]. In a 1992 study, Jones and Hunter reported directional effects of translation loads on the magnitude of the grip force [30]. Specifically, subjects used larger grip forces for distal loads that pulled away from the palm than for proximal loads that pushed towards the palm. It was postulated that the frictional condition at the finger-object interface was a function of loading direction due to the anisotropic properties of glabrous skin [30].

In 1996, Häger-Ross et al. specifically investigated the effects of load direction relative to gravity and hand orientation [31]. Subjects had to restrain an active object with two parallel, symmetric grip surfaces with the ulnar side of the hand facing either downward or upward. A motor was used to apply unpredictable loads in the distal, proximal, radial, and ulnar directions. Depending on hand orientation, the radial and ulnar load directions would be in the direction of gravity or opposed to gravity. Regardless of hand orientation, grip force response latencies were shorter for what were referred to as the "dangerous directions": distal loads as opposed to proximal loads, and directions with gravity instead of against gravity [31]. These findings were consistent with another study that varied load direction and support conditions for the hand and forearm [29]. Interestingly, the latencies for cutaneous afferent responses were similar for the distal and proximal load directions and, yet, the grip force response latencies were shorter for distal loads [31]. It was postulated that load directions relative to both gravity (with or against) and hand geometry (away or towards the palm) serve as "intrinsic task variables" for the central control of grasp stability when faced with unpredictable loads [31]. The fact that grip force response latencies and magnitudes were independent of load direction when digital sensibility was impaired suggests that digital sensory inputs play critical roles in reactive grip control [29].

1.4 Influence of Sensory Feedback on Fingertip Forces

1.4.1 Visual Feedback

Visual cues play a critical role in anticipatory parameter control [11], a control policy which proposes that fingertip forces are adapted to intrinsic object properties (e.g., shape [17], size [32], weight [33], surface curvature [19]) in anticipation of force requirements based on prior experience. During grasp experiments in which object shape (grip surface angle) was varied, it was found that visual feedback was used to adjust fingertip force in a feedforward manner from the onset of horizontal force generation [17]. Furthermore, the grip surface angle from the previous trial had no effect on the forces applied in the current trial when visual feedback was available, regardless of whether cutaneous afferent feedback was intact or not [17].

In a different study that varied the size and weight of objects, visual cues about object size influenced the scaling of fingertip forces during the loading phase even before additional somatosensory cues related to actual object weight became available with object lift-off [32]. Subjects used prior experience, as with common everyday objects, to scale fingertip forces after visually identifying the object and estimating the object's weight [33]. When size and weight were covaried and object density remained constant, visual cues about size were used as indirect measures of weight for anticipatory fingertip force scaling [32].

1.4.2 Digital Feedback

Digital sensibility plays a critical role in both the shaping of anticipatory grip adjustments that are informed by sensorimotor memories and the shaping of evoked or reactive grip responses to unpredictable load perturbations [25, 26]. Discrete event, sensory-driven control has been proposed as a control policy, used in addition to anticipatory parameter control, that uses discrete mechanical events encoded in the spatiotemporal patterns of sensory inputs to trigger pre-programmed corrective grip responses [11]. Due to their proximity to the finger-object interface, tactile receptors innervating the glabrous skin are believed to be particularly important for the adaptation of fingertip forces to intrinsic object properties [11]. It has even been said that "it appears that tactile afferents of the skin in contact with the object are the only species of receptor in the hand capable of triggering and initially scaling an appropriate change in grip force in response to an imposed change in load force" [34].

Multiple studies have addressed the importance of cutaneous afferent feedback to the adaptation of fingertip forces to variations in the frictional condition at the finger-object interface. A 1984 study using a lift-hold-replace task showed that the adaptation of grip forces and load forces to frictional condition was dependent upon input from cutaneous afferents [15]. Specifically, afferents innervating the glabrous skin areas that directly contacted the object were important for detecting "local" slip events and estimating the frictional condition [15]. It is believed that fast-adapting FA I units are primarily responsible for triggering and initially scaling the automatic, reactive grip force responses to ramped increases in load forces [35, 36].

In the absence of visual cues, cutaneous afferent feedback about object shape [17] and surface curvature [19] influenced fingertip forces soon after object contact. When grip surface angle was varied, fingertip force adjustments to the new surface angle emerged approximately 100 ms after object contact [17]. When subjects had to rely on cutaneous afferent feedback alone, fingertip forces prior to the 100 ms timepoint were influenced by the previous trial. When surface curvature was varied, fingertip force adjustments to the new surface curvature emerged 100–200 ms after object contact [19]. Again, initial fingertip forces were influenced by the force conditions of the previous trial. Once the new surface curvature was detected, tactile cues were used to adapt fingertip forces in a feedforward manner [19]. While both visual and tactile feedback have been associated with feedforward adjustments of fingertip force [17, 19], tactile feedback is especially critical for anticipatory grip force adjustments to frictional conditions [35, 37].

Nondigital sensory input from joint receptors, intrinsic and extrinsic hand muscles, and skin strain sensors located away from the fingertips may also contribute to grip responses. Häger-Ross and Johansson observed catch-up responses despite the use of three different support conditions (hand support, forearm support, no support) used to vary the transmission of the perturbations at the fingertips to other segments of the hand and forearm [29]. Nonetheless, it is believed that digital inputs are the principal sensory inputs used during reactive grip control [25, 26, 29, 34–36, 38]. Anesthetization of both digits in a precision grip weakened the effect of load rate on catch-up response strength, increased grip response latencies, and, in some cases, caused the absence of a grip force response altogether [26].

1.5 Motivation for the Following Study on Grip Responses to Unexpected Torque Loads

Prior studies on rotational slip involved expected, tangential torque loads as subjects lifted and tilted [18], lifted and held [19], or released one's grasp to allow the tilting of [14] passive objects whose center of mass location threatened rotational slip about the grip axis. In the work that follows here, torque loads were used to investigate grip responses to unexpected rotational perturbations of a grasped object about two different axes relative to the subject's hand: distal-proximal axis away from and towards the subject's palm, and the grip axis connecting the two fingertips. The objectives of this study were twofold: to determine if a catch-up response is elicited by *unexpected rotational perturbations*, and, if so, to determine if features of the catch-up response (onset latency, shape, duration, strength) are dependent on the direction, strength, and/or axis of the perturbation. To our knowledge, this is the first study to investigate grip responses to unexpected torque loads and to show characteristic, yet axis-dependent, catch-up responses for conditions other than pure linear slip.

2 Materials and Methods

2.1 Experiment

Eighteen unimpaired subjects (nine male, nine female) participated in the study under a protocol approved by the Arizona State University Institutional Review Board. All participants were naïve to the purpose of the study, naïve to the experimental apparatus, and gave their informed consent prior to participating in the study. Subjects had no known hand or neurological dysfunction. All subjects were right hand dominant and ranged in age from 19 to 38 years.

A six-camera motion-capture system (MX-T40 cameras, Vicon, Centennial, CO) and 3 mm diameter hemispherical markers (Mocap Solutions, Huntington Beach CA) were used to collect kinematic data from the thumb and index finger of each subject's dominant hand (24 markers in the form of 8 triads) and an instrumented test object (3 markers) (Fig. 1). Surface electromyography (EMG) was used to measure the activity of the first dorsal interosseous (FDI) muscle (BIOPAC EMG 100C, BIOPAC Systems, Goleta, CA).

The 194 g rigid test object had two parallel, symmetric grip surfaces spaced 39 mm apart. Fingertip forces and torques were measured independently for the thumb and index finger by six-axis force/torque transducers (Nano-25, ATI

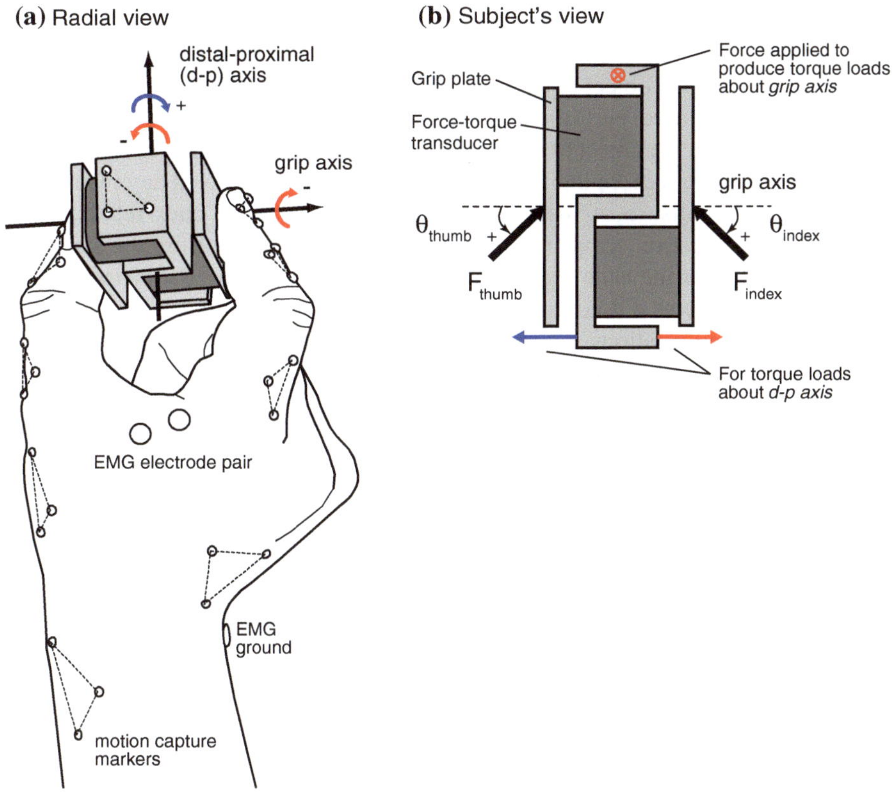

Fig. 1 Experimental set-up. **a** A radial view of the subject's hand is shown with triads of retro-reflective markers (indicated by dotted lines) that were used to track the motion of the thumb, index finger, and instrumented object. Surface EMG was used to record first dorsal interosseous activity. Directions of rotation (*red* = negative, *blue* = positive) were defined with respect to each of two hand-referenced, object-fixed axes: distal-proximal and grip axes. **b** A view from the perspective of the subject of the object shows the locations of the attachment points used to impose rotational perturbations with a cable, mass, and pulley system. The angle θ of the fingertip force vector was measured relative to the grip surface normal in the plane containing the grip axis. (Adapted from [12])

Industrial Automation, Apex NC) housed within the test object. Each aluminum grip plate was covered by a single layer of masking tape in order to minimize reflectivity during collection of motion capture marker data. Kinematic data were collected at 200 Hz while EMG and force/torque data were collected at 1.8 kHz by a 16-bit data acquisition board that was synchronized with the motion capture system.

Each subject sat upright with the dominant arm supported by a tabletop. With each subject's dominant hand and wrist unsupported at the end of the table, a vacuum positioning pillow (Versa Form, Sammons Preston, Bolingbrook, IL) and velcro straps were used to constrain the subject's forearm to the tabletop. Hand pronation and supination were restricted while wrist flexion/extension and radio-ulnar deviation were not. Subjects were instructed to hold the test object upright in midair using a precision grip with the thumb and index finger centered on its own grip plate and directly opposed to the other digit. Subjects were instructed to curl their other digits towards their palm as if making a fist. Subjects were told that a perturbation would jar the grasped object at a random, unannounced time point during each five second trial. A single perturbation occurred during every trial. Subjects were instructed to return the object to its initial orientation as soon as the perturbation was detected and to avoid dropping the object. Subjects were not given any instructions related to use or disuse of their wrists. To minimize visual cues related to the impending perturbation, subjects were instructed to look away from the test object and investigator during each trial. There were no auditory cues associated with the perturbation. Subjects did not have any practice trials prior to data collection.

Rotational perturbations were imposed about two different axes which were defined relative to the subject's hand, fixed relative to the object, and passed through the test object's center of mass (Fig. 1): distal-proximal ("d-p") axis away from and towards the subject's palm, and the grip axis which connects the fingertips. Both positive and negative rotations were imposed about the d-p axis while only negative rotations were imposed about the grip axis such that the top of the test object tilted away from the subject's hand. Step torque loads were imposed using a mass and pulley system attached to the object via lightweight, inextensible fishing line (200 lbf braided line, PowerPro, Irvine, CA). Each trial lasted five seconds, during which a mass was dropped vertically by 5 cm and remained hanging from the fishing line until the conclusion of the trial. No significant swinging of the mass was observed. The mass (100 or 150 g) combined with the off-center attachment points for the fishing line (34.3 mm above or below the object's center of mass, Fig. 1b) resulted in the following externally-applied torque loads, or *object-moments* [39], about the test object's center of mass: ± 33.6 or ± 50.5 mN-m about the distal-proximal axis, and -50.5 mN-m about the grip axis. Johansson et al. defined object-moment as the "product of the force of gravity acting on the object and the perpendicular distance between the grip axis and its CM [center of mass]" [39]. In this study, object-moment was defined as the external torque load imposed about the object's center of mass via a mass and pulley system.

Blocks of experimental conditions were created using different combinations of levels for three factors associated with each rotational perturbation: object-moment magnitude (small or large), axis of rotation (d-p or grip axis), direction of rotation (positive or negative). A total of six blocks were presented in the following order: +33.6 mN-m about the d-p axis, +50.5 mN-m about the d-p axis, −50.5 mN-m about the d-p axis, −33.6 mN-m about the d-p axis, −33.6 mN-m about the grip axis, and −50.5 mN-m about the grip axis. Subjects completed five trials for each block. Subjects received a mandatory 30 s rest period between each trial and were advised to notify the researcher if lengthier rest periods were necessary in order to avoid fatigue.

2.2 Data Analysis

Motion capture data for the hand and test object were post-processed using Vicon Nexus software to ensure complete marker sets for each trial. All data were then processed using MATLAB (Mathworks, Natick MA). Kinematic marker data were filtered using a fourth-order 30 Hz low pass Butterworth filter [40]. Triads of markers were used to establish local Cartesian reference frames for bodies of interest, such as the test object.

Angular orientation and angular velocity of the test object were used to verify the predominance of rotational motion in each perturbation and to determine the onset of object perturbation, which was defined as "t = 0 s." All trials were synchronized according to the onset of perturbation. Angular deviations of the object from its initial upright orientation were measured about the axis of interest (Fig. 1a) and averaged across trials for each block on a subject-specific basis. The peak angular deviation of the object was defined as the maximum magnitude of rotation of the object from its initial orientation and was extracted from the averaged angle data.

First dorsal interosseous surface EMG data were full-wave rectified and then filtered using a 50 Hz low-pass fourth-order Chebyshev filter [41]. The data were normalized using maximum voluntary contraction data collected at the start of each experimental session. For each subject, all trials for a given block condition were averaged to ensure that trends in muscle activation were consistent. To determine when FDI was active, activation threshold values were set at three standard deviations of the baseline noise prior to the onset of object perturbation. For each subject and block condition, the activation threshold was determined from the mean baseline noise data across all trials for that block. Each subject's FDI was considered to be active if its mean EMG activity (on a block-specific basis) exceeded the activation threshold and remained above threshold for a continuous period of at least 50 ms [41]. Visual inspection was used to verify that the activation thresholds resulted in accurate, reliable threshold crossings [42]. The time to initial FDI activation (time of the first upward threshold crossing relative to the onset of object perturbation) was used as the *onset latency of the catch-up response*. Due to poor surface EMG connectivity, EMG data are only only presented for 15–17 subjects, depending on the perturbation conditions.

Fingertip force and torque data were filtered using a fourth order, 30 Hz low-pass Butterworth filter [40]. Force components were separated into normal and tangential components relative to the grip surface (as in [37]) as opposed to "grip forces" and vertical lifting "load forces" (as in [13]). Normal and tangential force components were appropriate for this study in which object orientation was not always upright and external loads were neither purely translational nor aligned with gravity. Normal force components were the same as the grip forces referred to in [13]. Force data for individual trials were averaged for each block on a subject-specific basis to ensure consistent extraction of key features [18, 19, 27, 37]. For rotations about the d-p axis, each digit's mean force vector was used to calculate the *angle θ of the fingertip force vector with respect to the grip surface normal*, as viewed from the plane containing the grip axis (Fig. 1b). Positive angles indicated that a digit's force vector had a component that pointed in the radial direction towards the top of the object. Normal force rates were obtained by taking first-order differences of the normal force data for each individual trial while accounting for the force sampling rate, averaging those data according to experimental block conditions on a subject-specific basis, and smoothing with a moving boxcar average having a width of 50 ms. As in [23], the onset of muscle activity and peak normal force were used to mark the start and end of the catch-up response, respectively, in order to determine the *duration of the catch-up response*. Peak normal force rate was used to define the *strength of the catch-up response*.

2.3 Statistical Analysis

The data could not be transformed into normal distributions using square-root, log, or inverse sine functions. Thus, we used the Kruskal–Wallis test, the non-parametric counterpart to the one-way ANOVA, to evaluate the independent effects of three experimental factors: object-moment magnitude (small or large), axis of rotation (d-p or grip axis), and direction of rotation (positive or negative). The Kruskal–Wallis test was also used to test whether quantities of interest could be pooled across the thumb and index finger. For all statistical analyses, the alpha level was set at 0.05. Summary data are reported as either median ranges or median $\pm$ median absolute deviation (MAD) unless otherwise specified.

3 Results

Data for a representative subject for a single experimental block condition (-50.5 mN-m rotation about the d-p axis) illustrate common trends in object kinematics, first dorsal interosseous activity, and fingertip force components (Fig. 2). After the onset of perturbation (t = 0 s; "a" in Fig. 2), the object rotated about the d-p axis such that the top of the object turned towards the subject's

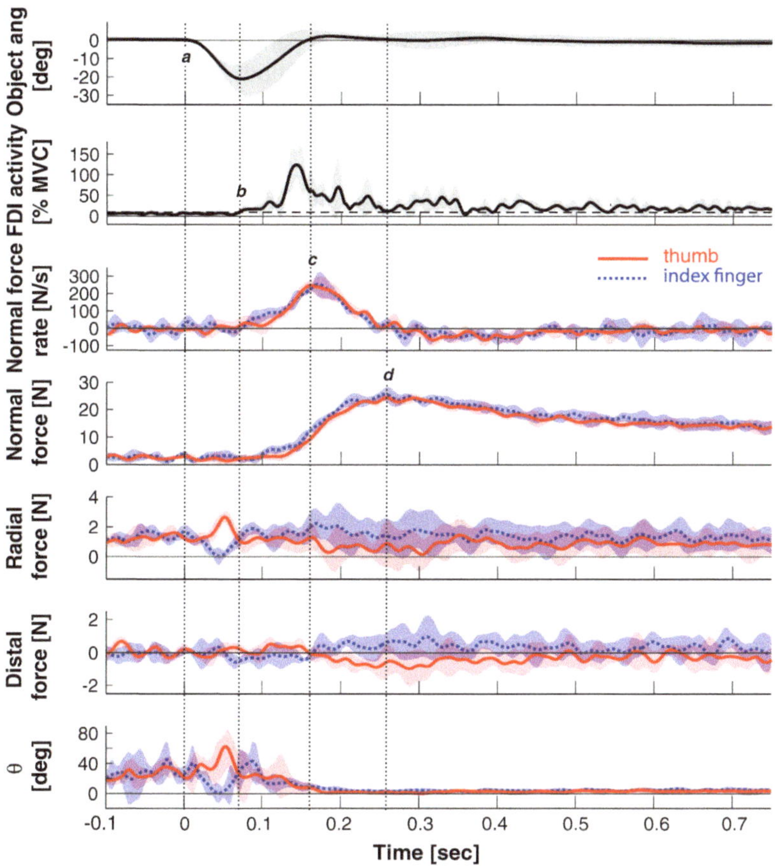

Fig. 2 Object angle, FDI activity, and fingertip forces, rates, and angles as a function of time. Mean data are reported for a representative subject for a single block of experiments (large object-moment magnitude, negative rotation about the distal-proximal axis) for angle of deviation of the object, FDI activity (normalized by maximum voluntary contraction), normal force rate, normal, distal, and radial force components, and angle θ of the fingertip force vector relative to the grip surface normal. The thumb (mean = *solid red line*) and index finger (mean = *dotted blue line*) data could be pooled. The shaded region (*red* and *blue* for thumb and index finger, respectively) around each mean trace represents ±SEM across five trials. Vertical dotted lines indicate times for: *a* onset of perturbation; *b* start of the catch-up response (time to FDI activation); *c* strength of catch-up response (peak normal force rate); and *d* end of the catch-up response (peak normal force). Each digit's fingertip forces are reported relative to an object-fixed reference frame in terms of normal components (+ refers to compressive forces along the grip axis) and tangential components (+ refers to forces in the radial direction that point towards the top of the object or to forces in the distal direction that point away from the subject along the d-p axis). A distinct unimodal catch-up response was observed in the normal force rate data for each digit. (Adapted from [12])

thumb. Shortly after the object reached its maximum angular deviation, the subject's FDI became active (onset of the catch-up response, "b" in Fig. 2) and the normal force rate for each digit began to increase until reaching a maximum value marking the strength of the catch-up response ("c" in Fig. 2). The thumb and index finger normal forces increased and decreased in parallel. A catch-up response was clearly visible in the normal force rate profiles of both the thumb and index finger (from "b" to "d" in Fig. 2). The subject had rotated the object back to its initial upright configuration by approximately 200 ms. The angle of the object appeared to have reached steady-state even though the catch-up response had not yet been completed, possibly because both digit's normal forces were scaled in parallel and did not change substantially the net torque about the object's center of mass.

3.1 Object Kinematics

For positive and negative rotations about the d-p axis, the object was rotated clockwise and counter-clockwise from the subject's perspective, respectively. Upon perturbation, the object was rotated away from its initial upright orientation to some peak angular deviation, and was then rotated back towards the upright orientation (Fig. 2). For rotational perturbations about the grip axis, the top of the object tilted away from the subject. With wrist movements limited, subjects could halt the tilting, but were unable to restore the object to its initial upright configuration.

Object-moment magnitude affected the peak angular deviation of the object for the d-p and grip axes for both rotation directions (largest p value was $p = 0.02$). As expected, peak angular deviations were greater for larger object-moment magnitudes (Table 1). Rotation axis also affected peak angular deviation of the object ($p = 5.3E-6$ and $3.9E-6$ for small and large object-moment magnitudes, respectively). The median values for negative rotations about the grip axis were 22–23° greater than those for negative rotations about the d-p axis (Table 1).

Object-moment magnitude did not affect the time to peak angular deviations of the object for rotations about the grip axis (p-value of 0.33), but did affect rotations about the d-p axis (largest p-value of 0.01). Time to peak angular deviation of the object was 10 ms greater for larger object-moment magnitudes (Table 1). Time to peak angular deviation was influenced by rotation axis (largest p-value of $1.2E-6$). Median values were 68–83 ms greater for the grip axis relative to the d-p axis (Table 1).

In general, direction of rotation did not affect object kinematics. For rotations about the d-p axis, peak angular deviation and time to peak angular deviation were slightly greater for negative rotations than positive rotations. While statistically significant, the actual differences in median values do not appear to be functionally significant (5° and 2–5 ms for peak angular deviation and time to peak angular deviation, respectively, Table 1).

3.2 Onset Latency of the Catch-Up Response

Surface EMG was used to investigate the timing of first dorsal interosseous activation relative to the onset of perturbation of the grasped object (Fig. 2). A burst of FDI activity was observed shortly after the onset of perturbation for all subjects

Table 1 Object kinematics and grip response events

Axis:	Distal-proximal		Grip
Direction:	Positive	Negative	Negative
Peak angular deviation of object [deg]	Small OMM 9.5 ± 1.9 Large OMM 12.7 ± 2.0	Small OMM 11.7 ± 1.9 Large OMM 17.8 ± 3.8	Small OMM 33.6 ± 5.2 Large OMM 40.5 ± 8.3
Time to peak angular deviation of object [ms]	Small OMM 55.0 ± 0.0 Large OMM 65.0 ± 5.0	Small OMM 60.0 ± 5.0 Large OMM 70.0 ± 5.0	137.5 ± 25.0
Onset latency of the catch-up response [ms] (*Time to FDI activation, "b" in* Fig. 2)	71.4 ± 9.2	83.9 ± 12.8	88.6 ± 9.2
Strength of the catch-up response [N/s] (*Peak normal force rate*)	Small OMM 79.5 ± 21.9 Large OMM 104.2 ± 27.0	83.4 ± 21.8	134.7 ± 50.8
Time to peak normal force rate [ms] (*"c" in* Fig. 2)	127.2 ± 15.5	152.2 ± 28.9	145.8 ± 16.7
Peak normal force [N]	13.0 ± 2.9	12.5 ± 3.3	19.2 ± 8.0
Time to peak normal force [ms] (*"d" in* Fig. 2)	309.2 ± 84.7	319.2 ± 80.5	282.5 ± 45.3
Duration of the catch-up response [ms] (*from "b" to "d" in* Fig. 2)	230.8 ± 74.7	221.9 ± 79.2	216.9 ± 50.8

Object kinematics and grip response events are summarized for all 18 subjects, pooled across both digits, and pooled by object-moment magnitude (OMM), unless otherwise specified. Data are reported as median ±MAD in the order of key events after the onset of perturbation and are separated according to axis and direction of rotation. (Reprinted with permission from [12])

and block conditions. Object-moment magnitude had no effect on the onset latency of the catch-up response (time to FDI activation) for any axis or direction of rotation (smallest p-value was $p = 0.11$ for positive rotations about the d-p axis). Axis of rotation also had no effect on the onset latency of the catch-up response (smallest p-value of 0.38 for the −50.5 mN-m object-moment magnitude) whose median values ranged from 71 to 89 ms (Table 1).

3.3 Fingertip Forces and Force Rates

Shortly after the onset of perturbation, thumb and index finger forces normal to the grip plate increased rapidly to maintain grasp of the object and eventually decreased to steady-state values (Fig. 2). A distinct, pulse-like catch-up response

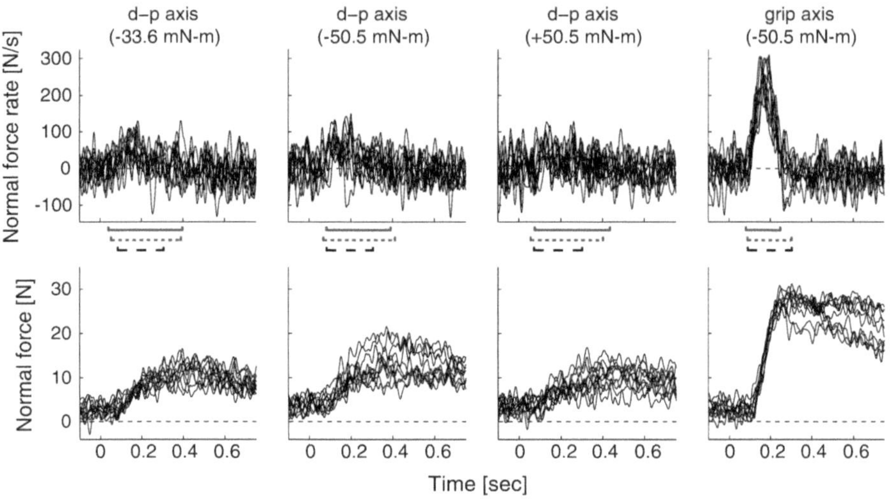

Fig. 3 Normal force rate and normal force as a function of time for different experimental conditions. Unimodal catch-up responses were observed in the normal force rates for all object-moment magnitudes, rotation directions, and rotation axes. Individual trial data (n = 5 trials) for one subject are shown here for both digits and five blocks. Catch-up responses were strongest for rotations about the grip axis. Positive normal forces correspond to compressive forces along the grip axis. *Solid red* and *dotted blue lines* beneath the normal force rates plots indicate the duration of the catch-up response for the thumb and index finger, respectively. Median values from Table 1 are indicated by *dashed black lines*. (Adapted from [12])

was observed in the normal force rate data for each digit (Fig. 2). In fact, catch-up responses were observed in the normal fingertip force data for all subjects, both digits, and all six block conditions. A comparison of catch-up responses across different object-moment magnitudes and axes of rotation demonstrates the consistency of the unimodal profile in the normal force rate (Fig. 3). Peak normal forces and peak normal force rates were, on average, greatest for rotational perturbations about the grip axis for all subjects, both digits, and both object-moment magnitudes (Fig. 4).

Peak normal force was affected by axis of rotation, but not by object-moment magnitude or direction of perturbation. Negative rotations about the grip axis resulted in the largest peak normal forces (19.2 ± 8.0 N, Table 1, Fig. 3). Time to peak normal force was not affected by object-moment magnitude, perturbation axis, or rotation direction. The smallest p-value of 0.053 resulted from a test on effects of perturbation axis (d-p or grip). Median values for time to peak normal force ranged from 283 to 319 ms (Table 1).

Radial, tangential force components revealed similar trends across all subjects regardless of object-moment magnitude or direction of rotation about the d-p axis. In particular, changes in radial force components from pre-perturbation baseline values were less variable immediately after the onset of perturbation and until the onset of the catch-up response (Fig. 5).

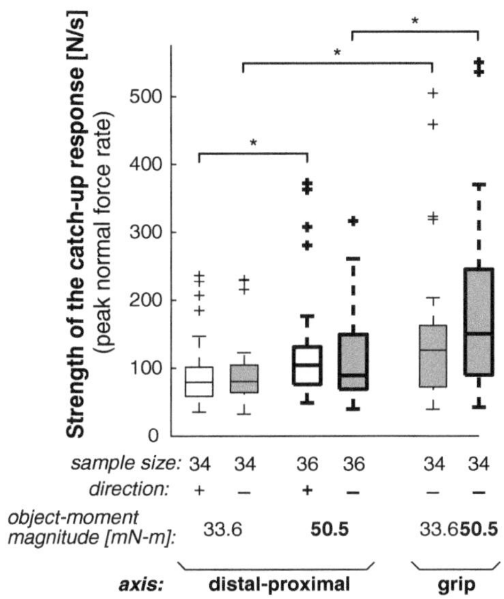

Fig. 4 Strength of the catch-up response (peak normal force rate). Each column represents strength of the catch-up response pooled across the thumb and index finger for 18 subjects. Thicker box plots are used for large object-moment magnitudes. Shaded boxplots are used for negative rotations. Catch-up responses were strongest for rotations about the grip axis. Asterisks indicate statistically significantly differences across groups. Each box plot indicates the 25th, 50th, and 75th percentiles. The whiskers indicate the 10 and 90th percentiles. Outliers (indicated by "+") had values that were more than 1.5 times the interquartile range from the top or the bottom of the box. (Reprinted with permission from [12])

3.3.1 Strength and Duration of the Catch-Up Response

For the most part, rotation direction and object-moment magnitude had no effect on the strength of the catch-up response (peak normal force rate), except for positive rotations about the d-p axis (p-value = 0.02), in which case the median value was 31 % larger for large object-moment magnitudes than for small object-moment magnitudes (Fig. 4, Table 1). Axis of rotation had the greatest effect on strength of the catch-up response, with the median value for the grip axis exceeding that of the d-p axis by 61.5 % for negative rotations (Fig. 4, Table 1).

The time to peak normal force rate was not affected by any experimental factor (object-moment magnitude, axis or direction of rotation) for either digit. The smallest p-value was 0.14 for positive rotations about the d-p axis. Median values for the time to peak normal force rate ranged from 127 to 152 ms (Table 1). Notably, the duration of the catch-up response was independent of all experimental factors as well as digit (smallest p-value was 0.13 for negative rotations about the d-p axis). Median values for the duration of the catch-up response ranged from 217 to 231 ms (Table 1).

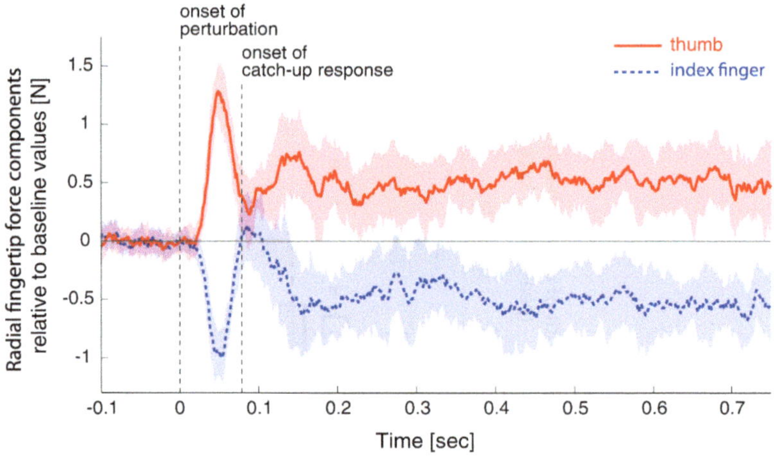

Fig. 5 Variability in radial fingertip force components. For clarity, radial force components relative to baseline values (prior to perturbation) are shown in order to account for trial-to-trial variations in baseline fingertip forces. Trends are shown for negative rotations about the distal-proximal axis, for all 18 subjects, for both object-moment magnitudes (n = 35 trials total), and both the thumb (*solid red line*) and index finger (*dotted blue line*). Thick lines represent median values and shaded regions represent the ±MAD ranges. Radial force components exhibited the least variability from the onset of perturbation (dashed vertical line at t = 0 s) to first dorsal interosseous activation at 83.9 ± 12.8 ms (dashed vertical line marking the onset of the catch-up response). (Reprinted with permission from [12])

3.3.2 Angle of the Fingertip Force Vector Relative to the Grip Surface Normal

The angle θ of each digit's fingertip force vector with respect to the grip surface normal (Fig. 1b) was used to visualize fingertip force responses to rotations about the d-p axis for large object-moment magnitude torque loads only. Due to unknown center of pressure locations,[1] each digit's fingertip force vector was drawn as if applied to the center of the grip surface for visualization purposes (Fig. 6).

Two types of force responses were observed based on the symmetry (or lack thereof) of the fingertip force angles across the thumb and index finger. The first type of response was characterized by an asymmetry in fingertip force angle about the vertical centerline of the object. Figure 6a shows one case in which the steady-state angle for the thumb was greater than that for the index finger for negative rotations about the d-p axis. This asymmetric relationship was reversed for positive rotations about the d-p axis. Typically, one digit's angle peaked while the other digit's angle dropped to near zero values at approximately 50 ms (Figs. 2,

[1] Center of pressure could not be determined due to the limited resolution of the force/torque transducers and low normal forces employed by subjects. A calibration experiment concluded that a minimum force of 20 N normal to the grip plate was necessary to calculate a digit's center of pressure to within 3 mm in the plane of the grip plate.

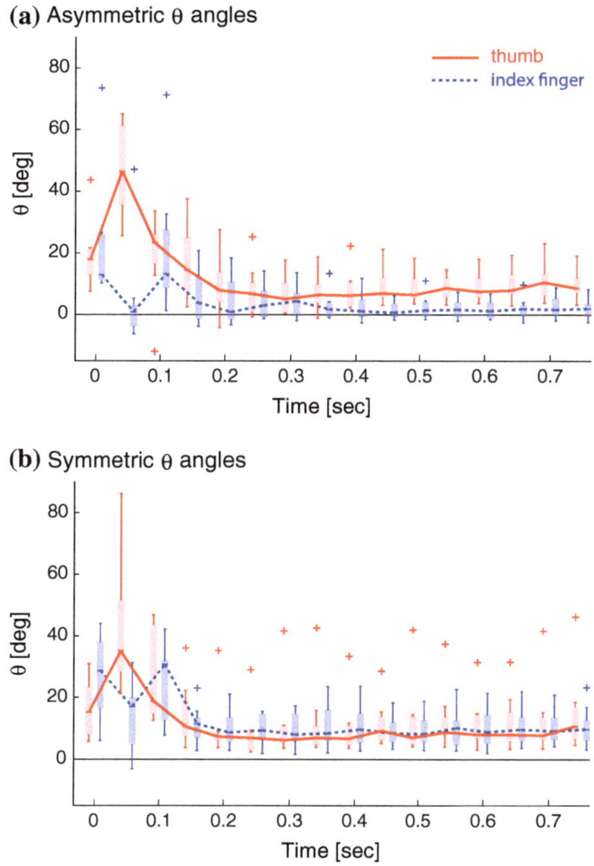

Fig. 6 Angle of fingertip force vector relative to the grip surface normal. The angle θ of each fingertip force vector was measured relative to the grip surface normal in the plane containing the grip axis. Data are shown for large object-moment magnitude, negative rotations about the distal-proximal axis for two groups of subjects who were separated based on the symmetry, or lack thereof, of forces applied by the thumb (*solid red lines* and *boxes*) and index finger (*dotted blue lines* and *boxes*) relative to the vertical centerline of the object. Boxplots are shown in 50 ms increments from perturbation onset ($t = 0$ s) until 750 ms. **a** Nine subjects used asymmetric force vectors about the centerline of the object (unequal steady-state θ angles) to correct the object rotation. Steady-state thumb and index fingertip force angles became statistically significantly different around 350 ms. **b** Nine subjects used symmetric force vectors (equal steady-state θ angles). Thumb and index fingertip force angles became statistically indistinguishable from one another around 100 ms

6a), regardless of direction of rotation. Steady-state thumb and index fingertip force angles became statistically significantly different around 350 ms ($p = 0.047$) and 400 ms ($p = 0.0496$) for negative and positive rotations, respectively, and remained disparate for the rest of the grip response with one angle near zero and one angle greater than zero (negative rotation case shown in Fig. 6a).

The second type of response was characterized by a symmetry in fingertip force angle about the vertical centerline of the object. Fingertip force angles for the thumb and index finger became statistically indistinguishable from one another around 100 ms and remained so for the rest of the grip response (negative rotation case shown in Fig. 6b).

Out of 18 subjects, twelve used asymmetric fingertip force angles across the two digits for positive rotations about the d-p axis and nine used asymmetric fingertip force angles for negative rotations about the d-p axis. Only eight subjects used asymmetric fingertip force angles for both positive and negative rotations about the d-p axis.

4 Discussion

4.1 Existence of the Catch-Up Response for Rotational Perturbations

Prior studies have described catch-up responses as a means for maintaining grasp stability when unexpected translational loads are imposed on a grasped object and linear slip is imminent [23, 24, 27, 29, 31]. We have shown that a unimodal catch-up response is also elicited for unexpected rotational perturbations of grasped objects that induce conditions other than pure linear slip or rotational slip at the fingerpad. The robust response was observed in all subjects, both digits, and irrespective of all three experimental factors: object-moment magnitude, axis of rotation, and direction of rotation. A characteristic unimodal catch-up response was always present in the normal force rate profile (Fig. 3).

Due to the method of perturbation, the external perturbations were not pure rotations but rather combinations of rotation and translation. Each perturbation was imposed by applying a force whose point of application was offset from the test object's center of mass (Fig. 1b). Nonetheless, a kinematic analysis confirmed that rotational effects dominated the (0, 100 ms) period immediately following the perturbation. Experiments are currently underway that use haptic devices to impart purely linear or rotational perturbations, randomize experimental conditions, and include rotations about a radial-ulnar axis.

4.2 Robust Timing of the Catch-Up Response

The temporal characteristics of the catch-up response were robust regardless of experimental conditions. Onset latencies of the catch-up response (71–89 ms, Table 1) were consistent with prior studies on linear slip that reported onset latencies of 80 $\pm$ 9 ms (mean $\pm$ std) [23], and 62 $\pm$ 9 ms and 74 $\pm$ 9 ms for catch-up responses to distal and proximal loads, respectively [31]. Our observations also fell within the 50-100 ms range reported for grip force responses to natural linear slips [35] and were not far from the 50–70 ms range reported for the latency of a "phasic burst of muscle activity" after the onset of unexpected force perturbations [25]. Peak normal forces ranged from 13 to 19 N (Table 1) and, yet, the time to achieve the peak normal force fell within a tight range (283–319 ms, Table 1). It should be noted that onset latency was based on FDI activation in the present study, but was based on changes in normal force rates in other studies [23, 31]. While exact onset threshold criteria for normal force rates were not provided in the literature, the time to FDI activation was essentially coincident with the initial increase in normal force rate (Fig. 2 in this work; Fig. 3 in [23]. Johansson et al. previously reported the duration of the catch-up response to be between 200 and 250 ms for linear slip conditions [23]. The duration of the catch-up responses observed in this study ranged from 217 ms (negative rotations about the grip axis) to 231 ms (positive rotations about the d-p axis; Table 1).

4.3 Axis of Rotation Affects Catch-Up Response Strength

Of the three experimental factors, axis of rotation relative to the hand had the greatest effect on the catch-up response. The robust temporal characteristics of the catch-up response suggests that the effects of axis of rotation manifested as changes in strength of the catch-up response (peak normal force rates) and are likely scaled according to load conditions [26]. The strength of the catch-up response for rotations about the grip axis were much larger than those for rotations about the d-p axis (Fig. 4, Table 1).

Axis of rotation critically affects the loading and slip conditions as well as the contributions of the digits to the passive resistance of the perturbations. With no part of the hand to physically oppose rotations about the grip axis, the rotational slip conditions resulted in the largest peak and time to peak angular deviations. Rotations about the grip axis are particularly difficult to counter because of the incompatibility of fingertip force/torque capabilities with the axis of rotation, which passes through both fingertips by definition. Normal grip forces are aligned with the axis of rotation and fingertips cannot actively produce tangential torques in the plane of the fingerpad. The strength of the catch-up response may have been scaled up to counter rotational slip, in particular (Figs. 3 and 4, Table 1).

Our finding that catch-up responses were strongest for rotations about the grip axis (Figs. 3 and 4, Table 1) is consistent with prior reports that grip forces are more closely regulated for rotational slip than linear slip [18, 19], and that grip responses to unanticipated loads vary with load (and slip) direction, with conservative grip responses being associated with "dangerous [loading] directions" [31]. In prior studies, stronger grip forces [30, 31] and shorter grip response latencies [29] were elicited by force loads away from the palm.

The idea that motion away from the hand constitutes a "dangerous direction" [31] applies to translational perturbations, but is not applicable to rotational perturbations that cause simultaneous motion of an object towards and away from the hand. While we did not observe axis-dependent changes in grip response latency [31], as measured by the time to initial FDI activation, we did observe axis-dependent effects on peak normal force and strength of the catch-up response. Based on these observations, we propose that the grip axis is a "dangerous axis" relative to the d-p axis, where critical differences exist in the conditions for slip and passive resistance of the digits. Rotations about the d-p axis be easier to counter because the object is rotated towards a fingertip that can immediately resist object motion via the passive mechanics of the hand.

4.4 Passive and Active Phases of the Grip Response

As with any unexpected loading of a hand-object system, grip responses will be comprised of a combination of passive and active components. While the non-invasive techniques used in this study preclude discussions regarding the neural pathways involved in the grip responses, some conclusions can be drawn about periods of the response which clearly precede the onset of FDI activity, as measured via surface EMG. The passive dynamics of the hand-object system are likely reflected in the reaction forces at the digit-object interface. The smooth, uniform changes in radial, tangential forces suggests that passive resistance contributes to at least the first 50 ms of the measured fingertip forces (Fig. 5). This was true for all subjects, both digits, and regardless of object-moment magnitude or direction of rotation about the d-p axis. After the FDI became active, the changes in radial forces exhibited increased variability about the median and were likely dominated by active muscle responses (Fig. 5).

Responses to perturbations may also have been influenced by joint stiffness and digital yield as affected by hand posture [26, 29], and orientation of the perturbation with respect to joints. Subjects were instructed to curl the middle, ring, and little fingers, which likely pre-tensioned the flexor digitorum profundus to resist flexion of the index finger. Rotations about the d-p axis might be resisted by the stiffnesses of entire digits while rotations about the grip axis might be resisted primarily by fingerpads. Passive resistance to perturbations could have arisen from natural limits on joint motion and skin stretch, viscoelastic musculoskeletal structures [43] and fingerpads [44], as well as passive joint torques [45, 46]. The temporal overlap and relative contributions of the passive and active components of the grip response remain unclear and require further investigation.

4.5 Independent Control of Fingertip Force Vectors

Edin et al. showed that digits could be independently controlled for a task in which subjects used a precision grip with symmetric contact points to lift an object having different frictional conditions at the parallel, vertical grip surfaces [37]. Specifically, the ratios of normal force to tangential force were found to be independent across the thumb and index finger [37]. Fu et al. showed that digit tangential forces could be independently controlled when subjects were tasked with simultaneously lifting and minimizing roll of an object using a precision grip with directly opposed, symmetric contact points [47].

The present study supports the idea of independent digit control and, importantly, suggests that the ability to independently control fingertip forces in an asymmetric manner can be used to compensate for torque loads that necessitate asymmetric digit responses during precision grip. Fingertip force vectors were projected onto the plane containing the grip axis in order to illustrate the asymmetry in the force response across digits for negative rotations about the d-p axis (Fig. 6). From an analysis of the angle θ of each fingertip force vector with respect to the grip surface normal, it appears that each digit's force vectors can be controlled independently and asymmetrically.

Two distinct types of grip responses were observed across the 18 subjects that could be differentiated by whether steady-state values for the angle θ were equal ("symmetric case") or unequal ("asymmetric case"). For the asymmetric case, one digit's force vector eventually pointed the top of the object while the other digit's force vector remained essentially normal to the grip surface (Fig. 6a). It would appear as if one digit (index finger for negative rotations about the d-p axis) played a supportive role as a pivot point while the other digit (thumb for negative rotations about the d-p axis) played a more active role to counter the external torque.

Physics dictates that fingertip forces must create a corrective moment about the object's center of mass to counteract an external torque load. The corrective moment can be created by modifying one's 3D fingertip force vector and/or digit placement. While subjects were instructed to grasp the object with both digits centered on each grip plate, it is possible that some subjects grasped the object with their digits in direct opposition to one another while others did not. Our observations suggest that selective digit placement and modulation of fingertip force direction are two strategies that could be used to counter an external torque load. However, conclusive remarks about digit roles (if any) cannot be made without further analyses of digit center of pressure.

4.6 FDI Activation Timing Suggests Automatic Grip Responses Distinct from Stretch Reflexes

Whether anticipatory or triggered by a perturbation stimulus, expressions of automatic or programmed motor responses have been found in the coordinated changes in fingertip forces during precision grip tasks [11, 16, 23, 24]. It is not entirely clear which neural pathways are involved in the automatic grip responses, but reflex loops have been considered. The identification, organization, labeling, and interpretation of reflex responses remain controversial in the neurophysiology community. While the application of surface EMG to a single muscle in this study is insufficient for resolving the long-standing controversies, we can still discuss our findings in the context of what is currently known about neural pathways associated with stretch reflexes and automatic grip responses.

Surface EMG was used to investigate the timing of first dorsal interosseous activation relative to the onset of perturbation of the grasped object (Fig. 2). The delays between the onset of perturbation and FDI activation overlap data previously reported by others for precision grip responses to unexpected loads [25, 35]. While it is physically plausible that the motions imposed on the thumb and index finger by the external torque loads could have triggered stretch reflexes, most FDI activation latencies observed in this study (71–89 ms, Table 1) exceeded those reported by others for tendon jerk and stretch reflexes in muscles of the hand. Forcible stretching of the flexor pollicis longus in the thumb has been observed to elicit 25–30 ms short-latency and 40 ms long-latency muscle responses [48]. Other studies have reported 25 ms tendon jerk latencies in the flexor pollicis longus and 40 ms stretch reflex responses in the flexor pollicis longus and flexor pollicis brevis in response to jerk-type perturbations of the thumb [49, 50]. FDI stretch reflex latencies have been described as having a short latency response of 32 ms and a long latency response of 55 ms [51]. If stretch reflexes were triggered, forces generated from long-latency stretch reflexes would likely be dwarfed by those resulting from supraspinally-mediated, automatic responses and/or voluntary responses [52, 53]. In a study that applied unexpected force loads directly to an object in precision grip or to the hand itself, it was hypothesized that cutaneous mechanoreceptors in the digital pulps play a larger role in automatic grip force adjustments than stretch reflexes [25].

5 Conclusion and Future Work

Previous studies [23, 24, 26] had shown that the catch-up response existed for unexpected translational perturbations of an object in a precision grip. We imposed unexpected rotational perturbations and found that catch-up responses can also be elicited by conditions other than pure linear slip or rotational slip at the fingerpad. The early period of the grip response appeared to be dominated by passive mechanics while the later period was dominated by a characteristic active catch-up response and independent (and often asymmetric) control of fingertip forces at steady-state. The uniform timing of the catch-up response was consistent with prior studies on grip responses to translational perturbations [25, 35]. The fact that qualititatively different load and slip conditions elicited similar catch-up responses suggests that a pre-programmed increase in normal (grip) force is a hallmark of human grip responses to unexpected perturbations.

Experiments are currently underway that use haptic devices to impart purely rotational perturbations and investigate perturbations about the radial-ulnar axis. While the effects of trial-to-trial learning were not studied here explicitly, it is possible that some learning took place due to the repetitive nature of the task. The use of haptic devices helps to minimize effects of learning and anticipation, by allowing the randomization of perturbation type (linear or rotational), axis, and direction. Digit placement and center of pressure analyses will also be possible with the use of force/torque transducers that are highly sensitive at low fingertip force magnitudes. Surface EMG of additional muscles besides FDI will enable investigations of coordinated muscle activity, activation latencies, and possible co-contraction or pre-stiffening strategies.

The results of this study could be used to inform the design of fingertip force control strategies for artificial hands having sensors that can detect the nature of perturbations imposed on the grasped object. The finding that grip responses to unexpected rotational perturbations can be axis-dependent has direct implications on the kinematic and kinetic control of high degree-of-freedom, anthropomorphic artificial hands. We previously showed that rotational perturbations elicited simultaneous ad/abduction and flexion/extension of the thumb carpometacarpal joint [54]. While the "catch-up response" appears to reflect a "grip harder" strategy, the act of gripping "harder" is not necessarily the result of pure flexion for human or artificial hands. In order to enact active grip responses beyond the passive responses of an artificial hand, one will need to consider the 3D motion and force production capabilities of each fingertip with respect to the load and slip conditions on the object to be stabilized. Stimulus-specific, automated grip strategies could help to reduce the cognitive burden on the user of a robotic or prosthetic hand.

Acknowledgments The authors gratefully acknowledge Kevin Bair, Nicholas Fette, and Ryan Manis for assistance with data collection and processing, Dr. Kevin Keenan for guidance in the EMG analysis, and Dr. Marco Santello, Dr. Stephen Helms Tillery, Dr. Marco Davare, and Qiushi Fu for technical discussions.

Funding This material is based upon work supported by the National Science Foundation under Grant No. 0954254. Any opinions, findings, and conclusions or recommendations expressed in this material are those of the author(s) and do not necessarily reflect the views of the National Science Foundation.

References

1. G. Bekey, R. Tomovic, Robot control by reflex actions, in *Proceedings of IEEE International Conference on Robotics and Automation*, vol. 3 (1986), pp. 240–247
2. M. Santello, M. Flanders, J.F. Soechting, Postural hand synergies for tool use. J. Neurosci. **18**(23), 10105–10115 (1998)
3. P.K. Allen, M.T. Ciocarlie, C. Goldfeder, H. Dang, Low-dimensional data-driven grasping, in *Proceedings of the Robotics Science and Systems Conference*, (Seattle, WA, 2009)
4. M.H. Lee, H.R. Nicholls, Tactile sensing for mechatronics—a state of the art survey. Mechatronics **9**(1), 1–31 (1999)
5. B.D. Argall, A.G. Billard, A survey of tactile human–robot interactions. Robotics Auton. Syst. **58**(10), 1159–1176 (2010)
6. H. Yousef, M. Boukallel, and K. Althoefer, Tactile sensing for dexterous in-hand manipulation in robotics, a review. Sens. Actuators A: Phys **167**(2), 171–187 2011
7. J. R. Napier, The prehensile movements of the human hand. J Bone Joint Surg Br **38-B**(4), 902–913 (1956)
8. D. Prattichizzo, J.C. Trinkle, in "*Grasping*," *in Springer Handbook of Robotics*, ed. by B. Siciliano, O. Khatib (Springer, Heidelberg, 2008), pp. 671–700
9. R.S. Johansson, J.R. Flanagan, in "*Tactile sensory control of object manipulation in humans*," in *Handbook of the Senses,* vol. 6, eds. by J.H. Kaas, E. Gardner. Somatosensation, (Academic Press, San Diego, 2008), pp. 67–86
10. R.S. Johansson, J.R. Flanagan, Coding and Use of Tactile Signals from the Fingertips in Object Manipulation Tasks. Nature Rev. Neurosci. **10**, 345–359 (2009)
11. R.S. Johansson, Sensory control of dextrous manipulation in humans, in *in Hand and brain: the neurophysiology and psychology of hand movements*, ed. by A.M. Wing, P. Haggard, J.R. Flanagan (Academic, San Diego, 1996), pp. 381–414
12. M. De Gregorio, V.J. Santos, Precision grip responses to unexpected rotational perturbations scale with axis of rotation. J. Biomech. **46**(6), 1098–1103 (2013)
13. G. Westling, R.S. Johansson, Factors influencing the force control during precision grip. Exp. Brain Res. **53**(2), 277–284 (1984)
14. H. Kinoshita, L. Bäckström, J.R. Flanagan, R.S. Johansson, Tangential torque effects on the control of grip forces when holding objects with a precision grip. J. Neurophysiol. **78**(3), 1619–1630 (1997)
15. R.S. Johansson, G. Westling, Roles of glabrous skin receptors and sensorimotor memory in automatic control of precision grip when lifting rougher or more slippery objects. Exp. Brain Res. **56**(3), 550–564 (1984)
16. R.S. Johansson, G. Westling, Programmed and triggered actions to rapid load changes during precision grip. Exp. Brain Res. **71**(1), 72–86 (1988)
17. P. Jenmalm, R.S. Johansson, Visual and somatosensory information about object shape control manipulative fingertip forces. J. Neurosci. **17**(11), 4486–4499 (1997)
18. A.W. Goodwin, P. Jenmalm, R.S. Johansson, Control of grip force when tilting objects: effect of curvature of grasped surfaces and applied tangential torque. J. Neurosci. **18**(24), 10724–10734 (1998)
19. P. Jenmalm, S. Dahlstedt, R.S. Johansson, Visual and tactile information about object-curvature control fingertip forces and grasp kinematics in human dexterous manipulation. J. Neurophysiol. **84**(6), 2984–2997 (2000)
20. P. Jenmalm, A.W. Goodwin, R.S. Johansson, Control of grasp stability when humans lift objects with different surface curvatures. J. Neurophysiol. **79**(4), 1643–1652 (1998)
21. J.R. Flanagan, M.C. Bowman, R.S. Johansson, Control strategies in object manipulation tasks. Curr. Opin. Neurobiol. **16**(6), 650–659 (2006)
22. A.M. Wing, J.R. Flanagan, Anticipating dynamic loads in handling objects, in *Proceedings of the ASME Dynamic Systems and Control Division*, vol. 64 (1998), pp. 139–143

23. R.S. Johansson, R. Riso, C. Häger, L. Bäckström, Somatosensory control of precision grip during unpredictable pulling loads. I. changes in load force amplitude. Exp. Brain Res. **89**(1), 181–191 (1992)
24. R.S. Johansson, C. Häger, R. Riso, Somatosensory control of precision grip during unpredictable pulling loads. II. Changes in load force rate. Exp. Brain Res. **89**(1), 192–203 (1992)
25. K.J. Cole, J.H. Abbs, Grip force adjustments evoked by load force perturbations of a grasped object. J. Neurophysiol. **60**(4), 1513–1522 (1988)
26. R.S. Johansson, C. Häger, L. Bäckström, Somatosensory control of precision grip during unpredictable pulling loads. III. Impairments during digital anesthesia. Exp. Brain Res. **89**(1), 204 (1992)
27. K.J. Cole, R.S. Johansson, Friction at the digit-object interface scales the sensorimotor transformation for grip responses to pulling loads. Exp. Brain Res. **95**(3), 523–532 (1993)
28. R.S. Johansson, K.J. Cole, Grasp stability during manipulative actions. Can. J. Physiol. Pharmacol. **72**(5), 511–524 (1994)
29. C. Häger-Ross, R.S. Johansson, Nondigital afferent input in reactive control of fingertip forces during precision grip. Exp. Brain Res. **110**(1), 131–141 (1996)
30. L.A. Jones, I.W. Hunter, Changes in pinch force with bidirectional load forces. J. Mot. Behav. **24**(2), 157–164 (1992)
31. C. Häger-Ross, K.J. Cole, R.S. Johansson, Grip-force responses to unanticipated object loading: load direction reveals body-and gravity-referenced intrinsic task variables. Exp. Brain Res. **110**(1), 142–150 (1996)
32. A.M. Gordon, H. Forssberg, R.S. Johansson, G. Westling, Integration of sensory information during the programming of precision grip: comments on the contributions of size cues. Exp Brain Res. **85**(1), 226–229 (1991)
33. A.M. Gordon, G. Westling, K.J. Cole, R.S. Johansson, Memory representations underlying motor commands used during manipulation of common and novel objects. J. Neurophysiol. **69**(6), 1789–1796 (1993)
34. V.G. Macefield, R.S. Johansson, Control of grip force during restraint of an object held between finger and thumb: responses of muscle and joint afferents from the digits. Exp. Brain Res. **108**(1), 172–184 (1996)
35. R.S. Johansson, G. Westling, Signals in tactile afferents from the fingers eliciting adaptive motor responses during precision grip. Exp. Brain Res. **66**, 141–154 (1987)
36. V.G. Macefield, C. Häger-Ross, R.S. Johansson, Control of grip force during restraint of an object held between finger and thumb: responses of cutaneous afferents from the digits. Exp. Brain Res. **108**(1), 155–171 (1996)
37. B.B. Edin, G. Westling, R.S. Johansson, Independent control of human finger-tip forces at individual digits during precision lifting. J. Physiol. (Lond.) **450**, 547–564 (1992)
38. G. Westling, R.S. Johansson, Responses in glabrous skin mechanoreceptors during precision grip in humans. Exp. Brain Res. **66**(1), 128–140 (1987)
39. R.S. Johansson, J.L. Backlin, M.K. Burstedt, Control of grasp stability during pronation and supination movements. Exp. Brain Res. **128**(1), 20–30 (1999)
40. K. Jordan, K. Newell, Task goal and grip force dynamics. Exp. Brain Res. **156**(4), 451–457 (2004)
41. P.W. Hodges, B.H. Bui, A comparison of computer-based methods for the determination of onset of muscle contraction using electromyography. Electroen Clin Neuro **101**, 511–519 (1996)
42. R.P. Di Fabio, Reliability of computerized surface electromyography for determining the onset of muscle activity. Phys. Ther. **67**(1), 43–48 (1987)
43. F.E. Zajac, Muscle and tendon: Properties, models. scaling, and application to biomechanics and motor control. Crit. Rev. Biomed. Eng. **17**(4), 359–411 (1989)
44. D.T. Pawluk, R.D. Howe, Dynamic lumped element response of the human fingerpad. J. Biomech. Eng. **121**(2), 178–183 (1999)

45. S.W. Lee, H. Chen, D.G. Kamper, Transmission of musculotendon forces to the index finger, in *Proceeding of the Robotics Science and Systems Conference*, Seattle, WA, 2009
46. P.H. Kuo, A.D. Deshpande, Contribution of passive properties of muscle-tendon units to the metacarpophalangeal joint torque of the index finger, in *Proceedingsof the IEEE International Conference on Biomedical Robotics and Biomechatronics*, 2010, pp. 288–294
47. Q. Fu, W. Zhang, M. Santello, Anticipatory planning and control of grasp positions and forces for dexterous two-digit manipulation. J. Neurosci. **30**(27), 9117–9126 (2010)
48. P.B. Matthews, Evidence from the use of vibration that the human long-latency stretch reflex depends upon spindle secondary afferents. J. Physiol. **348**(1), 383 (1984)
49. C.D. Marsden, P.A. Merton, H.B. Morton, Stretch reflex and servo action in a variety of human muscles. J. Physiol. **259**(2), 531–560 (1976)
50. P.B. Matthews, The contrasting stretch reflex responses of the long and short flexor muscles of the human thumb. J. Physiol. **348**(1), 545–558 (1984)
51. A.F. Thilmann, M. Schwarz, R. Töpper, S.J. Fellows, J. Noth, Different mechanisms underlie the long-latency stretch reflex response of active human muscle at different joints. J. Physiol. **444**(1), 631 (1991)
52. T.I.H. Brown, P.M.H. Rack, H.F. Ross, A range of different stretch reflex responses in the human thumb. J. Physiol. **332**(1), 101 (1982)
53. E. Bizzi, P. Dev, P. Morasso, A. Polit, Effect of load disturbances during centrally initiated movements. J. Neurophysiol. **41**(3), 542 (1978)
54. M. De Gregorio, V.J. Santos, Rotational object perturbations result in characteristic types of kinematic grip responses. *Proc Ann Mtg Amer Soc Biomech*, Providence, RI, 2010

Chapter 9
Hand Haptic Perception

Mahdi Tavakoli

Abstract Haptic perception encompasses tactile feedback and kinesthetic feedback. The haptic experience of touching an object by hand conveys information to the human about the object's material properties such as stiffness, texture, and weight and its shape properties such as size, orientation and curvature. In this chapter, we review how these properties are perceived through the sense of touch.

Keywords Tactile feedback · Kinesthetic feedback · Time-delayed haptic feedback · Object material properties · Object shape properties · Exploratory hand movements

1 Introduction

The sense of touch is the first sensory modality that develops and becomes functional in humans [1]. Touch feedback, which is also called haptic feedback, encompasses tactile (cutaneous) feedback relying on skin stimulation and kinesthetic (force) feedback involving muscle stimulation. For instance, in any haptics-based shape recognition task, active touch and contour following stimulate the kinaesthetic sense while passive pressure sensing is a form of tactile sense. Together, tactile feedback and kinaesthetic feedback influence the human's ability to distinguish objects.

The haptic experience of a human subject when touching an object includes sensations such as stiffness, texture, and weight. These sensations define the material properties of the object and our hands are adapted to best perceive them through touch and manipulation, surpassing vision in terms of discrimination

M. Tavakoli (✉)
Department of Electrical and Computer Engineering, University of Alberta,
Edmonton, AB, Canada
e-mail: mahdi.tavakoli@ualberta.ca

R. Balasubramanian and V. J. Santos (eds.), *The Human Hand as an Inspiration for Robot Hand Development*, Springer Tracts in Advanced Robotics 95,
DOI: 10.1007/978-3-319-03017-3_9,

accuracy. On the other hand, size, orientation and curvature define the shape (geometric) properties of an object and can be perceived by both touch and vision. Sections 2 and 3 discuss the haptic recognition of material properties and shape properties of an object, respectively.

Haptic exploration of an object is a task-dependent activity meaning that it necessitates highly-specialized exploratory hand movements for detecting various object properties. For instance, rubbing, pressing and lifting an object can provide information regarding object texture, stiffness and weight, respectively. In Sect. 4, we will contrast and compare such active haptics-based movements against passive stimulation of fingers in terms of detecting object material and shape properties.

2 Haptic Recognition of Object Material Properties

2.1 *Stiffness*

Different objects in our environment have different stiffnesses. The perception of object stiffness happens through evaluating the object deformation in response to a force applied on it by the hand (finger) or another object. While by touching a compliant object we can receive both the force and the deformation information required for stiffness discrimination, our vision can provide additional information concerning the deformation of the object. However, the utility of vision will be limited if critical movements of the task are orthogonal to the view or are occluded by the hand that is trying to sense the contact force. Moreover, vision cannot supply any information about the hand-object contact force. Thus, haptic interaction with an object is crucial to estimating its stiffness.

Srinivasan and LaMotte compared the ability of the human hand to discriminate the softness of objects when human subjects were given tactile information, kinesthetic information, or both [2]. The purpose was to isolate the components of haptic information that enables the human to make this discrimination. To do so, three experiments were performed: (a) active touch with the normal finger, (b) active touch under local cutaneous anesthesia of the finger, and (c) passive touch where the objects were brought in contact with the passive fingerpad of the subject using a mechanical device. Thus, in these experiments, the subjects received both tactile and kinesthetic feedback, only kinesthetic feedback, and only tactile feedback, respectively. Two types of specimens were used: (1) rubber-type objects with variable stiffnesses, and (2) rigid objects supported by springs of various stiffnesses (called spring cells). These specimens were chosen to represent compliant objects with deformable and with rigid surfaces, respectively. In the experiments, it was seen that the subjects demonstrated very good softness discriminability for the rubber objects in the active touch experiments, while they showed poorer discriminability for the compliance of spring cells. Another result was that to discriminate pairs of rubber objects, tactile information alone was sufficient but kinesthetic information alone was not. Nonetheless, for discriminating pairs of spring cells, tactile information alone was not sufficient and both tactile and kinesthetic information were necessary. Such a difference in the sufficiency of tactile information for successful stiffness discrimination can be traced to the mechanics of contact of the fingerpad, which affect tactile information, as explained in the following: As a subject indents an object, the compressive contact force applied by the fingerpad on the object increases, causing the fingerpad to deform. Only in the case of objects with deformable surfaces, the resulting deformation of the fingerpad depends on the object stiffness. This explains why tactile information alone is not sufficient for discriminating pairs of spring cells, which have rigid surfaces.

A human operator may try to perform stiffness discrimination using a teleoperated robot and through a haptic user interface that both controls the robot and displays the robot-object contact forces to the operator in the form of haptic (force) feedback. Since such a haptic telemanipulation system engages the operator's sense of touch, one would expect similar task performance as in direct touch. An interesting issue arises from the presence of a non-negligible time delay in the communication media between the user interface and the teleoperated robot, which happens in long-distance teleoperation. While the usefulness of haptic feedback in no-delay teleoperation has been established, e.g., in [3] and [4], the loss of temporal coincidence between the human operator's motions and the ensuing reflected forces in delayed teleoperation may confuse the operator so much so that the force feedback becomes useless or even misleading. To assess the value of providing haptic feedback to the user during delayed teleoperation, researchers have studied the effect of delay on the human's perception of the relative stiffness of virtual spring-like surfaces simulated by reflecting forces proportional to the user's virtual surface indentations. Subjected to a forced-choice paradigm (i.e., distinguish the stiffer of the two surfaces or identify them as having the same stiffness), users perceived the surfaces to be stiffer than actual under delayed force feedback and the stiffness overestimation increased for larger delays [5, 6]. The effect of crossing the boundary of a force field, where local stiffness is ill-defined, on the perception of delayed stiffness has also been studied [7]; note that stiffness is the local derivative of the force field. It has been found that subjects interacting with delayed force fields underestimate (overestimate) stiffness if they do not move (do move) across the boundary of the elastic field.

2.2 Texture

As a human draws a finger across the surface of an object, he/she receives information not only regarding the shape properties of the object, but also about the texture of the surface. Perception of texture is a multidimensional experience that encompasses roughness as its most prominent aspect. There are, however, other aspects of texture of a surface besides its roughness. Hollins et al. performed experiments to examine the dimensionality of surface texture perception [8]. In the experiments, 17 tactile stimuli were moved across the index fingers of the subjects in a direction perpendicular to its surface with constant speed. Out of the 17 stimuli, seven were thin or flexible materials such as wax paper, cardboard, smooth plastic or sand paper that were mounted on blocks of wood. Other stimuli were the surfaces of rigid objects such as a rubber eraser, Styrofoam, brick, or leather wallet. The stimuli presented to the subjects did not have any curvature and a random noise source masked out the faint sounds of contact between the finger and the stimuli. After each test, a subject was asked to sort the presented object into several categories on the basis of perceived similarity. Using multidimensional scaling methods and the data collected from the subjects, the study showed that subjects' judgments of surface texture can be represented in a perceptual space. Subjects' ratings of each stimulus fit into two (and possibly three) dimensions. The two robust and orthogonal dimensions were found to be roughness-smoothness and hardness-softness, and the third less robust dimension was judged to be compressional elasticity (springiness) of the surface—see Fig. 1. The warm-cold and sticky-slippery scales were not found to be independent of those for roughness and hardness. The study concluded that it is unlikely to have fewer than three dimensions in the perceptual space for texture.

Later, Picard et al. also investigated the perceptual dimensions of tactile stimuli [9]. Unlike Hollins et al. whose study involved passive stimulation of a static finger, Picard et al. allowed active exploration of objects by a dynamic finger. In the experiments, subjects were asked to use a lateral motion procedure to sort 24 car seat materials with different tactile properties on the basis of perceived similarity (free-sorting task). Their analysis showed that the tactile texture space did not exceed four dimensions: soft/harsh, thin/thick, relief, and hardness. While the first two dimensions may be qualified as separable, the last two dimensions were found to be related to the soft/harsh dimension. The hardness dimension was found to be close to the soft/harsh dimension as the two dimensions use similar exploratory hand movements to seek substance information about objects. This study did not identify roughness as a perceptual dimension although it is semantically close to harsh.

The roughness percept is generally believed to reflect the separation between raised elements that form the textured surface. For the underlying neural representations of roughness, readers are referred to [9]. Klatzky and Lederman studied the perception of roughness when a rigid structure was placed between the skin and the textured surface [10]. Subjects made roughness judgments through a stick-like probe held in the fingertips or a rigid fiberglass sheath mounted on the fingertip. Task performance under these rigid structures was compared to that with the bare finger (i.e., direct contact). A result of this study was that although discrimination was best with the finger, the rigid structures led to greater perceived roughness for the smoothest stimuli. The two experimental conditions in the above study (contact with a rigid structure versus direct touch) were different in the vibratory coding of roughness by the rigid structure. Vibration is highly important when a human operator uses a teleoperation system to explore a surface. For instance, the user needs to receive critical vibratory information associated with making contact with a rigid surface for teleoperation realism. The study by Klatzky and Lederman supports the use of vibro-tactile cues to display roughness when direct skin contact with an environment is not possible.

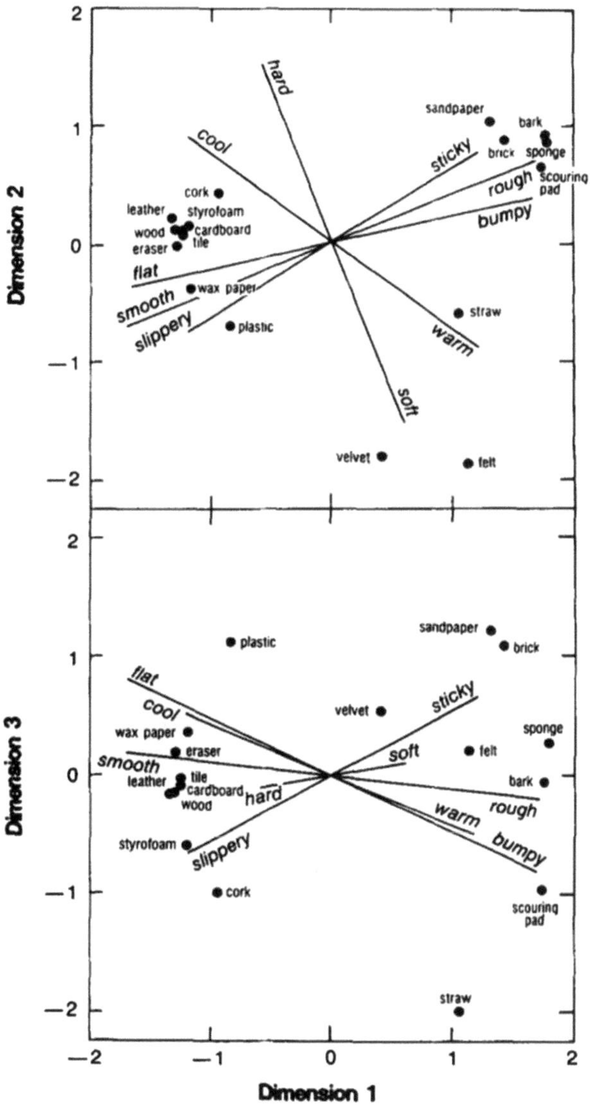

Fig. 1 Three-dimensional space viewed along the third dimension (*top*) and the second dimension (*bottom*) [8]

2.3 Weight

Ernst H. Weber (1795–1878) performed experiments that measured the sensitivity of the human hand in weight perception, which is limited in nature, and studied whether weight perception was more due to the tactile feedback resulting from holding an object or the kinesthetic feedback resulting from lifting an object. Subjects were made to lift different weights and asked if they could detect a difference between the two. Searching for the smallest perceivable difference—the "just-noticeable difference"—a 1 to 40 ratio between a standard weight and a different weight was found to be noticeable when subjects lifted weights (i.e., invoking both tactile and kinesthetic feedback). However, when the weights were rested on a subject's skin (i.e., providing tactile feedback only), the just-noticeable difference became a ratio of 1 to 30. This simply means that using both tactile and kinesthetic information results in more sensitivity to small weight differences.

Efficient manipulation of an object requires its accurate weight information, which as shown by Weber first requires manipulation (lifting) of the object. A practical solution to this paradox lies in using our past experience about object properties such as its size. Larger objects tend to weigh more (not considering environments such as under water or outer space). As such, research has shown that the size of an object influences the perceived weight of the object—this is referred to as size-weight illusion. As a result of the size-weight illusion, different sized objects of the same mass are perceived to have different masses. Ellis and Lederman conducted experiments to assess the extent to which haptic cues (physical weight) and visual cues (size/volume) about an object influence weight perception [11]. In a first set of experiments called haptics + vision experiments, subjects could see an object while picking it up to evaluate its heaviness. In a second set of experiments called haptics-only experiments, blindfolded subjects were asked to haptically estimate the weight of objects. Haptic cues alone were found necessary and sufficient to generate a size-weight illusion that has almost the same strength as that generated under both haptic and vision cues. In fact, as seen in Fig. 2, for objects of the same physical weight (904, 350, or 140 g), the perceived weight decreased as the volume increased under both haptics + vision and

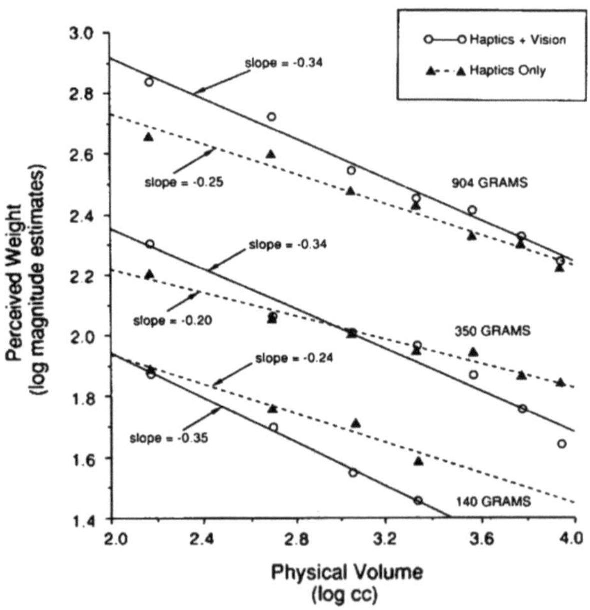

Fig. 2 Mean magnitude estimates of weight versus the physical volume for each modality [11]

haptics-only modalities. Such a strong size-weight illusion produced in the haptics-only condition showed that vision is not a necessary condition for this illusion to exist. The size-weight illusion is a primarily haptic phenomenon rather than a case of vision influencing haptic processing. The size-weight relationship can be used, for instance, in a robotic setting for lifting objects of unknown masses by finding an initial estimate of the object weight based on the size information obtained from the grasp action.

Later, Ellis and Lederman showed that weight perception is more broadly affected by a subject's expectation based on knowledge and past experience [12]. In experiments, golfers and non-golfers were presented with real and practice golf balls that had been tampered with to have different weights. It was found that golfers, who expect a weight difference between ball types, judged practice balls to weigh more than real golf balls of the same weight. On the other hand, non-golfers, who expect no weight difference between ball types, judged practice and real balls of equal weight to weigh the same.

3 Haptic Recognition of Object Shape Properties

3.1 Size

Humans can evaluate the size of an object using both vision and touch. They demonstrate a well-known tendency to overestimate the length of a vertical line compared to a horizontal line of the same length—this is referred to as the vertical-horizontal illusion and is very robust in vision [13]. The vertical-horizontal illusion has also been reported in the haptic modality in both blindfolded sighted and blind subjects [14], meaning that subjects overestimate vertical extents and underestimate horizontal extents when trying to judge sizes based on touch. As shown by Suzuki and Arashida, the vertical segment of an inverted T is perceived to be 1.2 times longer than the horizontal segment when using touch to judge the size [15]. Interestingly, the horizontal segment will be overestimated when the inverted-T figure is rotated by 90°, meaning that the segment that is divided into two parts (i.e., the horizontal segment in the inverted T and the vertical segment in the rotated inverted T) is underestimated [16]. The extent of the vertical-horizontal illusion in touch has been shown to depend on the object tracing motions made by the hand and the size and orientation of the explored object. For instance, Gentaz and Hatwell found an increase in the length overestimation with the inverted T when subjects used the index finger of the dominant hand to explore the object instead of free exploration by both hands [14].

Heller et al. studied whether the haptic horizontal-vertical illusion in the case of 2-D forms would generalize to 3-D objects [17]. They experimented with objects that had round or square bases and dowel rods projecting above them at heights equal to the widths of the horizontal bases—see Fig. 3. It was found that with free

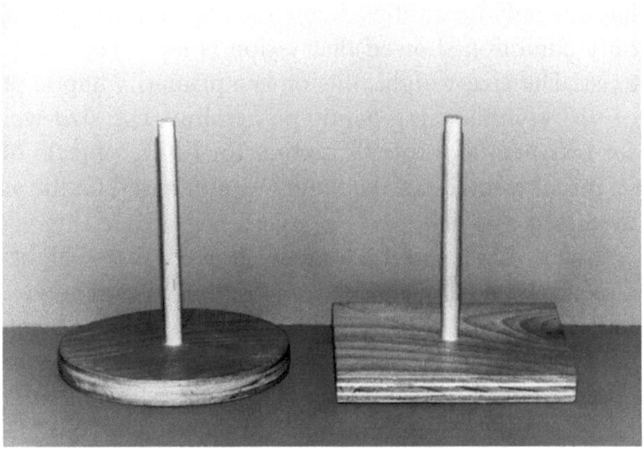

Fig. 3 3D objects with *round* and *square* bases and dowel rods projecting above them in [17]

haptic exploration to judge the size, the horizontals were overestimated by the subjects. This is referred to as negative illusion because it is contrary to the vertical-horizontal illusion.

3.2 Orientation

Proper spatial orientation of objects enables human subjects to recognize them through the sense of touch. In general, upright shapes are more easily recognizable than tilted shapes. Orientation is of particular importance in alphabet recognition, a prime example of which is reading Braille characters across the page from left to right based on the sense of touch. If a subject's hand is tilted, recognizing upright Braille characters becomes difficult [18]. Similarly, rotating the Braille characters can cause the reader to misinterpret the letters and words because of the misalignment between the finger and the characters [19]. Past research has shown a relationship between the orientation of the Braille page and the recognizability of Braille characters. Heller et al. measured the performance of both congenitally blind and blindfolded sighted individuals in reading non-rotated and rotated (by 180°) Braille letters [20]. While all subjects did worse on the 180° rotated Braille page, the blind subjects had less difficulty compared to the sighted subjects in terms of recognizing the rotated letters. Also, Ungar et al. found that while rotated Braille letters and words reduced the speed and accuracy of Braille recognition, experienced Braille readers could adjust to rotated characters in the presence of a context such as a set of words that are all in the same line [19]. Thus, for the highest speed and accuracy of character recognition through the sense of touch, Braille needs to be presented horizontally.

3.3 Curvature

Shape perception in 3D often happens through a combination of vision and haptics. The curvature of an object at any point on its surface is the reciprocal of the radius of curvature at that point. Similar to the overestimation of linear extents discussed in Sect. 3.1, Heller et al. have shown a haptic vertical-horizontal illusion in perception of convex curves [21]. Kappers et al. performed experiments to study the active haptic identification of 3D objects represented by quadric surfaces [22]. Each object was defined by a quantity describing its shape ("shape index") and a quantity describing its overall curvature ("curvedness")—see Fig. 4. In the experiments, both the shape index and curvedness were found to significantly impact the haptic shape identification. In fact, when the curvedness of test surfaces were kept constant, concave surfaces (negative shape index) led to a larger variation in the subjects' shape recognition response than convex ones (positive shape index). Also, it was found that surfaces with a high curvedness were identified more easily than those with a low curvedness. Further experiments with constant and with random curvedness yielded identical results (i.e., not knowing the curvedness had no influence on the subjects' response about the shape index), meaning that curvedness and shape index are not confounded from a psychophysical perspective.

4 Active Haptic Exploration Versus Passive Stimulation

To arrive at a basic understanding of how humans perceive objects, psychologist James Gibson carried out experiments using cookie cutters of different shapes such as square, star, etc. [23]. First, in a set of passive stimulation experiments, the

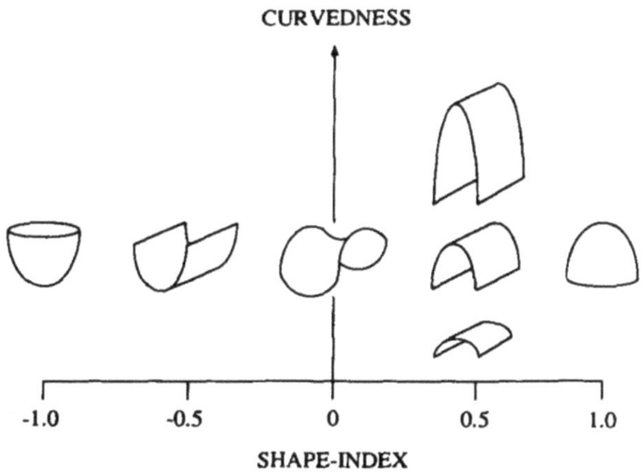

Fig. 4 The shape index can indicate a *concave spherical paraboloid* (−1), a *concave cylinder* (−0.5), a *hyperbolic paraboloid* (0), a *convex cylinder* (0.5), or a *convex spherical paraboloid* (1). The curvedness can range from a flat surface (0) to an extremely curved surface (infinity) [22]

cutters were pressed on the palm of a still hand of a blindfolded subject. Next, in a set of active exploration experiments, the subject was allowed to feel the cutter by his/her finger. The rate of correct identification of the shape of objects rose from 29 to 95 % from the passive stimulation to the active exploration conditions. This means that object shape perception is much more accurate when fingers are used to actively explore the object as the subject will receive feedback both from the fingers (cutaneous feedback) and from the arm and hand muscles (kinesthetic feedback). For further details on the advantages of active touch, the reader can refer to [24].

Lederman and Klatzky showed that in active exploration of objects, various subjects systematically performed appropriate hand movements (called exploratory procedures) depending on the task at hand [25]. The purpose of their experiments was to find links between desired knowledge about objects and the exploratory movements performed by subjects. After classifying the procedures for each task, they found that in free exploration a procedure is used that is necessary, sufficient, and optimal for the given task. Their experiments found several consistently applied associations between task and procedure including identifying object texture through lateral motion, identifying object hardness by applying pressure, identifying object temperature through static contact, identifying object weight by unsupported holding, identifying object global shape and volume through enclosure by fingers, and identifying object exact shape by following the object contours—see Fig. 5.

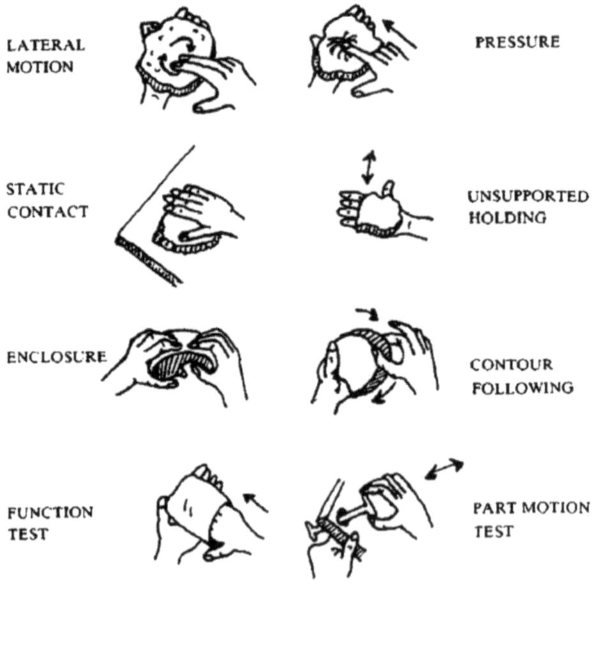

Fig. 5 Exploratory procedures (EPs) and associated property(ies) that each EP is optimal at providing. Adapted from [25] with permission of the authors

5 Concluding Remarks

In this chapter, we considered how an object's material properties such as stiffness, texture and weight are perceived through the sense of touch. For shape properties, we considered the effects of object size, orientation and curvature on object recognition. Finally, we briefly compared the effects of passive stimulation of fingers versus active movements of finger on the human's ability to detect object material and shape properties. The interested reader is also referred to [24] for additional studies.

Acknowledgments This research was supported by the Natural Sciences and Engineering Research Council (NSERC) of Canada.

References

1. G. Gottlieb, Ontogenesis of sensory function in birds and mammals, in *The Biopsychology of Development*, ed. by E. Tobach, L. Aronson, E. Shaw (Academic Press, New York, 1971)
2. M.A. Srinivasan, R.H. LaMotte, Tactual discrimination of softness. J. Neurophysiol. **73**, 88–101 (1995)
3. B. Hannaford, L. Wood, Performance evaluation of a 6 axis high fidelity generalized force reflecting teleoperator, in *Proceedings JPL/NASA Conference on Space Telerobotics*, Pasadena, CA, 1989, pp. 89–97
4. C.R. Wagner, R.D. Howe, Force feedback benefit depends on experience in multiple degree of freedom robotic surgery task. IEEE Trans. Rob. **23**(6), 1235–1240 (2007)
5. A. Pressman, L.J. Welty, A. Karniel, F.A. Mussa-Ivaldi, Perception of delayed stiffness. Int. J. Robot. Res. **26**(11–12), 1191–1203 (2007)
6. H. Ohnishi, K. Mochizuki, Effect of delay of feedback force on perception of elastic force: a psychophysical approach. IEICE Trans. Commun. **E90-B**(1), 12–20 (2007)
7. I. Nisky, F.A. Mussa-Ivaldi, A. Karniel, A regression and boundary-crossing-based model for the perception of delayed stiffness. IEEE Trans. Haptics **1**(2), 73–83 (2008)
8. M. Hollins, R. Faldowski, S. Rao, F. Young, Perceptual dimensions of tactile surface texture: a multidimensional scaling analysis. Atten. Percept. Psychophysics (formerly Perception Psychophysics) **54**(6), 697–705 (1993)
9. D. Picard, C. Dacremont, D. Valentin, A. Giboreau, Perceptual dimensions of tactile textures. Acta Psychol. **114**(2), 165–184 (2003)
10. R. L. Klatzky, S. Lederman, Tactile roughness perception with a rigid link interposed between skin and surface. Atten. Percept. Psychophysics (formerly Perception Psychophysics) **61**(4), 591–607 (1999)
11. R.R. Ellis, S.J. Lederman, The role of haptic versus visual volume cues in the size-weight illusion. Atten. Percept. Psychophysics (formerly Perception Psychophysics) **53**(3), 315–324 (1993)
12. R.R. Ellis, S.J. Lederman, The golf-ball illusion: evidence for top-down processing in weight perception. Perception **27**(2), 193–201 (1998)
13. S. Coren, J.S. Girgus, *Seeing is Deceiving: The Psychology of Visual Illusions* (Lawrence Erlbaum Associates Inc., Hillsdale, NJ, 1978)
14. E. Gentaz, Y. Hatwell, Geometrical haptic illusions: the role of exploration in the Müller-Lyer, vertical-horizontal, and Delboeuf illusions. Psychon. Bull. Rev. **11**(1), 31–40 (2004)

15. K. Suzuki, R. Arashida, Geometrical haptic illusions revisited: haptic illusions compared with visual illusions. Atten. Percept. Psychophysics (formerly Perception Psychophysics) **52**(3), 329–335 (1992)
16. S. Millar, Z. Al-Attar, Vertical and bisection bias in active touch. Perception **29**(4), 481–500 (2000)
17. M. Heller, D. Brackett, S. Salik, E. Scroggs, S. Green, Objects, raised-lines and the haptic horizontal-vertical illusion. Q. J. Exp. Psychol.Sect. A. Human Exp. Psychol. **56**(5), 891–907 (2003)
18. M.A. Heller, The effect of orientation on visual and tactual braille recognition. Perception **16**(3), 291–298 (1987)
19. S. Ungar, M. Blades, C. Spencer, Effects of orientation on Braille reading by people who are visually impaired: the role of context. J. Vis. Impairment Blindness **92**(7), 454–463 (1998)
20. M.A. Heller, J.A. Calcaterra, S. Green, F. Lima, The effect of orientation on Braille recognition in persons who are sighted and blind. J. Vis. Impairment Blindness **93**(7), 416–419 (1999)
21. M.A. Heller, A.M.L. Kappers, M. McCarthy, A. Clark, T. Riddle, E. Fulkerson, L. Wemple, A.M. Walk, A. Basso, C. Wanek, K. Russler, The effects of curvature on haptic judgments of extent in sighted and blind people. Perception **37**(6), 816–840 (2008)
22. A.M.L. Kappers, J.J. Koenderink, I. Lichtenegger, Haptic identification of curved surfaces. Atten. Percept. Psychophysics (formerly Perception Psychophysics) **56**(1), 53–61 (1994)
23. J.J. Gibson, Observations on active touch. Psychol. Rev. **69**(6), 477–491 (1962)
24. S. Ballesteros, M.A. Heller, Haptic object identification, in *Human Haptic Perception–Basics and Applications*, ed. by M. Grunwald (Birkhäuser, Basel, 2008)
25. S.J. Lederman, R.L. Klatzky, Hand movements: a window into haptic object recognition. Cogn. Psychol. **19**(3), 342–368 (1987)

Chapter 10
Classifying Human Hand Use and the Activities of Daily Living

Aaron M. Dollar

Abstract Humans use their hands for a large variety of tasks during daily life. In this chapter, a discussion of human hand use is presented, including classification schemes for grasping and manipulation behaviors. First, a simple classification of the Activities of Daily Living (ADLs) is presented, providing some structure to a terminology that is typically used in an ad hoc manner. Next, an overview of work related to classifications and taxonomies of static grasp types is presented, followed by a study investigating the frequency of use of various grasp types by a housekeeper and machinist. Finally, a taxonomy classifying hand-based manipulation is presented, providing a hand-centric and motion-centric categorization of hand use. These descriptions and classifications of hand use should prove useful to researchers interested in robotic manipulation, prosthetics, rehabilitation, and biomechanics.

Keywords Grasping · Manipulation · Activities of daily living · Robotics · Taxonomy

1 Introduction

Due to the complexity of the human hand and the immense variability in tasks that we use our hands for on a day-to-day basis, there has long been a desire for classification schemes to categorize these activities and the hand postures/movements

Manuscript received February 11, 2012. Portions of this work were previously published in [7-9]. This work was supported in part by the U.S. National Science Foundation under Grant IIS-0953856 and DARPA grant W91CRB-10-C-0141 (ARM-H).

A. M. Dollar (✉)
Department of Mechanical Engineering and Materials Science, Yale University,
New Haven CT 06511, USA
e-mail: aaron.dollar@yale.edu

R. Balasubramanian and V. J. Santos (eds.), *The Human Hand as an Inspiration for Robot Hand Development*, Springer Tracts in Advanced Robotics 95,
DOI: 10.1007/978-3-319-03017-3_10,

utilized in executing them. Initially, this interest was primarily centered in fields such as biomechanics, hand surgery, and rehabilitation [1–4]. With the advent of robotics into manufacturing tasks, the study of hand function received new life as researchers began to investigate human hands in order to shed light on the design and control of robotic end effectors.

In this chapter, I describe a few classification schemes related to hand use in everyday environments and tasks that should be useful to researchers interested in human manipulation, hand biomechanics, prosthesis and robotic end-effector design, and rehabilitation. In Sect. 2, I present a simple sub-classification of the Activities of Daily Living (ADLs) [5, 6] for the application of robotics in human environments, putting forth a standard categorization that allows robotic tasks to be discussed in terms of the analogous human tasks and their hierarchal classifications. While this is by no means an exhaustive classification, it does provide some structure to terminology that is primarily used in an ad hoc manner.

In Sect. 3, I briefly review the fairly deep literature related to grasp classification, with sub-categorizations according to the static placement of the fingers during power and precision grasps (instead of the movements of the fingers and hand during manipulation movements). This short section is followed by a description of a preliminary experimental examination (Sect. 4) of the frequency with which various grasp types are used in daily tasks, focusing on two subjects: a housemaid and a machinist. For each of these subjects, 4 h of video from a head-mounted camera is analyzed to determine how frequent each grasp type is used and for how long. For each subject, on the order of three thousand distinct grasp changes are observed over the 4 h period, a statistic indicative of the variability and frequency of hand use in human living.

Finally, in Sect. 5, I present a classification of hand and finger motions during in-hand manipulation tasks, focusing on the nature of the motion of the hand/fingers as well as contact with external objects. This taxonomy is hand- and movement-centric in nature (as opposed to object- and force-centric, for instance), and is some of the first extensive work on a topic that is sure to receive more attention as robotic and prosthetic hands become more dexterous.

In order to keep the tone of this chapter as an overview of hand classification, many of the details are not presented in substantial depth. Instead, the reader is asked to refer to the original publications [7–9] for a more extensive description of related work, details on experimental methodology, or in-depth description of the classifications presented here.

2 Activities of Daily Living

Many fields related to occupational therapy, rehabilitation, and gerontology use the term "ADLs" in evaluating the ability of a patient to perform self-maintenance and other daily tasks crucial for unassisted living [5, 6, 10–14]. The term is generally used broadly and qualitatively. Many different sub-categories of the ADLs have

Table 1 Activities of daily living

Domestic activities of daily living (DADLs)	
DADL1	Food preparation
DADL2	Housekeeping
DADL3	Laundry
DADL4	Telephone/computer/technology use
DADL5	Office tasks/writing
DADL6	Hobby/sport
Extradomestic activities of daily living (EADLs)	
EADL1	Transportation/driving
EADL2	Shopping
EADL3	Employment-related tasks/tool use
Physical self-maintenance (PSM)	
PSM1	Feeding/medicating
PSM2	Toileting
PSM3	Bathing
PSM4	Dressing
PSM5	Grooming
PSM6	Ambulation/transfer

been proposed to classify an individual's level of independence, including Physical Self-Maintenance (PSM) [10], ADLs [5], Instrumental Activities of Daily Living (IADLs) [12, 13], and mobility [12], among others. These categorizations of the ADLs were primarily developed to be used by a physician or occupational therapist to assist evaluation of human performance in daily tasks and determine, for instance, whether admission into a nursing home is justified for an elderly or disabled person.

Table 1 presents a new sub-classification of ADLs (drawn primarily from [11, 14]) designed for use with the application of robotics in domestic and work environments. These sub-categories are deemed "Domestic Activities of Daily Living (DADLs)", "Extradomestic Activities of Daily Living (EADLs)", and "PSM". In prior work, we discussed a variety of "Objects of Daily Living," putting forth a collection of objects identified as important from a number of sources related to prosthetics, rehabilitation, and robotics [7].

The first and cardinal category, "DADLs," contains subtasks spanning those regularly performed in human living environments. The majority of efforts related to assistive robotics focus on tasks in this category, particularly in Housekeeping and Food Preparation [15–17]. Typical approaches for assistance in this area consist as devices not intended to be utilized for tasks outside of this category. Exceptions, however, include work related to robotic wheelchairs and wheelchair-mounted manipulator arms (e.g. [18, 19]), which are frequently used outside of the home.

The second category, "EADLs," contains activities and tasks performed primarily outside of the home. Note that housekeeping activities, technology use, and office tasks are classified primarily as DADLs, even though they are often performed as employment-related tasks. Aside from wheelchairs and related technologies, robotics applications for these areas include driver assists (e.g. [20]) and cooperative robots for manufacturing tasks (e.g. [21, 22]).

Assistance with tasks related to the final category, "PSM," is one of the most important areas of need in assisted-living and hospital environments. However, this application generally requires physical contact between the robot and human and is sufficiently challenging such that many tasks will not likely be tractable in the near future. Exceptions include Feeding/Medicating, which have been assisted by wheel-chair mounted arms, as well as robotic orthoses [23] and prosthetics (e.g. [24]) for assistance during Ambulation/Transfer.

3 Human Grasp Classification

The first major attempt to organize human grasping behavior into distinct categories was by Schlesinger in 1919, which categorized grasps into six types: cylindrical, tip, hook, palmar, spherical, and lateral [2]. These grasps are primarily defined by the object that the hand interacts with. However, human grasps are often less dictated by size and shape of the object, but more by the tasks that need to be accomplished. In 1956, Napier suggested a scheme that would divide grasps into power and precision grasps [1]. Unfortunately not all the grasps fell cleanly into either of these two categories, with the lateral pinch in particular serving both power and precision functions.

In studying the grasps required for manufacturing tasks, Cutkosky in 1989 provided a much more comprehensive and detailed organization of human grasps (Fig. 1) [25]. This taxonomy was acquired through a set of observational surveys on professional machinists and is essentially an integration of the previous work done by Schlesinger and Napier. The taxonomy tree is organized such that it is first divided into power and precision grasps from left to right, and by task and geometry detail (precision) down the tree. A small number of successive taxonomies, built primarily from the Cutkosky taxonomy, have been proposed since (a comprehensive overview can be found in [26]).

A recent effort has resulted in what the author views as the most complete grasp taxonomy to date (Fig. 2) [26]. In this work, grasp types are organized primarily according to power, precision, and intermediate types, with sub-categorizations according to thumb position (abducted or adducted), and finger/palm contact type (palm, finger pad, and finger side). This work identifies 33 independent grasp types, which includes the 16 grasps from Cutkosky's taxonomy [25].

4 Grasp Frequency in Household and Machine Shop Tasks

Though there have been a number of efforts focused on classifying types of human grasps, no previous studies have examined the frequency of these grasps as they

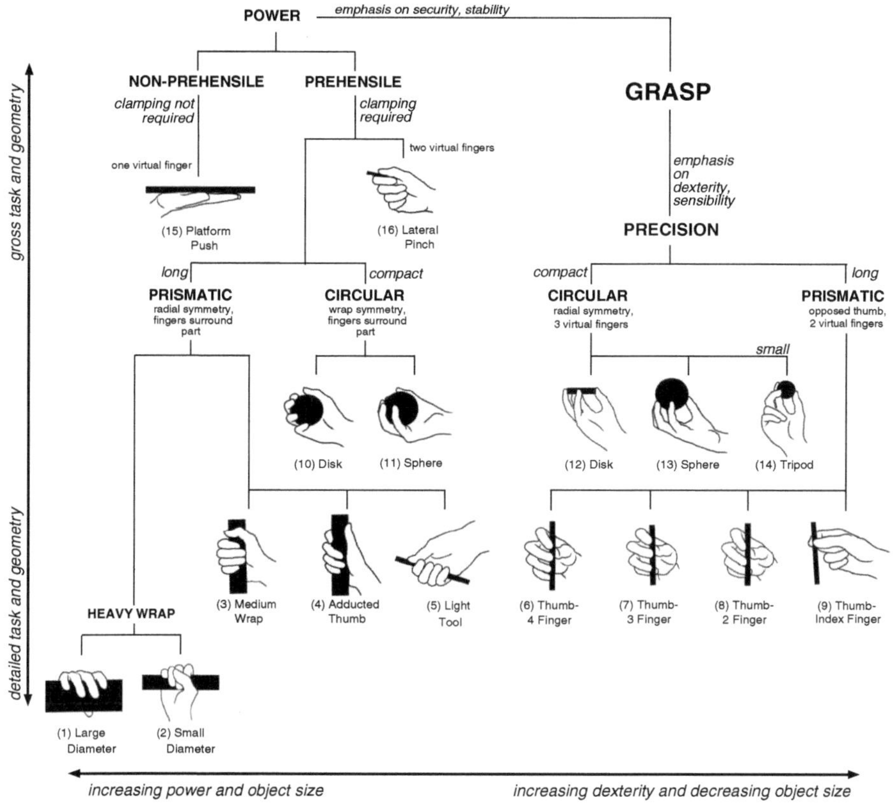

Fig. 1 Cutkosky grasp taxonomy. Adapted from [25]

are used in various settings. The frequency data is important as it will further clarify the relationship between task requirements and the various grasp types. Furthermore, it might serve to establish a sort of "prioritization" of grasp types according to the most frequently used in the examined daily activities. This is important to motivate the practical robotic and prosthetic hand design tradeoffs between complexity and performance. We begin by describing the experimental methodology, including details on the subjects used, apparatus, and protocol. We then present the results for two subjects, a professional housekeeper and a professional machinist, identifying the frequency of grasp type use for each. The complete version of this study is presented in [8].

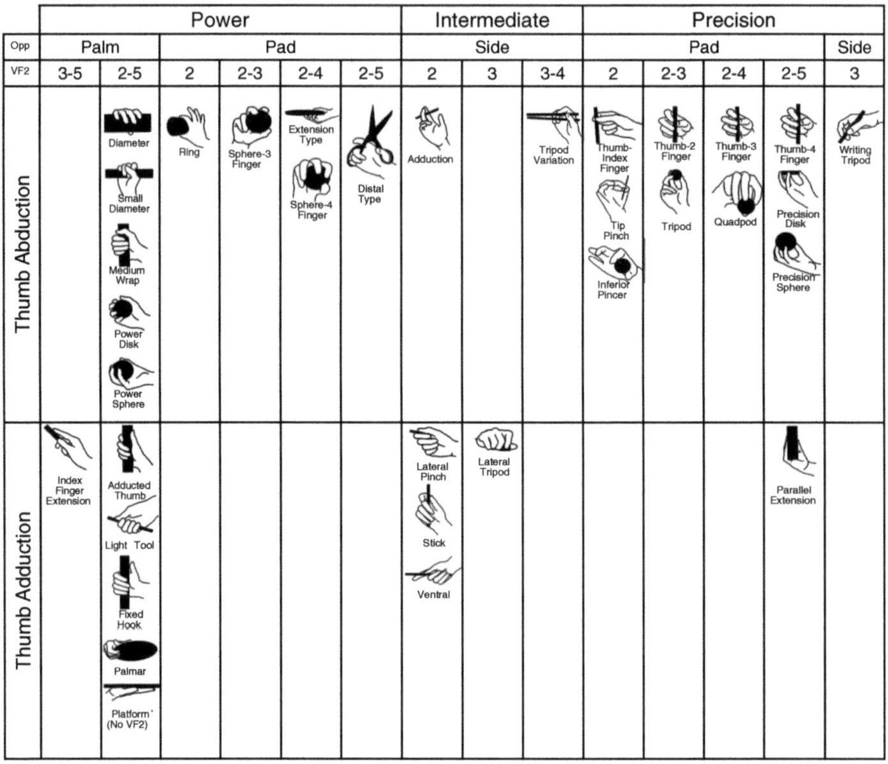

Fig. 2 Modified Feix grasp taxonomy [26]. Note that a few names are used from the Cutkosky taxonomy, such as for the thumb-n finger grasps. The platform grasp from the Cutkosky taxonomy is shown, although it does not have a second virtual finger

4.1 Experimental Procedure and Apparatus

Two subjects participated in the study presented in this section. The first, a 41 year old right-handed male, was a professional machinist who had worked in his profession for more than 20 years. The second, a 30 year old right-handed female, was a full-time house maid who had been working in that capacity for over 5 years. Neither subject had any injury or disability that would alter their grasping and manipulation ability from what would be expected as typical for their profession.

A total of at least 4 h of hand usage was analyzed for each subject, over multiple days. The days and times of recording were carefully chosen according to the subject's feedback such that there would be a wide range of tasks representative of the span of the job requirements performed throughout the total 4 h. Therefore, days and times for which the subject was performing a small number of tasks repetitively were not included.

The video hardware consisted of a tube camera with a wide-angle fisheye lens (2.5 mm, ~140° field of view) attached to a three-band head strap taken from a hiker's lamp. This setup allows the camera to rest on the subject's head without being intrusive or uncomfortable. The camera is connected to a mini digital video recorder and an external battery pack. Both the receiver and battery pack are worn in the back pocket of the subject. The setup is able to acquire video of sufficient quality for manual grasp classification. The overhead view was chosen after informal testing showed this to be the most useful for our purposes as it shows the entire workspace of both left and right arms in front of the body as well as enough of the surroundings to give the context of the grasps in addition to the grasp itself.

4.2 Results

The results below show the analysis of 4 h of video for each of the two subjects—house maid and machinist. During the 4 h analyzed, the subjects were performing a wide range of tasks associated with the regular demands of their profession. All data was manually recorded by a researcher trained in human grasp classification. The researcher went through the video and when the user changed their grasp (either acquiring a new object or releasing an existing object), the grasp type (according to the Cutkosky [25] and Feix taxonomies [25]), object and task being performed, and the time stamp associated with the change was recorded. Only the right (dominant) hand was examined in the present study. Approximately 2,500 and 2,000 grasp changes were made during the 4 h period by the house maid and machinist, respectively. The complete results can be found in [8].

Figure 3 shows the frequency data from the house maid (left) and machinist (right), with labels for all grasp types occurring at least 2 % of the time. Grasps followed by numbers in parentheses correspond to those identified in the Cutkosky taxonomy (Fig. 1) [25]. Unnumbered grasp types are labeled according to the terminology utilized in [26]. Figure 4 shows sample screen captures for the three most common grasps utilized by the house maid (top–medium wrap, index finger extension, and power sphere) and machinist (bottom–lateral pinch, light tool, and tripod) during the 4 h analyzed.

4.3 Discussion

From the results summarized in Fig. 3, it can be seen that only a small number of grasp types comprise the majority of those used. For the house maid, nearly 80 % of the time was spent utilizing six grasp types: medium wrap, index finger extension,

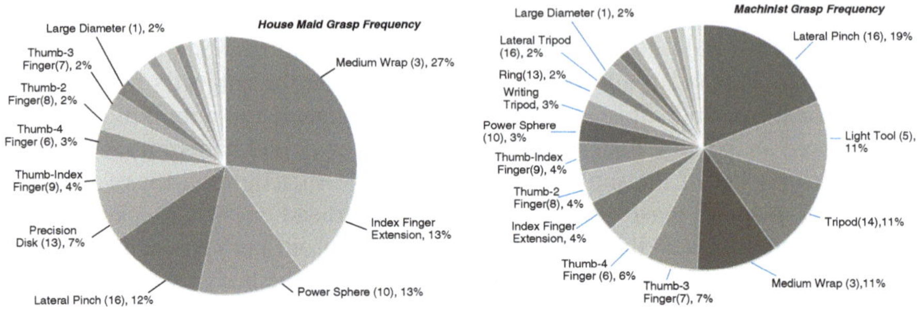

Fig. 3 Grasp frequency results for the House Maid (*left*) and Machinist (*right*), showing 4 h of work. Grasps occurring less than 2 % of the time are not labeled. The 'no grasp' case is not shown here

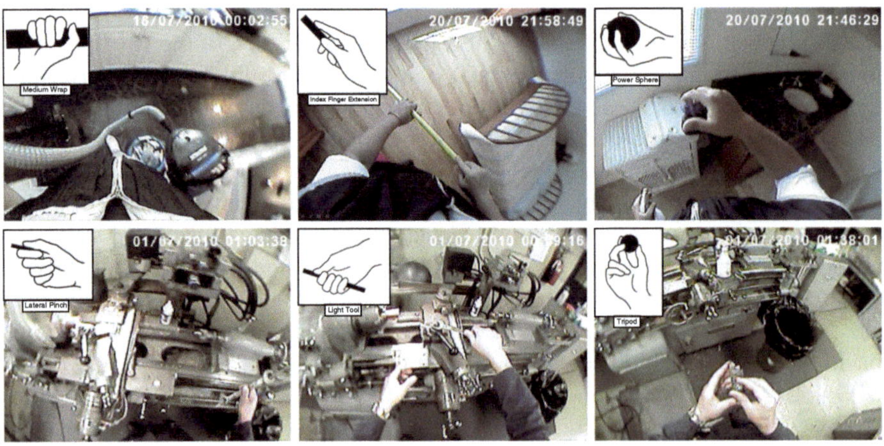

Fig. 4 Video stills from the House Maid (*top*) and Machinist (*bottom*) experiments showing the three most commonly-used grasps and example tasks performed using them

power sphere, lateral pinch, precision disk, and thumb-index finger. Nearly 80 % of the machinist's time grasping utilized nine grasps: lateral pinch, light tool, tripod, medium wrap, thumb-3, thumb-4, index finger extension, thumb-2, and thumb-index. Note that all 16 grasps identified in [25] occurred in both subjects' data, with the 'power disk' occurring least. However, two grasps frequently utilized by the subjects (index finger extension and writing tripod, >3 % for both subjects) are not represented in the Cutkosky taxonomy. It is also interesting to note that the house maid primarily used power grasps while the machinist used a balance of both. Furthermore, the machinist switched grasps more often (~2,500 in 4 h vs. ~2,000).

One particularly interesting question that was raised during our analysis related to how to classify grasps of non-rigid objects. The house maid in particular often used a rag or other cloth to wipe down surfaces for cleaning. We have classified these grasps primarily as 'power sphere', based on the observation that the subject utilized all five fingers in the grasp. However, a new subset of grasp types for compliant objects might be developed.

While 4 h is a fairly large amount of grasping data (>2,000 grasps per subject), these results will, of course, change to some extent based upon the specific subject being examined. Future work will involve completing the 8 h of video analysis for these two subjects, as well as investigating grasp behavior for additional professions that may be of interest to robotics, such as food preparation, machine maintenance, and others.

5 Human Manipulation Classification

This section presents a taxonomy for detailed classification of human and anthropomorphic manipulation behavior. This hand-centric, motion-centric taxonomy differentiates tasks based on criteria such as object contact, prehension, and the nature of object motion relative to a hand frame. A sub-classification of the most dexterous categories, within-hand manipulation, is also presented, based on the principal axis of object rotation or translation in the hand frame. Principles for categorizing complex, multi-faceted tasks are also presented, along with illustrative examples. Although illustrated with human hands, the taxonomy can easily be applied to any robot manipulator or end-effector. (Note that a more extensive version of this section can be found in [9].)

While the authors were unable to find any extensive classifications that differentiate the full range of human manipulation behaviors from one another, a number of related works should be mentioned. Elliott and Connolly described three general classes of within-hand (intrinsic) manipulation movements: simple synergies, reciprocal synergies, and sequential patterns [27]. In simple synergies, all digits involved move as one unit, such as while pinching or squeezing. In complex synergies, the fingers move together, but the thumb moves independently. In sequential patterns, the digits move independently in a repeatable sequence. Exner's classification [28] has been used fairly extensively in clinical settings, classifying within-hand manipulation into five categories: palm-to-finger translation, finger-to-palm translation, shift, simple rotation, and complex rotation. Gentile [29] proposes a task classification scheme based on environmental context and function of the action. It differentiates tasks according to whether it is being performed in addition to basic body stability or body transport motions.

A much different but related classification is the taxonomy of haptic disassembly tasks [30], which classifies tasks according to task type and type of force or torque required. The force classification differentiates between tasks where the force is aligned with the motion, such as pressing a button, and those where the force is not aligned, such as sanding a surface. Torque is differentiated by whether the torque axis passes through the grip space, expressing the difference between turning a screwdriver and a steering wheel.

Other work has used the term *manipulation primitives* to describe the steps used in a specific algorithm or by a specific robot, but work by Morris and Haynes [31] describes a more general definition based on motion in six possible degrees of freedom between two rigid bodies. Morrow and Khosla [32] later improved on the notation used and described a general architecture for executing manipulation task primitives based on sensorimotor primitives defined based on a specific robot. These efforts focus on object motions and degrees of freedom and therefore differ substantially from the current hand-centric taxonomy.

5.1 Human Manipulation Classification

Figure 5 presents our manipulation taxonomy (with terms defined in Fig. 6). Note that in creating this classification, we take a hand-centric view of the problem, as opposed to an object-centric view. The taxonomy therefore focuses on what the hand is doing during execution of the manipulation task, rather than how the object being contacted is moving during the task. As a result of this classification, a given movement of an object can be done from multiple locations on the tree (e.g. a low-friction knob could be turned with a single finger as a non-prehensile task, or with multiple fingers as a prehensile task). Object-centric classifications might be made in a manner similar to [32] described above. Note that this is also a motion-centric view of manipulation tasks, as opposed to a force-centric view (such as [30], as described above). However, the two are related by the Jacobian of the manipulator so that motions can occur in directions in which forces can be applied and vice versa. Tasks in which force is applied normal to the major direction of motion (e.g. writing on a board) would be considered two distinct simultaneous tasks, decomposed in a manner outlined in Sect. 5.1.2.

5.1.1 Further Sub-Categorization

Using the existing categorization criteria for the taxonomy, certain further sub-categories might be added depending on the specific application of the taxonomy. For example, for each sub-categorization that includes motion of the object, more specific details of the nature of that motion can be added. These motions might be broken up by rotations or translations about some fixed frame, as is presented for dexterous within hand manipulation in Sect. 5.2, or with regards to how many degrees of freedom in which the object can be actively manipulated in (similar to [32]). Alternatively, some type of classification related to the force, similar to [30], might be made.

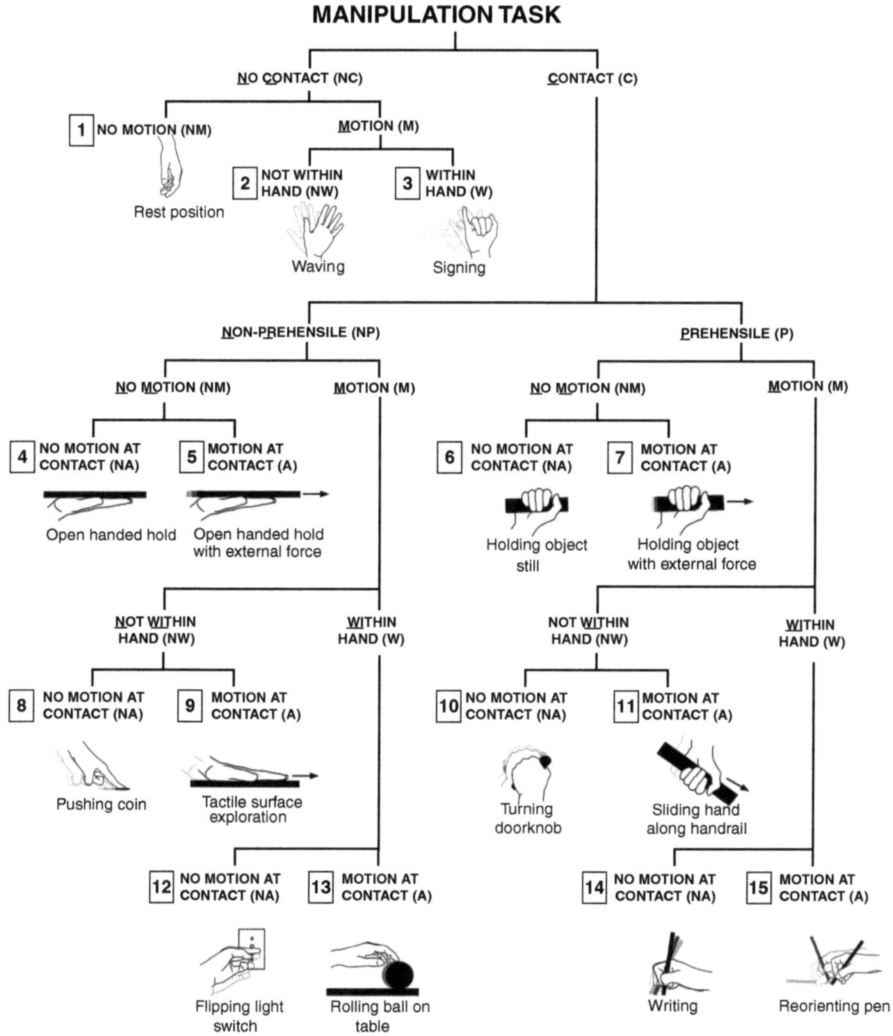

Fig. 5 Manipulation taxonomy. Any type of human or robotic manipulation task can be classified according to this taxonomy regardless of hand morphology. Example tasks are given for each leaf of the tree

5.1.2 Classifying Complex Tasks

The taxonomy above provides a structured way of classifying relatively simple tasks. More complex tasks have less obvious categorizations. There are three major types of complex manipulation tasks that require further discussion as to their categorization: time-separated sequences, simultaneous bi-manual tasks, and simultaneous within-hand tasks.

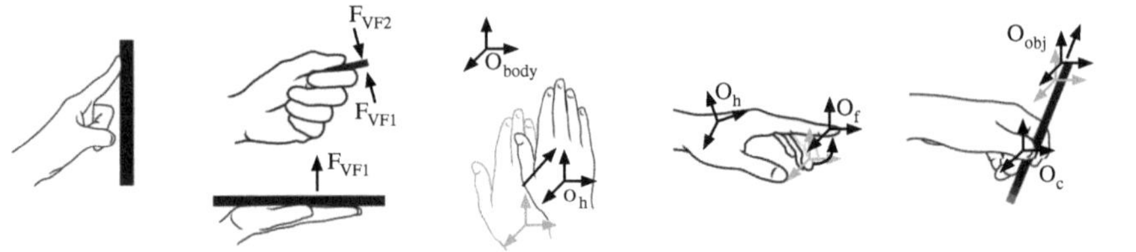

Contact
Hand is touching an external object or the environment.

Prehensile
Action of hand on object must be described with more than one virtual finger.

Motion
Any part of the hand moves relative to body fixed frame.

Within Hand
Points on the hand are moving relative to the hand base frame.

Motion at contact
Object reference frame moves relative to contact point frame(s).

Fig. 6 Explanation of important terms in the manipulation taxonomy

Time-separated motions, such as a long sequence of movements to accomplish an overall goal, should be classified as the sum of the discrete sub-components of the manipulation process. For instance, picking up a pen and writing with it might be decomposed into three sub-tasks: lifting the pen from the table (prehensile/motion/not within hand), rotating the pen into the writing position (prehensile/motion/within hand/motion at contact), and writing with it (prehensile/motion/within hand/no motion at contact).

Bi-manual tasks, where both hands are in use and *required* to perform the task, would be defined by the individual tasks being performed by each hand. Simultaneous use of the hands to perform independent tasks should not be considered 'bi-manual'.

Tasks where the hand performs two or more simultaneous functions (e.g. pulling a hand drill trigger, thumb-typing on smartphone, using calipers, writing on a chalkboard, etc.) are some of the most dexterous tasks regularly performed. We propose that these types of tasks should be treated as the sum of the distinct sub-tasks being performed. For example, the task of pulling the trigger on a power drill could be categorized as a prehensile, no motion task (grasping and holding the drill handle) combined with a non-prehensile/motion/within hand task (index finger compressing the trigger). Thumb-typing on a cell phone would be similarly considered as the sum of a prehensile, no motion task (holding the phone with the palm and fingers) combined with a non-prehensile/motion/within hand task (thumb pressing the keys). Writing on a chalkboard, where a force is being applied to the board (to maintain contact) and the chalk is being moved along the surface of the board would be considered as the sum of two prehensile/motion/within hand tasks, as applying force to the board and moving the chalk both require actuation in each direction.

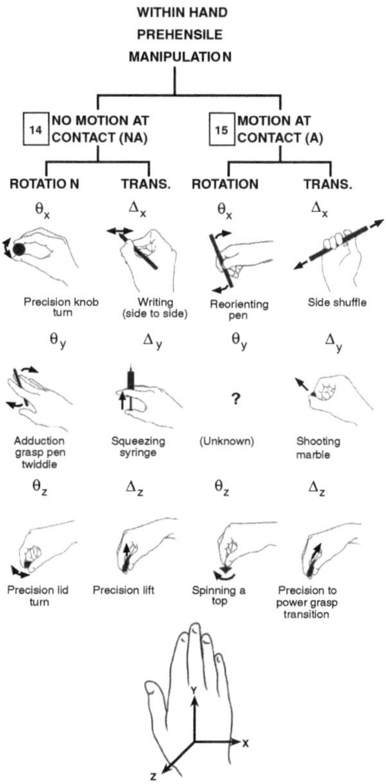

Fig. 7 Dexterous subclassification. Tasks are classified by rotations and translations along hand coordinate axes (*bottom*)

5.2 Dexterous, Within Hand Manipulation

The term "dexterous manipulation" is used frequently in the robotics community, but no widely accepted definition exists. Perhaps the most common use, however, refers to manipulations that would be categorized as prehensile, within hand manipulation according to the taxonomy presented above. Indeed, there is great interest in the field to impart greater dexterity to robotic and prosthetic hands in the form of "within hand" manipulation capability. Figure 7 shows the dexterous manipulation taxonomy. We sub-categorize the movements according to the axis about which the movement is primarily concentrated, with respect to a coordinate frame affixed to the back of the hand. Each movement subcategory ("no motion at contact" and "motion at contact") is therefore expanded to three rotational and three translational movements with respect to this coordinate frame (plus some positional offset).

Due to constraints imposed by hand kinematics, it is unlikely that any movement would precisely align with the fixed coordinate frame axes. Instead, these are intended to be approximate. For movements that are significantly askew from these axes, a linear combination of cardinal movements might be used to describe the task.

Based on the constraints inherent with human hand kinematics, it is difficult to affect dexterous translational movements in the x-direction or rotational movements about the y-axis. Indeed, there are few identifiable common dexterous manipulation tasks for those axes.

5.3 Conclusions and Future Work

This section provides a hand- and motion-centric categorization of human manipulation that might be applied in various ways. For example, the proposed classification scheme might enable better understanding of human hand use by emphasizing hand-centric differences between tasks that might appear similar if only object motion is considered. In some cases, a similar range of object motion might be accomplished through a "not within hand" or "within hand" strategy, with significant differences in the required hand dexterity. This might be used, for example, to help identify hand intensive tasks that a patient recovering from a hand injury should do carefully or sparingly.

The taxonomy might also guide the creation of a set of standard manipulation tasks for each leaf of the tree. This kind of set should include the most frequent tasks that humans perform, and span a wide variety of practical hand motions. Although creating a complete set may be difficult or impossible, even an incomplete set might be a powerful tool for evaluating manipulation performance. This standard set could be used to evaluate a human patient's manipulation ability or chart their progress during use of rehabilitative devices, or to compare the dexterity of robot hands. In either case, performance on the standard task set could be used to assign an overall dexterity score to a hand, providing a structured basis for comparing hand performance.

6 Overall Conclusion

In this chapter I presented an overview of work useful for classifying human hand use. These areas included a very brief overview of high-level tasks in the ADL—an area which would be appropriate for future additional categorization, both at the high level and low level. Next I described various work on grasp taxonomies, which have had a fairly extensive treatment in the literature. Following this, I presented a summary of some results on grasp use during daily activities, categorizing grasp use for 4 h of a professional housekeeper and a professional machinist. Finally, I presented a classification of within-hand, dexterous manipulation, from a hand-centric and motion-centric perspective. While the enormous range of uses of the human hand make it difficult to capture the whole range of function in a succinct format, I believe the work here represents a sufficient means of classification for much of the most important uses. However, there is of course much room for improvement and follow-on to these areas.

Acknowledgments The author would like to thank Ian Bullock, Josh Zheng, Sara De La Rosa, and Kayla Matheus for their work on the studies presented in this paper, Lael Odhner, Raymond Ma, and Leif Jentoft for their helpful discussions related to the manipulation taxonomy, and to Kayla Matheus for helping to create the hand drawings used in the figures.

References

1. J. Napier, The prehensile movements of the human hand. J. Bone Joint Surg. **38B**(4), 902–913 (1956)
2. G. Schlesinger, in *Der Mechanische Aufbau der Kunstlichen Glieder*, ed. by M. Borchardt et al. Ersatzglieder und Arbeitshilfen (Springer, Berlin, 1919)
3. L.M.K. Boelter, A.D. Keller, C.L. Taylor, V. Zahm, *Studies to Determine Functional Requirements for Hand and Arm Prosthesis* (Department of Engineering, UCLA, California, 1947)
4. C.L. Mackenzie, T. Iberall, *The Grasping Hand* (Elsevier/North-Holland, Amsterdam, 1994)
5. Staff of the Benjamin Rose Hospital, Multidisciplinary studies of illness in aged persons: II. a new classification of functional status in activities of daily living. J. Chronic Dis. **9**, 55–62 (1959)
6. S. Katz, A.B. Ford, R.W. Moskowitz, B.A. Jackson, M.W. Jaffe, Studies of illness in the aged: the index of ADL: a standardized measure of biological and psychosocial function. J. Am. Med. Assoc. **185**(12), 914–919 (1963)
7. K. Matheus, A.M. Dollar, Benchmarking grasping and manipulation: properties of the objects of daily living, in *Proceedings of the 2010 IEEE/RSJ International Conference on Intelligent Robots and Systems (IROS 2010)*, 2010
8. J.Z. Zheng, S. De La Rosa, A.M. Dollar, An investigation of grasp type and frequency in daily household and machine shop tasks, in *Proceedings of the 2011 IEEE International Conference on Robotics and Automation (ICRA)*, Shanghai, China, May 9–13, 2011
9. I.M. Bullock, A.M. Dollar, Classifying human manipulation behavior, in *Proceedings of the 2011 IEEE International Conference on Rehabilitation Robotics (ICORR)*, Zurich, Switzerland, June 29–July 1, 2011
10. M.F. Lowenthal, *Lives in Distress* (Basic Books, New York, 1964)
11. M.P. Lawton, E.M. Brody, Assessment of older people: self-maintaining and instrumental activities of daily living. Gerontologist **9**(3), 179–186 (1969)
12. S. Katz, Assessing self-maintenance: activities of daily living, mobility, and instrumental activities of daily living. J. Am. Geriatr. Soc. **31**(12), 721–727 (1983)
13. W.D. Spector, S. Katz, J.B. Murphy, J.P. Fulton, The hierarchical relationship between activities of daily living and instrumental activities of daily living. J. Chronic Dis. **40**(6), 481–489 (1987)
14. D. Galasko et al., An inventory to assess activities of daily living for clinical trials in Alzheimer's disease. Alzheimer Dis. Assoc. Disord. **11**(supplement 2), S33–S39 (1997)
15. A. Saxena, J. Driemeyer, A.Y. Ng, Robotic grasping of novel objects using vision. Int. J. Robot. Res. **27**(2), 157–173 (2008)
16. S. Srinivasa, D. Ferguson, M. Weghe, R. Diankov, D. Berenson, C. Helfrich, H. Strasdat, The Robotic busboy: steps towards developing a mobile robotic home assistant, in *International Conference on Intelligent Autonomous Systems*, 2008

17. H. Nguyen, A. Jain, C. Anderson, C.C. Kemp, A clickable world: behavior selection through pointing and context for mobile manipulation, in *Proceedings of IEEE/RJS International Conference on Intelligent Robots and Systems (IROS)*, 2008
18. H.A. Yanco, Integrating Robotic research: a survey of robotic wheelchair development, in *AAAI Spring Symposium on Integrating Robotic Research*, Stanford University, California, 1998
19. R. M. Alqasemi, E. McCaffrey, K. Edwards, R. Dubey, Analysis, evaluation and development of wheelchair-mounted robotics arms, in *Proceedings of IEEE International Conference on Rehabilitation Robotics*, 2005, pp. 469–472
20. C. Urmson et al., Autonomous driving in urban environments: boss and the DARPA urban challenge. J. Field Robot. **25**(8), 425–466 (2008)
21. M.A. Peshkin, J E. Colgate, W. Wannasuphoprasit, C.A. Moore, B. Gillespie, P. Akella, Cobot architecture. IEEE Trans. Robot. Autom. **17**, 377–390 (2001)
22. D. Shin, I. Sardellitti, O. Khatib, A hybrid actuation approach for human-friendly Robot design, in *Proceedings of the IEEE International Conference on Robotics and Automation*, Pasadena, CA, 2008
23. A.M. Dollar, H. Herr, Lower extremity exoskeletons and active orthoses: challenges and state of the art. IEEE Trans. Robot. Spec. Issue Biorobot. **24**(1), 144–158 (2008)
24. S.K. Au, M. Berniker, H. Herr, Powered ankle-foot prosthesis to assist level-ground and stair-descent gaits. Neural Netw. **21**, 654–666 (2008)
25. M.R. Cutkosky, On grasp choice, grasp models, and the design of hands for manufacturing tasks. IEEE Trans. Robot. Autom. **5**(3), 269–279 (1989)
26. T. Feix, Anthropomorphic hand optimization based on a latent space analysis, PhD Dissertation, Vienna University of Technology, 2011
27. J.M. Elliott, K. Connolly, A classification of manipulative hand movements. Dev. Med. Child Neurol. **26**, 283–296 (1984)
28. C.E. Exner, In-hand manipulation skills, in *Development of Hand Skills in the Child,* ed. by American Occupational Therapy Association (American Occupational Therapy Association, Rockville, 1992), pp. 35–45
29. J.H. Carr, R.B. Shepherd, J. Gordon, A.M. Gentile, J.M. Held, *Movement Science: Foundations for Physical Therapy in Rehabilitation* (Aspen Publishers, Inc., Rockville, 1987)
30. A. Bloomfield, Y. Deng, J. Wampler, P. Rondot, D. Harth, M. McManus, N. Badler, A taxonomy and comparison of haptic actions for disassembly tasks, in *Proceedings of the IEEE Virtual Reality Conference*, 2003, pp. 225–231
31. G. Morris, L. Haynes, Robotic assembly by constraints, in *Proceedings of IEEE International Conference on Robotics and Automation*, 1987, pp. 1507–1515
32. J.D. Morrow, P.K. Khosla, Manipulation task primitives for composing robot skills, in *Proceedings of International Conference on Robotics and Automation*, 1997, pp. 3354–3359

Part II
Human Hand-Inspired Robotic Hand Design and Control

Chapter 11
Design of Artificial Hands: A Review

Marco Controzzi, Christian Cipriani and Maria Chiara Carrozza

Abstract The human hand is capable of performing complex and useful tasks using an effective integration of mechanisms, sensors, actuators and control functions, and at the same time is also a cognitive instrument, allowing humans to develop a superior brain by interacting with the surrounding environment. The idea of developing a human-like artificial hand has always intrigued mankind, and to replicate it is still one of the main challenges of robotics, requiring large efforts, based on multidisciplinary knowledge ranging from engineering to neuroscience. This chapter is aimed at providing an overview of past and present artificial hands, developed in the frameworks of research projects in prosthetics and humanoid robotics.

Keywords Artificial hands · Robotics · Design issues · Grasp taxonomies · Review

1 Introduction

Duplicating the human hand is one of the most ancient and challenging objectives of robotics research, and in the last few decades a large number of research projects have been carried out to achieve this goal. The human hand is a complex integrated system with many components that provides subjects with high functionality and elegant behaviour. 35 intrinsic and extrinsic actuators (muscles) are engaged in a

M. Controzzi (✉) · C. Cipriani · M. C. Carrozza
Scuola Superiore Sant'Anna, Pisa, Italy
e-mail: marco.controzzi@sssup.it

C. Cipriani
e-mail: ch.cipriani@sssup.it

M. C. Carrozza
e-mail: chiara.carrozza@sssup.it

R. Balasubramanian and V. J. Santos (eds.), *The Human Hand as an Inspiration for Robot Hand Development*, Springer Tracts in Advanced Robotics 95,
DOI: 10.1007/978-3-319-03017-3_11, © Springer International Publishing Switzerland 2014

structure with 22 Degrees of Freedom (DoF) (articulations) and more than 17,000 sensors (mechanoreceptors) in the glabrous skin, make it capable of performing countless actions. Its versatile prehensile abilities allow for tasks ranging from fine digit manipulation to the handling of heavy objects. In addition, the hand is a marvellous cognitive instrument for sensing and exploring the surrounding world.

More than 2000 years ago Aristotle claimed: "*…the hand is a tool of tools…*", emphasising how humans were fascinated by this amazing limb [1].

Nowadays the main research fields in which robotic hands are investigated are upper limb prosthetics and humanoid robotics. Pioneering examples of articulated artificial hands were designed for upper-limb replacement due to war amputations. The first documented prosthetic hand user was, in 218 BC, the Roman emperor general Marcus Sergius, who during the second Punic War lost his right hand and after amputation was fitted with an iron-made prosthesis that enabled him to hold his shield and return to the battlefield [2]. In the early sixteenth century the German mercenary Goetz Von Berlichingen (1480–1562) built an incredible artificial limb for its time, capable of different grasp types due to the independent flexion of phalanges and thumb opposition, driven by racks and recalled by springs [3]. In the mid-sixteenth century Ambroise Paré (1510–1590), a French military surgeon, turned his attention to the design of artificial limbs able to replace those he had surgically removed, and devised a hand with fingers that could be operated independently with levers and gears contained therein. All of these devices were externally operated by gross movements of the healthy limb or chest. Only in the late nineteenth century was the first breakthrough in control input conceived, due to an idea by Dr. Vanghetti, later implemented by the surgeon E. F. Sauerbruch (1875–1951) who built, around the time of World War I, the first prosthetic hand to be operated by the muscles in the residual forearm in which ivory pins were surgically inserted [2]. A second breakthrough was achieved only fifty years later, when the first non-body powered myoelectric prosthesis that employed electric motors was developed. The first prosthesis of this kind was developed during the early 40s by Reinhold Reiter, a physicist working with the Bavarian Red Cross. He published his work in 1948 but it was not widely disseminated, and myoelectric control was destined to be addressed again in England, in the Soviet Union, and perhaps other places during the 1950s [4]. Today research in prosthetics is still very active. Recently two commercial prosthetic hands with greater DoF have been introduced to the market: the i-limb (developed by Touch Bionics in 2007) and BeBionic (developed by RSL Steeper in 2010) prostheses. Both hands are capable of different grasping patterns thanks to five individually-powered digits, but their functionality is limited by the passive movement of the thumb abduction/adduction joint. Despite massive efforts, current commercial prosthetic hands are very simple grippers with few active DoF and poor cosmetic appearance. The main limitations that affect the development of prosthetic hands are related to the number of communication channels of the current user-prosthesis interfaces: only few intentional commands generated by the amputee can be recognised for control of a prosthesis; this reason (together with others not mentioned here) prevents manufacturers from developing hands with a large number of DoFs.

The progress of the development of anthropomorphic hands for robots is in parallel to the development of humanoid robots. In the mid-eighteenth century brilliant engineers such as the French inventor Vaucanson (1709–1782) and the Swiss watchmaker Jaquet-Droz (1721–1790) developed the first models of mechanical androids. These precursors of today's robots were able to independently perform several human-like movements, usually artistic activities to charm audiences of nobility. Examples are the Vaucanson "*Automa ton flute*" flute-player whose leather-covered five-fingered hands opened and closed the holes of a flute, and the harpsichord "*Musician*" by Jacque-Droz [5].

Nowadays research into humanoid robotics holds the promise to produce exploitable results in the near future, and is growing in interest mainly due to the significant progress in the last decades in electronic integrated circuits and in applied computer science. The possibility of replacing man through remote manipulation in hazardous, strenuous, dirty and repetitive work is perhaps one of the most fascinating results that world expects from this research. Industrial robotic grippers that are specifically designed for constrained or restricted sets of operations in structured environments could in principle be successfully replaced by versatile hands able to grasp undefined shaped objects or tools, which the conventional gripper cannot, therefore decreasing cycle times, costs and power consumption.

Artificial hands can be exploited in neuroscience experiments where neurophysiological models of the human brain and body are investigated and a physical hand, with the aim to replicate the biological one, is needed, as for example in sensorimotor coordination [6], sense of body ownership [7, 8], brain plasticity [9] and others.

This chapter is aimed to provide an overview of past and present artificial hands by reviewing key design criteria and issues important for their development. Several excellent reviews addressing robotic hand design and control have been presented in the past [10–12]; the aim of the present review is to provide the reader with an updated perspective, including recent and ongoing important results achieved by several groups in the EU and US due to extensive research programs aimed to support the development of neuro-controlled upper limb prosthetics.

2 Human Hand Function and Synthesis

In order to understand the choice behind the design of an artificial hand, whatever application is envisaged, the functionalities of the human limb must be analysed. As recalled by Napier in 1956 the movements of the hand can be subdivided into two main groups: (1) *prehensile movements*—or movements in which an object is seized and held partly or wholly within the compass of the hand; and (2) *non-prehensile movements*—or movements in which no seizing or grasping is involved [13]. Since prehensile functions of the human hand are so manifold and more interesting than non-prehensile ones, this chapter is solely concerned with prehensile movements and will focus on the two main functional requirements with which an artificial hand should comply: grasp robustness and manipulative dexterity. Grasp robustness is defined as the action of the hand on an object consisting in preventing its motion relative to the hand itself considering external perturbations. Grasp robustness may be associated to the concept of whole-hand manipulation where to ensure grasp stability palmar areas (i.e. proximal phalanxes and the palm) of the hand are crucial. Manipulation dexterity is defined as the capability of changing the position and orientation of the manipulated object from a given reference configuration to a different one, arbitrarily chosen within the hand workspace [11, 14–16]. Usually manipulation dexterity implies contact and low-forces on the fingertips and distal phalanxes.

There are two main approaches to the study of grasping and manipulation: the empirical and the analytical approach. Empirical approaches study grasping by humans, trying to replicate the human knowledge by experience and will be briefly summarised in the next paragraph, whereas analytical approaches are based on physical models describing the manipulation action, for which the reader is invited to refer to Chap. 20. Besides analysis, both approaches can be used for the synthesis of artificial hands by engineers. As an example the pre-shaping or grasping-posture planning problem can be easily tackled by means of an empirical approach, whereas grasping optimisation problems require strict analytical treatment. For the design of hand prostheses, where requirements like multi-grasp ability, low weight and power consumption are more important than others, an empirical approach is the most obvious choice. In contrast, for the design of dexterous anthropomorphic hands or grippers aimed to manipulate objects or tools in structured environments, it becomes clear that an analytical approach is preferable. In short, the choice of the more suitable approach for the design and control of an artificial hand depends on the specific case, and in general it can be a combination of empirical and analytical methods.

2.1 Empirical Approach

The empirical approach to the duplication of the grasping and manipulation abilities in an artificial hand is based on observations of human behaviour and on the consequent efforts to mimic it. The human hand's prehensile capabilities, although highly versatile, can be classified by means of grasp taxonomies in order to focus the design on specific aspects, reducing the complexity.

The very first taxonomies of human grasps were proposed by Schlesinger in 1918 (summarised by Taylor and Schwarz in 1955 [17] and Griffiths in 1943 [18]). They associated grasps with the shape of the object; specifically six grasps/shapes were described: *cylindrical*, *tip* (or pincer), *hook*, *palmar*, *spherical* (or *ball*) and *lateral* grasps. McBride [19] partitioned prehensile functions depending on the areas of the hand engaged in the grasp: *whole hand* grasping, *palm and digit* grasping, *thumb and fingers* grasping. In 1956 Napier [13] showed that these classifications either lacked of practical utility due to reliance on the shape of the object held (as for Schlesinger's and Griffiths' taxonomies) or did not appear to have any functional or anatomical basis (e.g. McBride's taxonomy). As Napier argued, our experience in everyday tasks points out how the choice of grasp is less dictated by the size and the shape of the objects than by the tasks we want to accomplish. Even during the execution of a single task with a single object, the hand adopts different grips to adjust to changing force/torque conditions. Napier underlined the influence of the functional viewpoint (related to the grasp stability of the object held) in the choice of grasp posture, and deduced that two patterns appear to cover the whole range of prehensile activity of the human hand: *power grasp* and *precision grasp*. Where considerations of stability predominate (like holding a hammer) a power grasp is chosen. Where considerations of sensitivity and dexterity predominate (like threading a wire through a needle) a precision grasp is chosen.

The main feature of the power grasp is the ability to handle heavier weights compared to the precision grasp that allows the accurate grasping of small objects with low interaction forces. With a precision grip the contact points are located at the fingertips of the thumb, index and middle fingers, and the thumb works in opposition to the other fingers. Conversely, the power grasp is characterised by a wider contact surface with the object which is held in the hollow formed by the thumb, the palm and the fingers. Consequently, a fundamental role is played by the palmar areas of the hand (palm and proximal phalanxes) to enhance the stability of the grip.

In 1986, Iberall and co-workers analysed how the hand behaves in prehension by focusing on the fact that the posture involves at least two forces being applied in opposition to each other against the object and more general forces can be decomposed into external forces which equilibrate the forces and moments of a task [20]. Iberall et al. hypothesised three opposition axes (i.e. basic directions) along which the human hand can apply forces: *pad opposition*, *palm opposition* and *side opposition*. In 1985 Arbib et al. introduced the concept of virtual fingers, a natural abstraction to decrease system complexity by assembling more fingers belonging to the same side of opposition into a single entity [21]. In 1986 Cutkosky extended Napier's taxonomy building a hierarchical tree of grasps—as shown in Fig. 1—in which moving from left to right the grasp becomes less powerful and the grasped object becomes smaller [22]. In such a classification the heavy wrap grips are the most powerful and the least dexterous while the tripod and the thumb-index finger grips are the most precise. The main advantage of Cutkosky's taxonomy is that it allows us to understand the relationship between task requirements and the grasping "solution" adopted to meet those requirements: the task requirements and object attributes are combined to dictate the grasp choice.

All of these taxonomies are *discrete* classifications of the grasping capabilities of the hand. What could be referred as the first *continuous taxonomy* was described in 1998 by Santello et al. [23]. They reported a series of interesting experimental results on the analysis of human hand grasping postures, and demonstrated that the control of hand posture involves a few (two or three) postural synergies regulating

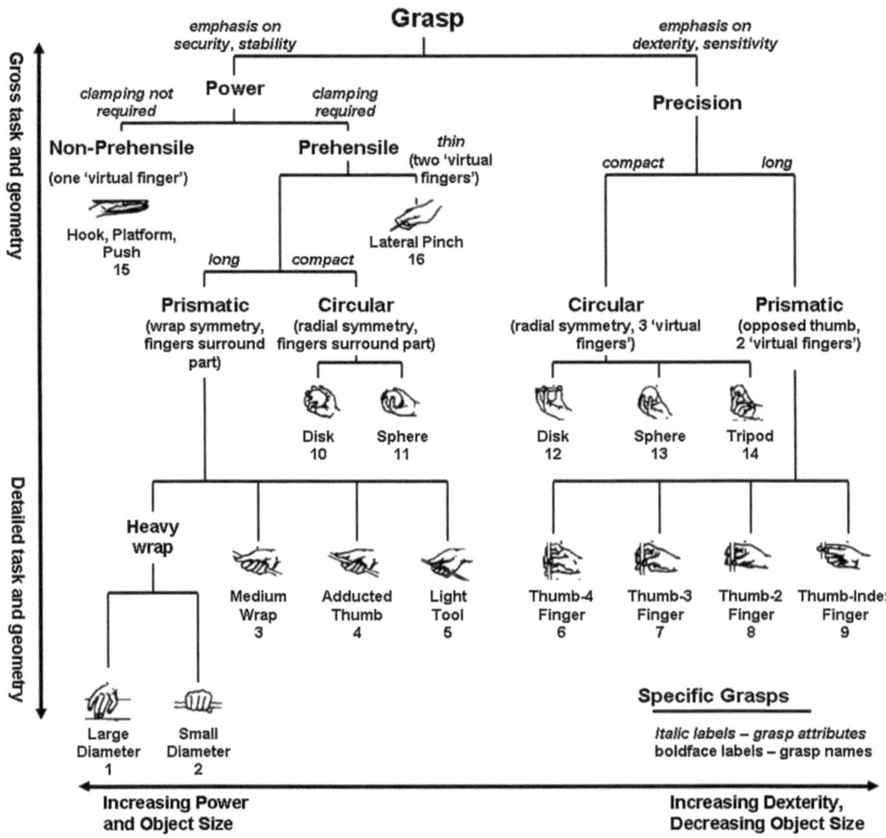

Fig. 1 Partial Cutkosky's grasp taxonomy from [22]. Moving from *left* to *right* the grasp becomes less powerful and the grasped object becomes smaller. The heavy wrap grips are the most powerful and least dexterous, while the tripod and the thumb-index finger grips are the most precise. However the trend is not strictly followed

the general posture of the hand, coupled with a finer control mechanism providing for small, subtle adjustments, thus revealing that prehensile movements of the human hand can be represented in a continuous reduced dimensionality space. This concept was exploited in a limited number of works with the aim of controlling artificial hands by means of a lower-dimension input space [24–28].

3 Hand Design Key Features

The choice of a specific design for an artificial hand must take into account the field of application and its peculiar requirements. In the authors' view the six most important features/issues that should be considered during the design and development phases are: (1) *kinematic architecture*, (2) *actuation principle*, (3) *actuation transmission*, (4) *sensors*, (5) *materials* and (6) *manufacturing method.* Starting from the definition of these features and the discussion of their critical issues, significant hands developed from the 60 s up to the present das will be presented, compared and discussed in the following section. Table 3 in the Appendix compares the general and technical information about a number of artificial hands. Since an exhaustive review cannot be accomplished here, we wish to apologise to those researchers engaged in important work that were not mentioned.

3.1 Hand Kinematics

The kinematics of the hand are related to the choice of required DoFs and degrees of control, DoC (i.e. controlled DoFs), thus it directly influences the performance of the hand.

The human hand's kinematics with its 22 DoFs engaged by a complex muscle-tendineous system is one of the main sources of inspiration for roboticists [29]. A DoF of a generic articulation is driven by the action of several muscles that produce the rotation of that specific joint and together operate on other joints [30].

Duplicating the complexity of the entire range of movement of the human hand is hard when all functional components must be fitted in the constrained volume imposed by the biological model, forcing engineers to find trade-offs for approximating the natural hand in relation to available technologies. In the past decades pioneering roboticists have approached the problem with a strong analytical approach. To perform basic prehension a minimum of 3 DoF are required under the assumption of a hand with rigid, hard finger, non-rolling and non-sliding contacts. Salisbury first postulated that to achieve dexterous manipulation a minimum of 9 DoC are needed [16], and in the past several hands have adopted such a configuration employing different transmission architectures: the Okada Hand [31], the Salisbury JPL Hand [32], the Karlsruhe Dexterous Hand [33], the High Speed Hand [34].

Morecki et al. showed that from a kinematical point of view a rigid body possessing n bidirectional DoFs can be completely controlled by $m = n + 1$ actuated tendons (only pulling actions); n tendons (or gears or belts) would only suffice in an endless routing in which each tendon is wrapped around two or more pulleys in an endless loop (Fig. 2) [35]. The first significant robotic hand using the $n + 1$ architecture was the Stanford/JPL Hand with 3 DoFs and 4 open-ended tendons, developed in the 80s by Salisbury and Craig [32]. In this hand arbitrary torques could be exerted on the three joints of each finger, by applying the appropriate combination of tensions in the cables. Examples of artificial hands with actuators equal to the number of DoFs, are the Belgrade/USC [36], DLR Hands I and II [37, 38], Gifu Hands II and III [39, 40], Barret [41], MANUS [42], Southampton-REMEDI [43], High Speed Hand [34], NTU Hand [44], i-limb prostheses [45], and BeBionic [46].

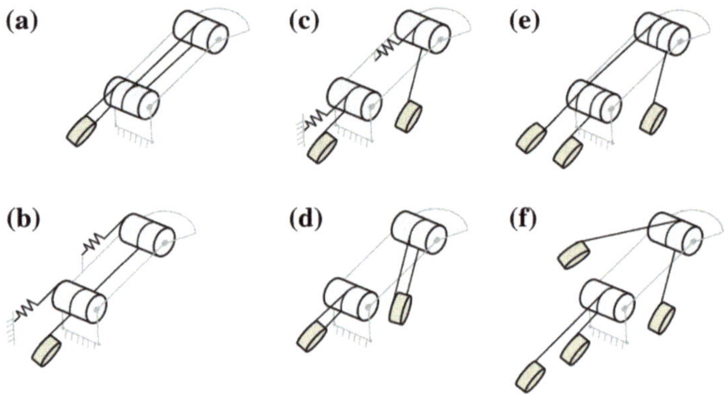

Fig. 2 Schematic representation of possible kinematic architectures in robotic hands: **a** M = N coupled joints, **b** M < N underactuated transmission, **c** M = N fully actuated open chain, **d** M = N fully actuated closed chain, **e** M = N + 1 fully controllable, **f** M = 2 N agonist/antagonist transmission

When m independently actuated tendons, with $m > n + 1$, are engaged in a system possessing n DoFs the actuation systems is referred to as redundant, and the resulting hand possesses an increased ability to apply arbitrary forces and motions to the object grasped. In such cases the evaluation of potential dexterity based on the kinematical structure can be addressed by investigating the manipulation ellipsoids, i.e. the ability of the finger in positioning/orientating the tip and in exerting forces/forces upon the object [47]. An example of a hand with redundant actuation is the UTAH/MIT Hand from the 80s by Jacobsen et al. [48], in which each finger comprised of 4 DoFs and 8 actuators ($m = 2n$). Other hands with redundant actuation are the Shadow [49], DIST [50], Hitachi [51], ACT [52], and DLR [53, 54].

Although a redundant number of actuators increases the potential dexterity, it also yields to bulkier and more complex systems and controls; hence in order to tackle this issue coupled transmission and underactuation have usually been exploited. Coupled mechanisms can be easily achieved by linking the motion of the joints in a finger or linking the motions of one finger to another (Fig. 2). The main drawback of this solution is that the kinematics of the joints is fixed by the design, and may yield unstable grasps. In contrast, coupled mechanisms are easy to control (i.e. kinematics is governed by linear equations) and are capable of producing large torques at the finger-tip. The most representative hands deploying coupled mechanisms are the commercial prostheses produced by Ottobock. In the Speed Hand one motor drives two fingers like a pinch gripper [55]; in the Michelangelo Hand two motors are used to flex/extend the fingers and rotate the thumb [56]. Underactuated mechanisms are systems with fewer inputs (DoC) than outputs (DoF); this is possible by employing differential mechanisms, mechanical limits and elastic elements. The great advantage of underactuated hands is that they allow automatic adaptation to the shape and contour of the objects, therefore increasing the contact area and hence stability [57]. Many underactuated and adaptive mechanisms have been developed in the past decades and been endowed in intrinsic prosthetic hands and robotic grippers. Examples are the Belgrade [58], MARCUS [59], Southampton [60], TBM [61] (prostheses), the Yokoi [62], SDM [63], and TUAT/Karlsruhe Hands [64] (grippers). In the Belgrade and Southampton Hands, for example, fingers were linked via whiffle-tree equalising mechanisms so that with a reduced number of motors multi-point adaptive grasps could be achieved.

By looking at the past, one of the representative underactuated mechanisms applied to this field is the Hirose's *soft* gripper, which is able to softly and gently conform to objects of any shape and hold them with uniform pressure [65]. In this device the gripping function is achieved by means of a mechanism consisting of multi-links and series of idle pulleys which can be simply actuated by a pair of wires. Based on such a principle several artificial hands were developed by Scuola Superiore Sant'Anna (Italy) in the last few decades, such as the RTR II [66], SPRING Hand [67], CyberHand [68], and SmartHand [69].

Smarter solutions could be inspired by the biological human hand. By observing the functionalities of the hand it is possible to divide the fingers into two types: one set of dexterous fingers (thumb, index and middle) that are able to accomplish manipulative tasks, and one set of underactuated fingers that are only involved in whole-hand grasp postures; such a choice was adopted e.g. in the design of the Robonaut Hand [69] (developed by NASA in 1999 as an end effector for a humanoid robot for extra-vehicular space activities), the i-Cub Hand [70], and the artificial hands developed within the RPP2009 (Revolutionary Prosthetics Program 2009) [71]. With regard to bioinspiration, the ACT-Hand project [72] aims to replicate most of the biomechanical aspects of the human limb, in terms of DoF/DoC it includes tendon insertion points and routing of the tendons including complex extensor mechanisms [72].

3.2 Actuation Principle and Mechanisms

One of the main bottle necks in the development of artificial hands with performance and size comparable to human ones is that current actuation technologies fail to provide efficient, high power density actuators [10]. In human beings muscles represent the active organs of the movement. As it can be easily gathered from the architecture of the hand discussed in the previous paragraph, a number of muscles participate to generate a single movement. In the hand the force produced by the muscles is transmitted to the bony segments by means of tendons. Tendons of the flexors are deployed to close the fingers, tendons of the extensors to reopen the hand, and their combined action allows the force and the stiffness of the hand to be modulated.

Table 1 Brief actuators comparison from [73]

Proprieties class	Power density ρ[W/Kg]	σ_{max} [MPa]	ε_{max}	E [GPa]	Efficiency
DC motors	100	0.1	0.4	*	0.6–0.8
Pneumatic	400	0.5–0.9	1	$5–9 \times 10^{-4}$	0.4–0.5
Hydraulic	2,000	20–70	1	2–3	0.9–0.98
SMA	1,000	100–700	0.07	30–90	0.01–0.02
Human muscle	500	0.1–0.4	0.3–0.7	0.005–0.09	0.2–0.25

Power density ρ = Power per unit of weight, σ_{max}= Maximum force exerted by the actuators per area, ε_{max}= Maximum run per length, *E* Actuator stiffness. Maximum stress and strain are indexes specifically designed for linear actuators. Units are expressed as follow: *W* Watt, *Kg* kilogram, *MPa* Mega Pascal, *GPa* Giga Pascal
*Depending on the gearhead

In the early stages of the design the requirements of the application (i.e. force, efficiency, displacement, specific power, noise, stiffness, size, mass and response time) must be matched by the specific performance of the actuator. To this aim it is helpful to define indexes for the comparison of the available actuators: (i) maximum actuation stress (i.e. the maximum value of actuation stress in a single stroke which produces maximum work output), and (ii) maximum actuation strain (i.e. the maximum value of actuation strain in a single stroke which produces maximum work output) are basic features of an actuator. In addition (iii) the power to weight ratio provides an idea of the power density of the specific class of actuator and it is useful, especially when lightweight solutions are sought. Table 1 presents a brief comparison of the main classes of actuators (including the biological muscles) based on values found in the literature. It is interesting to note that in terms of performance natural muscles have a very high power to weight ratio that can reach 500 W/Kg in young healthy subjects. A complete review useful for selecting the best actuator can be found in [73], whereas in [42] actuator types are compared specifically for prosthetic hand applications.

Most robotic hands developed are actuated by electrical brushed and brushless electric motors, mainly because these ensure the best performance compromise; examples of hands deploying these actuators are SmartHand [69], Speed Hand and Michelangelo by Ottobock [55] [56], Stanford JPL [32], Okada [31], Belgrade [58], Barrett [41], and others. The main advantage is obviously related to their ability to store electrical power in small batteries, in contrast—compared to other actuation types—they show the lowest power to weight ratio (the order of magnitude is 100 W/Kg).

Other designs involved pneumatic actuators such as the Utah/MIT [74], in which the mechanical energy of the output link was realised by means of potential and kinetic energy of the working fluid, representing a light-weight solution with high power. Because of gas compressibility pneumatic actuators show an inherent compliance (i.e. low *stiffness* (E) in Table 1) which is an advantage in terms of safety during human-robot interaction but, since it is difficult to modulate, it becomes a disadvantage during the execution of precision tasks. The Shadow Hand

[49] employed a particular lightweight pneumatic actuator belonging to the Pneumatic Artificial Muscle (PAM) class: the McKibben artificial muscle. This actuator, invented in 1958 by Richard H. Gaylord [75] and successively employed by McKibben in prosthetics as reported in [76], consists of an expandable inner elastic camera surrounded by a covered, braided mesh shell and closed by two ends, one being the air input and the other the force attachment point. When the inner tube is pressurised the muscle inflates, tries to contract and exerts a force on the end point [77]. *Power density* ρ of pneumatic artificial muscles exceed values of 1 kW/Kg. The difficulty in controlling the stiffness of the actuator is usually overcome by hydraulic actuation. In the latter the working fluid is liquid (instead of gas) and hence the compliance can be modulated by means of servo-valve control. The main drawbacks of hydraulic actuators are the high working pressure (about 30 MPa) and the consequential high stress that parts are subjected to. Examples of artificial hands employing such technology are the Karlsruhe Hand [78] and the Vanderbilt gas actuated arm [79].

Alternative actuators like shape memory alloy (SMA) were also investigated as they are particularly promising for the design of artificial hands; examples are the Hitachi robot hand [51] and the hand developed at the State University of New Jersey [80]. SMA are usually made of nickel-titanium alloys, and are driven by a temperature-induced phase change, produced by heating above the transformation temperature of their electrical resistance, which produces a significant shear strain. Although electrical heating can be achieved rapidly cooling is the main limitation on operating frequency, and additionally SMA usually present a relatively low efficiency [81, 82].

With regard to prosthetics, the multi-functionality of the hand strongly depends on the performance of the actuator unit, since this will affect the autonomy of the entire system. One important consideration concerns the need of non-back-drivable mechanisms. A transmission is defined as non-back-drivable when motion can only be transmitted from the input to the output axis and not vice versa. Such a property enables the actuator to deliver a stall torque without energy consumption, therefore the possibility to switch off the power, once a desired position of the hand or grasp stability is achieved. Examples of hands with non-back drivable actuation units are Speed Hand and Michelangelo [55], and SmartHand [69]. Generally speaking, in artificial hands for robotics applications non-back-drivability is important for safety reasons: indeed electrical supply or battery failures should not cause the (potentially dangerous) release of grasped objects or tools due to back-drivable transmission [83]. Another aspect is related to the ability of humans to co-contract their muscles, changing the stiffness of a given joint. In classical robotics application actuators are preferred to be as stiff as possible in order to carry out precise positioning movements, whereas a high compliance rate is required to absorb shocks and to safely interact with users. Within this framework, variable stiffness actuators (VSAs) are currently of growing interest in the robotics community [84], utilising both passive mechanical stiffness and actively controlled stiffness, especially for robotic hand actuation [85], they are capable to prevent damages if shocks occur. The last prototype developed by the German Aerospace Center DLR [86] is one of the main representative devices of this change.

3.3 Anthropomorphism

The word *anthropomorphism* means the attribution of human qualities to nonhumans, and in this specific case to a robotic device, i.e. the capability of a robotic end-effector to mimic the human hand, partly or totally, as far as concerns shape, size, weight, consistency, kinematics and contact surfaces [12]. In robotics an anthropomorphic design might be a desirable goal when it is required that the robot presents human-like appearance and behaviour, e.g. if working in a man orientated-environment, cooperating with humans or being tele-operated by humans. Conversely the *minimalist* approach chooses the simplest mechanical structure, the minimum number of actuators, the simplest set of sensors, etc., that fulfil the task requirements [11].

Anthropomorphism is mostly influenced by the choice of the number of fingers: a three-fingered artificial hand although being able to perform as a wide range of complex tasks as a human hand, might not be perceived as anthropomorphic due to its cosmetic appearance. The Stanford/JPL Hand was designed by using the minimalist approach with the aim to investigate finger-tip dexterous manipulation. Hence it has just three fingers with sensors on the tips [32]. Other examples of three-fingered hands are the Okada [31], the Barrett [41], and the High Speed Hand [34]. Prosthetic hands must be designed following the anthropomorphic approach, as cosmetic appearance is an important issue for user's acceptability. Regarding the choice of the number of fingers the state of the art shows that the trend is to follow a bio-inspired approach, employing five fingers, as in the Robonaut [69], the Gifu Hand [39], and the Shadow Hand [49]. Four-fingered hands, such as the Utah/MIT Dextrous Hand [48], DIST Hand [50], and DLR Hands [37, 38], represent a compromise between the extra mechanical complexity and cost of a five-fingered design and the added functionality over a minimalist three-fingered design.

Weight is an additional requirement affecting anthropomorphism, and when considering a prosthesis the weight should be based on the amputee's natural limb (i.e. about 400 g in a male depending on subject weight [87]) in order to find the orthopedic balance between the prosthesis and natural limb, as shown by several examples developed in the past decades: the TBM (280 g) [61], RTR II (350 g) [66], Southampton REMEDI Hand (400 g) [43], and SmartHand (520 g) [69].

In humanoid robots and industrial applications, the weight of the robotic hand should be the lowest possible in order to minimise power consumption, but in general the weight is higher than a prosthetic hand because of the large number of actuators, such as in the Stanford/JPL (1,100 g) [32], the Gifu Hand (1,400 g) [39], the DLR/HIT Hand sold by Schunk GmbH (2,200 g for the first version, and 1,500 g for the last) [38], and the Shadow Hand (3,900 g) [49].

3.4 Transmission

In designing artificial hands a wide range of transmission systems such as tendons, linkages, gear trains, belts or flexible shafts can be employed, but a large variety of design considerations and constrains can guide designers to select the optimal transmission mechanism. Important design goals are to minimise friction, backlash and inertia while trying to maintain small overall size and weight. Low friction and inertia and virtually no-backlash are especially sought if fingers are required to perform dexterous manipulation tasks, indeed non-linear effects due to friction or backlash could make it difficult or impossible to accurately control the movements.

In artificial hands the most commonly implemented transmissions are flexible tendons running into sheaths, in analogy with tendons in the human hand. The main advantage of a tendon-sheath transmission is that it allows actuators to be located remotely from joints, reducing the dimensions and weight of the fingers. In opposition, friction between tendon and sheath that occurs in curves introduces non-linear effects and reduces efficiency. Several artificial hands like the Stanford/JPL Hand [32], DIST Hand [50], LMS Hand [88], SmartHand [69] and the Vanderbilt Hand [79] are driven by Teflon-coated cables in flexible conduits.

Effects of friction can be drastically reduced by replacing sheaths with idle pulleys at the expense of system and control complexity or by flat bends instead of tendons in order to increase the strength and stiffness of the transmission system, as in the Utah/MIT Hand [48]. An immediate drawback of flexible cables or flat bend transmission is that they can only be used to pull, so in order to achieve the active two-way control of a generic joint a pair is required (hence increasing the complexity). Moreover, if a high stiffness is required, flexible cable must be constantly pre-loaded, inducing dangerous pre-stress on the components.

Linkages or trains of gears give the best stiffness proprieties to the transmission, low maintenance is required, and they allow bidirectional control of the joint. Nevertheless their employment substantially increases the weight, complexity and sometimes dimensions of the hand. An example of a hand driven by linkages is the Belgrade/USC [36]. Gear train transmission in which bevel gears were combined to reproduce the 2 DoFs flexion/extension and the adduction/abduction movements of the metacarpophalangeal joints (MCP) of the human finger can be found in the DLR/HIT Hand [37] and in the Gifu Hand [39]. The two gears on the flexion axis were driven by two different actuators, and their combined control allowed the generation of a variety of movements ranging from pure flexion/extension to pure adduction/abduction, with the possibility of combined motions with desired torque and speed.

In the Robonaut Hand [89] different power transmissions were adopted in order to achieve optimal performance. The actuation unit is located in the forearm of the robot and the mechanical power is transmitted through a flexible drive train from the forearm to the hand, where a leadscrew assembly converts the rotary motion of the flexible shaft to linear motion of a short cable fixed on the first phalanx of the finger, meanwhile other phalanxes are linked by linkages and cams [69].

In prosthetics it is preferred that the transmission is stiff during the flexion of the fingers, meaning that if the opening is forced the fingers do not move from their position, while they should be compliant during extension, avoiding damage that can occur during impacts of the dorsum of the prosthesis against obstacles. The actuation unit must be as simple as possible since it is driven by a small number of input signals directly controlled by the human subject, so in practice the employment of VSA is not possible. In order to achieve this requirement (stiff closure and compliant opening) the transmission of the Michelangelo Hand that drives the flexion/extension of the index and middle fingers are composed by linkages in which the bearings at the end points of the links are enclosed in a compliant material (a simple rubber) and surrounded by pre-loaded cables [56]. Such a link is capable of transmitting high tensile loads due to the embedded cable, whereas the rubber confers the desired compliant behaviour in bending and compression loads. In this framework an interesting transmission concept, alternative to traditional ones, was the COBOT architecture employed in DARPA's Revolutionary Prosthetic 2009 extrinsic hand, consisting in a single 40 W DC motor to power the prosthetic's various movements from the main shaft to each DoF of the hand by means of many Continuously Variable Transmissions (CVTs).

Another aspect concerning the transmission system is related to the ability of the mechanism to convert the motion of the actuation unit into complex or different movements. This possibility can reduce the weight and size and is especially interesting for prosthetic hands. In the Michelangelo Hand [56] the flexion/extension of all fingers is driven by a main drive composed by a rotational disk in which the linkages that are attached to the fingers are secured at different points. As a consequence of the arrangement of the linkages on the disk, the two different rotational directions of the main drive (clockwise and anticlockwise) produce the flexion of the fingers. The thumb abduction is realised by means of a micro motor embedded in the proximal phalanx of the thumb. The hand is capable of three grasp postures: neutral, lateral, and cylindrical.

Pons et al. [42] designed a thumb mechanism that employed one single motor connected to a so-called Geneva wheel that allowed the thumb to flex/extend and abduct/adduct simultaneously.

A brief comparison based on the discussed parameters, between the main transmission mechanisms that can be employed in artificial hands, is shown in Table 2.

Table 2 Brief transmission comparison

Proprieties class	Weight	Dimension	Stiffness	Backlash	Inertia	Friction	Noise	Reliability	Push and pull
Tendon/idle pulleys	●●●●○	●●●●○	●○○○○	●●●●○	●●●●○	●●●●●	●●●●●	●○○○○	No
Tendon/sheath	●●●●●	●●●●●	●○○○○	●●●●○	●●●●●	●○○○○	●●●●○	●○○○○	No
Linkages and cams	●●○○○	●○○○○	●●●●●	●○○○○	●●○○○	●●○○○	●●●○○	●●●●○	Yes
Cylindrical gear	●○○○○	●●○○○	●●●●○	●●○○○	●○○○○	●●●●○	●○○○○	●●●●●	Yes
Bevel gear	●○○○○	●●○○○	●●●●●	●●●○○	●○○○○	●●●●○	●●●○○	●●●●●	Yes
Flat bends and belt	●●●●○	●●●○○	●●●○○	●●●●●	●●●●○	●●●○○	●●●●○	●●○○○	No
Flexible shaft	●●●○○	●●●●○	●●○○○	●○○○○	●●●○○	●●○○○	●●○○○	●●●○○	Yes

3.5 Sensors

Sensors in a robotic hand allow it to assess exteroceptive data, such as information about the interaction between the object and the hand and between the object and the environment (force, object slip and temperature sensors), and proprioceptive data, which are related to the measurement of the status of the hand, such as the fingers configuration or joints torques/speeds (position and torque sensors). Therefore sensors play a unique role in the perception, and without the information given by them the potential dexterity of a given hand's kinematics architecture comes to naught since the hand can be only used as an open-loop device [11, 12]. In general sensors should be lightweight, robust, small, with low drift, and low power consumption. The characteristic of the sensors strongly depends on the information it is intended to monitor: for example an ideal force sensor employed in an artificial hand would have high sensitivity in the lower range of forces and a monotonic characteristic, while a position sensor requires a linear characteristic [90].

The reader can find a detailed review of sensors for artificial hands in [90] and (with a focus on the touch sensors) in [91], while here we restrict the discussion to two type of sensors: force sensors and position sensors.

Most force sensors are developed using strain gauges mounted on a deformable structure. For example the tendon's tension in the Stanford/JPL Hand was achieved by means of strain gauges glued at the base of the idle pulley support [92], while in the SmartHand the ends of each tendon were fixed to a micromechanical structure in order to obtain a cantilever elastically strained by the cable. Two silicon strain gauges were glued onto the sensor cantilever: one is the varying resistor, the other is a dummy resistor used for temperature compensation [69]. In the Utah/MIT Hand, cable tendon tensions (up to 130 N) were measured at the wrist by Hall sensors in charge of monitoring the deflection of idler pulleys on which the tendons were wrapped [93].

Considering the direct measurement of touch, a matrix of miniaturised tri-axial sensors combining micromechanical systems (MEMS) and piezoresistors [94] were successfully integrated in the fingertip of a dexterous finger [95]. Similarly, fingertips of the Stanford/JPL Hand were covered by tactile sensors composed by a 7×8 tactile elements (tactels) on the cylindrical portion, and 1×8 elements under the hemispherical tip [92], while a capacitance sensor array using floating electrodes and embedded in rubber layers has been used in the Utah/MIT Hand [93]. The fingers and palm of the Gifu Hand II were covered with a distributed tactile sensor with 624 detecting points. The tactile sensor had a grid pattern electrode and used conductive ink in which the electric resistance changes in proportion to the pressure on the top and bottom of a thin film. The maximum load supported was about 74 kN/m, the resolution of measurement was 8 bits, with a sampling cycle 10 ms/frame [96]. Other examples of sensors that can be used in artificial hands for monitoring the interaction between hand and object are based on optical effects [97] and force sensitive resistors (FSR) [98].

The measurement of the configuration of the finger is usually indirectly achieved with optical encoder since they are integrated into the actuator, robust and miniaturised with the only disadvantage of a relative high cost compared with other solutions. A direct measure of the joint's angle of the finger is achievable through Hall effect sensors that sense the magnetic field produced by permanent magnets located on the movable phalanx, such as in the SmartHand [69] and Utah/MIT Hand [93]. In the DLR I Hand the actual position of each DoF was acquired by means of the combination of two different sensors: the motor's position was monitored by means of Hall effect sensors whereas the joints configuration was observed by sensors based on a one-dimensional PSD (Position Sensing Device) allowing consideration of the tendon hysteresis [99].

3.6 Materials, Compliance and Integration (Manufacturing)

In the design of the hand components, since it is an end-effector and due to the movements of the arm it is subject to high speeds, the Young's modulus to density is one of the best performance indices to maximise. Other constraints could be considered, like minimum wall thickness (aluminium alloys or composites), bio-compatibility (titanium alloys), and corrosive environments (inox steels) [100]. In all cases the design and fabrication of a prototype using traditional machinery techniques is a rather long and expensive process, and often the employment of rapid prototyping techniques provides several advantages towards the development of a functional prototype. One of the main advantages is represented by the chance to develop parts with complex geometry that could not be manufactured using traditional techniques.

The choice of the material usually affects most of the features of a mechanical component, e.g. weight, compliance, strength, etc. As in the design in all engineering fields there are many factors that should be considered when choosing materials and/or manufacturing processes for robotic hands. These topics are out of the scope of this chapter, which instead will provide a quick overview of the materials and manufacturing techniques chosen for some of the state of the art robotic hands.

Won et al. [101] developed a five finger anthropomorphic hand by using a Selective Laser Sintering technique (SLS), which allows joints to be manufactured in one step without requiring assembly. The main drawback of SLS is the lack of mechanical proprieties (such as stiffness and strength), moreover it fails when dimension tolerances are narrowed. Dalley et al. [102] developed a transradial anthropomorphic hand, in which the parts were physically realised in high-strength, nickel-coated thermoplastic using an additive directly incorporated during the manufacturing process; this method combines the flexibility of the rapid technique with the strength/stiffness of the metallic material. The main drawbacks of this method are the low fatigue resistance, and poor surface finish and geometrical tolerance that often requires re-machining by traditional methods.

Dollar and Howe [63] showed how a further integration of sensors, electronics, and actuation could be obtained using polymer-based Shape Deposition Manufacturing (SDM). Another advantage of the SDM technique is the possibility of simultaneously creating rigid links and compliant joints of the fingers, providing the hand with passive mechanical compliance useful for grasping in unstructured environments.

The use of compliant materials in the manufacturing of joints allows a number of components to be avoided (such as pulleys, axis, torsion springs, and so on) and the ability to embed the extension system, disguised as releasing springs, in the structure thus resulting in a reduction in joint size. Examples of artificial hands exploiting compliant joints based on these principles were developed at the University of Bologna [103], Stanford [104], Genoa [105], Scuola Superiore Sant'Anna [106], and University of Iowa [107]. In the last few decades researchers have investigated new materials with proprieties close to the human skin. The necessity of compliant materials for fingers is better understood by considering the physiology of the human finger, which can be summarised as a multi-layer structure where the layers have different mechanical properties. This unique structure results in a peculiar force-deformation characteristic under external loads and plays a fundamental role during the action of precision grasping and manipulation. Indeed the human skin modulates the contact forces and hence their physical transmission and transduction into neural signals by mechanoreceptors, due to its viscoelasticity. The outer layer is composed by the skin which plays a key role in the sense of touch and in the mechanics of contact, then a thick layer is composed by the soft tissue, which is responsible for strain dissipation during force interaction so to reduce wear and possible damage, meanwhile the bone represents the stiff core of the finger and, only in the fingertip, nail as a hard shell suppresses excessive deformation of soft tissue and enlarges the range of the possible friction coefficient [108]. Examples of bio-inspired fingertips for robotic hands were developed by the University of Leeds [109] and Bologna [110].

4 Open Issues and Conclusion

The human hand is powerful organ, able to perform surprising actions ranging from power grasps to fine manipulation tasks. The chance of duplicating it and its extraordinary capabilities has been a fascination for humans for years. Today, the employment of an end-effector able to replicate the same tasks that only the human hand is able to realise, enables notable benefits to be earned in terms of the flexibility of manufacturing lines: most of the tasks required to develop the current product may be accomplished by the same robot and, assisted by humans, it may learn new tasks making the same robot suitable for future developments of different products. In addition, humanoid robots for providing entertainment and services to human subjects have been emerging on the commercial market in the last few years, and the possibility of robots that act and look like humans, working in cooperation with them, is now concrete. Despite the advantages that only a robotic hand can offers, today's industry is far from the development of an end-effector with proprieties similar to the human hand, and it is still convenient to choose grippers that are developed to solve a limited set of tasks.

So, why there are not robotic hands in today's factories?

By reviewing the state of the art of existing robotic hands it is clear how, although many artificial hands have been developed, in all the examples that were analysed in this chapter, the designers are always forced to reach a compromise between the wide range of constraints and the limitations of available technologies. The main practical reason that limits the diffusion of robotic hands in factories is related to the complex architecture of this device, which has been demonstrated by the low reliability of the prototypes so far developed. The good news is that the growing computing power of current systems allows the management of complex kinematical architecture and sensors in real time, while manufacturing technologies are in continuous evolution, and these are promising for future applications of these devices in practical environments.

In the prosthetic field, in the last few years two multi-articulated prosthesis have been introduced to the market: the i-limb (developed by Touch Bionics in 2007) and BeBionic (developed by RSL Steeper in 2010) prostheses.

Thanks to the five individually-powered digits and the passive movement of abduction/adduction of the thumb, these hands are capable of different grasping patterns, nevertheless they still use traditional two-input EMG interfaces to simultaneously open and close all fingers. Advances in the design of future prosthetic hands are fixed by the bandwidth of the interface that is currently very poor, disallowing the direct control of DoF architectures. While the research strives to increase the bandwidth of the human machine's interfaces, hands will be improved in terms of cosmesis (reducing the drawback of a chilly device that could disturb the amputee), robustness of the mechanisms, and the ability of the hand to perform more complex automatic grasps, both useful to extending the autonomy of the amputee in the execution of their daily activities.

Appendix

See Table 3:

Table 3 Some examples of artificial hands

Name	Year	Fingers	Hand kinematics		Actuation type	Sensors		
			Joints/DoC	N° actuators		Position	Force/torque	Contact
Human hand		5	22/18	38	Intrinsic and extrinsic muscles	About 17,000 mechanoreceptors for touch, pressure, pain and temperature detection, and musculo-tendinous receptors for proprioceptive information		
Belgrade	1969	3	13/1	1	Remote DC motors	–	–	Finger tip tactile sensors
Okada hand	1978	3	9/9	9	Remote DC motors	Encoders	Torque sensors	–
Stanford/JPL hand	1981	3	9/9	12	Remote DC motors	Encoders	Tendon tension sensors based on strain gauges	Finger tip force sensors 8 × 8 tactile sensors array
Utah/MIT Hand	1982	4	16/16	32	Remote pneumatic actuators	Encoders and joints angle sensors based on Hall effect	Tendon tension sensors based on Hall effect	Capacitive tactile sensors
The Hitachi Robot hand	1984	4	12/12	48	Remote SMA	–	–	–
Belgrade/USC hand	1988	5	18/4	4	Remote DC motors	Encoders	–	Tactile sensors
Barrett hand	1988	3	18/4	4	Brushless DC motors	Encoders	Torque sensors based on strain gauges	–
UB hand II	1992	5	13/13	13	Remote DC motors	Encoders and joints angle sensors based on Hall effect	–	6-axis sensors
NTU hand	1996	5	17/17	17	Micro DC motors	Joint position sensors	–	Tactile sensors
DLR hand I	1997	4	16/12	12	Micro Brushless DC motor	Hall motor sensors motors, PSD-LED joint sensors	Torque sensors based on strain gauges	Tactile sensors
DIST hand	1998	4	16/16	20	Remote Brushless DC Motors	Encoders and joints angle sensors based on Hall effect	–	3-axis fingertip force sensors

(continued)

Table 3 (continued)

Name	Year	Fingers	Hand kinematics		Actuation type	Sensors		
			Joints/DoC	N° actuators		Position	Force/torque	Contact
LMS hand	1998	4	17/16	16	Remote DC motors	Encoders and potentiometers	Force obtain by means of tendons elongation	–
Robonaut	1999	5	20/12	12	Remote DC motors	Encoders, joint sensors	Force sensors	FSR tactile sensors
MANUS	1999	5	9/2	2	Brushless DC motors	Encoders and joints angle sensors based on hall effect	Sensors based on hall effect	–
DLR hand II	2000	4	17/13	13	Brushless DC motors	Encoders, potentiometers	Torque sensors based on strain gauges	Tactile sensors
TUAT/Karlsruhe hand	2000	5	17/1	1	Remote ultrasonic motor	–	–	–
Southampton REMEDI hand	2001	5	13/6	6	DC Motors	Encoders	–	–
High speed hand	2001	3	8/8	8	DC Brushless motors	Encoders	Sensors based on strain gauges and 6-axis force sensors	Tactile sensors
GIFU hand II and III	2001	5	20/16	16	DC Brushless motors	Encoders	–	Tactile sensors (859 detecting point)
Shadow hand	2002	5	23/23	36	Remote Pneumatic actuators (McKibben muscles)	Joints angle sensors based on Hall effect	Pressure sensors	Tactile sensors
RTR II	2002	3	11/2	2	DC Motors	Joints angle sensors based on Hall effect and encoders	Sensors based on strain gauges	FSR sensor

(continued)

Table 3 (continued)

Name	Year	Fingers	Hand kinematics		Actuation type	Sensors		
			Joints/ DoC	N° actuators		Position	Force/torque	Contact
Sensorhand speed	2002	3	3/1	1	DC Motor	Encoder	Sensor based on strain gauges	SUVA sensor
UBH III	2003	5	20/20	20	DC Brushless motors	Bending sensors (Piezoresistive effect)	Sensors based on strain gauges	Intrinsic tactile sensors
IOWA-hand	2004	5	11/11	11	-	–	–	–
Karlsruhe hand	2005	5	15/15	15	Pneumatic actuators (flexible fluid actuators)	Joint sensors	–	Tactile sensors
I-Limb	2007	5	11/5	5	DC motors	Encoder	–	–
Vanderbilt Hand	2007	5	17/6	6	Remote Gas actuators	–	Tension sensors	–
SmartHand	2008	5	16/4	11	DC Motors	Joints angle sensors based on hall effect and encoders	Sensors based on strain gauges	LED and photo detector
SDM hand	2008	4	8/1	1	Remote DC Motor	Joints angle sensors based on Hall effect	–	LED and photo detector
BeBionics	2010	5	11/5	5	DC Motors	Encoders	–	–

Weight [Kg]	Dimension /HH	Transmission	Force [N]/speed [s]	Purpose	References
0.4	1	Net of tendons and sheats	400/0.25	Multipurpose	[87]
–	1	Linkages	–/–	Prosthetics	[58]
–	1	Tendons	–/0.2	Object-handling system for manual industry	[31]
5.5	1.2	Tendons, pulleys and sheats	45/–	Machine dexterity	[32]
3.2	2	Tendons, idle pulleys	32/0.1	machine dexterity, fingertips, phalanges, palm manipulation	[48]
4.5	<1	Tendons, pulleys and sheats	20/–		[51]

(continued)

Table 3 (continued)

Weight [Kg]	Dimension /HH	Transmission	Force [N]/speed [s]	Purpose	References
–	1.1	Linkages	22/2	Prosthetics	[36]
1.2	1	Gears	15/1	Multipurpose end-effector	[41]
–	1	Tendons, pulleys and sheats	–/–	Machine dexterity	[111]
1.57	>1	Spur gears	10 / 1	Industrial and prosthetics applications	[44]
1.8	2	Tendons, and pulleys	–/0.5	Grasping control and telemanipulation	[37, 99]
<1	1	Tendons, pulleys and sheats	–/–	Grasping control and telemanipulation	[50]
1	–	Tendons, pulleys and sheats	–/–	Grasping control and telemanipulation	[88]
1.2	–	Linkages, flexible shafts, and cams	–/–	Space operations	[69]
1.2	1.2	Linkages, and tendons	40/–	Prosthetics	[42]
–	1.5	Bevel gears	40/–	Space operations	[38]
0.125	1	Linkages	–/–	Humanoid robot	[64]
1	0.4	Linkages	9.2/0.8	Prosthetics	[43]
0.8	1	Gears	28/0.1	Catching and dexterous manipulation by vision	[34]
1	1.4	Gears	5/0.1	Tele-operation and dexterous manipulation	[39, 96]
4	1	Tendons, pulleys and sheats	16/0.5	Tele-operation and dexterous manipulation	[49]
0.35	1	Tendons, pulleys and sheats	–/1.5	Prosthetics	[66]
0.4	1	Linkages	100/1	Prosthetics	[55]
–	1	Tendons and sheats	10/0.85	Humanoid robot	[112]
0.09*	1	Tendons and sheats	–/–	Prosthetics	[107]
0.15	1	Direct drive	12/0.1	Prosthetics and Humanoid	[78]
0.518	1	Flexible rackets		Prosthetics	[45]
1.6 (whole arm)	1	Tendons and sheats	150/0.5	Prosthetics and humanoid	[79]
0.560	1	Tendons, pulleys and sheats	45/1.2	Prosthetics	[69]
0.16*	1	Tendons, pulleys and nylon conduits	–/–	Multipurpose end-effector	[63]
0.539	1	Linkages	–/–	Prosthetics	[46]

* Actuation unit not included

References

1. Aristotle, *De Anima*, 350 BCE, III 8, 431 b 30 432a 3
2. R.H. Meier, D.J. Atkins, *Functional restoration of adults and children with upper extremity amputation* (Demos Medical Publishing, New York, 2004)
3. D.S. Childress, Artificial hand mechanisms, in *Mechanisms Conference and International Symposium on Gearing and Transmissions, San Francisco California,* Oct 1972
4. D.S. Childress, Historical aspects of power limb prostheses. Clin Prosthet Orthot **9**(1), 2–13 (1985)
5. M.E. Rosheim, *Robot evolution: The development of Anthrobotics* (Wiley, New York, 1994)
6. B.B. Edin, G. Westling, R.S. Johansson, Independent control of human finger-tip forces at individual digits during precision lifting. J Physiol **450**, 547–564 (1992)
7. B. Rosén, H.H. Ehrsson, C. Antfolk, C. Cipriani, F. Sebelius, G. Lundborg, Referral of sensation to an advanced humanoid robotic hand prosthesis. Scand. J. Plast. Reconstr. Surg. Hand Surg. **43**, 260–266 (2009)
8. H.H. Ehrsson, A. Fagergren, T. Jonsson, G. Westling, R.S. Johansson, H. Forssberg, Cortical activity in precision- versus power-grip tasks: An fMRI study. J. Neurophysiol. **83**, 528–536 (2000)
9. H. Yokoi, R. Katoh, A. Hernandez Arieta, T. Miyamoto, K. Ikoma, T. Onishi, W. Yu, T. Arai, Cerebral reorganization by adaptable robot hand. Neurosc. Res. **58**(supplement 1), 29 (2007)
10. J.L. Pons, R. Ceres, F. Pfeiffer, Multifingered dextrous robotics hand design and control: A review. Robotica **17**(6), 661–674 (1999)
11. A. Bicchi, Hands for dextrous manipulation and robust grasping: A difficult road towards simplicity. IEEE Trans Robot Autom **16**(6), 652–662 (2010)
12. L. Biagiotti, F. Lotti, C. Melchiorri, G. Vassura, *How Far Is the Human Hand? A Review on Anthropomorphic Robotic End-Effectors* (DEIS–DIEM, University of Bologna, Bologna, 2004)
13. J.R. Napier, The prehensile movements of the human hand. J. Bone Joint Surg. Br. **38-B**(4), 902–913 (1956)
14. R.M. Murray, Z. Li, S.S. Sastry, *A Mathematical Introduction to Robotic Manipulation* (CRC Press, Boca Raton, 1994)
15. M.A. Okamura, N. Smaby, M.R. Cutkosky, An overview of dexterous manipulation, in *Proceedings of the 2000 IEEE International Conference on Robotics and Automation, Symposium on Dexterous Manipulation*, 2000, pp. 255–262
16. M.T. Mason, J.K. Salisbury Jr, *Robot Hands and the Mechanics of Manipulation* (MIT Press, Cambridge, 1985)
17. C.L. Taylor, R.J. Schwarz, The Anatomy and Mechanics of the Human Hand. Artif Limbs **2**, 22–35 (1995)
18. H.E. Griffiths, Treatment of the injured workman. *Lancet* **241**(6250), 729–733 (1943)
19. E.D. McBride, *Disability Evaluation* (T.B. Lippincott Co., Philadelphia, 1942)
20. T. Iberall, G. Bingham, M.A. Arbib, Opposition space as a structuring concept for the analysis of skilled hand movements, in *Experimental Brain Research Series 15—Generation and Modulation of Action Patterns*, ed. by H. Heuer, C. Fromm (Springer, Berlin, 1985), pp. 158–173
21. M.A. Arbib, T. Iberall, D.M. Lyons, Coordinated control programs for movements of the hand, in *In Hand Function and the Neocortex*, ed. by A.W. Goodwin, I. Darian-Smith (Springer, Berlin, 1985), pp. 111–129
22. M.R. Cutkosky, On grasp choice, grasp models, and the design of hands for manufacturing tasks. IEEE Trans Robotics and Automation **5**(3), 269–279 (1989)
23. M. Santello, M. Flanders, J.F. Soechting, Postural hand synergies for tool use. J Neurosci **18**(23), 10105–10115 (1998)

24. C.Y. Brown, H. Asada, Inter-finger coordination and postural synergies in robot hand via mechanical implementation of principal components analysis, in *Proceedings of the IEEE/RJS International Conference on Intelligent Robots and Systems*, 2007, pp. 2877–2882
25. A. Tsoli, O.C. Jenkins, Robot grasping for prosthetic applications, in *Proceedings of the International Symposium on Robotic Research*, 2007
26. J. Rosell, R. Suárez, C. Rosales, J. A. García, A. Pérez, Motion planning for high DoF anthropomorphic hands, in *Proceedings of IEEE International Conference on Robotics and Automation*, 2009, pp. 4025–4030
27. M.T. Ciocarlie, P.K. Allen, Hand posture subspaces for dexterous robotic grasping. Int J Robot Res **28**(7), 851–867 (2009)
28. G. Matrone, C. Cipriani, M. C. Carrozza, G. Magenes, Real-time myoelectric control of a multi-fingered hand prosthesis using principal components analysis. J. NeuroEng, Rehabilit. **9**(40) (2012)
29. I.A. Kapandji, *The Physiology of the Joints—Upper Limb* (Churchill Livingstone, Edinburgh, 1982)
30. L. Barbieri, M. Bergamasco, Nets of tendons and actuators: an anthropomorphic model for the actuation system of dexterous robot hands, in *Proceedings of the IEEE International Conference on Advanced Robotics*, 1991, pp. 357–362
31. T. Okada, Object-handling system for manual industry. IEEE Trans. Syst. Man Cybern. **9**(2), 79–89 (1979)
32. J.K. Salisbury, J.J. Craig, Articulated hands: force control and kinematic issue. Int J Robot Res **1**(1), 4–17 (1982)
33. G. Wöhlke, A programming and simulation environment for the Karlsruhe dexterous hand. J Robot Auton Syst **9**, 243–262 (1990)
34. A. Namiki, Y. Imai, M. Ishikawa, M. Kaneko, Development of a high-speed multifingered hand system and its application to catching, in *Proceedings of the 2003 IEEE/RSJ International Conference on Intelligent Robots and Systems (IROS 2003)*, vol. 3, 27–31 Oct 2003, pp. 2666–2671
35. A. Morecki, Z. Busko, H. Gasztold, K. Jaworek, Synthesis and control of the anthropomorphic two-handed manipulator, in *Proceedings of the 10th International Symposium on Industrial Robots*, 1980
36. G.A. Bekey, R. Tomovic, I. Zeljkovic, *Control architecture for the Belgrade/USC hand. Dextrous robot hands* (Springer, New York, 1990)
37. J. Butterfass, G. Hirzinger, S. Knoch, H. Liu, DLR's multisensory articulated hand. Part I: Hard- and software architecture, in *Proceedings of the 1998 IEEE International Conference on Robotics and Automation, Leuven, Belgium*, May 1998
38. J. Butterfass, M. Grabenstein, H. Liu, G. Hirzinger, DLR-Hand II: next generation of dexterous robot hand, in *Proceedings of the IEEE International Conference on Robotics and Automation*, 2001, pp. 109–114
39. H. Kawasaki, T. Komatsu, K. Uchiyama, Dexterous anthropomorphic robot hand with distributed tactile sensor: Gifu hand II. IEEE/ASME Trans. Mechatron. **7**(3), 296–303 (2002)
40. T. Mouri, H. Kawasaki, S. Ito, Unknown object grasping strategy imitating human grasping reflex for anthropomorphic robot hand. J Adv Mech Des Syst Manuf **1**(1), 1–11 (2007)
41. W. Townsend, The Barrett hand grasper—programmable flexible part handling and assembly. Ind Robot **27**(3), 181–188 (2000)
42. J.L. Pons, E. Rocon, R. Ceres, D. Reynaerts, B. Saro, S. Levin, W. Van Moorleghem, The MANUS-HAND dextrous robotics upper limb prosthesis: mechanical and manipulation aspects. Auton Robot **16**, 143–163 (2004)
43. C.M. Light, P.H. Chappell, Development of a lightweight and adaptable multiple-axis hand prosthesis. Med Eng Phys **22**, 679–684 (2000)
44. L. Lin, H. Huang, NTU hand: a new design of dexterous hands. J. Mech. Des. **120**(2), 282–293 (1998)
45. Touch EMAS Ltd. http://www.touchbionics.com

46. RSL Steeper BeBionic. http://www.bebionic.com
47. T. Yoshikawa, Manipulability of robotic mechanisms. Int. J. Robot. Res. **4**, 3–9 (1985)
48. S.C. Jacobsen, E.K Iversen, D.F. Knutti, R.T. Johnson, K.B. Biggers, Design of the Utah/MIT dexterous hand, in *Proceedings of the 2001 IEEE International Conference on Robotics and Automation*, San Francisco, CA, USA, 1986, pp. 1520–1532
49. R. Walker, Developments in dextrous hands for advanced robotic applications, in *10th International Symposium on Robotics and Applications (ISORA 2004)*, Seville, Spain, 28 June–1 July 2004
50. A. Caffaz, G. Cannata, The design and development of the DIST-hand dextrous gripper, in *Proceedings of the IEEE International Conference on Robotics and Automation*, 1998
51. Y. Nakano, M. Fujie, Y. Hosada, Hitachi's robot hand. Robot. Age **6**(7), 18–20 (1984)
52. A.D. Deshpande, Z. Xu, M.J.V. Weghe, B.H. Brown, J. Ko, L.Y. Chang, D.D. Wilkinson, S.M. Bidic, Y. Matsuoka, Mechanisms of the anatomically correct testbed hand. IEEE/ASME Trans. Mechatron. **1**(18), 238–250 (2013)
53. M. Grebenstein, P. Van Der Smagt, Antagonism for a highly anthropomorphic hand-arm system. Adv. Robot. **22**, 39–55 (2008)
54. M. Grebenstein, G. Hirzinger, R. Siegwart, Antagonistically driven finger design for the anthropomorphic DLR hand arm system, *Humanoids*, 2010
55. Otto Bock Healthcare, Italy, MN, 2011. http://www.ottobock.com
56. G. Puchhammer. Hand prosthesis and force transmission device, US Patent 2008/0319553 A1, 25 Dec 2008
57. T. Laliberte, L. Birglen, C.M. Gosselin, Underactuation in robotic grasping hands. Mach. Intell. Robot. Control **4**(3), 1–11 (2002)
58. M. Rakic, An automatic hand prosthesis. Med. Electron. Biol. Eng. **2**, 47–55 (1964)
59. P.J. Kyberd, O.E. Holland, P.H. Chappel, S. Smith, R. Tregdigo, P.J. Bagwell, M. Snaith, Marcus: a two degree of freedom hand prosthesis with hierarchical grip control. IEEE Trans Rehabilit Eng **3**(1), 70–76 (1995)
60. P. Kyberd, C. Light, P.H. Chappell, J.M. Nightingale, D. Whatley, M. Evans, The design of anthropomorphic prosthetic hands: a study of the Southampton hand. Robotica **19**, 593–600 (2001)
61. N. Dechev, W.L. Cleghorn, S. Naumann, Multi finger, passive adaptive grasp prosthetic hand. Mech. Mach. Theory **36**, 1157–1173 (2001)
62. H. Maekawa, K. Yokoi, K. Tanie, M. Kaneko, N. Kimura, N. Imamura, Development of a three-fingered robot hand with stiffness control capability. Mechatronics **2**(5), 483–494 (1992)
63. A.M. Dollar, R.D. Howe, The highly adaptive SDM hand: design and performance evaluation. Int. J. Robot. Res. **29**, 595 (2010)
64. N. Fukaya, S. Toyama, T. Asfour, R. Dillmann, Design of the TUAT/Karlsruhe humanoid hand, in *Proceedings of the 2000 IEEE/RSJ International Conference on Intelligent Robots and Systems*, 2000
65. S. Hirose, Y. Umetami, The development of soft gripper for the versatile robot hand. Mech. Mach. Theory **13**, 351–359 (1977)
66. B. Massa, S. Roccella, M. C. Carrozza, P. Dario, Design and development of an underactuated prosthetic hand, in *Proceedings of the ICRA'02. IEEE International Conference on Robotic Automation*, 2002, vol. 4, pp. 3374–3379
67. M.C. Carrozza, C. Suppo, F. Sebastiani, B. Massa, F. Vecchi, R. Lazzarini, M.R. Cutkosky, P. Dario, The SPRING Hand: development of a self-adaptive prosthesis for restoring natural grasping. Auton. Robots **16**(2), 125–141 (2004)
68. M.C. Carrozza, G. Cappiello, S. Micera, B.B. Edin, L. Beccai, C. Cipriani, Design of a cybernetic hand for perception and action. Biol. Cybern. **95**(6), 629–644 (2006)
69. C. Cipriani, M. Controzzi, M. C. Carrozza. The SmartHand transradial prosthesis. J. NeuroEng. Rehabil. **8**, 29 (2011)
70. A. Schmitz, U. Pattacini, F. Nori, L. Natale, G. Metta, G. Sandini, Design, realization and sensorization of a dextrous hand: The iCub design choices, in *10th IEEE-RAS International Conference on Humanoid Robots*, Nashville, Tennessee, 6–8 December 2010

71. Defense Advanced Research Project Agency (DARPA), *DARPA initiates revolutionary prosthetics programs*, Press release of 8 February 2006, Arlington, VA, USA, 2006
72. X. Zhe, E. Todorov, B. Dellon, Y. Matsuoka, Design and analysis of an artificial finger joint for anthropomorphic robotic hands, in *2011 IEEE International Conference on Robotics and Automation (ICRA)*, 9–13 May 2011, pp. 5096–5102
73. J.E. Huber, N.A. Fleck, M.F. Ashby, The selection of mechanical actuators based on performance indices. Proc. R. Soc. A **453**, 2185–2205 (1997)
74. F. Smith, S. Jacobsen, D. Potter, C. Davis, Miniature high performance servovalves, *International Fluid Power Exposition and Technical Conference*, Chicago, 24–26 March 1992
75. R.H. Gaylord, Fluid actuated motor system and stroking device, US Patent 2,844,126, 1958
76. V.L. Nickel, J. Perry, A.L. Garrett, Development of useful function in the severely paralyzed hand. J. Bone Joint Surg. **45-A**, 933 (1963)
77. B. Tondu, P. Lopez, Modeling and control of McKibben artificial muscle robot actuators. Control Syst. Mag. **20**(2), 15–38 (2000)
78. S. Schulz, C. Pylatiuk, M. Reishl, J. Marti, R. Mikut, G. Bretthauer, A hydraulically driven multifunctional prosthetic hand. Robotica **23**, 293–299 (2005)
79. K. Fite, T. J. Withrow, K. W. Wait, M. Goldfarb, A gas-actuated anthropomorphic transhumeral prosthesis, in *Proceedings of the 2007 IEEE International Conference on Robotics and Automation*, 2007
80. K. DeLaurentis, C. Pfeiffer, C. Mavroidis, Development of a shape memory alloy actuated hand, in *Proceedings of the 7th International Conference on New Actuators*, 2000, pp. 281–284
81. D. Reynaerts, H. Van Brussel, A high performance actuators for robot hands. Journal de Physique **IV 01**(C4), 157–162 (1991)
82. M. Bergamasco, F. Salsedo, P. Dario, Shape memory alloy micromotors for direct-drive actuation of dexterous artificial hands. Sens. Actuators **17**(1–2), 115–119 (1989)
83. M. Controzzi, C. Cipriani, M.C. Carrozza, Miniaturized non-back-drivable mechanism for robotic applications. Mech. Mach. Theory **45**(10), 1395–1406 (2010)
84. R. Ham, T. Sugar, B. Vanderborght, K. Hollander, D. Lefeber, Compliant actuator designs. IEEE Robot. Autom. Mag. **16**(3), 81–94 (2009)
85. T. Wimbock, Ch. Ott, A. Albu-Schaffer, A. Kugi, G. Hirzinger, Impedance control for variable stiffness mechanisms with nonlinear joint coupling, in *IEEE International Conference on Intelligent Robotic Systems*, 2008, pp. 3796–3803
86. M. Grebenstein, A. Albu-Schaffer, T. Bahls, M. Chalon, O. Eiberger, W. Friedl, R. Gruber, S. Haddadin, U. Hagn, R. Haslinger, H. Hoppner, S. Jorg, M. Nickl, A. Nothhelfer, F. Petit, J. Reill, N. Seitz, T. Wimbock, S. Wolf, T. Wusthoff, G. Hirzinger, The DLR hand arm system, in *2011 IEEE International Conference on Robotics and Automation (ICRA)*, 9–13 May 2011, pp. 3175–3182
87. R. Tubiana. *The Hand* (W. B. Saunders Company, West Washington Square, 1981)
88. D. Chaigneau, M. Arsicault, J.-P. Gazeau, S. Zeghloul, LMS robotic hand grasp and manipulation planning (an isomorphic exoskeleton approach). Robotica **26**, 177–188 (2008)
89. C. S. Lovchik, M. A. Diftler, The Robonaut hand: a dexterous robot hand for space, in *Proceedings of the 1999 IEEE International Conference on Robotics & Automation*, Detroit, Michigan, May 1999
90. P.H. Chappell, Making sense of artificial hands. J. Med. Eng. Technol. **35**(1), 1–18 (2011)
91. R.S. Dahiya, G. Metta, M. Valle, G. Sandini, Tactile sensing—from humans to humanoids. IEEE Trans. Rob. **26**(1), 1–20 (2010)
92. R. Fearing, Some experiments with tactile sensing during grasping, in *Proceedings of the 1987 IEEE International Conference on Robotics and Automation*, March 1987, vol. 4, pp. 1637–1643
93. D. Johnston, P. Zhang, J. Hollerbach, S. Jacobsen, A full tactile sensing suite for dextrous robot hands and use in contact force control, in *Proceedings of the 1996 IEEE International Conference on Robotics and Automation*, Minneapolis, Minnesota, April 1996

94. L. Beccai, S. Roccella, A. Arena, F. Valvo, P. Valdastri, A. Menciassi, M.C. Carrozza, P. Dario, Design and fabrication of a hybrid silicon three-axial force sensor for biomechanical applications. Sens. Actuators, A **120**, 370–382 (2005)
95. C.M. Oddo, M. Controzzi, L. Beccai, C. Cipriani, M.C. Carrozza, Roughness encoding for discrimination of surfaces in artificial active touch. IEEE Trans. Rob. **27**(3), 522–533 (2011)
96. H. Kawasaki, T. Komatsu, K. Uchiyama, Dexterous anthropomorphic robot hand with distributed tactile sensor: Gifu hand II. IEEE/ASME Trans. Mechatron. **7**(3), 296–303 (2002)
97. A. Persichetti, F. Vecchi, M. C. Carrozza, Optoelectronic-based flexible contact sensor for robot finger application, in *10th International Conference on Rehabilitation Robotics (ICORR 2007)*, 13–15 June 2007
98. A. Tura, C. Lamberti, A. Davalli, R. Sacchetti, Experimental development of a sensory control system for an upper limb myoelectric prosthesis with cosmetic covering. J. Rehabil. Res. Dev. **35**, 14–26 (1998)
99. H. Liu, P. Meusel, J. Butterfass, G. Hirzinger, DLR's multisensory articulated hand—Part II: The parallel torque/position control system, in *Proceedings of the 1998 IEEE International Conference on Robotics & Automation*, Leuven, Belgium, 1 May 1998
100. W.D. Callister Jr, *Materials science and engineering: an introduction* (Wiley, New York, 1997)
101. J. Won, J.K. DeLaurentis, C. Mavroidis, Fabrication of a robotic hand using rapid prototyping, in *ASME Design Technical Conference*, DECT00/MECH-14203, December 2000
102. S.A. Dalley, T.E. Wiste, T.J. Withrow, M. Goldfarb, Design of a multifunctional anthropomorphic prosthetic hand with extrinsic actuation. IEEE/ASME Trans. Mechatron. **14**(6), 699–706 (2009)
103. F. Lotti, P. Tiezzi, G. Vassura, A. Zucchelli. Mechanical structures for robotic hands based on the "compliant mechanism" concept, in *7th ESA Workshop on Advanced Space Technologies for Robotics and Automation (ASTRA 2002)*, Noordwijk, The Netherlands, 19–21 November 2002
104. R. Doshi, C. Yeh, M. LeBlanc, The design and development of a gloveless endoskeletal prosthetic hand. J. Rehabil. Res. Dev. **35**(4), 388–395 (1998)
105. E. Ullmann, F. Cepolina, M. Zoppi, Upper limb prosthesis for developing countries, in *Proceedings of the IEEE International Conference on Intelligent Manipulation and Grasping (IMG 04)*, Genova, Italy, 1–2 July 2004, ISBN 88 900 426-1-3
106. M.C. Carrozza, G. Cappiello, G. Stellin, F. Zaccone, F. Vecchi, S. Micera, P. Dario, A cosmetic prosthetic hand with tendon driven under-actuated mechanism and compliant joints: Ongoing research and preliminary results, in *Proceedings of the 2005 IEEE International Conference on Robotics and Automation (ICRA 2005)*, 18–22 April 2005, pp. 2661–2666
107. J. Yang, E.P. Pitarch, K. Abdel-Malek, A. Patrick, L. Lindkvist, A multi-fingered hand prosthesis. Mech. Mach. Theory **39**, 555–581 (2004)
108. K. B. Shimoga, A.A. Goldenberg, Soft materials for robotic fingers, in *Proceedings of the 1992 IEEE International Conference on Robotic and Automation (ICRA)*, Nice, France, May 1992
109. F. Shao, T.H.C. Childs, B. Henson, Developing an artificial fingertip with human friction properties. Tribol. Int. **42**, 1575–1581 (2009)
110. P. Tiezzi, G. Vassura, Experimental analysis of soft fingertips with internal rigid core, in *Proceedings of the 12th International Conference on Advanced Robotics (ICAR '05)*, 18–20 July 2005, pp. 109–114
111. C. Melchiorri, G. Vassura, Mechanical and control features of the UB hand version II, in *Proceedings of IEEE/RSJ International Conference on Intelligent Robots and Systems, IROS*, 1992
112. F. Lotti, P. Tiezzi, G. Vassura, L. Biagiotti, C. Melchiorri, UBH 3: an anthropomorphic hand with simplified endo-skeletal structure and soft continuous fingerpads, in *ICRA'04, IEEE International Conference on Robotics and Automation*, New Orleans, LA, USA, 26 April–1 May, 2004

Chapter 12
Optimizing the Topology of Tendon-Driven Fingers: Rationale, Predictions and Implementation

Joshua M. Inouye, Jason J. Kutch and Francisco J. Valero-Cuevas

Abstract Tendon-driven mechanisms in general, and tendon-driven fingers in particular, are ubiquitous in nature, and are an important class of bio-inspired mechatronic systems. However, the mechanical complexity of tendon-driven systems has hindered our understanding of biological systems and the optimization of the design, performance, control, and construction of mechatronic systems. Here we apply our recently-developed analytical approach to tendon-driven systems [1] to describe a novel, systematic approach to analyze and optimize the routing of tendons for force-production capabilities of a reconfigurable 3D tendon-driven finger. Our results show that these capabilities could be increased by up to 277 % by rerouting tendons and up to 82 % by changing specific pulley sizes for specific routings. In addition, we validate these large gains in performance experimentally. The experimental results for 6 implemented tendon routings correlated very highly with theoretical predictions with an R^2 value of 0.987, and the average effect of unmodeled friction decreased performance an average of 12 %. We not only show that, as expected, functional performance can be highly sensitive to tendon routing and pulley size, but also that informed design of fingers with fewer tendons can exceed the performance of some fingers with more tendons. This now enables the systematic simplification and/or optimization of the design and construction of novel robotic/prosthetic fingers. Lastly, this design and

J. M. Inouye
Department of Biomedical Engineering, University of Southern California, CA 90089, USA
e-mail: jinouye@usc.edu

J. J. Kutch
Division of Biokinesiology and Physical Therapy, University of Southern California, CA 90089, USA
e-mail: kutch@usc.edu

F. J. Valero-Cuevas (✉)
Department of Biomedical Engineering and the Division of Biokinesiology and Physical Therapy, University of Southern California, CA 90089, USA
e-mail: valero@usc.edu

R. Balasubramanian and V. J. Santos (eds.), *The Human Hand as an Inspiration for Robot Hand Development*, Springer Tracts in Advanced Robotics 95,
DOI: 10.1007/978-3-319-03017-3_12,

analysis approach can now be used to model complex biological systems such as the human hand to understand the synergistic nature of anatomical structure and neural control.

Keywords Biologically-inspired robots · Mechanism design

1 Introduction

Bio-inspired robotic hands employ multiple robotic fingers for dexterous grasping and manipulation tasks [2–13]. Bio-inspiration can refer to their tendon-driven nature, but also the asymmetry of the routings, the variation in moment arm sizes, and the non-uniform distribution of maximal tendon tensions. Robotic finger kinematics may be anthropomorphic or they may be less complex to simplify the construction and control of the fingers. Moreover, two fundamental classes of actuation are typically used: (i) those that use remote actuation (e.g., motors outside the fingers which actuate tendons, cables, or gears) and (ii) those that use internal actuation (e.g., motors inside the fingers). Tendon-driven limbs and fingers are ubiquitous in vertebrates, and such bio-inspired tendon-driven actuation has proven engineering design advantages such as light weight, low inertia, small size, backdrivability, and design flexibility [14]. However, the mechanical complexity of tendon-driven systems (e.g., the large number of design parameters) has precluded the development of modeling, design, and analysis tools to optimize their performance, control, and construction. In this paper, we analyze, optimize, and test alternative implementations of a 3D tendon-driven robotic finger. We validate this approach with physical hardware implementations from the functional perspective of maximizing the set of feasible endpoint static forces.

Many considerations go into the design of robotic fingers and hands, such as force and velocity production, control, ease of construction, design simplicity, and cost. Adequate force-production capabilities are a necessary element of the multidimensional design puzzle: according to Firmani, "The knowledge of maximum twist and wrench capabilities is an important tool for achieving the optimum design of manipulators" [15]. In fact, if a finger cannot produce sufficient endpoint force while meeting other critical design requirements such as size and number of motors (for example in space, hazardous or surgical applications), then the mechatronic system is useless regardless of the attributes of the controller or ease of manufacturing. Therefore, as a demonstration of our novel modeling, analysis and optimization approach, we concentrate on the kinetostatic (endpoint force-production) capabilities for robotic fingers.

Several studies have analyzed the kinetostatic performance of tendon-driven and torque-driven manipulators [15–23] (determining the kinetostatic capabilities given design parameters), and several others have addressed their optimization or synthesis (specifying the design parameters given desired capabilities) [24–32].

These studies are based on mathematical theory. The fabrication of robotic fingers has been widely accomplished for robotic hands [2–13]. Experimental testing of kinetostatic performance can be found in the biomechanics literature [33, 34], but these do not implement a system whose parameters can be altered. We combine these three areas of *theory, fabrication, and testing* to optimize and validate hardware implementations of alternative robotic finger designs.

2 Nuances of Biological and Engineering Systems that are Tendon-Driven

Understanding the mechanical properties of tendon-driven plants is critical to understanding the actual problem confronting both brain-body dynamics and the plant-controller engineering design problem. Elsewhere we have shown the importance of understanding the nuances of tendon-driven systems that cannot be understood when applying the common torque-driven mathematical approach to the actuation of serial kinematic chains (i.e., limbs [35]). However, a brief summary is warranted because the present work is motivated by the need to understand well (i) the consequences of alternative tendon routings (i.e., system topologies) and their optimization; and (ii) the need to validate the analytical and computational approaches we have developed and promoted in prior work.

The nuances of tendon-driven systems include:

- *Topology matters*: When thinking of the structure and function of anatomical and engineering systems, it is important to distinguish between the topology of the system and its specific parameter values [36]. The topology consists of the type and connectivity of elements of any particular systems, which in this case include the number and type of kinematics degrees of freedom, and the way in which tendons cross and actuate joints. The specific parameters then include the actual number of limb segments, degrees of freedom, the kinematic connectivity across limb segments and the specifics of how many tendons there are, which kinematic degrees of freedom they cross and actuate, and the moment arm at each degree of freedom (assuming only rotational degrees of freedom). We have shown, for example, that apparently similar topologies can have very different mechanical capabilities [36], and that some topologies can be used to satisfy various functional goals [37]. Therefore the topology and parameters values of a tendon-driven system must be considered carefully.
- *Agonists, antagonists and co-contraction*: The perspective and language that emerges from analyzing single joint systems such as the elbow or knee are very often not valid in the case of multi-articular, multi-tendon systems [35]. More specifically, the idea that there is an agonist tendon that produces torque in the desired sense, which is opposed by an antagonist tendon on the "other side" of the joint does not extrapolate well. In the case of two or more kinematics degrees of freedom, each tendon produces a specific combination of joint

torques; and all constitute a basis for mechanical actions. The positive addition of their actions (muscles can only pull, not push) creates the set of feasible outputs; be they in the joint torque space (feasible torque set) or the end-point space (feasible wrench set). Thus the routing of each tendon and the strength of each muscle can dramatically affect the size and shape of the feasible actions of the system. In the multi-joint case, therefore, mechanical actions are a vector addition of these basis vectors, and it is no longer possible to label some muscles as agonists or antagonists. This is because all active basis vectors contribute to the task equally. That is, the loss of any one muscle will affect the ability to implement that solution. Similarly, the vector addition of all contributing vectors at times requires the simultaneous cancellation of mechanical actions at a joint [34]. This can lead to activity in muscles on opposite sides of a joint, which has often been called co-contaction by extrapolation from the single joint case. However, such simultaneous activity is not an optional strategy in the sense that co-contraction is usually thought of. Rather, such vector addition that engages muscles on either side of a joint is at times simply what is required to accomplish the task [38] and is not a always a decision the nervous system makes to, for example, increase joint impedance, etc.
- *Muscle Redundancy and Synergies*: More recently we have called attention to the fact that muscle redundancy, while mathematically undeniable, does not necessarily imply robustness to muscle dysfunction [39]. That is, the solution space for a given task has a specific structure given by the constraints of the task [34]. This structure then defines which muscles must be active at a level higher than zero (and are therefore necessary); and which muscles are needed at moderate or high levels of activation (and therefore the task is sensitive to weakness or dysfunction of those muscles) [39]. Lastly, because the structure of the solution space is given by the interaction among the topology of the plant and the constraints of the task, it is not surprising (in fact, it is expected) that EMG and other recordings often exhibit a lower dimensional structure [40]. This structure depends directly on how many tendons there are, how they are routed, how they can be used to fulfill the constraints of the task and is not necessarily the consequence of a specific choice of neural control strategy. Disambiguating mechanically necessary versus neurally optional features of muscle coordination required using the techniques presented here to identify the solution space for a task given a specific tendon topology.

Therefore, our work here is motivated by the need to be able to test the functional and control consequences of any topology for a tendon-driven systems, be it biological or mechatronic. Moreover, the predictions and interpretations mentioned above depend to a certain extent on the validity of such linear methods to find and describe the structure of the solution space for static force production. Therefore, it is of critical interest to be able to validate these methodology predictions experimentally.

3 Finger Construction

We had several design requirements when designing a reconfigurable robotic finger as a test bed for analysis, optimization, and testing. They were

(1) Ability to arbitrarily change tendon routing (i.e., the joints each tendon crosses, and whether they produce positive or negative torque at each joint).
(2) Ability to vary pulley sizes (i.e., moment arms of the tendons).
(3) Low friction.
(4) Sufficient and well arranged degrees-of-freedom (DOFs) to allow three-dimensional endpoint motion and force production.
(5) Robust, durable, rigid.

We designed the finger in SolidWorks 2010 (Dassault Systémes), as shown in Fig. 1. The actual reconfigurable finger is also shown. It was constructed with one ad-abduction DOF and two flexion–extension DOFs. The primary materials were aluminum (for the links and terminating pulleys), turcite (for the spacers and rotating pulleys), and ball bearings with extended inner rings (mounted on all pulleys and link axes). All of the pulleys were custom-machined and two sizes were constructed for reconfigurability: a radius of 8.0 mm for the large pulleys and 4.4 mm for the small pulley, as shown in Fig. 2. There were multiple pulleys that had to be added between the axes to ensure reconfigurability of the tendon routing (i.e., that each tendon could rout on either side of every joint and no bowstringing of the tendons would occur).

The selection of link lengths and pulley sizes was otherwise fairly arbitrary, and since our study did not involve optimization of incremental changes in these parameters (except the 2 pulley sizes), we simply constructed it to have reasonable size that could be fabricated and tested.

4 Methods

After construction of the finger with the desired capabilities, we were then able to analyze and optimize tendon routing and pulley sizes based on the actual kinematics and reconfiguration options of the finger.

4.1 Force Polytope Analysis

Quantification of the force-production capabilities of a robotic finger (or manipulator) can be accomplished by determination of the feasible force set (or force polytope) of the finger. This convex set encloses all feasible forces that the fingertip can exert given kinematic parameters, tendon routing and pulley sizes, and

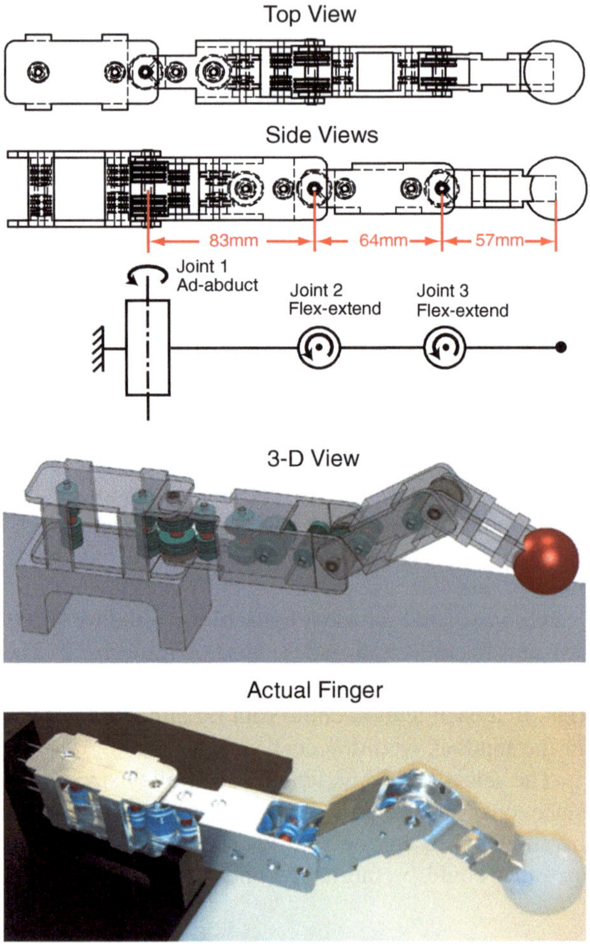

Fig. 1 2-D and 3-D views of finger model in solidworks, and the actual finger

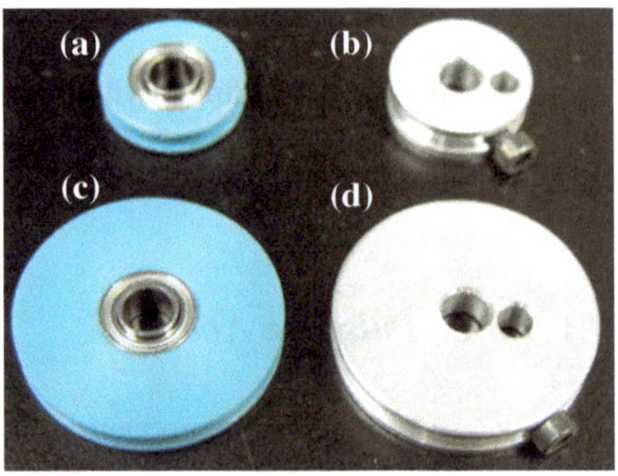

Fig. 2 Pulleys used in finger design. **a** Turcite rotating small pulley. **b** Aluminum terminating small pulley. **c** Turcite rotating large pulley. **d** Aluminum terminating large pulley

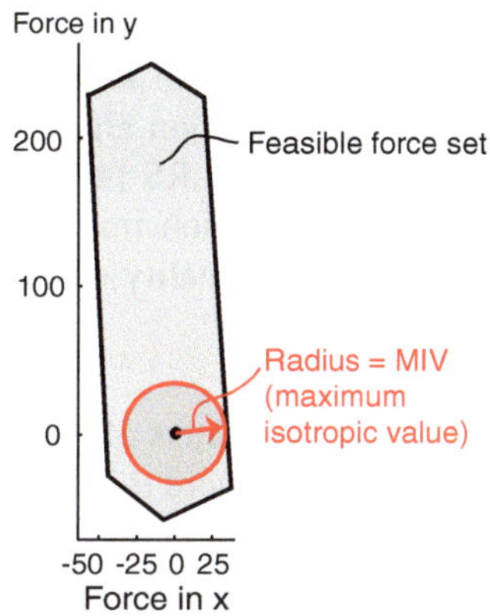

Fig. 3 Illustration of calculation of MIV (maximum isotropic value) from feasible force set

maximal tendon tensions. A quality metric that can be assigned to this set is known as the maximal isotropic value (MIV) [18]. Since we are not assuming any specific task that this finger must perform, then we chose to use this metric. We could have used any other metric instead of the MIV. Further comments can be found in the discussion section. The MIV is the radius of the largest ball, centered at the origin, that the feasible force set can contain, as illustrated in Fig. 3 for a 2-D feasible force set. A finger can exert at least that many units of force in any direction.

We can use an activation vector, $\vec{a}$, to represent the degree to which a tendon is activated. Each element of $\vec{a}$ ranges between 0 (no activation, zero force) and 1 (full activation, maximal force). Further discussion may be found in [35]. If we define F_0 as a diagonal matrix of maximal tendon tensions, R as the moment arm matrix (or structure matrix) relating tendon tensions to joint torques, and J as the posture-dependent Jacobian relating joint velocities to fingertip velocities, then we can get the fingertip force vector $\vec{f}$ from tendon activations [34] if the Jacobian is square and invertible:

$$\vec{f} = J^{-T} R F_0 \vec{a} = A\vec{a} \tag{1}$$

For a given fixed finger posture, the J^{-T}, R, and F_0 matrices can be grouped into a linear mapping from activations into fingertip force, which we call an action matrix A [34, 35]. Each column of A represents the force vector each tendon produces at the fingertip in that posture if fully activated. The collection of all such forces (i.e., all columns of matrix A) forms a set of output force basis vectors. Linearity of this mapping holds true for static forces because the Jacobian and moment arms remain constant. The Minkowski sum of these basis vectors forms the feasible force set of the fingertip, and can be computed by taking the convex hull of the points generated by mapping each vertex of the activation hypercube (i.e., each vertex of the unit hypercube in the positive orthant) to fingertip wrench space via the action matrix A [34].

There are two ways to describe a convex hull: (i) a set of vertices and (ii) a set of linear inequalities. Vertex enumeration methodologies can calculate one description given the other. The Qhull software package uses the Quickhull algorithm [41] and is used to perform the MIV calculations in this study. Other vertex enumeration algorithms that can perform these calculations easily include CDD [42] and LRS [43].

The description involving a set of linear inequalities (similar to a linear programming inequality constraint formulation) takes the form

$$Ax \leq b \tag{2}$$

where A is a matrix of constants defining the inequalities, x is a vector of variables of length d, where d is the dimensionality of the convex hull, and b is a vector of constants. If we denote A_i as the ith row of A, then the linear inequality $A_i x \leq b_i$ defines a halfspace, which also defines a facet of the convex hull. The perpendicular (i.e., shortest) Euclidean distance (or offset) of this facet from the origin, in general, will be given by

$$\frac{b_i}{\|A_i\|_2} \tag{3}$$

The Qhull output, however, automatically sets each $\|A_i\|_2$ equal to 1, so the ith offset from the origin is simply the signed constant b_i. Calculation of the MIV in this study involves simply finding the minimum of b corresponding to the feasible force set.

For our finger, the Jacobian is a 3×3 matrix which is square and invertible in our experimental postures, R is a $3 \times \ell$ matrix (ℓ is the number of tendons, which is 4, 5, or 6), and F_0 is an $\ell \times \ell$ diagonal matrix of maximal tendon tensions.

4.2 Evaluating Tendon Routings

The construction of the finger allowed for various moment arm matrices to be implemented which had 4, 5, or 6 tendons. These designs are known as N + 1, N + 2, and 2N designs, where N is the degrees of freedom of the finger. We enumerated all possible moment arm matrices beginning with the "base" matrices shown in Fig. 4. The N + 2a and N + 2b designs differ only in that the second tendon terminates at the first joint in the N + 2a designs and at the second joint in the N + 2b designs. We replaced each '#' with either a 1 or -1 (in accordance with the sign of the moment exerted on a joint when the corresponding tendon is under tension; see Fig. 1 for definition of joint axes) in a full combinatoric search and then checked the controllability (i.e., that all of the joints could be actuated independently in torque and motion) conditions as described in [22]. We then calculated the MIV for these routings using the large pulleys in the main posture: 0° at joint 1, $-45°$ at joint 2, and $-45°$ at joint 3, as shown in Fig. 5. To make comparisons feasible across finger designs with different number of tendons, we used a uniform

Fig. 4 Base moment arm matrices used when finding admissible and unique tendon routings

N+1 design: 24 admissible, 3 unique

$$R = \begin{bmatrix} \# & \# & \# & \# \\ 0 & \# & \# & \# \\ 0 & 0 & \# & \# \end{bmatrix}$$

N+2a design: 88 admissible, 6 unique

$$R = \begin{bmatrix} \# & \# & \# & \# & \# \\ 0 & 0 & \# & \# & \# \\ 0 & 0 & 0 & \# & \# \end{bmatrix}$$

N+2b design: 296 admissible, 11 unique

$$R = \begin{bmatrix} \# & \# & \# & \# & \# \\ 0 & \# & \# & \# & \# \\ 0 & 0 & 0 & \# & \# \end{bmatrix}$$

2N design: 872 admissible, 20 unique

$$R = \begin{bmatrix} \# & \# & \# & \# & \# & \# \\ 0 & 0 & \# & \# & \# & \# \\ 0 & 0 & 0 & 0 & \# & \# \end{bmatrix}$$

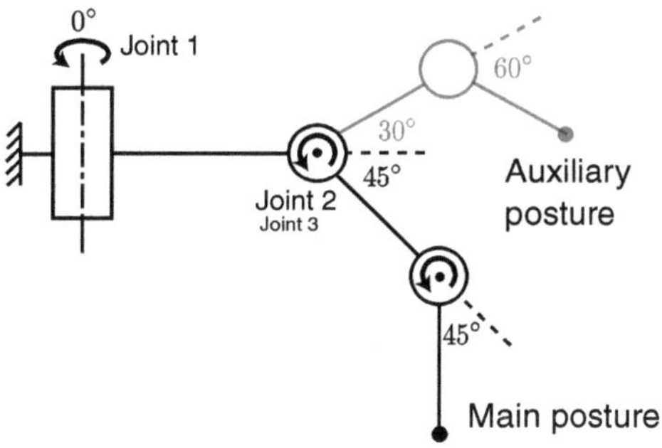

Fig. 5 Finger posture used in computations and experimental testing

maximal tendon tension distribution, with the sum being constrained[1] to 60N (i.e., for designs with 4, 5, and 6 tendons, the maximal tensions were 15, 12, and 10N, respectively). We found that many of the admissible routings produced the exact same MIVs and feasible force set volumes, likely corresponding with structurally isomorphic routings [22]. The number of routings that produced unique MIVs was a very small subset of the admissible routings, as can be seen from the numbers in Fig. 4. In cases of optimization of a more complex finger or manipulator where the number of unique MIVs may be orders of magnitude higher, methods for selection of a subset for further optimization such as in [46] may be used.

[1] The sum of maximal tendon tensions being equal is an important constraint due to the size, weight, and motor torque (and therefore tendon tension) limitations inherent in dexterous hands. For example, the torque capacity of motors is roughly proportional to motor weight, and minimization of weight was an important consideration in the design of the DLR Hand II [44]. In addition, the maximal force production capabilities of McKibben-style muscles are roughly proportional to cross-sectional area [45]. Since the actuators typically will be located in the forearm, then the total cross-sectional area will be limited to the forearm cross-sectional area. In this study, for simplicity and without affecting the generalizability of our approach or results, we do not consider alternative constraints on the actuation system (e.g., electrical current capacity, tendon velocities, etc).

The total number of routings producing distinct feasible force sets[2] was 40. For each of those routings, we calculated the MIV for all combinations of large and small pulleys. For example, the N + 1 design has 9 moment arm values. Therefore, there are 2^9 combinations of large and small pulleys for that case. Taking the combination with the highest MIV for each routing gave 40 moment-arm-optimized routings. Therefore, we had 40 unoptimized routings and 40 optimized routings. Out of these 80, we chose 6 different routings to test experimentally in a fashion that permitted testing of a large range of MIVs, and included the design with the highest predicted MIV. Otherwise the selection was arbitrary.

4.3 Experimental Testing of Tendon Routings

For each of the tendon routings tested, we first arranged the pulleys and strings (0.4 mm braided polyester twine) to match the desired configuration. We then mounted the finger onto a base that was part of a motor array system as shown in Fig. 6. The DC motors were coupled to capstans on which the string wound. Each string was then routed around pulleys that were attached to load cells (Interface SML 25, Scottsdale, AZ) which provided force measurements for the closed-loop controller implemented in Realtime LabView. The endpoint of the finger was fixed to a custom made gimbal which constrained translational motion but not rotational motion (we did not want the fingertip to be over-constrained). The gimbal was attached to a 6-axis load cell (JR3, Woodland, CA). The sampling rate and control loop frequency were both 100 Hz.

A small pretension of 1N was applied to each string to remove slack and prevent it from falling off of the pulleys. Then each vertex of the activation hypercube (as described in the previous section) was applied to the strings (in addition to the pretension) in ramp-up, hold, and ramp-down phases to find the feasible force set [39]. As in prior work, [39], vertices of this experimental feasible force set were determined from the hold phases and then used to find the MIV using Qhull as described earlier. The experimental MIV could then be compared with the theoretical MIV (from computational results).

To compare the shapes of the experimental and theoretical feasible force sets, we first normalized the volume of the experimental feasible force set to make it equal to the volume of the theoretical feasible force set. We then calculated the mean Euclidean distance of each vertex from the theoretical feasible force set from

[2] Due to the nature of our full combinatoric search, moment arm matrices that produced mirrored feasible force sets about a plane passing through the origin (which would have the same MIV) were discarded and also those moment arm matrices that were produced by a rearrangement of the columns. For example, in Fig. 4, interchanging columns 5 and 6 does not change the feasible force set, it only reverses the "numbering" of the tendons. But in the full combinatoric search, both of these numberings would be different matrices producing identical feasible force sets.

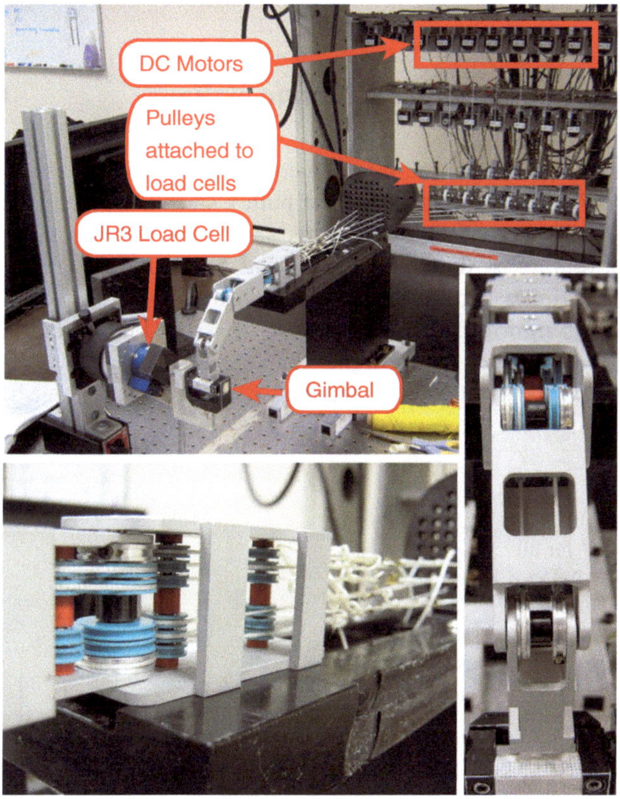

Fig. 6 Experimental system for feasible force set testing

the corresponding vertex in the experimental feasible force set. We did this because there was always friction loss in the experiment.[3] In addition, we calculated the average angle between the two vectors (starting and the origin and ending at the corresponding vertex) formed from corresponding vertices.

We tested the finger in the main posture (for which we optimized MIV) and an auxiliary posture (to validate the predictions more fully). These postures are shown in Fig. 5. For each design and posture, we did three repetitions of tests. Since there were 6 designs, 2 postures, and 3 repetitions, we conducted a total of 36 tests.

[3] Take an extreme case in which friction loss was 50 % exactly for every tendon. The theoretical feasible force set is a unit cube. While the shape of the experimental feasible force set would be also an exact cube, it would be 50 % contracted in every direction and therefore the corresponding vertices would be far from each other. If we normalized the volume, the corresponding vertices would be in the same location, and the mean distance (in shape similarity) would be zero.

5 Results

5.1 Calculating Maximum Isotropic Values

The 40 unique unoptimized and 40 unique optimized routings produced the MIVs shown in Fig. 7a, b. Optimization of the pulley sizes increased the average MIV from 0.60 to 0.78N, a 30 % increase as shown in Fig. 7a, and the maximum increase for a routing by this optimization was 82 %. It is interesting to note that this force-production capability increase is achieved by simply decreasing specific pulley sizes in an informed manner. We can see from Fig. 7b that designs with 4 tendons could not produce MIVs higher than the best designs with 5 or 6 tendons. However, the best design with 4 tendons did have a higher MIV than many

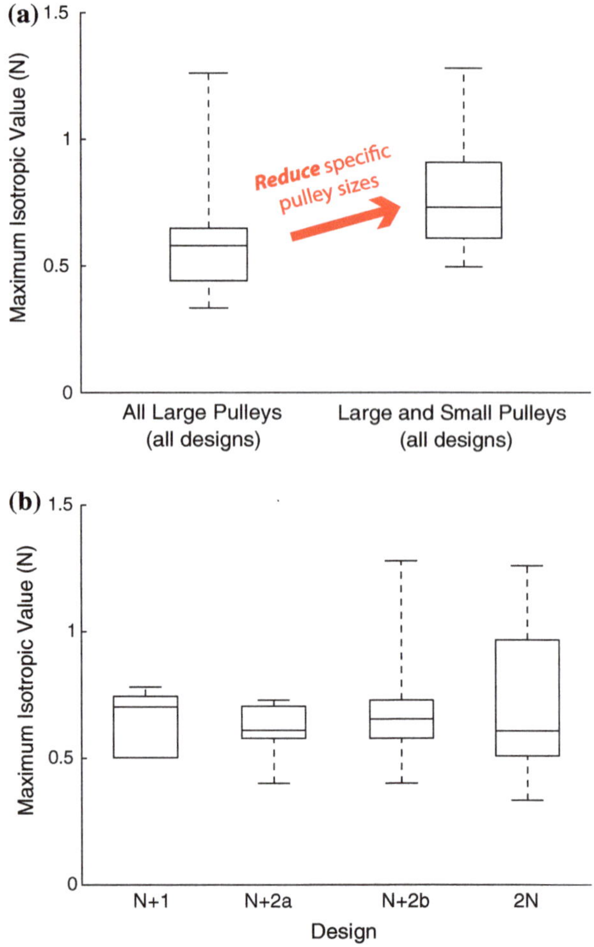

Fig. 7 Maximum isotropic values for various routings. **a** Boxplot of MIV for all designs before and after pulley-size optimization. **b** Boxplot of MIV versus design (includes optimized and unoptimized pulley sizes)

alternative routings that had more tendons. In addition, the maximal increase from only rerouting tendons (no pulley size optimization) was 277 % (i.e., the increase from the worst admissible routing to the best admissible routing for a given number of tendons).

5.2 Theoretical Predictions Versus Experimental Results

The experimental results and the routings tested are shown in Fig. 8. The data points shown in Fig. 8b are averages of the three test repetitions in both the main posture and the auxiliary posture for each of the designs. The average standard deviation from the three test repetitions was very low at 0.0090N, showing that the

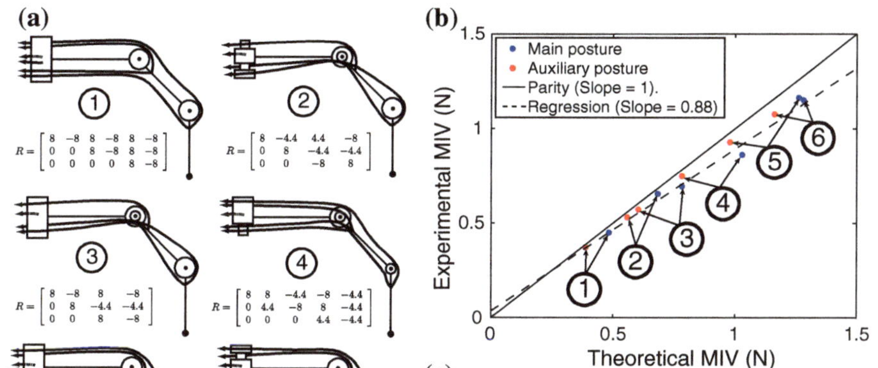

(c)

	Main posture					
	MIV			Shape Similarity		
Design	Theoretical (in N)	Experimental (in N)	Error (in %)	Mean Angle (in degrees)	Mean Distance (in N)	Normalization Factor
1	0.48	0.45	-6.69	4.68	0.25	1.10
2	0.68	0.66	-4.01	3.06	0.32	1.07
3	0.78	0.69	-11.15	2.57	0.27	1.09
4	1.03	0.86	-16.12	2.56	0.11	1.10
5	1.26	1.16	-7.63	6.23	0.33	1.14
6	1.28	1.15	-9.99	4.72	0.26	1.13

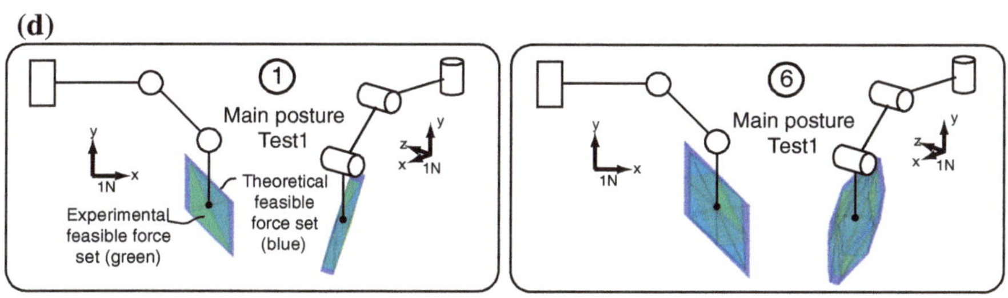

Fig. 8 Results from experimental testing of various routings. **a** The 6 different routings tested. Shown to scale. *R* matrix values are in mm. **b** Experimental versus theoretical MIV. Parity line is where experimental MIV would be exactly equal to theoretical MIV (intercept of 0, slope of 1). Regression has an R^2 value of 0.987. **c** Table of averages from 3 tests for each design in main posture. **d** 3-D visualization of experimental and theoretical feasible force sets for designs 1 and 6

results for each design/posture combination were extremely repeatable. We see a consistent linear relationship between theoretical and experimental MIVs with an R^2 value of 0.987. This result shows that the theoretical calculations are very good at predicting actual performance. The slope of the line is 0.879, which we interpret to represent an average loss of performance of about 12 % from theoretical predictions, likely due mainly to friction in the system. We show experimental and theoretical feasible force sets for one test of designs 1 and 6 in the main posture in Fig. 8d, and we can see visually that the shape of the theoretical feasible force sets was extremely similar to those of the experimental feasible force sets, although the experimental ones were contracted to an extent.

We can see several interesting features in Fig. 8. First of all, in Fig. 8a, we see that routings 1 and 5 are identical except that 2 of the signs in the moment arm matrix are reversed (i.e., 2 of the tendons are switched from one side of the ad-abduction joint to the other). However, we can see in Fig. 8b and the table in Fig. 8c that the MIV of routing 5 is more than twice that of routing 1 both theoretically and experimentally. Figure 9 emphasizes this large change in MIV for a small, intelligent (but perhaps counterintuitive before performing the analyses) change in tendon routing. Secondly, the MIVs of routings 5 and 6 are very similar, but routing 6 has one less tendon. Thirdly, routings 2 and 3 have two fewer tendons than routing 1 but still outperform it in terms of MIV. Figure 8d demonstrates visually that the experimental feasible force sets corresponded very closely with the theoretical feasible force sets in shape, and that the size was similar but contracted by a small amount due to friction. While the side views of both of these feasible force sets look similar, the isometric views show clearly that the feasible force set of routing 1 is quite thin along one direction (which results in a low MIV) and the feasible force set of routing 6 is much more expanded in all directions (so the MIV is much higher).

In Fig. 8b, we see that the data points lie underneath but fairly close to the parity line (if the theoretical and experimental MIVs were identical, the data points would lie exactly on the parity line). As would be expected, none of them are above the parity line. In the table in Fig. 8c, we see that the error in the prediction of MIV ranged roughly between 4 % and 16 % which is the percentage MIV below the parity line where the data points are located. As far as shape goes, we see that the average difference in angles between corresponding vertices of the feasible force sets was between 2.56° and 6.5°. A small amount of angular error would be expected due to error in the positioning and alignment of the JR3 axes relative to the finger axes, so this contributes to the average difference in angles. The mean distance of corresponding vertices is less than 0.35N. The normalization factors were less than 1.15, and it is the factor by which the experimental feasible force set had to be expanded in every direction to have the same volume as the theoretical feasible force set. It roughly corresponds with the MIV error.

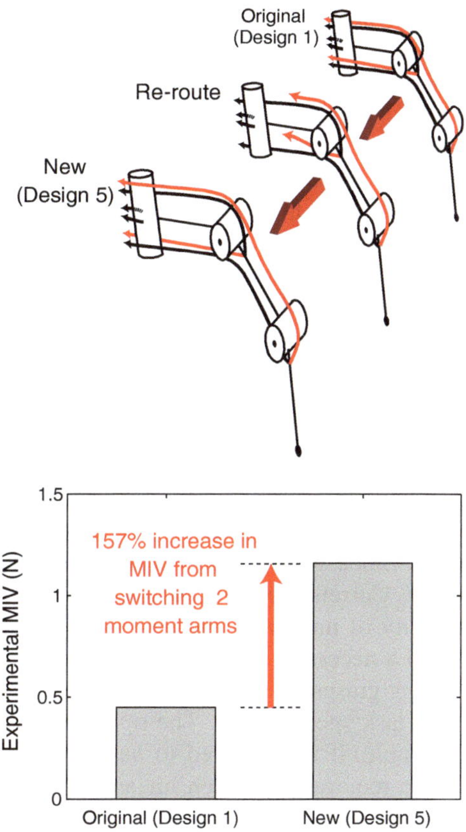

Fig. 9 Illustration of a simple (but intelligent) tendon re-routing that drastically increases MIV

6 Discussion

In this work, we have investigated the effect of various tendon routings on the set of feasible forces that can be exerted by a robotic finger both computationally and in a physical system. We see that routing has a very dramatic effect on the shape and size of the feasible force set. We also see that computational predictions are quite accurate and that they can be useful when making informed design decisions. Therefore the main conclusions of this study are twofold:

(1) Different routings in robotic fingers can result in extremely different force-production capabilities.
(2) Theoretical feasible force set analyses predict experimental force-production performance quite well and therefore they are a useful design tool.

One application of this work is to design tendon-driven fingers and manipulators for a given task that are optimized in terms of minimal size, weight, complexity, and cost. Since tendon-driven systems are linear for fixed postures, if we double all the moment arms (or the maximal tensions of all tendons), we double the size of feasible force set in every direction. Our results showed that the MIV for routing 6 was more than 100 % greater than that of routing 1. Therefore, one could either reduce all of the maximal tendon tensions or all of the moment arms in routing 6 by 50 % and still have a greater MIV than routing 1. In a robotic hand system, if the maximal tendon tensions were cut in half (by implementing smaller motors), then the weight of the actuators could be roughly halved and cost would be reduced. This would be very desirable, in general. If tendons are driving a minimally-invasive surgical instrument, then the moment arms could be halved and therefore the diameter of the instrument would be halved (which actually would reduce the cross-sectional area by 75 %!). The instrument would then be much smaller and could be much better suited for certain surgical procedures.

We have used MIV as the fitness metric for our analyses since no prior assumption of task specification was made. We acknowledge that in general, the MIV would typically be more practically used for a tendon-driven, multi-purpose manipulator rather than a robotic finger. These analyses and optimizations that we described apply equally to tendon-driven manipulators and fingers regardless of size. Since we were only testing one finger in this paper, we decided to use the MIV. Current work has accomplished design optimization and validation for grasp quality of multiple fingers of identical design to the one in this chapter [47].

If a necessary task or set of tasks is known (e.g., to have high flexion force for strong grasps) then the analyses could assign a fitness metric to a routing based on that task specification. The optimization could then be based on that metric. For example, if it is desired to have a very strong flexion force with low extension force requirements, then linear programming can easily be used to determine the maximal force possible in the flexing direction(s) after the feasible force set has been calculated (e.g., using the generic procedure outlined in [24]). We have previously described that the feasible force sets of the human fingers are asymmetrically biased towards endpoint forces in the flexion direction than in the extension which is anatomically reasonable for grasping tasks [34, 48–50]. If strong grasp and minimal size/weight/cost is desired for a set of fingers, then analyses like those used in [51] can be used to design an optimized tendon-driven robotic hand.

We have investigated force-production capabilities in this paper, but there are many other considerations that go into the design of a robotic hand. Other significant considerations include the robustness and effectiveness of control algorithms, passive stiffness characteristics, sensitivity to friction and positioning errors, and maximal endpoint velocities. We acknowledge that force-production capabilities are only one piece of the design puzzle for optimized robotic fingers.

For reasons of practicality, we only analyzed and constructed routings where the tendons routed around every joint that they passed (i.e., that the structure matrix is pseudo-triangular, as in [22]) and where there were only two sizes of pulleys that could be chosen. Routings can, however, be designed where tendons pass through the center of joints [52], or where moment arms can have many feasible magnitudes. This opens up the design space even more, and exhaustive searches like the ones we performed in this study may be more laborious, or even not be feasible given the exponential growth of design options. In addition, tendon-driven fingers or manipulators with more than 3 degrees of freedom will tend to suffer from the curse of dimensionality in the design space, and a designer may have to use various optimization algorithms [1] in a search for a "good enough" design which could then be selected for physical construction. Alternatively, a designer could come up with a handful of feasible, physically-realizable routings and then search in the vicinity of that region of the parameter space to determine feasible improvements with affordable computational cost [53].

The optimization process we used in this study only addressed this realization of a robotic finger. If a general robotic finger or manipulator has more joints or tendons, the dimensionality of the design space increases dramatically and finding a globally optimum solution for a specific fitness metric (of which any task-specific metric may be used, not only the general MIV metric) may be computationally infeasible due to the curse of dimensionality. Other custom or typical optimization algorithms could be used to find solutions with a high fitness. Furthermore, in the case when the optimization may involve link lengths and D-H parameters in addition to tendon routing, number of tendons, and pulley sizes, then finding a locally optimal solution or just a good-enough solution could still be very useful. The main purpose of this study was to investigate the correlation of predictions with experiments, as opposed to identifying a general optimization method for tendon-driven robotic fingers and manipulators.

Tendon friction was a significant factor in our experiment, (as it is for any tendon driven system), especially for the tendons at the last joint that had to wrap around as many as 12 pulleys. The main source of friction seemed to be the pulleys, as general observation of the data indicated that tendons attached to the ad-abduction joint (which wrapped around 4 pulleys) suffered from very little friction loss (less than a few %) while the tendons that attached to the last joint (which wrapped around 12 pulleys and routed through the fairly complicated tendon redirection between the first and second joints) suffered from as much as 20 % friction loss.

Future work will extend this experimental validation approach to routings of multiple fingers for optimized grasp quality. In addition, this work is easily applicable to refine the design of generic tendon-driven manipulators. Furthermore, investigation of the control and structure of biological tendon-driven systems is now made possible using a similar framework.

7 Conclusions

We conclude from this validation that these computational methods are effective at predicting the performance of drastically different tendon-driven robotic finger (or manipulator) designs, and are therefore a useful design tool. Various benefits of fully utilizing this design tool include:

(1) *Minimization of weight*: if a superior design has a force-production performance twice that of an inferior design, the superior design's actuators only need to be half the strength of the inferior design's to match the inferior design's performance, which in general corresponds to a large reduction in weight of the actuators.
(2) *Minimization of size*: if a superior design has a force-production performance twice that of an inferior design, the superior design's moment arms only need to be half the size of the inferior design's to match the inferior design's performance, which could be used to half the overall thickness of the finger (or manipulator, or minimally-invasive surgical device).
(3) *Minimization of number of tendons (and therefore actuators)*: If a design with less tendons (such as an N + 1 design) can be synthesized with the same force-production performance as that of one with more tendons (such as a 2N design), then the actuator system can be simplified and less space to rout the tendons inside the finger (or manipulator, or minimally-invasive surgical device) is needed.

Acknowledgments The authors gratefully acknowledge the help of Dr. Manish Kurse in providing the data acquisition routine for the experimental procedure, and Dr. Veronica Santos for construction of the gimbal used in the experiments.

References

1. J.M. Inouye, J.J. Kutch, F.J. Valero-Cuevas, A novel synthesis of computational approaches enables optimization of grasp quality of tendon-driven hands. IEEE Trans. Rob. 1–9 (2012)
2. S. Jacobsen, E. Iversen, D. Knutti, R. Johnson, K. Biggers, Design of the utah/mit dextrous hand. IEEE Int. Conf. Rob. Autom. **3**, 1520–1532 (1986)
3. J.K. Salisbury, J.J. Craig, Articulated hands: Force control and kinematic issues. Int. J. Robot. Res. **1**(1), 4 (1982)
4. Shadow Dexterous Hand, Shadow Robot Company
5. M. Grebenstein, A. Albu-Schffer, T. Bahls, M. Chalon, O. Eiberger, W. Friedl, R. Gruber, U. Hagn, R. Haslinger, H. Hppner, The dlr hand arm system, Submitted to ICRA, vol. 11
6. R.O. Ambrose, H. Aldridge, R.S. Askew, R.R. Burridge, W. Bluethmann, M. Diftler, C. Lovchik, D. Magruder, F. Rehnmark, Robonaut: Nasa's space humanoid. IEEE Intell. Syst. Appl. **15**(4), 57–63 (2000)
7. B.M. Jau, Dexterous telemanipulation with four fingered hand system, in *Proceedings of IEEE International Conference on Robotics and Automation*, vol. 1, pp. 338–343, 1995
8. B. Massa, S. Roccella, M.C. Carrozza, P. Dario, Design and development of an underactuated prosthetic hand, in *Proceedings of IEEE International Conference on Robotics and Automation (ICRA'02)*, vol. 4, pp. 3374–3379, 2002
9. L.R. Lin,s H.P. Huang, Mechanism design of a new multifingered robot hand, in *Proceedings of 1996 IEEE International Conference on Robotics and Automation*, vol. 2, pp. 1471–1476, 1996
10. H. Kawasaki, T. Komatsu, K. Uchiyama, Dexterous anthropomorphic robot hand with distributed tactile sensor: Gifu hand II. IEEE/ASME Trans. Mechatronics **7**(3), 296–303 (2002)

11. A. Namiki, Y. Imai, M. Ishikawa, M. Kaneko, Development of a high-speed multifingered hand system and its application to catching, in *Proceedings of 2003 IEEE/RSJ International Conference on Intelligent Robots and Systems (IROS 2003)*, vol. 3, pp. 2666–2671, 2003
12. I. Yamano, T. Maeno, Five-fingered robot hand using ultrasonic motors and elastic elements, in *Proceedings of the 2005 IEEE International Conference on Robotics and Automation (ICRA 2005)*, pp. 2673–2678, 2005
13. I. Gaiser, S. Schulz, A. Kargov, H. Klosek, A. Bierbaum, C. Pylatiuk, R. Oberle, T. Werner, T. Asfour, G. Bretthauer, A new anthropomorphic robotic hand, in *Prooceedings of 8th IEEE-RAS International Conference on Humanoid Robots*, pp. 418–422, 2008
14. J.L. Pons, R. Ceres, F. Pfeiffer, Multifingered dextrous robotics hand design and control: a review. Robotica **17**(6), 674 (1999)
15. F. Firmani, A. Zibil, S.B. Nokleby, R.P. Podhorodeski, Wrench capabilities of planar parallel manipulators. part I: wrench polytopes and performance indices. Robotica **26**(06), 791–802 (2008)
16. S. Bouchard, C.M. Gosselin, B. Moore, On the ability of a cable-driven robot to generate a prescribed set of wrenches, in *Proceedings of the ASME International Design Engineering Technical Conferences, Mechanics and Robotics Conference, Citeseer,* 2008
17. P. Chiacchio, Y. Bouffard-Vercelli, F. Pierrot, Force polytope and force ellipsoid for redundant manipulators. J. Robot. Syst. **14**(8), 613–620 (1997)
18. R. Finotello, T. Grasso, G. Rossi, A. Terribile, Computation of kinetostatic performances of robot manipulators with polytopes, in *Proceedings 1998 IEEE International Conference on Robotics and Automation*, vol. 4, pp. 3241–3246, 1998
19. M. Gouttefarde, S. Krut, Characterization of parallel manipulator available wrench set facets. Adv. Robot Kinematics Motion Man Mach. 475–482 (2010)
20. A. Zibil, F. Firmani, S.B. Nokleby, R.P. Podhorodeski, An explicit method for determining the force-moment capabilities of redundantly actuated planar parallel manipulators. J. Mech. Des. **129**, 1046 (2007)
21. L.W. Tsai, Design of tendon-driven manipulators. J. Mech. Des. **117**, 80 (1995)
22. J.J. Lee, L.W. Tsai, The structural synthesis of tendon-driven manipulators having a pseudotriangular structure matrix. Int. J. Robot. Res. **10**(3), 255 (1991)
23. J.J. Lee, Tendon-driven manipulators: analysis, synthesis, and control. Ph.D Dissertation, University of Maryland, 1991
24. J.L. Fu, N.S. Pollard, On the importance of asymmetries in grasp quality metrics for tendon driven hands, in *Proceedings of 2006 IEEE/RSJ International Conference on Intelligent Robots and Systems*, pp. 1068–1075, 2006
25. D.Z. Chen, J.C. Su, K.L. Yao, A decomposition approach for the kinematic synthesis of tendon-driven manipulators. J. Robot. Syst. **16**(8), 433–443 (1999)
26. Y.J. Ou, L.W. Tsai, Kinematic synthesis of tendon-driven manipulators with isotropic transmission characteristics. J. Mech. Des. **115**, 884 (1993)
27. Y.J. Ou, L.W. Tsai, Isotropic design of tendon-driven manipulators. J. Mech. Des. **118**(3), pp. 360–366 (1996)
28. J.B. Sheu, J.J. Huang, J.J. Lee, Kinematic synthesis of tendon-driven robotic manipulators using singular value decomposition. Robotica **28**(01), 1–10 (2009)
29. J. Angeles, *On the Optimum Dimensioning of Robotic Manipulators* (McGill University, Montreal, 2004)
30. M.M. Aref, H.D. Taghirad, S. Barissi, Optimal design of dexterous cable driven parallel manipulators. Int. J. Robot. Theory Appl. **2**(4), 43–51 (2009)
31. D. Chablat, J. Angeles, On the kinetostatic optimization of revolute-coupled planar manipulators. Mech. Mach. Theory **37**(4), 351–374 (2002)
32. W.A. Khan, J. Angeles, The kinetostatic optimization of robotic manipulators: the inverse and the direct problems. J. Mech. Des. **128**, 168 (2006)
33. L. Kuxhaus, S.S. Roach, F.J. Valero-Cuevas, Quantifying deficits in the 3d force capabilities of a digit caused by selective paralysis: application to the thumb with simulated low ulnar nerve palsy. J. Biomech. **38**(4), 725–736 (2005)

34. F.J. Valero-Cuevas, F.E. Zajac, C.G. Burgar, Large index-fingertip forces are produced by subject-independent patterns of muscle excitation. J. Biomech. **31**(8), 693–704 (1998)
35. F.J. Valero-Cuevas, A mathematical approach to the mechanical capabilities of limbs and fingers. Prog. Mot. Control 619–633 (2005)
36. F.J. Valero-Cuevas, V.V. Anand, A. Saxena, H. Lipson, Beyond parameter estimation: extending biomechanical modeling by the explicit exploration of model topology. IEEE Trans. Biomed. Eng. **54**, 1951–1964 (2007)
37. J.M. Inouye F.J. Valero-Cuevas, Asymmetric routings with fewer tendons can offer both flexible endpoint stiffness control and high force-production capabilities in robotic fingers, in *Proceedings of the Fourth IEEE RAS/EMBS International Conference on Biomedical Robotics and Biomechatronics (BioRob)*, Rome, 2012
38. F.J. Valero-Cuevas, An integrative approach to the biomechanical function and neuromuscular control of the fingers. J. Biomech. **38**, 673–684 (2005)
39. J.J. Kutch, F.J. Valero-Cuevas, Muscle redundancy does not imply robustness to muscle dysfunction. J. Biomech. **44**(7), 1264–1270 (2011)
40. J.J. Kutch, F.J. Valero-Cuevas, Challenges and new approaches to proving the existence of muscle synergies of neural origin. PLoS Comput. Biol. **8**(5), http://www.hubmed.org/display.cgi?uids=22570602. Accessed May 2012
41. C.B. Barber, D.P. Dobkin, H. Huhdanpaa, The quickhull algorithm for convex hulls. ACM Trans. Math. Softw. **22**(4), 469–483 (1996)
42. K. Fukuda, A. Prodon, Double description method revisited. Comb. Comput. Sci. **1120**, 91–111 (1996)
43. D. Avis, in *Polytopes—combinatorics and computation* ed. by G. Kalai, G. Ziegler. A Revised Implementation of the Reverse Search Vertex Enumeration Algorithm, vol. 29 (Birkhauser-Verlag, DMV Seminar Band 29, 2000), pp. 177–198
44. J. Butterfafl, M. Grebenstein, H. Liu, G. Hirzinger, Dlr-hand II: next generation of a dextrous robot hand, in *Proceedings 2001 ICRA IEEE International Conference on Robotics and Automation*, vol. 1 (IEEE, 2001), pp. 109–114
45. N.S. Pollard, R.C. Gilbert, Tendon arrangement and muscle force requirements for humanlike force capabilities in a robotic finger. Environment **17**, 14 (2002)
46. G. Taguchi, S. Konishi, *Orthogonal Arrays and Linear Graphs: Tools for Quality Engineering* (American Supplier Institute, Allen Park, 1987)
47. J.M. Inouye, Bio-inspired tendon-driven systems: computational analysis, optimization, and hardware implementation. Ph.D. Dissertation, University of Southern California, 2012
48. J.M. Inouye, J.J. Kutch, F. Valero-Cuevas, Quantitative prediction of grasp impairment following peripheral neuropathies of the hand, in *Proceedings of the 35th Annual Meeting of the American Society of Biomechanics*, Long Beach, CA, 2011
49. F.J. Valero-Cuevas, J.D. Towles, V.R. Hentz, Quantification of fingertip force reduction in the forefinger following simulated paralysis of extensor and intrinsic muscles. J. Biomech. **33**(12), 1601–1609 (2000)
50. F.J. Valero-Cuevas, V.R. Hentz, Releasing the A3 pulley and leaving flexor superficialis intact increases pinch force following the zancolli lasso procedures to prevent claw deformity in the intrinsic palsied finger. J. Orthop. Res. **20**(5), 902–909 (2002)
51. J.M. Inouye, J.J. Kutch, F. Valero-Cuevas, A novel methodology to compare grasp quality: application to two dominant tendon-driven designs, in *Proceedings of the 35nd Annual Meeting of the American Society of Biomechanics*, Long Beach, CA, 2011
52. M. Grebenstein, M. Chalon, G. Hirzinger, R. Siegwart, Antagonistically driven finger design for the anthropomorphic dlr hand arm system, in *Proceedings of IEEE-RAS International Conference on Humanoid Robots (HUMANOIDS)*, 2010
53. F.J. Valero-Cuevas, H. Hoffmann, M.U. Kurse, J.J. Kutch, E.A. Theodorou, Computational models for neuromuscular function. IEEE Rev. Biomed. Eng. 2, 110–135 (2009)

Chapter 13
Dynamic Simulation of the Hand

Shinjiro Sueda and Dinesh K. Pai

Abstract Robots and simulations provide complementary approaches for exploring different aspects of hand modeling. Robotic systems have the advantage in that all physical interactions, such as contact between tendons and bones, are automatically and correctly taken into account. Conversely, software simulations can more easily incorporate different material models, muscle mechanics, or even pathologies. Previously, no software for dynamic simulation could efficiently handle the complex routing and contact constraints of the hand. We address these challenges with a new simulation framework well suited for modeling the hand. We use the spline basis as the system's dynamic degrees of freedom, and place them where they are most needed, such as at the pulleys of the fingers. Previous biomechanical simulation approaches, based on either line-of-force or solid mechanics models, are not well-suited for the hand, due to the complex routing of tendons around various biomechanical constraints such as sheaths and pulleys. In line-of-force models, wrapping surfaces are used to approximate the curved paths of tendons and muscles near and around joints, but these surfaces affect only the kinematics, and not the dynamics, of musculotendons. In solid mechanics models, the fiber-like properties of muscles are not directly represented and must be added on as auxiliary functions. Moreover, contact constraints between bones, tendons, and muscles must be detected and resolved with a general purpose collision scheme. Neither of these approaches efficiently handles both the *dynamics* of the musculotendons and the complex routing *constraints*, while our approach resolves these issues.

Keywords Biomechanics · Computer simulation · Dynamics · Constraints · Finger · Hand

Manuscript received April 19, 2005; revised January 11, 2007.

S. Sueda · D. K. Pai (✉)
Sensorimotor Systems Laboratory, Department of Computer Science,
University of British Columbia, Vancouver, Canada
e-mail: pai@cs.ubc.ca

R. Balasubramanian and V. J. Santos (eds.), *The Human Hand as an Inspiration for Robot Hand Development*, Springer Tracts in Advanced Robotics 95,
DOI: 10.1007/978-3-319-03017-3_13, © Springer International Publishing Switzerland 2014

1 Introduction

Recent advances in robotics and biomechanics have furthered the understanding of the dynamic anatomy and function of the hand. There remains, however, a critical lack of a robust, efficient software simulation framework specialized for the human hand. Simulations and robots provide complementary approaches for exploring different aspects of hand modeling. If the end goal is to design an industrial, haptic, or prosthetic hand, then having a physical robotic hand is a requirement. Alternatively, if the goal is to study the effects of changing model parameters, such as tendon routing or muscle activation mechanics, then a software simulator is much more efficient. Most importantly, using subject specific model parameters is far easier to achieve with a software simulator. For example, pathologies and their corresponding surgical results can be simulated, by removing a tendon or by applying a tendon transfer.

There are several general-purpose musculotendon simulators developed for biomechanics research. These simulators can be broadly categorized into two types: line-of-force models and solid mechanics models. Neither of these approaches can fully handle the complex routing constraints present in the hand and the dynamic coupling of musculotendons and bones.

1.1 Line-of-Force Models

In line-of-force models [1–5], muscles and tendons are simulated quasistatically as kinematic paths, with no mass and inertia, which do not fully account for the shape of the biomechanical structures. The inertia of the musculotendons is lumped onto the bones in a canonical skeletal configuration, and is assumed to be constant during skeletal movement. This may be a reasonable assumption depending on the question being asked—important biomechanical properties, such as moment arm and isometric forces, are of this type. However, some questions, such as those concerning the force exerted during a grasp maneuver, or the motion of the fingertips during a tapping sequence, are better answered with a dynamic simulator, since in these situations, the errors in inertia due to lumping can be large, and can vary with hand posture and joint configuration [6].

Perhaps more importantly for hand simulation, handling of routing constraints, including branching and kinematic loops, is difficult with these methods. The complex routing constraints of tendons around various biomechanical structures such as sheaths and pulleys are non-trivial, and cause the moment arms of the musculotendons in the hand to depend on the posture [7, 8]. Some of the extrinsic tendons of the hand, such as the extensor digitorum communis, originate from a common muscle in the forearm, and furthermore, many of these tendons are connected transversely through inter-tendinous bands. These kinds of branching structures are difficult to simulate with line-of-force models, because musculotendons are idealized as abstract forces acting on the bones, and each muscle–tendon pair is represented by a single entity. Furthermore, the kinematic loop formed by the lumbrical muscle between the flexor and extensor is non-trivial to handle. The lumbrical muscle is unique in that it originates not from a bone but from a tendon (flexor digitorum profundus). The resulting coupling between the contractile forces of the flexor digitorum profundus and the lumbrical muscles cannot be captured properly by line-of-force models.

1.2 Solid Mechanics Models

Simulators based on solid mechanics models, such as the finite element or finite volume method [9–11], are also not ideal for hand simulation, since the musculotendons of the hand are thin and anisotropic, and would require many disproportionately small elements. Furthermore, solid mechanics models require general collision detection and resolution algorithms, which are expensive and not easy to use with deformable bodies. (See [12] for an alternative approach.) Various dynamic models [13, 14] from the computer graphics community could potentially be used for musculotendon simulation, but these models were designed for use in free-floating configurations, and do not work well in the highly-constraining situations present in the hand.

1.3 Strands

The challenges discussed above motivated us to develop a novel and efficient biomechanical *strand simulator*, which can simulate thin, strand-like soft tissues with complex routing constraints, such as tendons, muscles, and ligaments.[1] Following the terminology of Pai [15], we use the term *strands* to indicate that these physical primitives are not just space curves but also have mass, elasticity, and other physical properties that influence their dynamics. Our focus is on the *large-scale, hyperelastic* behavior of muscles and tendons responsible for skeletal movement, rather than on the low-level details of the muscle mechanics, such as hysteresis and residual force enhancement. In this work, we simply treat fiber level muscle mechanics as a black-box function that takes as input the current state (position, velocity, activation, etc.) of the muscles and returns as output the computed contractile and passive elastic forces. Our decision to use strands is also motivated by the anatomical structure of real muscle tissue. Muscles consist of

[1] Portions of this work are based on an earlier work: Musculotendon Simulation for Hand Animation, in ACM Trans. Graph., vol. 27, no. 3, (2008) © ACM, 2008. http://doi.acm.org/10.1145/1360612.1360682.

fibers, curved in space, which are bundled into groups called fascicles. When a muscle is activated, the fibers contract and transmit a contractile force directly along each fiber. By using strands in our muscle simulation, we are able to directly model this behavior. Strands allow us to define smooth curves to represent tendons, muscles, or even the individual fascicles of each muscle.

What is the most appropriate representation of the strand for hand simulation? What are the best degrees of freedom (DoF) for the musculotendon strands of the hand? Choosing the most appropriate generalized coordinates for the application is the key to the robustness, efficiency, and overall usability of the model. One can decide to use, for example, splines [16, 17] Cosserat frames [15], or quaternion segments [13]. For hand simulation, we require a representation that gives us the ability to handle complex constraints, and to properly transmit force around these constraints. Line-of-force models, solid mechanics models, or other strand models from computer graphics cannot satisfy these criteria completely.

To answer the questions above, we first make two important observations about hand simulation. First, for hand simulation, we know a priori where the routing *constraints* are going to be. We do not need a general formulation of contact and its associated cost in simulation and modeling time. Second, for hand simulation, we know a priori the regions on the musculotendon strands that are going to require many *degrees of freedom*, and the regions that do not. Therefore, we can reduce the degrees of freedom as much as possible while still maintaining mass, inertia, and other material properties of the musculotendons of the hand, by selectively inserting degrees of freedom in strategic regions along the strand.

The strand framework based on *splines* provides both of these important advantages. With cubic B-splines, we can have smooth routing constraints at various predefined locations along the strand, and we can selectively insert more degrees of freedom in some regions of the strand, such as near joints or pulleys. It is well suited for the hand and other parts of the body with complex routing constraints, but, just like line-of-force simulators, it can easily be used for many other musculoskeletal systems in the body. The strand framework gives the added benefit of the coupling between the dynamics of bones, muscles, and tendons, since the musculotendons are always simulated *along with* the bones—they have mass, inertia, and momentum, and interact with the bones through contact constraint forces.

2 Rigid Body Representation

We use rigid bodies expressed in maximal coordinates for the bones. Following the notation of Cline and Pai [18], we use bold letters, $\boldsymbol{x}$, to denote vectors and points in $\mathbb{R}^3$, and sans serif letters, q, to denote generalized coordinates and related quantities. Everywhere possible, q refers to the generalized coordinates of a body, and $\boldsymbol{x}$ refers to a point on the body in $\mathbb{R}^3$. Time derivatives are indicated with a dot, $\dot{\boldsymbol{x}} \equiv d\boldsymbol{x}/dt$, and material derivatives are indicated with a prime, $\boldsymbol{x}' \equiv d\boldsymbol{x}/du$.

The configuration of a rigid body is represented by the usual 4×4 transformation matrix consisting of rotational and translational components,

$$ {}^{0}_{i}\mathsf{E} = \begin{pmatrix} {}^{0}_{i}\Theta & {}^{0}\boldsymbol{p} \\ 0 & 1 \end{pmatrix}. \tag{1} $$

The leading subscripts and superscripts indicate that the coordinates of rigid body (or frame) i are defined with respect to the world frame, 0. Thus each column of ${}^{0}_{i}\Theta$ corresponds to the frame's basis vectors, e_k, expressed in world coordinates, and ${}^{0}\boldsymbol{p}$ is the position of the frame's origin expressed in world coordinates. Given a local position ${}^{i}\boldsymbol{x}$ on a rigid body, its world position is

$$ {}^{0}\boldsymbol{x} = {}^{0}_{i}\mathsf{E}\,{}^{i}\boldsymbol{x}, \tag{2} $$

where we have omitted the homogeneous coordinates for brevity. Unless otherwise stated, we assume that the reference frame is the world frame, and use a trailing subscript to indicate the frame of a rigid body, as in E_i. With this notation, E_i transforms a position from the local space of the ith rigid body to world space. Although there are other choices for the parameterization of the rotational component of the rigid body, such as unit quaternions and Euler angles, we use rotation matrices because they allow us to transform position and velocity vectors from one coordinate frame to another very efficiently with a matrix vector multiplication. For other types of simulations where the number of rigid bodies is greater, other choices may be more appropriate.

The spatial velocity ${}^{i}\phi_i$ of a rigid body ${}^{0}_{i}\mathsf{E}$ describes the motion of the rigid body at time t. The spatial velocity is composed of the angular component, ${}^{i}\boldsymbol{\omega}_i$, and the linear component, ${}^{i}\boldsymbol{v}_i$, both expressed in body coordinates, and is defined by

$$ {}^{i}_{0}\mathsf{E}\,{}^{0}_{i}\dot{\mathsf{E}} = \begin{pmatrix} [{}^{i}\boldsymbol{\omega}_i] & {}^{i}\boldsymbol{v}_i \\ 0 & 0 \end{pmatrix}, \tag{3} $$

where the 3×3 matrix, $[\boldsymbol{a}]$, is the cross-product matrix such that $[\boldsymbol{a}]\boldsymbol{b} = \boldsymbol{a} \times \boldsymbol{b}$. ${}^{i}\boldsymbol{\omega}_i$ and ${}^{i}\boldsymbol{v}_i$ are repackaged into the 6-vector ${}^{i}\phi_i$.

$$ {}^{i}\phi_i = \begin{pmatrix} {}^{i}\boldsymbol{\omega}_i \\ {}^{i}\boldsymbol{v}_i \end{pmatrix}. \tag{4} $$

The spatial velocity transforms from one frame to another according to the adjoint of the coordinate transform, which is defined from the rigid transform ${}^{0}_{i}\mathrm{E}$.

$$ {}^{0}_{i}Ad = \begin{pmatrix} {}^{0}_{i}\Theta & 0 \\ [{}^{0}_{i}\boldsymbol{p}]\,{}^{0}_{i}\Theta & {}^{0}_{i}\Theta \end{pmatrix}. \tag{5} $$

The spatial velocity of the ith rigid body in world coordinates is then

$$ {}^{0}\phi_i = {}^{0}_{i}Ad\,{}^{i}\phi_i. \tag{6} $$

If we ignore the angular component of a spatial velocity, it is easy to see that a 3D velocity is transformed simply with the rotation matrix,

$$ {}^{0}\dot{\boldsymbol{x}} = {}^{0}_{i}\Theta {}^{i}\boldsymbol{v}_i. \tag{7} $$

Again, we suppress the leading superscript for brevity and write ϕ_i, assuming that all spatial velocity quantities are expressed in local coordinates.

If a rigid body is moving with spatial velocity, ϕ_i, the world velocity of a point, ${}^{i}\boldsymbol{x}$, affixed to the rigid body is computed as

$$ \begin{aligned} {}^{0}\dot{\boldsymbol{x}} &= \Theta_i(-[{}^{i}\boldsymbol{x}]\; I)\; \phi_i \\ &= \Gamma\; \phi_i, \end{aligned} \tag{8} $$

where the 3×6 matrix, $\boldsymbol{\Gamma}$, the material Jacobian, transforms the local spatial velocity of the rigid body, ϕ_i, into the velocity of a local point on the rigid body in world coordinates, ${}^{0}\dot{\boldsymbol{x}}$. Its transpose, a 6×3 matrix, transforms a point force in world space, $\boldsymbol{f}_0$ into a local wrench acting on the rigid body.

$$ \mathsf{f}_i = \Gamma^T \boldsymbol{f}_0. \tag{9} $$

The Newton–Euler equations of motion of a rigid body can be written in a compact form as

$$ \begin{aligned} \mathrm{M}_i\; \dot{\phi}_i &= [\text{Coriolis forces}] + [\text{body forces (e.g., gravity)}] \\ &= [\phi_i]^T\; \mathrm{M}_i\; \phi_i + B(\mathsf{E}_i). \end{aligned} \tag{10} $$

Here, M_i is the spatial inertia of the rigid body, and $[\phi_i]$ is the spatial cross product matrix,

$$ [\phi_i] = \begin{pmatrix} [\boldsymbol{\omega}_i] & 0 \\ [\boldsymbol{v}_i] & [\boldsymbol{\omega}_i] \end{pmatrix}. \tag{11} $$

Expressing the spatial velocity of a rigid body in local coordinates has the advantage in that the mass matrix is diagonal and can be precomputed at the beginning of the simulation. We use the following time-stepping discretization at the velocity level to obtain the discrete impulse-momentum equations at time $t^{(k)}$.

$$ \mathrm{M}_i\; \phi_i^{(k+1)} = \mathrm{M}_i\; \phi_i^{(k)} + h\left(\left([\phi_i]^{(k)}\right)^T \mathrm{M}_i\; \phi_i^{(k)} + B\left(\mathsf{E}_i^{(k)}\right)\right), \tag{12} $$

where $h = t^{(k+1)} - t^{(k)}$ is the time step size.

Joint constraints between rigid bodies are implemented using the adjoint formulation [19], from which we can easily derive various types of joints simply by dropping different rows in the 6×6 adjoint matrix. Given two rigid bodies, i and k, and a joint frame defined with respect to the first body, ${}^{i}_{j}\mathsf{E}$, we constrain the rigid bodies' spatial velocities, ϕ_i and ϕ_k, with respect to the joint frame. Using Eq. 6, the relative velocity at joint j is given by

$$\begin{aligned} \delta\phi_j &= {}^j_i Ad\ \phi_i - {}^j_k Ad\ \phi_k \\ &= \begin{pmatrix} {}^j_i Ad & -{}^j_i Ad\ {}^i_k Ad \end{pmatrix} \begin{pmatrix} \phi_i \\ \phi_k \end{pmatrix}. \end{aligned} \tag{13}$$

For a rigid joint, we want the relative velocities to be zero, so we set $\delta\phi = 0$. From this, we can derive different types of joints, by dropping various rows of the constraint equation: for example, the top three rows (corresponding to the three rotational DoFs) for a ball joint, or the third row (corresponding to the rotation about the z-axis) for a hinge joint.

Writing these joint constraint equations in the form $G\phi = 0$, we simultaneously obtain the velocity that satisfies the constraints at the time $t^{(k+1)}$ and the contact impulse $G^T\lambda$ by solving the Karush–Kuhn–Tucker (KKT) system

$$\begin{pmatrix} M & G^T \\ G & 0 \end{pmatrix} \begin{pmatrix} \phi^{(k+1)} \\ \lambda \end{pmatrix} = \begin{pmatrix} \mathsf{f}^{(k)} \\ 0 \end{pmatrix}, \tag{14}$$

where λ is the vector of Lagrange multipliers for the constraints. The joint reaction forces are then $-G^T\lambda/h$.

The rigid body configuration $\mathsf{E}_i^{(k+1)}$ can be obtained by integrating $\phi_i^{(k+1)}$. We must be careful here, because E_i belongs to a non-Euclidean space (SE(3), the Special Euclidean group in three dimensions). We use a first order implicit discretization, with the time step h:

$$\mathsf{E}_i^{(k+1)} = \mathsf{E}_i^{(k)} \exp\left(h \begin{pmatrix} \left[\omega_i^{(k+1)}\right] & v_i^{(k+1)} \\ 0 & 0 \end{pmatrix} \right). \tag{15}$$

The matrix exponential can be computed efficiently using Rodriguez' formula [19].

3 Strand Framework

The *Spline* strand is an extension of the physically-based spline models previously used in computer graphics and computer aided design [16], with the addition of muscle activation dynamics (Sect. 3.1), and simple yet robust sliding and surface constraints (Sect. 3.2). The spline strand, or just "strand" for the rest of this chapter, is a cubic B-spline curve with mass, elasticity, and other physical properties that influence its dynamics. It is parameterized by $n \geq 4$ control points, and has $3n$ degrees of freedom, corresponding to the x, y, and z coordinates of the n control points. The state of the system is given by the stacked positions and velocities of the rigid bodies and strand control points. These are the generalized coordinates and velocities of the system, respectively:

$$\begin{aligned} \mathsf{q} &= \left[\cdots\ \mathrm{E}_i\ \cdots\ \boldsymbol{q}_j\ \cdots\right]^T \\ \dot{\mathsf{q}} &= \left[\cdots\ \phi_i\ \cdots\ \dot{\boldsymbol{q}}_j\ \cdots\right]^T. \end{aligned} \tag{16}$$

Using the notation from Sect. 2, $\mathsf{E}_i \in SE(3)$ and $\phi_i \in se(3)$ are the configuration and the spatial velocity of the ith rigid body. $\boldsymbol{q}_j \in \mathbb{R}^3$ and $\boldsymbol{q}_j \in \mathbb{R}^3$ are the position and the velocity of the jth spline control point of the strand.

For each generalized coordinate, we construct an impulse-momentum equation which, when discretized at the velocity level, is

$$\mathrm{M}\ \dot{\mathsf{q}}^{(k+1)} = \mathrm{M}\ \dot{\mathsf{q}}^{(k)} + h\,\mathsf{f} - G^T\lambda, \tag{17}$$

where M is the block-diagonal generalized mass matrix of rigid bodies and strand control points, h is the step size, f is the generalized force, and $G^T\lambda$ is the constraint force (Sect. 3.2). The generalized force includes the body forces for rigid bodies [18], and the passive and active forces for strands (Sect. 3.1).

3.1 Strand Forces

The generalized force term in Eq. 17 is composed of four terms, and the dynamics equation for the control points of the strands is given by

$$M\dot{\boldsymbol{q}}^{(k+1)} = M\dot{\boldsymbol{q}}^{(k)} + h\big(\boldsymbol{f}_d + \boldsymbol{f}_g + \boldsymbol{f}_p + \boldsymbol{f}_a\big) - G^T\lambda, \tag{18}$$

where $\boldsymbol{f}_d$ is the Rayleigh damping force, $\boldsymbol{f}_g$ is the gravity force, $\boldsymbol{f}_p$ is the passive elastic force, and $\boldsymbol{f}_a$ is the active force. The last term, $G^T\lambda$, is the constraint force term, which will be discussed in Sect. 3.2. The mass matrix, M, derived from the kinetic energy of the spline, can be precomputed at the beginning of the simulation, and is sparse, due to the local support structure of cubic spline curves (See Fig. 1). For the derivation of $\boldsymbol{f}_g$ and M, we refer the reader to [20]. In the rest of this section, we derive the passive and active forces for the musculotendon strands.

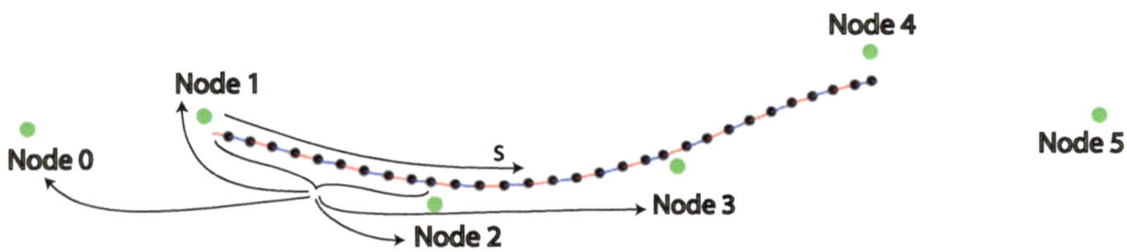

Fig. 1 A cubic B-spline strand, with the control points shown in green. Any point on the strand is defined by four control points and a scalar parameter s. For example, the first portion of the strand only depends on nodes 0 through 3. The black circles illustrate the material points sampled at $s = 0.1, 0.2, 0.3, \ldots\ldots$ For the computation of the dynamics, we integrate along s, rather than taking discrete samples

As an elastic solid, a muscle is made of a complex nonlinear viscoelastic material [21], and modeling and understanding its properties is an active area of research with divided views [22, 23]. The main contentious issue is the behavior of the muscle in the "descending limb" of the force–length curve.

The force exerted by a muscle is the sum of the passive and active forces. In the Hill-Zajac model [21], the passive force is assumed to be due to a hyperelastic material model, while the active force is computed as a function of the muscle's activation level, a, maximal isometric force, f_0, current strain, ε, and current strain rate, $\dot{\varepsilon}$.

$$f_a = a \cdot f_0 \cdot FL(\varepsilon) \cdot FV(\dot{\varepsilon}). \tag{19}$$

The Force–Length (FL) and Force–Velocity (FV) curves are measured empirically from isolated muscle experiments. First, the muscle's passive elasticity is measured by gradually increasing its length and tabulating the generated tension. Then the experiment is repeated with the muscle activated; the muscle at rest is stretched to the desired length, then maximally activated, and then the tension is measured. Depending on the muscle specimen, the resulting FL relationship often shows an area of negative slope—as the length is increased, the force decreases. The order of operation here is critical, and is the source of the contention. If the muscle is activated first and then stretched to the desired length, it results in a significantly different FL relationship and does not show any areas of negative stiffness ([23], p. 92). This "history dependency" is also present in the ascending arm and the plateau of the FL curve, but is most prominent in the descending arm.

The slope is usually interpreted as stiffness, and this can cause instability in the muscle response. The reason for this instability can be demonstrated by imagining the muscle as being composed of two springs in series (Fig. 2b). If the stiffness is positive, then any perturbation away from equilibrium causes the springs to exert restoring forces. If the stiffness is negative, however, the resulting spring forces cause further perturbation away from equilibrium, since the lengthened spring becomes weaker and the shortened spring becomes stronger.

Incorporating proper muscle mechanics is a key component of building an accurate musculoskeletal simulator. However, understanding and modeling a robust and accurate muscle model is not the focus of this work. One approach is to

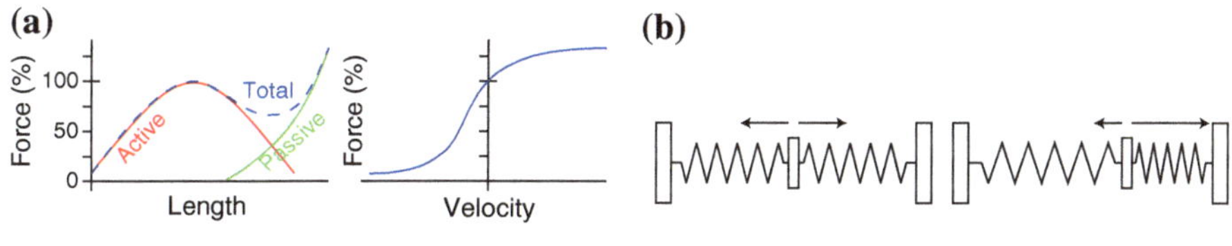

Fig. 2 **a** Typical Force–Length and Force–Velocity curves. As a muscle is stretched, the total force increases, then decreases, and then increases again. If the muscle is being shortened, its force decreases, and if the muscle is lengthened, its force increases. **b** If the springs have negative stiffness, then any perturbation away from the equilibrium results in forces that push the system further away from equilibrium

replace the area with negative slope with a straight horizontal line. Instead we assume that the active force takes on the standard Hill-Zajac form shown in Eq. 19, and treat $FL(\varepsilon)$ and $FV(\dot{\varepsilon})$ as customizable black-box functions defined by the user.

With the customizable FL and FV curves, we are able to insert any muscle mechanics model into a strand simulation, as long as it fits Eq. 19. Depending on the application, we can choose different models. This is similar to the approach taken by general line-of-force musculoskeletal simulators [1], with the main difference being that with the spline basis, the strand can have a variable strain value along the length of the strand, rather than having a single value for the entire musculotendon.

In order to compute the strain and strain rate $(\varepsilon, \dot{\varepsilon})$, we need to derive the strand's kinematic path and its derivative. The kinematic path of a strand is a cubic B-spline curve:

$$\begin{aligned} \boldsymbol{x}(s,t) &= \sum_{i=0}^{3} b_i(s)\boldsymbol{q}_i(t) \\ b_0(s) &= \left(-s^3 + 3s^2 - 3s + 1\right)/6 \\ b_1(s) &= \left(3s^3 - 6s^2 + 4\right)/6 \\ b_2(s) &= \left(-3s^3 + 3s^2 - 3s + 1\right)/6 \\ b_3(s) &= s^3/6, \end{aligned} \tag{20}$$

where $\boldsymbol{q}_i(t)$ denote the control points of the strand, with velocities $\dot{\boldsymbol{q}}_i(t)$. The cubic B-spline basis functions, $b_i(s)$, depend on the material point, s, along the spline (which will be integrated out in the final equations of motion). Although a strand can have an arbitrary number of control points, a material point on a strand only depends on four control points, due to the local support of the B-spline basis. Thus, to traverse along a strand with n control points, we sum the contribution from an overlapping set of four control points at a time: 0 through 3, 1 through 4,…, and $n-3$ through n. The velocity and the tangent vectors of a point $\boldsymbol{x}(s,t)$ can be obtained in a similar manner.

$$\begin{aligned} \dot{\boldsymbol{x}}(s,t) &\equiv \frac{d\boldsymbol{x}}{dt} = \sum_{i=0}^{3} b_i(s)\boldsymbol{q}_i(t) \\ \boldsymbol{x}'(s,t) &\equiv \frac{\partial \boldsymbol{x}}{\partial s} = \sum_{i=0}^{3} b'_i(s)\dot{\boldsymbol{q}}_i(t), \end{aligned} \tag{21}$$

where $b'_i(s)$ are the derivatives of the B-spline basis functions. The strain and the strain rate at a material point s on the strand are given by

$$\begin{aligned}\varepsilon(s,t) &= \frac{\|\boldsymbol{x}'(s,t)\|}{\|\boldsymbol{x}_0'(s)\|} - 1 \\ \dot{\varepsilon}(s,t) &= \frac{\|\dot{\boldsymbol{x}}'(s,t)\|}{\|\boldsymbol{x}_0'(s)\|},\end{aligned} \tag{22}$$

where $\boldsymbol{x}_0'(s) = \boldsymbol{x}'(s,0)$, computed at $t = 0$. $\varepsilon(s)$ is defined at an infinitesimal material point $\boldsymbol{x}(s)$, and this ability to represent varying strain values along the musculotendon strand is one of the strengths of our framework.

For the passive and the active forces, the user specifies two FL functions, $FL_p(\varepsilon)$ and $FL_a(\varepsilon)$, that return the stress as a function of strain. The work function at a material point s is then $W = \int FL(\varepsilon)d\varepsilon$, which is integrated along the strand to give us the strain potential energy, $V = \int_0^1 Wds$. Finally, the generalized force acting on the control points is derived by taking the derivative of the elastic energy. For the passive force, the generalized forces acting on the first four control points are

$$\boldsymbol{f}_p = A_{CS} \sum_{i=0}^{3} \int_0^1 FL_p(\varepsilon) b_i' \frac{\boldsymbol{x}'}{\|\boldsymbol{x}'\|} ds, \tag{23}$$

where A_{CS} is the cross-sectional area of the strand. If the strand contains $n > 4$ control points, the contributions to the other control points are computed by incrementing the range of the summation from 0 through 3 to $n - 3$ through n. The integral can be evaluated accurately using a numerical integration scheme. Note that we do not require the pennation angle, since the geometric paths of the strands automatically account for the arrangement of fascicles in the muscles. However, it can be added if required, for instance, when approximating a pennate muscle with a single strand, with the appropriate scaling of the physiological cross-sectional area.

We assume that the active force, $\boldsymbol{f}_a$, is linear in the activation levels, and can be expressed as a matrix–vector product, $\boldsymbol{f}_a = Aa$, where a is the vector of muscle activation levels between zero (no activation) and one (full activation). The matrix A, which is of size (#DoF $\times$ #muscles), is the ''activation transport'' matrix that takes as input the activations of the muscles and returns as output the corresponding forces on the strand DoFs (Fig. 3). This separation of force and activation will be used in Sect. 4 for the derivation of the controller. The active force is derived in the same manner as the passive force, but with FL_a instead of FL_p:

$$\boldsymbol{f}_a = \underbrace{\left(f_0 FV(\dot{\varepsilon}) A_{CS} \sum_{i=0}^{3} \int_0^1 FL_a(\varepsilon) b_i' \frac{\boldsymbol{x}'}{\|\boldsymbol{x}'\|} ds \right)}_{A} a. \tag{24}$$

The average strain rate of the whole strand, computed using the velocities from the last time step, is used to scale the force by the force–velocity factor, $FV(\dot{\varepsilon})$.

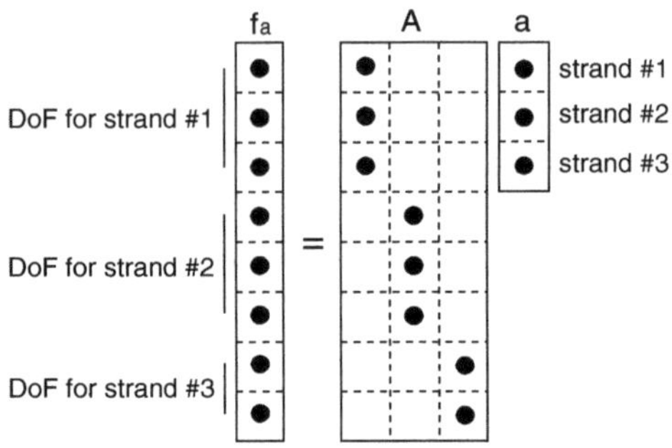

Fig. 3 A schematic of the activation transport matrix, A, for a simplified system. There are three musculotendons, each with its own activation signal: a. The activation transport matrix multiplies these activation levels, and returns the forces, f_a, on the degrees of freedom of the strands

3.2 Constraints

Constraint handling is an important component of a biomechanical simulator, used for musculotendon origins/insertions and for tendon routing. Although wrapping surfaces implemented in general biomechanical simulators [1, 24] are effective for kinematic constraints, they do not work for dynamic constraints, and are also limited to simplified geometries, such as spheres and cylinders.

Tendon routing is particularly difficult, and often ignored by existing biomechanical simulators. In our system, there are two types of constraints for tendon routing: sliding and surface constraints, as shown in Fig. 4. A sliding constraint is used to thread a tendon through a specific point in space on a bone, and surface constraints are used to approximate the contact constraint between the surface of the bone and the tendon.

Collision detection and resolution for a general multi-body simulator with deformable strands are non-trivial. There are, however, two key assumptions in our application that simplify the constraint formulation. First, since tendons and muscles stay in contact with surrounding tissue and do not come apart, *we deal only with equality constraints*; inequality constraints, which are more difficult to solve numerically, do not need to be included. Second, because we know where the contact regions are going to be, *general-purpose collision detector is not required*. Because strands are based on spline curves, keeping track of contacting

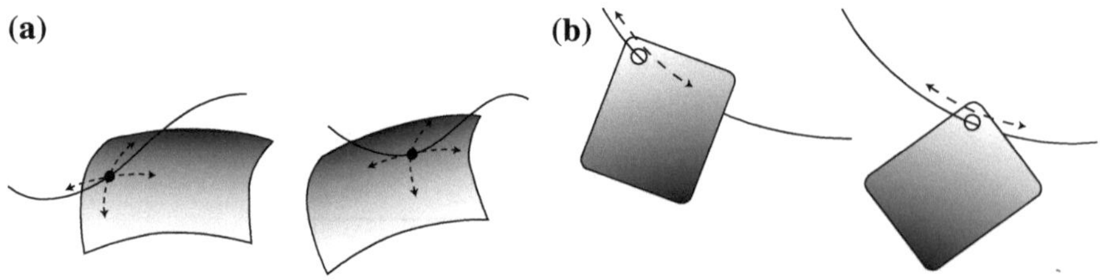

Fig. 4 **a** Surface constraint and **b** sliding constraint. With a surface constraint, the constraint point moves on the rigid body surface, whereas with a sliding constraint, the constraint point moves along the strand

points is computationally inexpensive. Potential contact points are first predetermined along each strand, and after each time step, these closest points are updated using the Newton–Raphson method. Contact points on rigid bodies for surface constraints are tracked in a similar manner. First, each rigid body is wrapped with a cubic tensor-product surface, and the contact point is tracked on the surface at each time step using the Newton–Raphson method.

The constraints in our system are formulated at the velocity level. Let $g(\mathsf{q})$ be a vector of position-level equality constraint functions, such that when each constraint is satisfied by the generalized coordinates, q, $g(\mathsf{q}) = 0$ By differentiating g with respect to time, we obtain a corresponding velocity-level constraint function that is consistent with our discretization.

$$\frac{dg(\mathsf{q})}{dt} = \frac{\partial g(\mathsf{q})}{\partial \mathsf{q}} \dot{\mathsf{q}} = 0. \tag{25}$$

Denoting the gradient of g by the constraint matrix G, we obtain the constraint equation $G\dot{\mathsf{q}} = 0$.

Fixed Constraints: We use fixed constraints for musculotendon origins and insertions, as well as for attaching several strands together to form branching structures. For example, if we want to constrain a point on a strand, $\boldsymbol{x}$, to a point on a rigid body, $\boldsymbol{x}_0$, we set their relative velocities to be equal.

$$\begin{aligned} \dot{\boldsymbol{g}} &= \dot{\boldsymbol{x}} - \dot{\boldsymbol{x}}_0 \\ &= \begin{pmatrix} b_0(s) & b_1(s) & b_2(s) & b_3(s) & -\Gamma \end{pmatrix} \begin{pmatrix} \dot{\boldsymbol{q}}_0 \\ \dot{\boldsymbol{q}}_1 \\ \dot{\boldsymbol{q}}_2 \\ \dot{\boldsymbol{q}}_3 \\ \phi \end{pmatrix}. \end{aligned} \tag{26}$$

Both $\dot{\boldsymbol{x}}$ and $\dot{\boldsymbol{x}}_0$ are linear with respect to the rigid body and spline DoFs of the system, as given by Eqs. 8 and 21.

Surface Constraints: Surface constraints are similar to the usual rigid body contact constraints. A fixed point on a strand is constrained to lie on a point on the surface of a rigid body. Let $\boldsymbol{x}$ denote the 3D position of the strand point to constrain, and $\boldsymbol{x}_0$ and $\boldsymbol{n}$ denote its corresponding contact point and normal on the rigid body. Equating the relative velocities along the normal gives

$$\dot{g} = \boldsymbol{n}^T(\dot{\boldsymbol{x}} - \dot{\boldsymbol{x}}_0). \tag{27}$$

The constraint point on the strand is fixed, and the point on the surface is updated before each step, by finding the closest point on the tensor-product surface attached to the rigid body (Fig. 4a).

Sliding Constraints: In most situations, such as in the carpal tunnel, tendons are confined to slide axially but not laterally. We can achieve this behavior by adding an additional dimension to the surface constraint. Given the tangent vector of the point to constrain on a strand, we generate the normal, $\boldsymbol{n}_1$, and binormal, $\boldsymbol{n}_2$, vectors, and apply the constraint with respect to both of these vectors.

$$\begin{aligned} \dot{g}_1 &= \boldsymbol{n}_1^T(\dot{\boldsymbol{x}} - \dot{\boldsymbol{x}}_0) \\ \dot{g}_2 &= \boldsymbol{n}_2^T(\dot{\boldsymbol{x}} - \dot{\boldsymbol{x}}_0). \end{aligned} \tag{28}$$

Unlike the surface constraint, the constraint point on the rigid body is fixed, and the point on the strand is updated before each step (Fig. 4b).

3.3 Time Integration

The strand framework is agnostic to the choice of integrator. Using the mass matrix M from [20], the force vector f from Sect. 3.1, and constraint matrix G and vector g from Sect. 3.2, we can use any integrator we choose to step the state of the system forward in time. In our current implementation, we take the two step approach commonly used in the computer graphics community. We solve for the constrained velocities, and then update the positions using these new velocities. Using the assumption that muscles are always in close contact, we know the active constraints at the current time step, and we obtain a linear KKT system:

$$\begin{pmatrix} M & G^T \\ G & 0 \end{pmatrix} \begin{pmatrix} \dot{q} \\ \lambda \end{pmatrix} = \begin{pmatrix} M\dot{q}_0 + hf \\ 0 \end{pmatrix}, \tag{29}$$

where $\dot{q}_0$ is the velocity at the previous time step and λ is the vector of Lagrange multipliers for the constraints. This KKT system contains both rigid bodies and strands, whereas the KKT system in Eq. 14 contains only rigid bodies. The position update is performed trivially for all the strand control points: $\boldsymbol{q} = \boldsymbol{q}_0 + h\dot{\boldsymbol{q}}$, where h is the time step size. For rigid bodies, we use the Rodrigues' formula [19].

Because the constraints are solved at the velocity level, the system may drift away from the constraint manifold over time. To fight this drift, we add a post-stabilization step [18] after taking a position step.

$$\begin{pmatrix} M & G^T \\ G & 0 \end{pmatrix} \begin{pmatrix} \Delta q \\ \lambda \end{pmatrix} = \begin{pmatrix} 0 \\ -g \end{pmatrix}. \tag{30}$$

The generalized positions are updated similarly as before: $\boldsymbol{q} = \boldsymbol{q}_0 + \Delta\boldsymbol{q}$ for strand control points and with the Rodriguez' formula for rigid bodies.

4 A Controller for Musculoskeletal Systems

Computing the activation levels required by the musculotendons to produce desired hand motion is difficult because of the redundancy of the muscles and the complexity of the routing of the tendons. Even simple tasks, such as moving a finger from one position to another, require the coordinated activation of several muscles acting as synergists and/or antagonists. Here, we present a tracking controller that has been used successfully for controlling the hand [17], as well as the eye and the jaw [25, 26].

For the purpose of our simulation, the job of the controller is to compute the activation levels of the muscles, which innervates the muscles to derive the skeleton to follow some target trajectory. We assume that the simulator has the following form; any strand formulation should work, as long as the contractile force is linear in the activation level, a:

$$\mathsf{M}\dot{\mathsf{q}} = \mathsf{f} + Aa, \tag{31}$$

where M is the system mass matrix, $\dot{\mathsf{q}}$ is the system velocity vector, f is the system force vector, and A is the activation transport matrix that maps the activation vector, a, to the appropriate DoFs (see Fig. 3 and Eq. 24). The KKT system from Eq. 29 can be converted to this form, by first extracting out the active force, $f_a = Aa$, from the rest of the forces, assuming that the active force is linear in the activation levels, and then combining the velocities and Lagrange multipliers into a single vector, $\dot{\bar{q}}$.

$$\underbrace{\begin{pmatrix} M & G^T \\ G & 0 \end{pmatrix}}_{\overline{\mathsf{M}}} \underbrace{\begin{pmatrix} \dot{q} \\ \lambda \end{pmatrix}}_{\dot{\bar{\mathsf{q}}}} = \underbrace{\begin{pmatrix} f + Aa \\ 0 + 0 \end{pmatrix}}_{\bar{\mathsf{f}} + \bar{\mathsf{A}}a}. \tag{32}$$

The length of the vector $\dot{\bar{\mathsf{q}}}$ is the number of DoFs in the system *plus* the number of constraints. Thus, for a system with rigid bodies and strands we have,

$$\dot{\bar{q}} = (\,\phi_0 \quad \cdots \quad \phi_n \quad \dot{\boldsymbol{q}}_0 \quad \cdots \quad \dot{\boldsymbol{q}}_m \quad \lambda_0 \quad \cdots \quad \lambda_l\,)^T. \tag{33}$$

The controller computes the activation levels, a, from a set of target velocities specified by the user, for example, from motion-capture data, key-framed animations, or even from the output of other skeletal controllers. If the input target is a sequence of rigid body configurations rather than velocities, it can be converted into the required form by computing the spatial velocities needed to move the bodies from their current configurations to the target configurations in a time step, using the matrix logarithm.

$$\phi = \left]\log\left(\mathsf{E}^{-1}\mathsf{E}_{target}\right)\right[, \tag{34}$$

where E is the current rigid body configuration, and E_{target} is the target rigid body configuration. The unbracket operator, $]A[$, extracts the spatial velocity vector from the argument matrix (See Eq. 3).

The input data specify the target velocities of rigid bodies, v_x. This can be either a 3D point velocity or a 6D spatial velocity, and the total size of v_x is $(3 \times \#\text{point targets} + 6 \times \#\text{spatial targets})$. Let us consider the following two examples:

(1) If there are n rigid bodies in the simulator, and if all of them have spatial velocity targets, then v_x would be a vector of size $6n$ composed of n target spatial velocities stacked on top of each other: $v_x = \left(\phi_0^{target} \cdots \phi_n^{target}\right)^T$.
(2) If there is a single 3D point target velocity (e.g., at the tip of the end effector), then v_x would be a vector of size 3 consisting of the target velocity vector: $v_x = \mathbf{v}_n^{target}$.

We then require that the controller computes the muscle activations to match the resulting system velocities to the desired velocities. That is,

$$\Gamma_x \dot{\bar{q}} = v_x, \tag{35}$$

where the entries of the Jacobian Γ_x contain the 6×6 identity matrix for spatial velocity targets and the 3×6 matrix Γ from Eq. 8 for point velocity targets. For Examples (1) and (2) from above, we have

$$\begin{aligned} \begin{pmatrix} I & \cdots & 0 & 0 & \cdots & 0 & 0 & \cdots & 0 \\ \vdots & \ddots & \vdots & \vdots & \ddots & \vdots & \vdots & \ddots & \vdots \\ 0 & \cdots & I & 0 & \cdots & 0 & 0 & \cdots & 0 \end{pmatrix} \dot{\bar{q}} &= \begin{pmatrix} \phi_0^{target} \\ \vdots \\ \phi_n^{target} \end{pmatrix}, \\ \begin{pmatrix} 0 & \cdots & \Gamma & 0 & \cdots & 0 & 0 & \cdots & 0 \end{pmatrix} \dot{\bar{q}} &= v_n^{target}. \end{aligned} \tag{36}$$

Since we only have rigid body targets, the entries in Γ_x for strand control points and Lagrange multipliers are always zero.

Substituting Eq. 31 into Eq. 35 and eliminating $\dot{\bar{q}}$, we arrive at the *targeting constraint equation*

$$H_x a + v_f = v_x, \tag{37}$$

where $H_x = \Gamma_x \overline{M}^{-1} \overline{A}$ and $v_f = \Gamma_x \overline{M}^{-1} \bar{\mathrm{f}}$. The matrix H_x can be interpreted as the effective inverse inertia experienced by the muscle activation levels in order to produce the target motion. Multiplying H_x by a gives the velocity of targets due to muscle activations, and the "free velocity" vector, v_f, is the velocity of the targets due to the non-active forces acting on the system. Thus, we are looking for the activation levels, a, that zero out the difference between the desired target velocities and the sum of active and passive velocities: $a = H_x^{-1}\left(v_x - v_f\right)$.

In some cases, it is possible to solve for the activations using the targeting Eq. 37 as a hard constraint. In most cases, however, the dynamics of the system cannot *exactly* follow the requested targets. For example, the bone joint constraints may prevent certain poses, or muscles may not be strong enough to produce the required force. So, we convert the targeting equation into a quadratic program (QP) instead. The linearity assumption in Eq. 31 allows us to obtain a convex quadratic program shown in Eq. 38. Without this assumption, we would instead obtain a general optimization problem that may or may not be convex.

To make the dynamics follow the target trajectory as closely as possible, the controller may sometimes return activations that switch spastically. In order to prevent this, we add a damping term to the objective function, using finite differencing to approximate the derivative of the activation. In addition, to minimize the total activation, we add an "activation energy" term to the objective. Putting this all together, the activation QP is

$$\begin{aligned} \min_{a} \quad & w_a\|a\|^2 + w_x\left\|\left(H_x a + v_f\right) - v_x\right\|^2 + w_d\|a - a_0\|^2 \\ \text{s.t.} \quad & 0 \le a \le 1, \end{aligned} \tag{38}$$

where w_a, w_x, and w_d are the blending weights, and a_0 is the activation vector from the previous time step. The first term minimizes the total activation and also adds regularization to the quadratic problem, whereas the second and third terms function as a spring-and-damper controller to guide the dynamics of the system toward the target motion. For easy motions that are not over-constrained, w_a and w_d can be set to zero to achieve close tracking of the target. However, for some motions, especially those that involve configurations that are almost singular, such as full flexion or extension, regularization (w_a) and damping (w_d) help stabilize the solution. For the simulations in Fig. 5, we used $(w_a, w_x, w_d) = (1, 1, 0.01)$.

The weight for the first term, w_a, need not be a scalar; each muscle can have its own weight, and can include such terms as the muscle's physiological cross sectional area and average muscle stress [27]. Herzog [28] takes a similar approach, and includes an inverse factor for the corresponding moment arm of the muscle so that muscles with large moment arms are recruited preferentially.

Although the original dynamics equation can be large (albeit very sparse), the dimensionality of the activation QP is the number of muscle strands, which is small in general (<100). The main computational cost is in the construction of the quadratic matrix, H_x, which requires $min(\#targets, \#muscles)$ sparse solves on the KKT matrix from Eq. 29. However, since the LU factors of the KKT matrix are required for solving the forward model, the only additional cost is the back-solve, which is relatively cheap. In practice, we note that the controller adds no significant computational cost to the overall system.

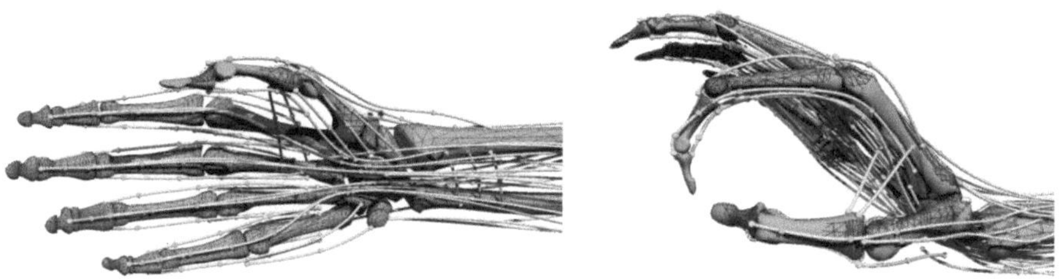

Fig. 5 Hand simulation with strands. The inverse solver computes the activation levels required to make the hand follow the input trajectory shown in wireframe from the initial hand pose shown in the left image to the final hand pose shown in the right image

5 Results

5.1 Whole Hand Simulation with Activation Controller

To show the scalability of the strand framework, we simulate the hand with 54 musculotendons and 17 bones (Fig. 5). The bone and muscle meshes were purchased from Snoswell Design in Adelaide, and the musculotendon paths were constructed based on standard textbook models in the literature [29]. For the test motions, the skeleton was rigged and animated using a commercial 3D software, and the resulting motions were imported into our simulator, implemented in Java. The activations for an motion sequence of several seconds are computed within a few minutes. The degrees of freedom of the skeleton are flexion/extension and abduction/adduction of the fingers, thumb, and wrist, and pronation/supination of the forearm. Our controller is able to compute the muscle activation levels for motions involving all of these ranges of motion.

5.2 Mass and Inertia

In a recent article, Pai [6] showed that even for a simple mono-articular limb, ‘‘errors in inertia due to [mass] lumping can be quite large and variable, changing with body posture and coupling the velocities of different joints.’’ This was shown with a simple simulator based on joint angles with and without mass-lumping. Here, we demonstrate that the result of our more general simulator matches that of Pai [6]. In Fig. 6, the simulation results of a mono-articular limb with and without mass-lumping are shown. The weight and length of the two segments are 4.2 kg

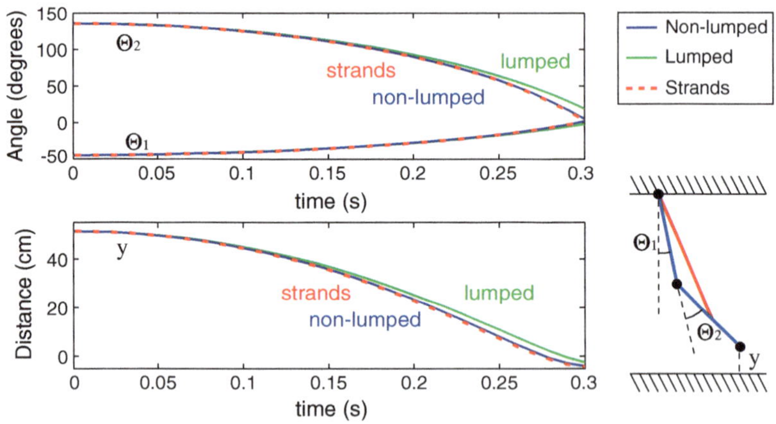

Fig. 6 Comparison of movement with the data from Pai [6], Fig. 2. A mono-articular limb was simulated for 300 ms with and without mass-lumping. The more general strand framework closely matches the output of the simpler model

and 43 cm respectively. With mass-lumping, the mass of the muscle, shown in red in the bottom right figure, is lumped to the two bones, shown in blue, at some default pose, and the system inertia is assumed to be fixed over time, even though it should change due to the movement of the muscle with respect to the degrees of freedom of the system. Without mass-lumping, the continuous change in the effective mass of the degrees of freedom of the system is taken into account.

With the strand simulator, the two bones are modeled as rigid bodies, and the muscle strand is modeled as a straight strand between the rigid bodies; and the generalized mass matrix of the system is recomputed at every time step to reflect the movement of the strand with respect to the rigid bodies. The limb, starting from the initial pose ($\Theta_1 = -45°, \Theta_2 = 135°$), falls due to gravity. In the top graph, the two joint angles are plotted over time. The solid blue (non-lumped) and green (lumped) curves show the results from Pai [6], and the overlaid, dotted red curve shows the result from the strand simulator. The strand simulator closely matches the non-lumped model of Pai [6] with the final mean error of 0.7° for Θ_1 and Θ_2, compared to the error of 14° between the lumped and non-lumped models. Similarly, if the height of the end point (foot) is plotted over time, the strand simulator matches the result from Pai [6] with the final error of 1.4 mm, compared to the error of 29 mm that would result from mass-lumping. The small discrepancy between the non-lumped models is likely due to the different integrators used; Pai [6] used a higher order integrator (RK45), whereas we used the linearly-implicit Euler integrator.

5.3 *Complex Routing Constraints*

Branching, kinematic loops, and slack are difficult to model using line-of-force models, but can be modeled naturally with the strand framework. With a simple simulation example of the lumbrical in Fig. 7, we show how the lumbrical can change the transmission of force of the flexor muscle across multiple joints. The simulation model consists of three rigid bodies and three strands. Both the extensor (EDC) and the flexor (FDP) insert into the distal bone. The lumbrical (LUM) originates from the FDP and inserts into the EDC. From the finger pose shown in Fig. 7a, the simulation is run with the FDP active and the lumbrical inactive

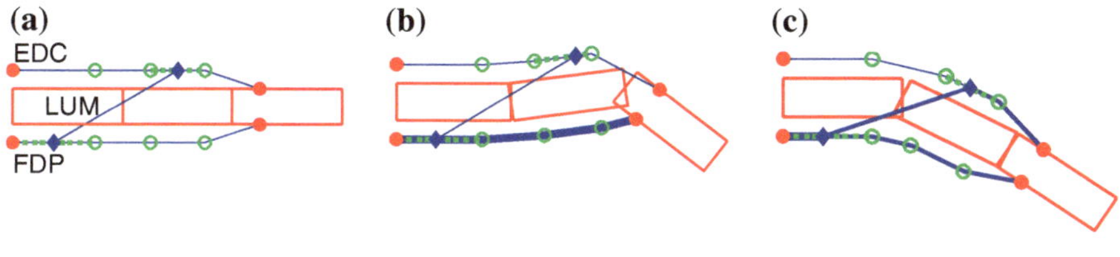

Fig. 7 **a** Simplified lumbrical simulation model, with a kinematic loop. The effect of activating the FDP differs depending on whether the lumbrical is coactivated. **b** Only the FDP is activated, and the distal joint flexes. **c** Both the FDP and LUM are activated, and the proximal joint flexes

(Fig. 7b), and with both the FDP and the lumbrical coactivated (Fig. 7c). When the FDP is activated in isolation, the distal joint is flexed, and the proximal joint is slightly hyper-extended. When the lumbrical is coactivated, however, the tension in the FDP is distributed to the distal part of EDC, and only the proximal joint flexes.

6 Conclusion

We proposed a new simulation framework well suited for modeling the fingers and the hand, which have been challenging to simulate with previous biomechanical simulators, due to the complex routing of tendons around various constraints such as sheaths and pulleys. Neither line-of-force models nor solid mechanics models efficiently handle both the *dynamics* of the musculotendons and the complex routing *constraints*. Our approach resolves these issues, by using the spline basis as the dynamic degrees of freedom (DoFs) of the system, and placing these DoFs at strategic locations along muscles and tendons.

We showed two small examples that are difficult to model with other simulators. In Sect. 5.2, we demonstrated the effect of mass and inertia on a simple 2-link example, and how our simulator correctly accounts for the change in inertia incurred by the change in joint angles. In Sect. 5.3, we showed a simulation with a lumbrical-like muscle that originates and inserts into another musculotendon, forming a kinematic loop. Our approach can efficiently simulate such configurations that are more challenging with other approaches. These are both important issues for hand simulation, because (1) the inertia of the phalanges are heavily dependent on the musculotendons they are connected to, and (2) the tendon routing in the hand is highly complex.

In a recent work, we have further improved the handling of constraints between musculotendons and bones, by adding a mixed formulation of Eulerian and Lagrangian approaches into the strand framework [30]. This extension allows us the handle the constraint independently from the initial discretization of the strand, by decoupling the the material and spatial descriptions of the strand.

References

1. S.L. Delp, F.C. Anderson, A.S. Arnold, P. Loan, A. Habib, C.T. John, E. Guendelman, D.G.Thelen, OpenSim: open-source software to create and analyze dynamic simulations of movement. IEEE Trans. Biomed. Eng. **54**(11), 1940–1950 (2007). http://www.ncbi.nlm.nih.gov/pubmed/18018689
2. M. Damsgaard, J. Rasmussen, S. Christensen, E. Surma, M. Dezee, Analysis of musculoskeletal systems in the AnyBody Modeling System. Simul. Model. Pract. Theory **14**(8), 1100–1111 (2006)
3. H. Lipson, A relaxation method for simulating the kinematics of compound nonlinear mechanisms. ASME J. Mech. Des. **128**, 719–728 (2006)

4. E. Johnson, K. Morris, T. Murphey, A variational approach to strand-based modeling of the human hand, in *Algorithmic Foundation of Robotics VIII*, ed. by G. Chirikjian, H. Choset, M. Morales, T. Murphey. Series Springer Tracts in Advanced Robotics, vol. 57 (Springer, Berlin, 2009), pp. 151–166
5. W. Tsang, K. Singh, F. Eugene, Helping hand: an anatomically accurate inverse dynamics solution for unconstrained hand motion, in *Proceedings of the ACM SIGGRAPH/Eurographics Symposium on Computer Animation* (2005), pp. 319–328
6. D.K. Pai, Muscle mass in musculoskeletal models. J. Biomech. **43**(11), 2093–2098 (2010)
7. S.W. Lee, D.G. Kamper, Modeling of multiarticular muscles: importance of inclusion of tendon-pulley interactions in the finger. IEEE Trans. Biomed. Eng. **56**(9), 2253–2262 (2009)
8. A. Deshpande, R. Balasubramanian, J. Ko, Y. Matsuoka, Acquiring variable moment arms for index finger using a robotic testbed. IEEE Trans. Biomed. Eng. **57**(8), 2034–2044 (2010)
9. J. Teran, S. Blemker, V.N.T. Hing, R. Fedkiw, Finite volume methods for the simulation of skeletal muscle, in *Proceedings of the ACM SIGGRAPH/Eurographics Symposium on Computer Animation* (2003), pp. 68–74
10. S.S. Blemker, S.L. Delp, Three-dimensional representation of complex muscle architectures and geometries. Ann. Biomed. Eng. **33**(5), 661–673 (2005)
11. K.R. Kaufman, D.A. Morrow, G.M. Odegard, T.L.H. Donahue, P.J. Cottler, S. Ward, R. Lieber, 3d model of skeletal muscle to predict intramuscular pressure, in *American Society of Biomechanics Annual Conference* (2010)
12. D.I.W. Levin, J. Litven, G.L. Jones, S. Sueda, D.K. Pai, Eulerian solid simulation with contact. ACM Trans. Graph. **30**(4), 36:1–36:9 (2011)
13. J. Spillmann, M. Teschner, An adaptive contact model for the robust simulation of knots. Comput. Graph. Forum **27**(2), 497–506 (2008)
14. M. Bergou, B. Audoly, E. Vouga, M. Wardetzky, E. Grinspun, Discrete viscous threads. ACM Trans. Graph. **29**(4), 116:1–116:10 (2010)
15. D.K. Pai, Strands: interactive simulation of thin solids using cosserat models. Comput. Graph. Forum **21**(3), 347–352 (2002)
16. A. Theetten, L. Grisoni, C. Andriot, B. Barsky, Geometrically exact dynamic splines. Comput. Aided Des. **40**(1), 35–48 (2008)
17. S. Sueda, A. Kaufman, D.K. Pai, Musculotendon simulation for hand animation. ACM Trans. Graph.**27**(3), 83:1–83:8 (2008)
18. M.B. Cline, D.K. Pai, Post-stabilization for rigid body simulation with contact and constraints, in *Proceedings of the IEEE International Conference on Robotics and Automation*, vol. 3 (2003), pp. 3744–3751
19. R.M. Murray, Z. Li, S.S. Sastry, *A Mathematical Introduction to Robotic Manipulation* CRC Press, Boca Raton, FL, USA (1994)
20. Y. Remion, J. Nourrit, D. Gillard, Dynamic animation of spline like objects, in *Proceedings of WSCG Conference* (1999), pp. 426–432
21. F. Zajac, Muscle and tendon: properties, models, scaling, and application to biomechanics and motor control. Crit. Rev. Biomed. Eng. **17**(4), 359–411 (1989)
22. D. Morgan, New insights into the behavior of muscle during active lengthening. Biophys J. **57**(2), 209–221 (1990)
23. M. Epstein, W. Herzog, *Theoretical Models of Skeletal Muscle* (Wiley, New York, 1998)
24. B. Garner, M. Pandy, The obstacle-set method for representing muscle paths in musculoskeletal models. Comput. Methods Biomech. Biomed. Eng. **3**(1), 1–30 (2000)
25. Q. Wei, S. Sueda, D.K. Pai, Physically-based modeling and simulation of extraocular muscles. Prog. Biophys. Mol. Biol. **103**(2), 273–283 (2010)
26. I. Stavness, A. Hannam, J. Lloyd, S. Fels, Predicting muscle patterns for hemimandibulectomy models. Comput. Methods Biomech. Biomed. Eng. **13**(4), 483–491 (2010)
27. R.D. Crowninshield, R.A. Brand, A physiologically based criterion of muscle force prediction in locomotion. J. Biomech. **14**(11), 793–801 (1981)

28. W. Herzog, Individual muscle force estimations using a non-linear optimal design. J. Neurosci. Methods **21**(2–4), 167–179 (1987)
29. K.L. Moore, A.F. Dalley, *Clinically Oriented Anatomy*, 4th edn., ed. by P.J. Kelly (Lippincott Williams & Wilkins, Philadelphia, 1999)
30. S. Sueda, G.L. Jones, D.I.W. Levin, D.K. Pai, Large-scale dynamic simulation of highly constrained strands. ACM Trans. Graph. **30**(4), 39:1–39:9 (2011)

Chapter 14
Why Humans can Manipulate Objects Despite a Time Delay in the Nervous System

Takahiro Inoue and Shinichi Hirai

Abstract A difference between the human and robots from the viewpoint of motor control is a sampling time in everyday movements. That is, the sampling time for controlling conventional robot systems is required to become approximately 1 ms. On the other hand, the motor control based on human nervous system permits considerably large time-delay due to the transmission latency on afferent/efferent pathways of the central nervous system. To date, it has been difficult for the robot to acquire dexterous tasks and precise movements as long as the robot system has an unexpected large delay in terms of sensory information. Based on the above observation, this chapter provides a new control strategy to accomplish precise orientation control of a target object grasped by a robotic hand consisting of two degrees-of-freedom (DOFs). The controller, named as serial two-phased (STP) controller, proposed in this study can realize secure grasping and manipulation in the case that a large visual feedback delay induced by the low specification of a camera is hiding in the control loop. Finally, through several simulations, we indicate that the closed loop dynamics designed by integrating the STP controller and a soft-fingertip structure is robust even in at most 100 ms-delay relatin to the updating of camera images.

Keywords Visual feedback · Delay · Control · Soft fingertip · Manipulation

T. Inoue
Okayama Prefectural University, Okayama, Japan
e-mail: inoue@ss.oka-pu.ac.jp

S. Hirai (✉)
Ritsumeikan University, Kusatsu, Japan
e-mail: hirai@se.ritsumei.ac.jp

R. Balasubramanian and V. J. Santos (eds.), *The Human Hand as an Inspiration for Robot Hand Development*, Springer Tracts in Advanced Robotics 95,
DOI: 10.1007/978-3-319-03017-3_14,

1 Introduction

We humans can grasp and manipulate objects with outstanding dexterity thanks to our highly developed brain, binocular vision, and the abundance of motor and sensory nerves in our hands and fingers. Human hands have high degrees-of-freedoms, from which sophisticated movements are generated by the muscles. The human retina in the eye includes photoreceptors, which provide visual feedback during grasping and manipulation. Human hands and fingers have sensory receptors in their skin, such as Meissner's corpuscles and Merkel disk receptors, which provide tactile and haptic feedback during grasping and manipulation. For their past, most robots exhibit a high performance and an enormously high-speed and precise motion compared with the everyday movements of human beings. Multi-fingered robot hands with vision and tactile sensors have become available recently. However, there still exists a large gap between human hands and robot hands. Robot hands cannot perform such dexterous manipulation as humans can, even although robots outperform humans in terms of control.

Current technology enables robots to accomplish 1 ms-periodic loop control as a result of the high performance of computers. In human sensation, sensory signals from visual and tactile receptors are sent to the cerebellum or the cerebrum, and muscle motion is commanded via the nervous system. The human nervous system has a relatively large latency of several tens of milliseconds, which results in slow signal processing; for example, a human cannot see motion over about 30 Hz. Despite such slow signal processing, humans exhibit high dexterity in object manipulation. Conversely, if the intrinsic neurophysiological latency that is expressed as a sum of central motor conduction time (CMCT) [1] and neuromuscular transmission delay is applied to a controller designed for a robot, it is clear that certain fatal disadvantages occur in the robot control system. Figure 1 shows a summary of the neurophysiological latency that stems from the central nervous system (CNS) [2]. Humans exhibit a high degree of dexterity in object manipulation despite a latency that is too large for current-generation robots to accept. We have to tackle this paradox to determine the source of dexterity in human grasping and manipulation so that robot hands can perform the same dexterous manipulation as humans do.

This chapter describes a novel and simple control law, and demonstrates that stable and dexterous soft-fingered manipulation can be achieved, even under a delay of up to 100 ms resulting from the updating of camera images that are utilized for visual feedback by the robot.

1.1 Related Work

Finite element (FE) analysis is often used when studying the deformation of objects, and it can be used to describe the x deformation of a hemispherical soft fingertip exactly [3–5]. However, although FE analysis can be used to simulate

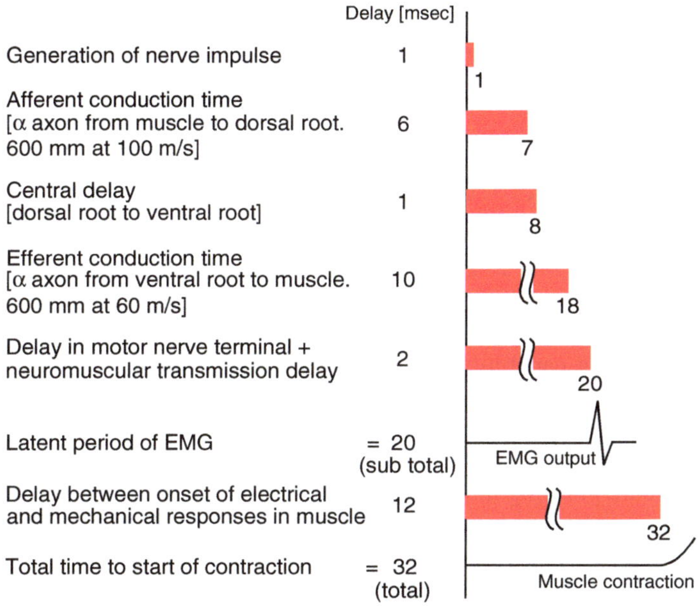

Fig. 1 Delay in latency from stimulation to onset of CMAP (compound muscle action potential) [21] is summarized with the exception of the optic [2]. This neurophysiological latency of human motor control is a result of the sum of the nerve propagation delay, the neuromuscular transmission delay, and the muscle fiber propagation delay [22]

grasping and manipulation numerically, it cannot be applied to a theoretical analysis of grasping and manipulation due to its complex formulation. In other words, FE analysis yields a *procedural* deformation model, which enables us to simulate the deformation of objects, but it cannot be applied to theoretical analysis. According to the principle of Occam's razor, we should choose a simple model to analyze and explain grasping and manipulation by soft fingertips. The Hertzian contact model provides a simple closed—form description of the contact between two quadratic surfaces of elastic objects [6], however, because the surfaces are assumed to be open ended, it cannot be applied to a hemispherical elastic fingertip with a rigid backplate. Arimoto et al. formulated the dynamics of pinching by a pair of soft fingertips [7] and used a *radially distributed deformation model* to analyze the mechanics of a soft fingertip [8]. Based on the concept of *stability on a manifold*, they showed theoretically that a 2-DOF (degrees-of-freedom) finger and a 1-DOF finger can together realize secure grasping and posture control [9, 10]. However, from our observations below, in addition to being able to grasp a rigid object, a pair of 1-DOF fingers with soft hemispherical fingertips can control the orientation of the object, which calls for a new model. Control laws for soft—fingered grasping and manipulation have been proposed in [11, 12] based on a radially distributed model. The proposed laws require an estimation of physical properties. This makes the proposed control laws sensitive to the delay in the control loop. Recall that humans can grasp and manipulate an object despite a relatively large delay of several tens of milliseconds in the nervous system, and it is clear that this calls for a control law that is robust against such a delay.

2 Mechanical Structure of the Human Hand

Human fingers have quite an interesting mechanical structure, as shown in Fig. 2a, in which a clearly parallel space for object pinching appears due to the passively deformable surface of the skin of the fingertips. Such a space configuration of the fingertip indicates a capability for extremely easy grasping. In other words, the large area of contact clearly contributes to stable and dexterous manipulation. In the case of the two-fingered manipulation shown in Fig. 2b, the geometrically opposing structure between the thumb and index finger plays a critical role in realizing a secure grasping motion, resulting in the construction of a highly simple and straightforward control strategy for dexterous object manipulation if a robotic hand were designed using soft materials for the fingers.

In what follows, we demonstrate a quite interesting physical characteristic of pinch motion by the the human fingers on the basis of a simple vibration test. Figure 3a shows a small wooden stick, on which a 3-axis acceleration sensor is mounted, grasped by the thumb and the index finger. In this test, an enforced rotational displacement around the pinching point is given to the stick, and its oscillating motion is shown after the external displacement is released. Figure 3b indicates a strong attenuation of the vibrating region. Note that no conscious manipulating forces by the two fingers were applied to the stick.

This vibration test implies that some sort of intrinsic mechanical property associated with human fingers contributes to stable grasping and ideal attenuation. This observation leads to the conclusion that a more precise fingertip model and

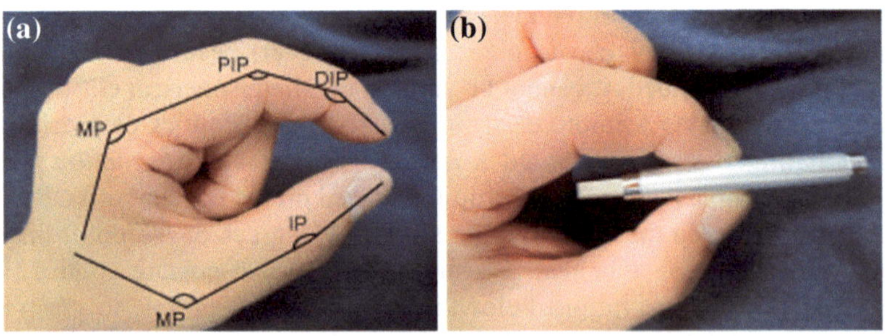

Fig. 2 This shows a classic geometrical configuration of two-fingered grasping by the index finger and the thumb, in which anatomical terms associated with the finger joints of the human hand are included. The index finger includes metacarpophalangeal (MP), proximal interphalangeal (PIP), and distal interphalangeal (DIP) joints. The thumb includes MP and interphalangeal (IP) joints. **a** Fingertip configuration. **b** Object grasping

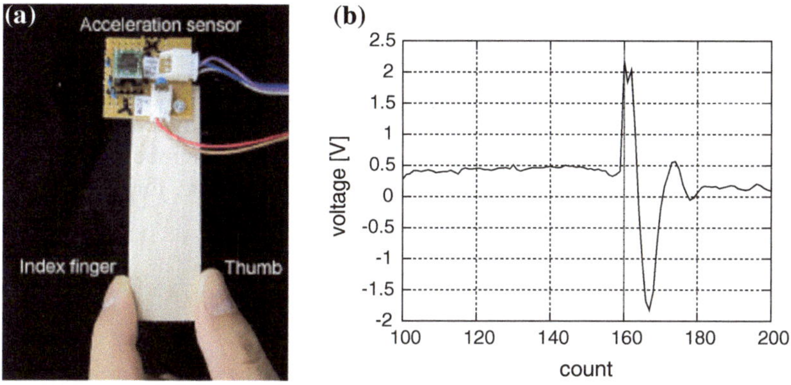

Fig. 3 In this simple test, a 3-axis acceleration sensor (KXM52) is mounted on the end of a wooden stick. The oscillating trajectory of acceleration was obtained when the stick was forcibly swung by the initial rotation and its rapid release. It can be seen that the initial value of the sensor and the steady-state one obtained are different from each other around the vibration. This discrepancy results from the viscoelastic property of the structure of the fingertip. Note that no additional or conscious manipulating forces by the two fingers were applied to the stick. **a** A Wooden stick. **b** Output of voltage

formulation are needed to enable a robotic hand to skillfully grasp an object. In this case, a method of straightforward expression, including not only mathematical perspectives but also the static and dynamic characteristics of the fingertip, is rather important to describe the natural vibration. Furthermore, we note that the simplification of the control law for dexterous manipulation can be realized by means of the physical properties of the human finger.

3 Simple Model of the Human Fingertip

Recently, there exist several related researches focusing on mechanical modeling of the human finger pad and verification of dynamic force response which described with an instantaneous force response and a reduced relaxation function [13–15]. Although these studies validated that the experimental data obtained by sinusoidal displacement input fit well with the elastic force model, the load configuration to the finger was restricted only to the normal contact on the finger pad. As a result, obvious dependence with the variation of contact direction had not been found. In what follows, the modeling process of soft finger pad, which is simply defined as a hemispherical solid shape, is presented. In this formulation, we first indicate the dependency on contact direction between the soft fingertip and a flat surface.

3.1 Normal Deformation Model

Let us recall the formulation process of a soft-contact model of a soft fingertip, which was developed completely in [16, 17]. We treat the fingertips as if they had a hemispherical shape and were composed of an infinite number of virtual linear springs that stand vertically, as shown in Fig. 4. Elastic force F and elastic potential energy P are described as

$$F = \frac{\pi E d^2}{\cos\theta_p}, \tag{1}$$

$$P = \frac{\pi E d^3}{3\cos^2\theta_p}, \tag{2}$$

where d is the maximum displacement of the finger, E is the Young's modulus of its material, and θ_p is the relative angle between the finger base and the contacting object. In order to verify the soft fingertip model, we validated the locus of both equations. Figure 5 shows good results, in which it can seen that there is a minimal point of each physical quantity, F and P, with respect to the contacting angle between the object and the soft fingertip. From these results, when considering the case of two-fingered grasping, we can see that a couple force induced by the elastic moment of both fingers arises and contributes to stable grasping. In addition, with the human fingers, as we grasp an object more forcefully, the stability of the grasping increases more and more.

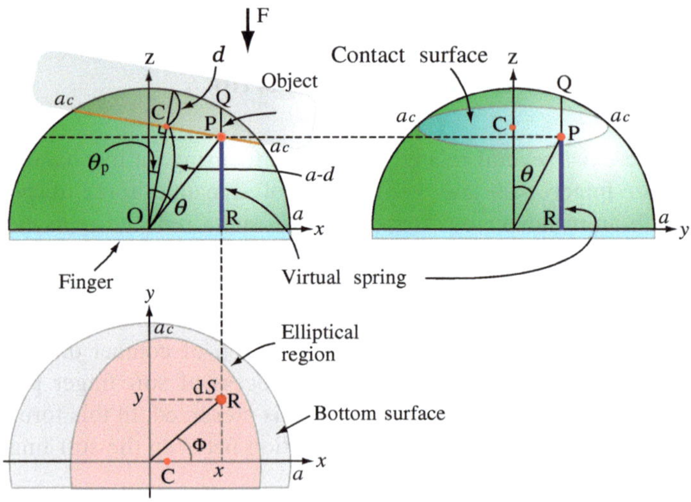

Fig. 4 Model of normal deformation of a soft fingertip

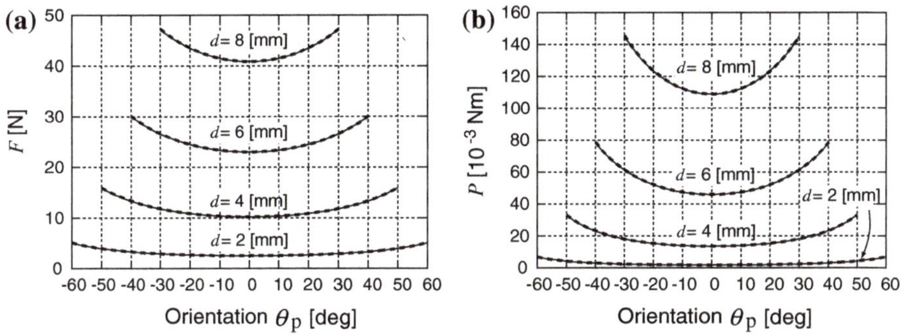

Fig. 5 In both figures, a unique minimal value exists when $\theta_p = 0$. That is, the elastic force and its potential energy depend on the contacting angle of a grasped object. **a** Elastic force. **b** Elastic energy

3.2 Model with Normal and Tangential Deformation

While the normal deformation model can express a minimal point of the elastic potential energy that was revealed in the practical compression test, the model lacks the tangential deformation that would appear in real manipulation. Let us construct a more precise fingertip model that includes tangential deformation along the tangential direction shown in Fig. 6. This figure shows an extension of the normal deformation model of a fingertip, in which a virtual spring having spring constant k is placed vertically inside the fingertip. The spring can be compressed and deformed in the lateral direction, which corresponds to its bending motion. We assume that each constant related to compression and bending is equal in this model. This extended fingertip model can be finally expressed as

$$P\left(d_n, d_t, \theta_p\right) = \pi E\left\{\frac{d_n^3}{3\cos^2\theta_p} + d_n^2 d_t \tan\theta_p + d_n d_t^2\right\}. \tag{3}$$

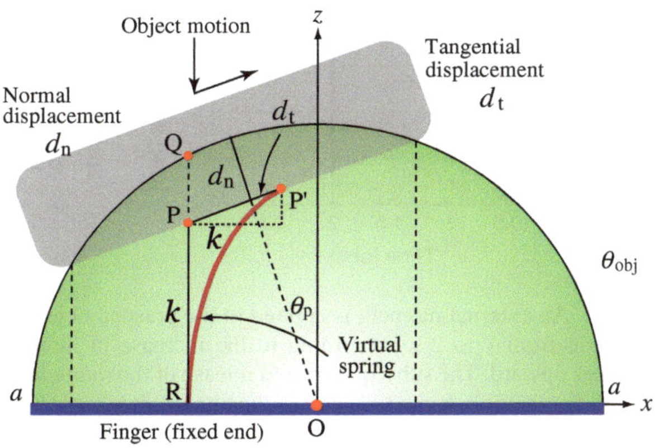

Fig. 6 Model of normal and tangential deformation of a soft fingertip

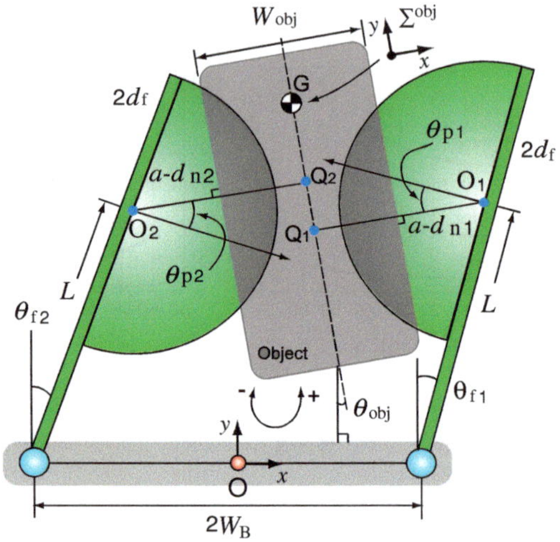

Fig. 7 Soft-fingered manipulation by two fingers

See [16] for detailed deviation. Note that this formula is an equation having three independent variables, d_n, d_t, and θ_p, while the previous one-dimensional model had only two variables, d_n and θ_p.

Now, let us verify whether the oscillating motion of a grasped object can be seen when this fingertip model is used in the same vibration test. Let us consider an extremely simple robotic hand shown in Fig. 7, which has two degrees-of-freedom, and then, one whose structure has the minimal degrees-of-freedom to complete successful grasping by two fingers. As in the previous test by the human hand (Fig. 3b), the grasped object oscillates according to the release of the initial external moment as shown in Fig. 8a. Note that factitious external moment has been applied

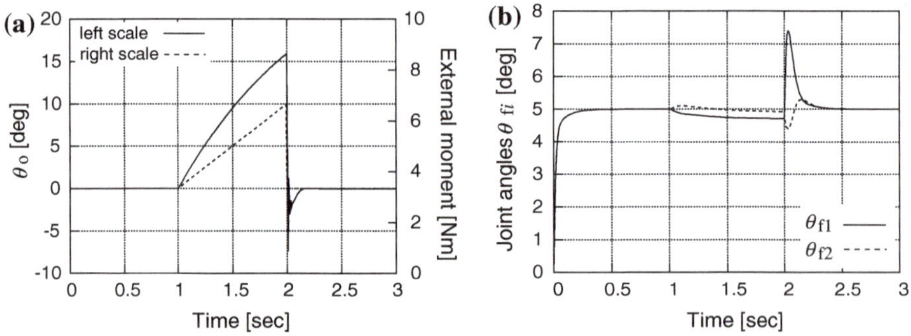

Fig. 8 An external moment is applied to the grasped object as per the dotted line throughout the duration from 1 to 2 s. According to the increase in moment, the object orientation gradually inclines upward. The subsequent rapid release of the moment provokes a relatively high-frequency oscillating motion. It can be seen that the finger joints settle back to the desired joint angle (5°) after the release because of the application of the joint PI controller. **a** Object vibration. **b** Joint PI control

to the object at 1 s, and it increases up to 2 s. Finally, the moment was suddenly released at 2 s. Both the force and elastic energy equations (Eqs. (1) and (2)) are both breakthrough physical models in the sense that the fingertip model can duplicate similar vibration. We applied a simple PI controller for articular joint control in this simulation, in which the desired values of the joint were both 5°, as shown in Fig. 8b.

4 Control Law for Soft-Fingered Manipulation

4.1 Lagrange's Equations of Motion

Recalling the simple 2-DOF model of the robotic hand shown in Fig. 7 from our previous studies [16, 17], the Lagrangian of the hand can be written as

$$\mathcal{L} = K - P + \sum_{i=1}^{2} \lambda_{\mathrm{n}i} C_{\mathrm{n}i}, \tag{4}$$

where K and P respectively mean the kinetic energy of the total system and the potential energy including not only gravitational potential but also elastic energy induced by the deformation of the soft fingertip. Therefore, K and P can be finally described as

$$P = \sum_{i=1}^{2} P_i(d_{\mathrm{n}i}, d_{\mathrm{t}i}, \theta_{\mathrm{p}i}) + M_{\mathrm{o}}\, g\, y_{\mathrm{o}} + \sum_{i=1}^{2} M_{\mathrm{f}i}\, g\, L \cos\theta_{\mathrm{f}i},$$

$$K = \frac{1}{2} M_{\mathrm{o}}\left(\dot{x}_{\mathrm{o}}^2 + \dot{y}_{\mathrm{o}}^2\right) + \frac{1}{2} I_{\mathrm{o}} \dot{\theta}_{\mathrm{o}}^2 + \frac{1}{2}\sum_{i=1}^{2} I_{\mathrm{f}i}\dot{\theta}_{\mathrm{f}i}^2 + \frac{1}{2} m \sum_{i=1}^{2} \dot{d}_{\mathrm{n}i}^2 + \frac{1}{2} m \sum_{i=1}^{2} \dot{d}_{\mathrm{t}i}^2,$$

where the parameters of the system are defined in Table 1. In addition, the last term of the right-hand side of Eq. (4) corresponds to the virtual energy due to the constraint forces, $\lambda_{\mathrm{n}i}$, which appear upon the contact between the object and the soft fingertips. Differentiating geometric constraints, $C_{\mathrm{n}i}$, with respect to system variables, the direction of $\lambda_{\mathrm{n}i}$, dynamically changing during the manipulation motion, can be clarified. As a result, the equations of motion of the whole system are described as

$$\frac{d}{dt}\frac{\partial \mathcal{L}}{\partial \dot{\boldsymbol{q}}} - \frac{\partial \mathcal{L}}{\partial \dot{\boldsymbol{q}}} = \sum_{i=1}^{2} \lambda_{\mathrm{t}i} \frac{\partial \dot{C}_{\mathrm{t}i}}{\partial \dot{\boldsymbol{q}}} + \boldsymbol{f}_{\mathrm{ext}} + \boldsymbol{u}, \tag{5}$$

where an external force vector and a control input vector are newly added. Note that the input, $\boldsymbol{u}$, corresponds to the input torque applied to the joint angle of the finger. The first term on the right-hand side of Eq. (5) denotes generalized forces in terms of the *Pfaffian* constraint, $\dot{C}_{\mathrm{t}i}$. In addition, $\lambda_{\mathrm{t}i}$ is the constraint forces tangential to the grasped object.

Table 1 Definitions of parameters

Parameter	Definition
$2W_B$	Base width of hand
a	Fingertip radius
L	Length of each finger
d_f	Thickness of finger
W_o	Width of object
M_o	Mass of object
M_{fi}	Mass of ith finger
m	Mass of fingertip
θ_{fi}	Joint angle of ith finger
θ_o	Orientation angle of object
θ_{pi}	$(\theta_{fi} + (-1)^i \theta_o)$
d_{ni}	Maximum displacement of ith fingertip
d_{ti}	Tangential displacement of ith fingertip
x_o, y_o	Position of object
I_o	Inertia of object
I_{fi}	Inertia of finger

4.2 Simple Control Law for Stable Grasping

The results of Figs. 3, 5 and 8 indicate that the flexibility of soft fingertips contributes to secure grasping, as the elastic energy is at its minimum during manipulation. As a result, it is easy to understand that the controller must be designed so as not to destroy the intrinsic physical property of the soft fingertips, which acts to realize stable grasping.

Applying a simple PI controller, which is expressed by Eq. (6), to the robot dynamics described in Eq. (5), the time trajectories of the object and the joints are obtained as shown in Fig. 9.

$$u_i = -K_P\left(\theta_{fi} - \theta_{fi}^{d}\right) - K_I \int_0^t \left(\theta_{fi} - \theta_{fi}^{d}\right) d\tau. \tag{6}$$

In the above equation, K_P and K_I denote proportional and integral gains. As a matter of fact, a controller using Eq. (6) was used in the vibration test shown in Fig. 8. In this case, after having applied the external force continuously to the object, we kept the force at rest for 1 s. Figure 8b shows that both joints converge to the desired angle, indicating that the error in the finger joints does not remain not only during the release of force, but also during the duration of the constant external force. Thus, restricted to stable grasping, neither the object information nor the real-time measurements of grasping forces are necessary, in designing a controller for robotic hands. Without the mechanical flexibility of the fingertip, a simplification of the controller design could not be achieved.

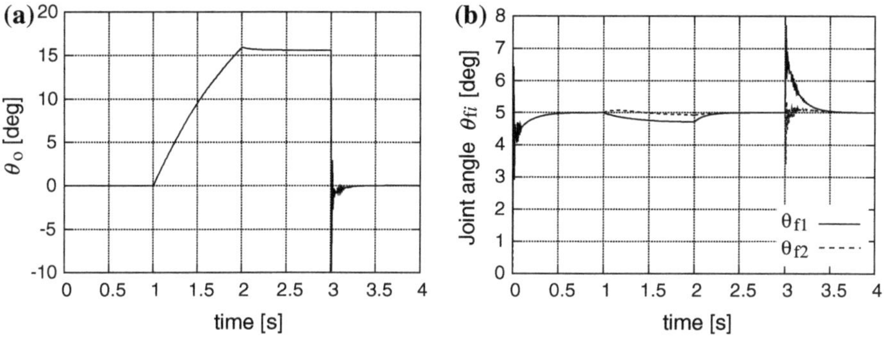

Fig. 9 In addition to the vibration test of Fig. 8, this shows the dynamic behavior of object orientation and the joint angles. Unlike the previous test, we kept the external force at rest for 1 s before releasing. Likewise, the desired value of the joint was set to 5°. **a** Object vibration. **b** Joint PI control

4.3 Controller for Dexterous Manipulation

Let us introduce a very simple controller for achieving precise object manipulation, which was named *serial two-phased (STP)* controller in [18]

$$\theta_{fi}^{d} = -(-1)^{1} K_{I} \int_{0}^{t} \left(\theta_{o} - \theta_{o}^{d}\right) d\tau, \tag{7}$$

$$u_{i} = -K_{P}\left(\theta_{fi} - \theta_{fi}^{d}\right) - K_{D}\dot{\theta}_{fi} + \tau_{b}, \tag{8}$$

where K_P, K_D, and K_I denote proportional, differential, and integral gains, respectively. This controller can perform precise orientation control of a grasped object even when the robotic hand has the minimal degrees-of-freedom shown in Fig. 7. The block diagram of the STP controller can be simplified as Fig. 10. In addition, the biased torque, τ_b, has a positive constant value and acts to prevent the motor torque u_i produced in PD control remaining negative. An inner local loop with respect to joint angle, θ_{fi}, located within the block diagram, the desired value is merely a pseudo value of the joint. That is, it is not necessary that the joint variable converges to its desired value even when the object orientation has attained a given posture, θ_o^d.

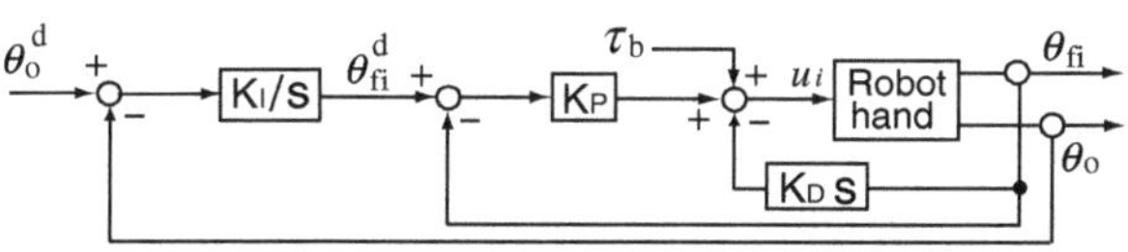

Fig. 10 Block diagram of a *serial two-phased (STP)* controller capable of achieving robust convergence of object orientation, when grasped by two soft fingertips. The characteristic of this controller is that desired trajectory of the joint angle in the second stage (Eq. (8)) is serially coupled and remains constant when the object orientation goes to convergence

5 Simulation

5.1 Simulation of a Pair of 1-DOF Fingers

Figure 11 shows the simulation results when the desired angle of the object is set to be a step input such that

$$\theta_o^d = \begin{cases} 3^\circ & (0\,\mathrm{s} \le t < 1\,\mathrm{s}), \\ 8^\circ & (1\,\mathrm{s} \le t < 2\,\mathrm{s}), \\ -5^\circ & (2\,\mathrm{s} \le t < 3\,\mathrm{s}), \end{cases} \tag{9}$$

In addition, we set the constant force at $\tau_b = 30$ Nm and assume that this manipulating motion is implemented on a vertical plane with gravitational force (Fig. 11d). Figure 11a–c depict the trajectory of the grasped object (x_o, y_o, θ_o) with respect to time. We can see that the STP orientation controller works well so that θ_o robustly converges to the desired step trajectory (Fig. 11c). In particular, it is important to produce the desired finger angles that satisfy the equivalent positive and negative values of the two fingers. This simple control structure of the first stage takes advantage of the natural rolling of the target object along both spherical surfaces of the fingertips. At the same time, the position of the object goes to a stable equilibrium point together with the orientation convergence, as shown in Fig. 11a, b. In other words, the equilibrium point of (x_o, y_o, θ_o) corresponds to a local minimum of elastic potential energy with constraints (LMEEwC [16]) during the manipulation. Figure 11e, f show the trajectories of both joint angles along with the desired angle, θ_{fi}^d, which is dynamically produced at the first stage of the controller expressed by Eq. (7). Here we find that there are very large errors of θ_{f1} and θ_{f2}. Despite this, the object orientation shows an exact convergence. In fact, these discrepancies of both fingers play a significant role in achieving orientation control of the grasped object.

5.2 Simulation of a Pair of 1-DOF Fingers Under Time Delay

Now, let us investigate a case in which a delay in updating information obtained by image processing exists in a visual feedback control system. Usually, conventional vision systems have an inevitable time-delay of 33 ms as a result of the video frame rate.

For instance, as shown in Fig. 12, the time to access the corresponding area of the memory, in which a processed image for the visual feedback is saved, becomes extremely large. As illustrated Fig. 12, let T_c be the robot control period, T_p be the memory accessing time for image processing, T_m be the sampling time for motor control, and T_u be the update timing of computed visual information from a captured image. The continuous updating manner in the upper figure is executed in the case that artificial update delay does not exist, i.e., $T_u = T_c$. On the other hand,

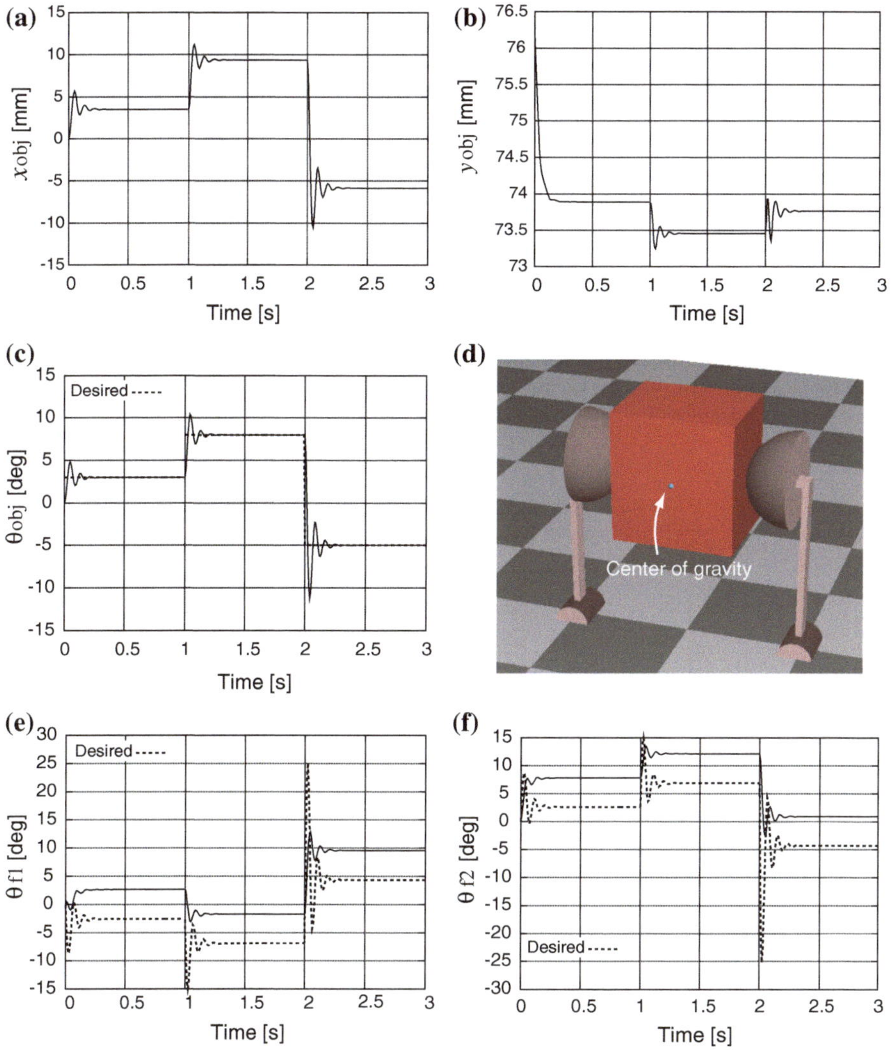

Fig. 11 Simulation results show that posture control of an object grasped by a minimal-DOF robotic hand can be achieved, although each of the joint angles contains a steady-state error. **a** x_0 versus time. **b** y_0 versus time. **c** θ_0 versus time. **d** Snapshot of the manipulation. **e** θ_{f1} versus time. **f** θ_{f1} versus time

if the image-update delay is incorporated into the robot, the periodic time for the motor control becomes extremely short. That is, the robot control period coincides with the period of motor control, i.e., $T_c = T_m$. As a result, $T_u \neq T_c$ is satisfied. In this simulation, we obtain successful results for object orientation control when the time delay, T_u, is increased up to 99 ms.

Figure 13 shows that the periodic time of image updating is assumed to be 33 ms. See [19] in detail. Note that an improved trajectory was obtained by only

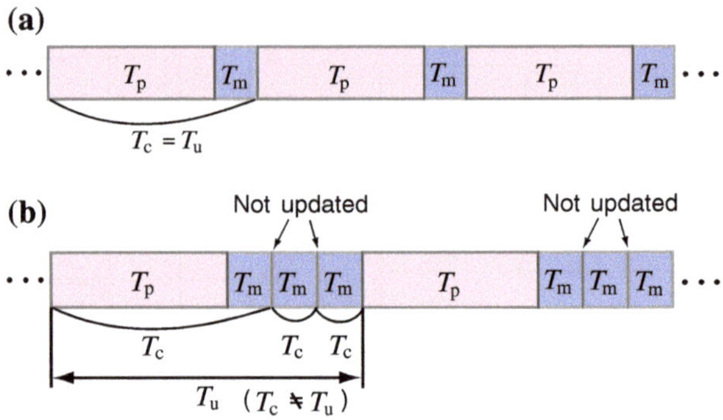

Fig. 12 A conceptual diagram of the time-delayed robotic system in visual feedback control. **a** Continuous updating of a camera image. **b** Irregular updating of a camera image

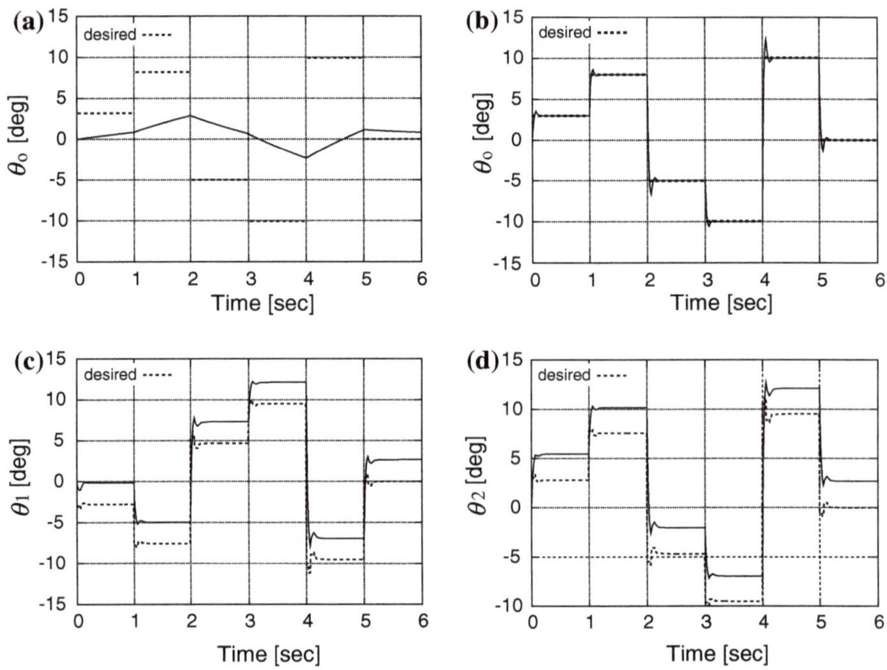

Fig. 13 This result shows an improved trajectory for object orientation, in which the periodic time of image updating is assumed to be 33 ms. In this result, we set the integral gain to $K_I = 0.01$ in the case of failure (a) and to $K_I = 1$ in the case of success (b). Note that the joint angle does not converge to each desired trajectory produced in Eq. (7). That is, the desired angle, θ_{fi}^{d}, corresponds to the virtual desired trajectory. **a** Before improvement. **b** After improvement. **c** θ_{f1} versus θ_{f1}^{d}. **d** θ_{f2} versus θ_{f2}^{d}

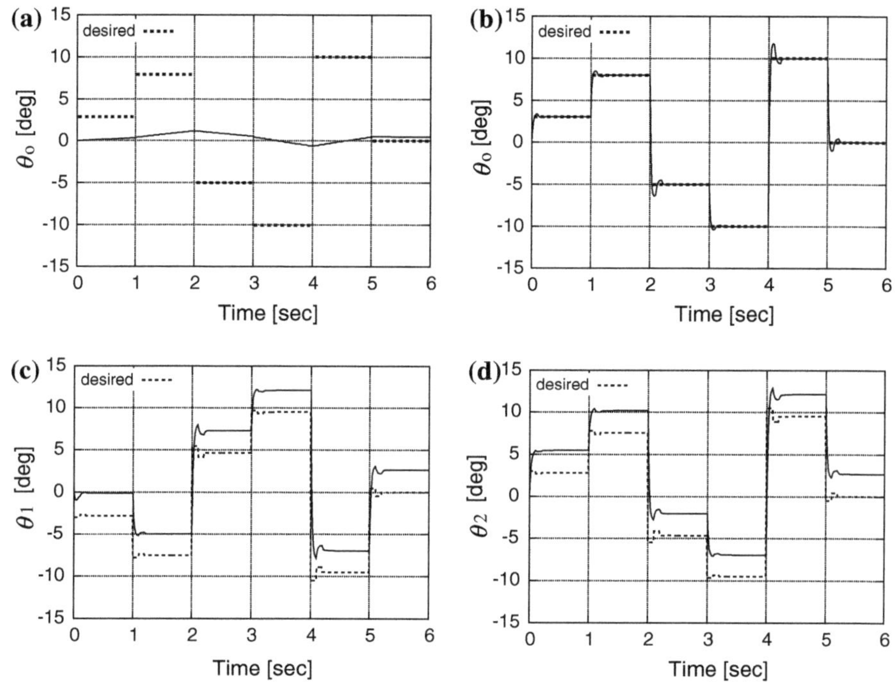

Fig. 14 This result shows an improved trajectory of object orientation, in which the periodic time of image updating is assumed to be 99 ms. In this result, we set the integral gain to $K_{\mathrm{I}} = 0.01$ in the case of failure (a) and to $K_{\mathrm{I}} = 1$ in the case of success (b). **a** Before improvement. **b** After improvement. **c** θ_{f1} versus θ_{f1}^{d}. **d** θ_{f2} versus θ_{f2}^{d}

changing the integral gain from 0.01 to 1. In addition, we show another result obtained when the updating delay is 99 ms (Fig. 14). Also in this case, the dynamic response of object orientation was drastically improved solely by modifying the integral gain from $K_{\mathrm{I}} = 0.01$ to $K_{\mathrm{I}} = 1$. This successful result stems from the fact that the passive deformation of the soft fingertips contributes to stable rotation of the object grasped by them. As stated previously, it is found that admissible deviations between the joint angle and the desired appear clearly at every step.

5.3 Simulation of a Robotic Hand with Multi-DOF Fingers

Next, let us describe the Lagrangian of the 5-DOF robotic hand system shown in Fig. 15. It can then be expressed using the mass of the link m_{ij}, inertia of the link I_{ij}, acceleration of gravity g, and Young's modulus E of the soft fingertip, as follows:

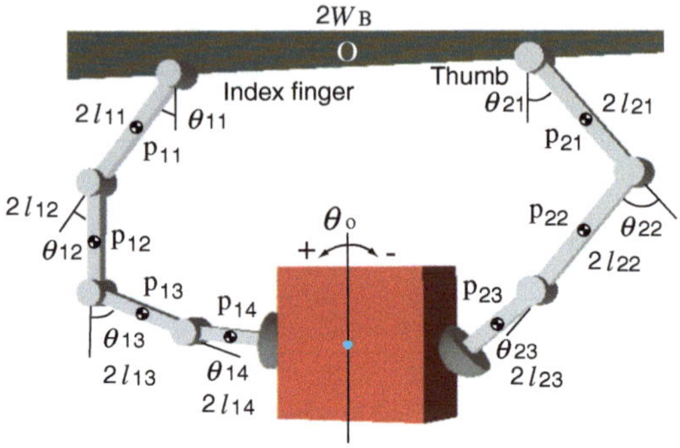

Fig. 15 Structure of a 5-DOF robotic hand

$$
\begin{aligned}
\mathcal{L} = &\frac{1}{2}\sum_{j}^{3} m_{1j}\dot{p}_{1j}^{2} + \frac{1}{2}\sum_{j}^{2} m_{2j}\dot{p}_{2j}^{2} + \frac{1}{2}I_{11}\dot{\theta}_{11}^{2} + \frac{1}{2}I_{12}\left(\dot{\theta}_{11} + \dot{\theta}_{12}\right)^{2} \\
&+ \frac{1}{2}I_{13}\left(\dot{\theta}_{11} + \dot{\theta}_{12} + \dot{\theta}_{13}\right)^{2} + \frac{1}{2}I_{21}\dot{\theta}_{21}^{2} + \frac{1}{2}I_{22}\left(\dot{\theta}_{21} + \dot{\theta}_{22}\right)^{2} \\
&- g\sum_{j}^{3} m_{1j}p_{1jy} - g\sum_{j}^{2} m_{2j}p_{2jy} \\
&- \pi E\sum_{i}^{2}\left\{\frac{d_{\mathrm{n}i}^{3}}{3\cos^{2}\theta_{\mathrm{p}i}} + d_{\mathrm{n}i}^{2}d_{\mathrm{t}i}\tan\theta_{\mathrm{p}i} + d_{\mathrm{n}i}d_{\mathrm{t}i}^{2}\right\} + \sum_{i}^{2} C_{\mathrm{n}i}\lambda_{\mathrm{n}i},
\end{aligned}
\tag{10}
$$

where the unexplained parameters are detailed in [20]. In the above equation, $\theta_{\mathrm{p}i} = \sum_{k=1}^{\tau}\theta_{ik} + (-1)^{i}\theta_{\mathrm{o}}$ with $i = 1$ and $r = 4$ for the index finger while $i = 2$ and $r = 3$ for the thumb. Note that the final joints in each finger, θ_{14} and θ_{23}, were introduced only for grasping an object as perpendicularly as possible, as shown in Fig. 15a. Therefore, these joints are assumed to be fixed at constant angle, that is, $\theta_{14} = 15°$ and $\theta_{23} = 10°$ are fulfilled during this simulation study.

Let us consider a case in which a 5-DOF robotic hand performs orientation control of a grasped object. We attempt to carry out the same object orientation control using only one finger of the robot. That is, while the STP controller is applied only to the index finger, a traditional and straightforward PD controller for joint angles is implemented for the thumb. This novel controller can be represented as a two-phase structure:

$$\theta_{11}^{\mathrm{d}} = K_{\mathrm{I}}\int\left(\theta_{\mathrm{o}}^{\mathrm{d}} - \theta_{\mathrm{o}}\right)\mathrm{d}t, \tag{11}$$

$$u_{11} = -K_{\mathrm{P1}}\left(\theta_{11} - \theta_{11}^{\mathrm{d}}\right) - K_{\mathrm{D}}\dot{\theta}_{11}, \tag{12}$$

$$u_{21} = -K_{\mathrm{P2}}\left(\theta_{21} - \theta_{21}^{\mathrm{d}}\right) - K_{\mathrm{D}}\dot{\theta}_{21}. \tag{13}$$

Not having a pseudo desired value for the joint angle, which is generated at the first stage, the second stage controller for the thumb (Eq. (13)) is not engaged in

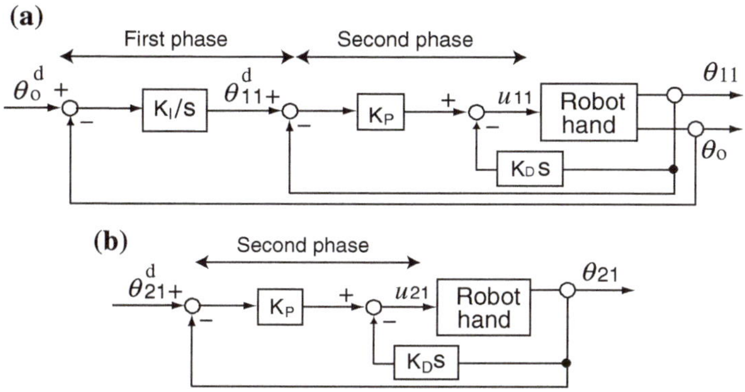

Fig. 16 In these two block diagrams, the *upper* figure corresponds to the STP controller, and the control law in the *lower* figure is a simple PD controller applied to the thumb. The first stage of the STP controller is eliminated in the *lower* figure. **a** For index finger. **b** For thumb

the orientation control of the object. In this case, θ_{21}^d can arbitrarily be set to be a neighborhood value from the initial joint θ_{21} of the thumb. The structure of the controller design can be clearly seen from the block diagrams shown in Fig. 16. It is obvious that the controller for the thumb does not act directly for object orientation control.

In addition, we assume that all joints except the actuated joints, θ_{11} and θ_{21}, satisfy a set of constraint conditions related to angular velocity. These constraints are expressed as

$$\dot{\theta}_{11} = \dot{\theta}_{12}, \quad \dot{\theta}_{12} = \dot{\theta}_{13}, \quad \dot{\theta}_{21} = \dot{\theta}_{22}. \tag{14}$$

This assumption results from the experimental knowledge that the relative angular velocities among the distal, middle, and proximal phalanxes are nearly identical to each other, as shown in Fig. 17.

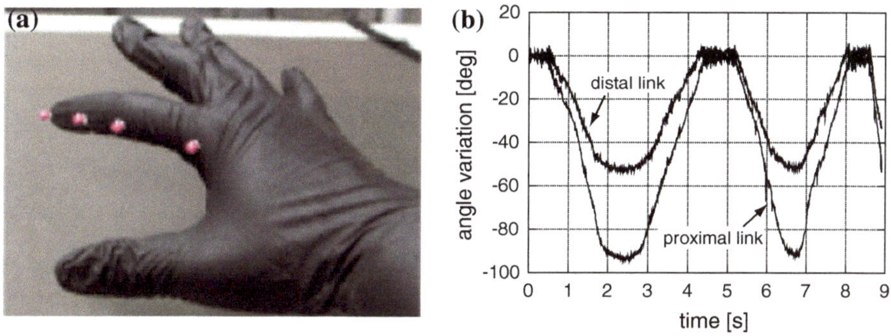

Fig. 17 By capturing the movement of the index finger on a 300-fps high-speed camera, it is shown that the relative relationship that exists among the rotation angles associated with the DIP, PIP, and MP joints have a certain motional regularity. **a** Human hand. **b** Angular relation

Figure 18 shows a simulation result in which the dotted line of the upper figure is the desired orientation of the object. Note that the time delay due to the image processing is not implemented in this case. It is clearly indicated that the object trajectory converges to the desired one with no error in each time step. Figure 18b, c show that extremely large errors between $(\theta_{11}, \theta_{21})$ and $(\theta_{11}^{d}, \theta_{21}^{d})$ remain throughout the object orientation control. This result implies that a discrepancy between the two joints is not important in object orientation control. As a result, the discrepancy can be defined as an admissible error. Hence, we conclude that the orientation of the grasped object can be precisely controlled by just one finger in the sense that θ_{o}^{d} remains constant in the controller for the thumb (Fig. 19).

One reason we consider such a finger movement is that the index finger and the thumb of the human hand manipulate an object smoothly by making extension and flexion movements alternately at the fingers. Furthermore, each of the fingers can

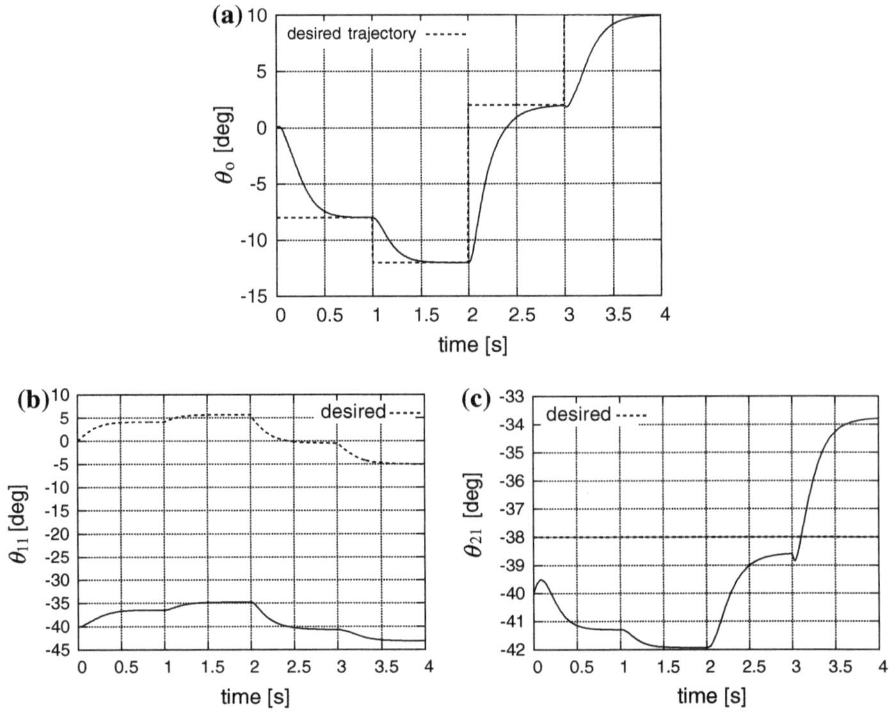

Fig. 18 This shows successful trajectory tracking of object orientation, when an objcet is grasped by two fingers. In this case, no time delay is incorporated. No error remains in the time steps of the trajectory. The STP controller proposed in this section may generate extremely large errors during manipulation; however, these discrepancies are not fatal errors in achieving orientation control. **a** Trajectory of the object orientation. **b** Error of θ_{11}. **c** Error of θ_{21}

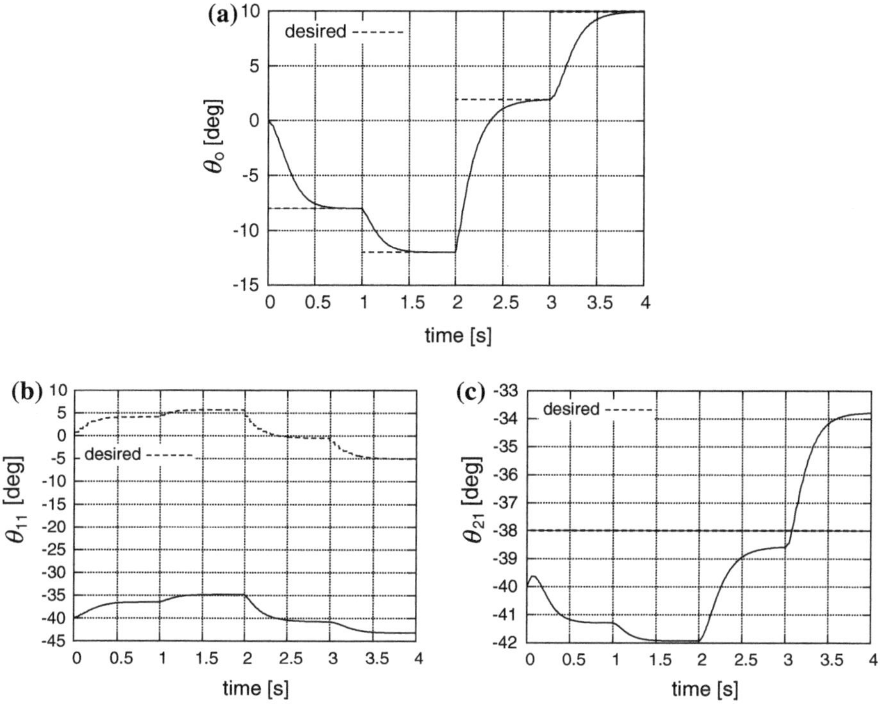

Fig. 19 This shows successful trajectory tracking of object orientation, where the time delay due to the image update becomes 50 ms. No error remains in the time steps of the trajectory as well as the result of Fig. 18. **a** Trajectory of the object orientation. **b** Error of θ_{11}. **c** Error of θ_{21}

roll the object dexterously, while the other finger acts solely to maintain stable grasping. This natural movement has not been achieved to date by conventional robot mechanisms and their control methods.

Finally, we show a simulation result in which 50 ms delay, T_u, exists within the control loop. It is obviously found that the trajectory of the object orientation converges robustly to the desired. In this case, this good result was obtained by only increasing the integral gain, K_I, up to 500 times from 0.0002 which was used in the previous simulation indicated in figure: simulation-results. In fact, the time interval T of Runge-Kutta method used in this analysis is 0.1 ms, which corresponds to the robot control period in practical experiments. Therefore, we know that the successful trajectory can be achieved when performing 500-times increase of the integral gain according to 500-times delay arising between the update delay T_u and the control period T, where $T = T_c$ and $T_u \neq T_c$.

6 Experimental Results

6.1 A Pair of 1-DOF Fingers Under Time Delay

As well as the simulations, we give the same task to a soft-fingered robotic hand that is configured so that two finger have an opposed structure, as shown in Fig. 11d. In this experiment, we utilize a CCD camera capable of capturing a sequence of grayscale images at 200 fps. Continuously, the grabbed images are processed to compute object position and orientation at intervals of 5 ms. Therefore, we intentionally slow the updating of the object information used for feedback control, that is, the update timing becomes once per 20 times to simulate a 100 ms updating delay.

Figure 20 shows an experimental result depicting the desired trajectory of the object orientation, the delayed response, and the improved response are depicted. It is obvious that the orientation trajectory of the grasped object has been dramatically improved just by changing the integral gain. However, the improved trajectory tends to become a step-like response because of the large time delay.

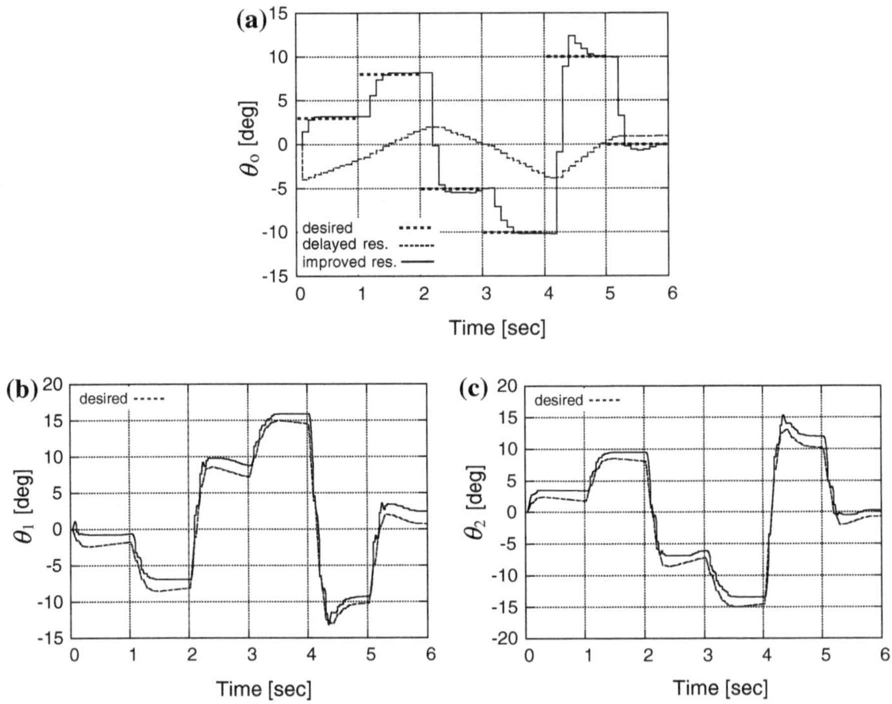

Fig. 20 This shows the experimental result of the orientation trajectory of a grasped object when the updating delay is equivalent to 100 ms. The gain parameters were set as $K_P = 60$, $K_D = 0.001$. As well as the simulation result, the orientation trajectory has been dramatically improved just by changing K_I from 0.0008 to 0.008. **a** θ_0. **b** θ_{f1} and θ_{f1}^d **c** θ_{f2} and θ_{f2}^d

In addition, it has been clearly clarified that a discrepancy in joint angles remains throughout the manipulation. In other words, this consistent error does not have to be eliminated provided the object orientation converges to the desired trajectory, that is, $\theta^{\mathrm{d}}_{\mathrm{f}i}$ corresponds to the apparent desired trajectory in the STP controller.

Next, we show another simulation and experimental results in Fig. 21, where the sinusoidal desired input for the object orientation, $\theta^{\mathrm{d}}_{\mathrm{o}}$, is given in Eq. (7), and

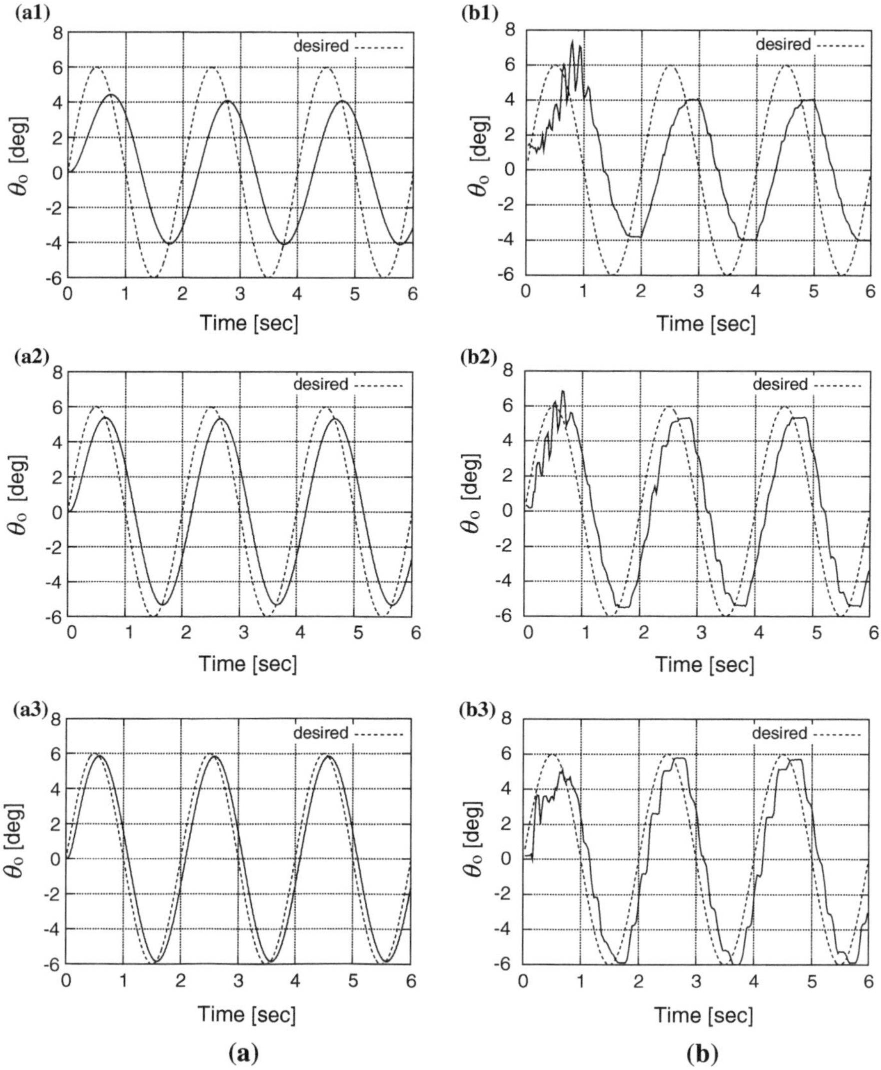

Fig. 21 Sinusoidal wave input for the desired object orientation, $\theta^{\mathrm{d}}_{\mathrm{o}}$, is fed into the robotic system. In simulations (a–1)–(a–3), the performance of tracking control is gradually improved as the integral gain increases. In experiments (b–1)–(b–3), the same trend occurs, yet the response in the initial state clearly indicates an oscillatory appearance. **a** Simulation.**b** Experiment

the updating delay is equivalent to 25 ms consistently. It is clearly indicated that the performance of tracking control is gradually improved as the integral gain increases in both the simulation and the experiment. In the experimental result, the step-like response starts to increase gradually. This result stems from the fact that nonlinearity due to Coulomb friction of the finger joint appears obviously when the joint torque increases according to the change in integral gain. In addition, the response in the initial state of the experimental results clearly indicates oscillatory appearance. The reason for this is that the response becomes particularly susceptible to static friction when the angular velocity of the joint is reduced.

6.2 Multi-Fingered Hand

We have designed a 5-DOF wire-driven robotic hand as shown in Fig. 22a. This robot has 8 DC motors, and two motors drive each angular joint, as illustrated in Fig. 23. The index finger has 6 pulleys for idle revolution around the MP and PIP joints, and 2 pulleys fixed at the PIP and DIP joints, as shown in Fig. 22b. A pair of pulleys is positioned to implement a figure-eight structure capable of achieving the coupled movements of the DIP and PIP joints of the human finger. The thumb

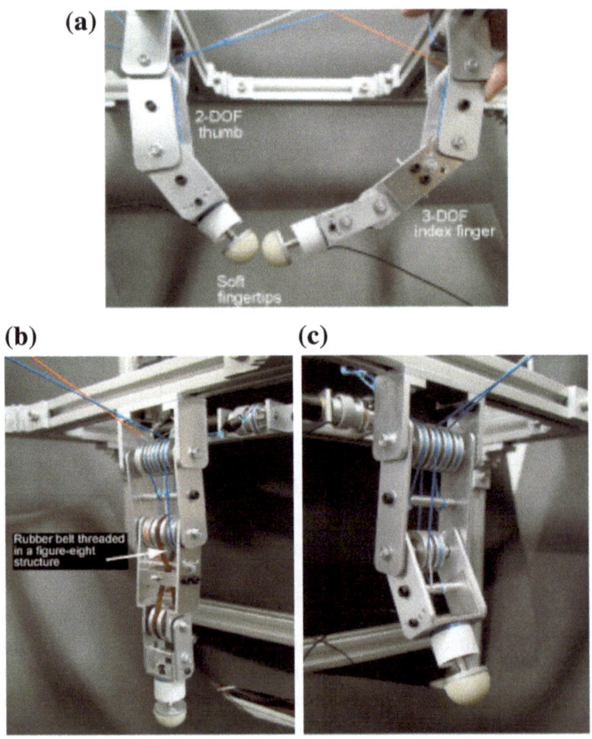

Fig. 22 This index-finger robot has a figure-eight structure achieved by running a cylindrical rubber belt between the DIP joint and the PIP joint. This system has 8 DC motors, 12 idle-revolution pulleys, 2 fixed pulleys, and 2 soft fingertips at the ends of both fingers. **a** Overall view of 5-DOF robotic hand. **b** Index finger robot. **c** Thumb robot

robot has 6 idle-revolution pulleys for wires, but the figure-eight structure is not used in this design. At the ends of both of the robot fingers, hemispherical soft fingertips are mounted, as shown in Fig. 22a. As shown in Fig. 23, the odd-number motors (1, 3, 5, 7) drive the finger so that its joint moves in the direction in which the grasping force increases, while the even-number motors (2, 4, 6, 8) move the finger joint in the opposite direction. We have used high-torque motors for the inward motion rather than the motors used for the outward motion of the finger. In this experiment, we use a more simpler integral controller than that used in the simulation (Eqs. 11), (12), and (13)) for realizing orientation control of the grasped object, which is expressed as

$$u_{12} = -(-1)^i K_I \int (\theta_o^d - \theta_o)\mathrm{dt} + \tau_{i2}, \tag{15}$$

where the last term is referred to as *biased torque*, which works so as to maintain the initial secure grasping before the actual control task. Note that the left-hand side of Eq. (15) corresponds to the torque applied to the two joints, θ_{12} (PIP) and θ_{22} (IP), as shown in Fig. 23. Therefore, both fingers are sustained by the motors (1, 2, 5, 6) to which only biased torque is applied. Figure 24 shows an experimental result, in which the desired object trajectory is given at intervals of 5 s. In addition, the update delay of image processing is estimated as 25 ms because of the electrical property of a camera and the computer performance, which was used in this experiment. It clearly shows that the actual trajectory robustly converges to the desired value, and as a result, indicates that the simple integral controller applied to the IP and PIP joints performs well to decrease orientation errors in each time step.

We should emphasize that the straightforward control law expressed in Eq. (15) does not involve the Jacobian matrix, despite the fact that the first term of the right-hand side of Eq. (15) is described as a workspace control. This non-Jacobian control method naturally enables the elimination of the joint angles that are sensed in real time, that is, encoder sensing is not necessary for robotic manipulation. Eventually, this result suggests that a model-less and grasping-force-less control can be realized if certain mechanical constraints associated with the musculo-skeletal structure are involved in the robot mechanism.

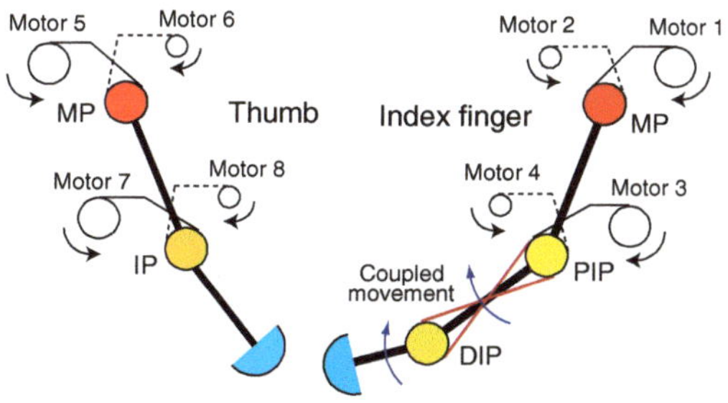

Fig. 23 Schematic view showing motor configuration and wire connections. The odd number motors drive the finger toward grasping the object, and the even-number motors move the finger in the opposite direction

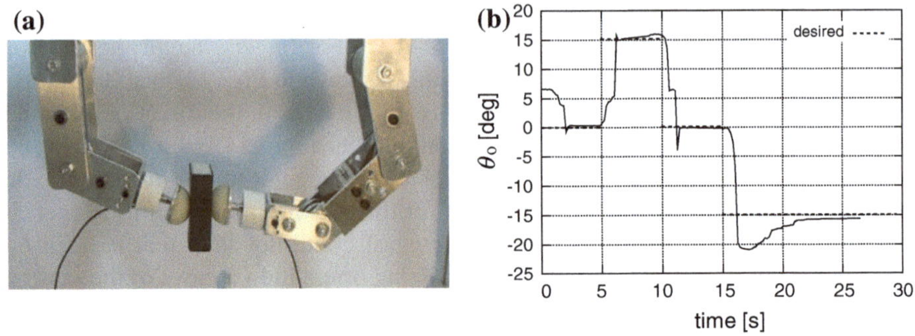

Fig. 24 This photograph shows dexterous manipulation by a 5-DOF robotic hand in which object orientation converges to the desired trajectory, while the error remains small. **a** Manipulation test. **b** Result

7 Concluding Remarks

Through our research into soft-fingered manipulation, we have proposed a simple object orientation controller that consists of two-phased controllers serially connected with each other, called an STP controller expressed in Eqs. (7) and (8). The first stage acts as a robust integral controller from which virtual (pseudo) desired trajectories of joint angles are generated. It is not necessary for the actual joint angle to converge to the virtually generated desired value provided the object is guided to the desired orientation. This is why the desired joint angle is termed a *virtual trajectory*.

In addition, we have clarified that an STP controller with soft fingertips works well in a case where a large time delay exists within the visual feedback robotic system. It had also been shown that the method of gain tuning for improving delayed responses is very simple and useful in a lot of practical situations. We have proposed a straightforward control model capable of realizing precise and secure manipulation by means of a 5-DOF two-fingered robotic hand. This control law does not involve Jacobian matrices and the grasping forces that have been conventionally required in robotic manipulation. In particular, it has been clearly indicated that even a novel controller, Eq. (15), in which encoder measurements are not used, works well for achieving object orientation control in the experiment. These results imply that various inner models, such as robot dynamics and kinematics, are, in fact, not necessary if the robot system has a similar structure and performance to that of the human fingers. That is, an antagonistic wire-driven mechanism and a soft-finger structure enable mimicking of biomechanical characteristics. We conclude that such a human-like robot can comply with nonlinear properties (e.g., gravitational force and friction) without a complicated robot model, and namely that a complete sensor-based control scheme could be constructed in environmental robotics.

References

1. C.W. Hess, K.R. Mills, N.M.F. Murray, Measurement of central motor conduction in multiple sclerosis by magnetic brain stimulation. Lancet **328**(8503), 355–358 (1986)
2. O. Holmes, *Human Neurophysiology* (Unwin Hyman, Crows Nest, 1990)
3. N. Xydas, I. Kao, Modeling of contact mechanics and friction limit surfaces for soft fingers in robotics, with experimental results. Int. J. Robot. Res. **18**(8), 941–950 (1999)
4. N. Xydas, M. Bhagavat, I. Kao, Study of soft-finger contact mechanics using finite elements analysis and experiments, in *Proceedings of IEEE International Conference on Robotics and Automatation* (2000), pp. 2179–2184
5. I. Kao, F. Yang, Stiffness and contact mechanics for soft fingers in grasping and manipulation. IEEE Trans. Robot. Autom. **20**(1), 132–135 (2004)
6. J.L. Johnson, *Contact Mechanics* (Cambridge University Press, Cambridge, 1985)
7. S. Arimoto, K. Tahara, M. Yamaguchi, P. Nguyen, H.Y. Han, Principle of superposition for controlling pinch motions by means of robot fingers with soft tips. Robotica **19**(1), 21–28 (2001)
8. P. Nguyen, S. Arimoto, Performance of pinching motions of two multi-dof robotic fingers with soft-tips, in *Proceedings of IEEE International Conference on Robotics and Automation* (2001), pp. 2344–2349
9. Z. Doulgeri, J. Fasoulas, S. Arimoto, Feedback control for object manipulation by a pair of soft tip fingers. Robotica **20**(1), 1–11 (2002)
10. J. Fasoulas, Z. Doulgeri, Equilibrium conditions of a rigid object grasped by elastic rolling contacts, in *Proceedings of IEEE International Conference on Robotics and Automation* (2004), pp. 789–794
11. S. Arimoto, R. Ozawa, M. Yoshida, Two-dimensional stable blind grasping under the gravity effect, in *Proceedings of IEEE International Conference on Robotics and Automation* (2005), pp. 1196–1202
12. M. Yoshida, S. Arimoto, J.-H. Bae, Blind grasp and manipulation of a rigid object by a pair of robot fingers with soft tips, in *Proceedings of IEEE International Conference on Robotics and Automation* (2007), pp. 4707–4714
13. D.T.V. Pawluk, R.D. Howe, Dynamic lumped element response of the human finger pad. ASME J. Biomech. Eng. **121**(2), 178–183 (1999)
14. D.T.V. Pawluk, R.D. Howe, Dynamic contact of the human finger pad against a flat surface. ASME J. Biomech. Eng. **121**(6), 605–611 (1999)
15. J.Z. Wu, R.G. Dong, W.P. Smutz, S. Rakheja, Dynamic interaction between a finger pad and a flat surface: experiments and analysis. Med. Eng. Phys. **25**(5), 397–406 (2003)
16. T. Inoue, S. Hirai, *Mechanics and Control of Soft-fingered Manipulation* (Springer, London, 2008)
17. T. Inoue, S. Hirai, Elastic model of deformable fingertip for soft-fingered manipulation. IEEE Trans. Robot. **22**(6), 1273–1279 (2006)
18. T. Inoue, S. Hirai, A two-phased object orientation controller on soft finger operations, in *Proceedings of IEEE International Conference on Intelligent Robots and Systems* (2007), pp. 2528–2533
19. T. Inoue, S. Hirai, Robotic manipulation with large time delay on visual feedback systems, in *Proceedings of IEEE/ASME International Conference on Advanced Intelligent Mechatronics* (2010), pp. 1111–1115
20. T. Inoue, D. Takizawa, S. Hirai, Modelless and grasping-forceless control by robotic fingers capable of mechanically coupled movement, in *Proceedings of IEEE/RSJ International Conference on Intelligent Robots and Systems* (2010), pp. 5875–5879
21. K.R. Mills, Power spectral analysis of electromyogram and compound muscle action potential during muscle fatigue and recovery. J. Physiol. **326**, 401–409 (1982)
22. S.R. Devasahayam, Signals and Systems in Biomedical Engineering (Plenum Publishing Corporation, New York, 2000)

Chapter 15
Control for Multi-Finger Hands

Suguru Arimoto and Morio Yoshida

Abstract This chapter attempts to develop a complete set of dynamics of 2-dimensional grasping under rolling contact constraints and propose a control methodology for preserving stability of a grasp under arbitrarily given object and fingerend shapes. The computational model is derived by means of moving frame coordinates under the postulate that two contact points on each contour curve share a single common point and have the same tangent. A control signal for blind grasping is given by referring to the fingers-thumb opposability of human hands. It is shown theoretically and computationally that the control stabilizes motion of the fingers-object system toward its optimal composition. Together with a historical note on blind grasping, a mathematical modeling for numerically simulating 3D grasping is discussed.

Keywords Multi-finger hand · Stable grasping · Blind grasping · Fingers-thumb opposition · Optimal composition

1 Introduction

This chapter discusses a fundamental problem of design of control signals preserving stability of a grasp or pinching (precision prehension) of arbitrary rigid objects subject to rolling contacts by using a pair of multi-joint robot fingers. First, modeling and control of 2D object grasping and manipulation under rolling contacts are discussed under the assumption that motion of the whole fingers-object system

S. Arimoto (✉) · M. Yoshida
Research Organizaion of Science and Technology, Ritsumeikan Univ, 525-8577, Kusatu, Shiga, Japan
e-mail: s_n_riemann@zb.ztv.nt.jp

M. Yoshida
e-mail: yoshida-@fc.ritsumei.ac.jp

R. Balasubramanian and V. J. Santos (eds.), *The Human Hand as an Inspiration for Robot Hand Development*, Springer Tracts in Advanced Robotics 95,
DOI: 10.1007/978-3-319-03017-3_15,

is confined to the horizontal plane. Even in this restricted situation, finding a robust control signal for preserving stability of a grasp is not obvious if the object to be grasped is changeable and hence its shape and location of the mass center are unknown. Nevertheless, human grasps and manipulates an arbitrary object with his or her eyes closed, once preshaping of the set of fingers against the object was successful beforehand. This talent of mankind comes from anthropological and neuro-physiological characteristics. The former claims that fingers-thumb opposability is one of authentifications of humankind discriminated from other primates (see [1]). The latter does that fast voluntary movements of human limb are generated in a feedforward manner by using spring-like forces of muscles evoked by neuromotor signals.

The Sect. 2 shows a design methodology of control signals for stabilization of a grasp of 2D objects without knowing object kinematics and without using visual or tactile sensing. Section 3 is devoted to the derivation of the Euler–Lagrange equation of motion of the fingers-object system under the situation that fingertips have an arbitrary but smooth and convex contour curve and the object has an arbitrary smooth contour. Section 4 shows stability of a grasp when such a blind control signal is supplied to finger-joint actuators. It is shown that the stable steady grasp is achieved in a geometric composition of the system when the line connecting the two fingertip contours is parallel to the line connecting the two contact points. Section 5 shows a numerical simulation result on stabilization of a 2D blind grasp by using a pair of planar fingers with redundant joints. Another simulation result on orientation control of a 2D object on the basis of the principle of superpose is presented. Section 6 presents a mathematical modeling of 3D object grasping with multi-contacts of rolling by a multi-finger hand in the circumstances that the 3D object has smooth surfaces and the fingerends have also smooth and convex surfaces. In a final section a short historical note on blind grasping is presented.

2 Control Signals for Blind Grasping

As pointed out by John Russell Napier (1917–1987) in his book [1], the most important movement of the human hand is opposition, that he defined in the following.

Opposition is a movement by which the pulp surface of the thumb is placed squarely in contact with—or diametrically opposite to—the terminal pads of one or all of the remaining digits.

Then, it is important to consider what kind of control signals for finger joints can generate such an opposable movement and securely grasp a rigid object (see Figs. 1 and 2). In the simplest case of grasp of a 2D rectangular object by a pair of multi-joint robot fingers commonly with a circular fingerend (see Fig. 3), a physical insight into the fingers-thumb opposition suggests that the principal term of control signals must be composed in the form

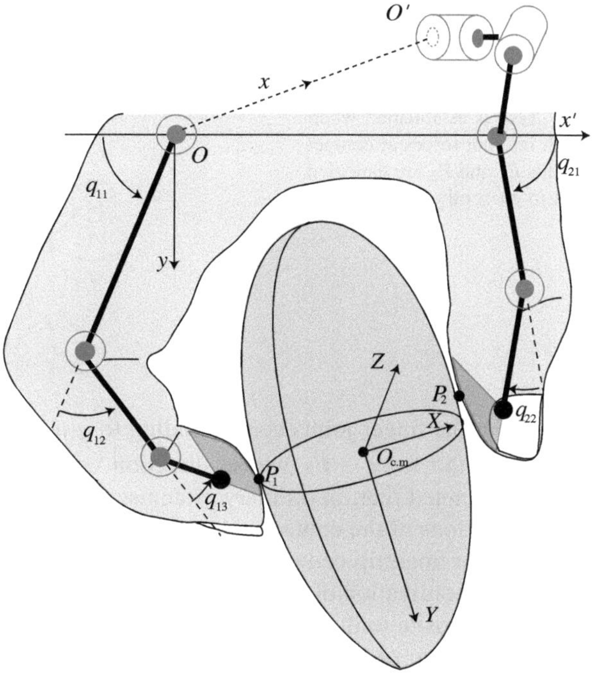

Fig. 1 Grasping of a 3D rigid object of an arbitrary shape on the basis of fingers-thumb opposition typically

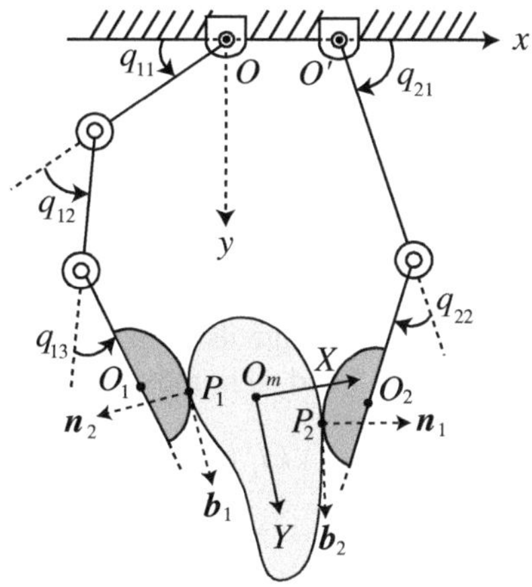

Fig. 2 2D grasping of a 2D rigid object with arbitrary smooth contours by means of multi-joint planar robot fingers with non-spherical fingertips

$$\boldsymbol{u}_i = (-1)^i \frac{f_d}{2r} J_i^{\mathrm{T}}(\boldsymbol{q}_i) \begin{pmatrix} x_1 - x_2 \\ y_1 - y_2 \end{pmatrix}, \ i = 1, 2 \tag{1}$$

where r denotes the radius of the fingertip hemispheres, $J_i(q_i)$ signifies the Jacobian matrix of the position vector $\boldsymbol{r}_i = (x_i, y_i)^{\mathrm{T}}$ of O_i with respect to finger joint vectors $\boldsymbol{q}_i = (q_{i1}, q_{i2}, \cdots)^{\mathrm{T}}$, and $i = 1$ corresponds to the left-hand finger and $i = 2$ the right-hand finger (see Fig. 3). In order to press the object from the left the reaction

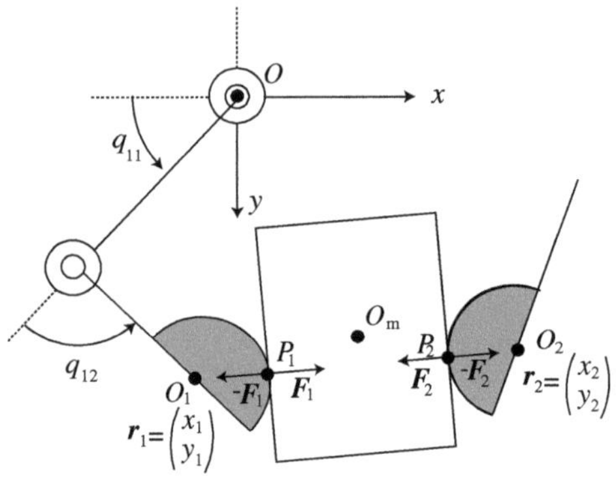

Fig. 3 Physical conditions of fingers-thumb opposable motion in the case of a 2D ball-plate grasp. Stability of the grasp is established when both internal forces at contact points P_1 and P_2 are canceled out to each other

torques at the finger joints corresponding to $q_{1j}(j = 1, \cdots)$ should be generated to withstand the force $-\boldsymbol{F}_1$ whose direction coincides with straight line $\overline{O_1O_2}$. It should be noted that, in ordinary circumstances of grasping, it is difficult to know exact locations of the contact points between the fingertips and object. Instead, the locations of fingertip centers O_1 and O_2 can be assumed to be known. In order to maintain coordination of pressing forces from the left and right, the direction of $\mathbf{F}_2$ must be opposite to that of $\mathbf{F}_1$ and the magnitudes of $\mathbf{F}_1$ and $\mathbf{F}_2$ must be equal for synchronization of both the fingerend motions and stabilization of the grasp. The control signals of (1) reflect these requirements.

What change of the principal component of control torque signals described by (1) is needed to cope with a general case when the fingerend is not spheric and the object sides are not a flat plane (see Fig. 2)? If the 2D fingerend has a contour curve with a variable curvature then the center of the curvature depends on the location of the contact point. That is, the center of the curvature of the fingerend contour must be considered to be unknown and therefore it is uncertain whether the straight line connecting fingerend centers O_1 and O_2 (see Fig. 4) represents the opposability. Notwithstanding this difficult situation, it is fortunately possible to consider the control torque signals for a secure grasp of 2D objects with smooth curved contours (see Figs. 2 and 4) in the following form:

$$u_i = (-1)^i kJ_i^{\mathrm{T}}(q_i)(r_1 - r_2) - c_i\,\dot{q}_i - \alpha_i k\{p_i(t) - p_i(0)\}\; e_i,\; i = 1,2 \qquad (2)$$

where $p_1 = q_{11} + q_{12} + q_{13}$, $p_2 = q_{21} + q_{22}$, $e_1 = (1,1,1)^{\mathrm{T}}$, $e_2 = (1,1)^{\mathrm{T}}$, and $\alpha_i > 0$ and $c_i > 0$ for $i = 1,2$. The first term of $\boldsymbol{u}_i$ in (2) represents a main component of the control signal reflecting the fingers-thumb opposability, where the constant $k > 0$ with physical unit Nm^{-1} should be common to both fingers and $\boldsymbol{r}_i$ denotes the position vector of fingertip center O_i expressed in terms of the inertial frame coordinates $O - xy$, i.e., $r_i = (x_i, y_i)^{\mathrm{T}}$. The second term of $\boldsymbol{u}_i$ expresses the angular velocity feedback regulating damping effects of overall motion of the system. The third term plays an important role in suppressing excess movements of each finger joint toward some equilibrium position of the system. Thus, the control torque input of (2) supplied to finger joint actuators is called a

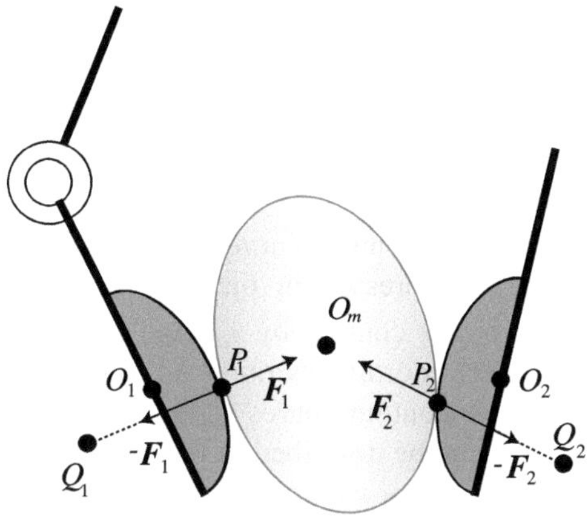

Fig. 4 Physical relations of a 2D grasp in not a ball-plate case. The curvature center Q_i of the fingertip contour curve at the contact point P_i is not fixed. What geometric relations among fingertip centers O_1 and O_2 and contact points P_1 and P_2 lead to a stable grasp?

blind control signal for blind grasping, since the right hand side of $\boldsymbol{u}_i$ in (2) can be computable in real time by using only the knowledge of finger kinematics and measurement data of finger joint angles and angular velocities. In other words, $\boldsymbol{u}_i$ of (2) does not need any knowledge of object kinematics and geometry.

In what follows, we will show how effective the control signal of (2) is for blind grasping even in the situation of existence of rolling contact constraints. Further, it is implicitly assumed that 2D fingerend rigid bodies have a smooth convex contour curve and 2D rigid objects to be grasped have unknown but smooth contour curves. At first, 2D grasping is treated by assuming that overall motions of the fingers-object system are confined to the horizontal plane, that is, all rotational axes of the finger joints and the object around its mass center are in the direction of z-axis perpendicular to the xy-plane (see Fig. 2). Then, the ultimate goal of this chapter is to show that the blind control signal is enable to stabilize the closed-loop dynamics of the fingers-object system and maneuver it toward an equilibrium composition of the system in such a manner that the straight line connecting the two contact points P_1 and P_2 is moving to be in parallel to $\overline{O_1O_2}$, the straight line between the fingerend centers O_1 and O_2 (Fig. 6). To verify this, it is important to derive dynamics of interactive motion of the fingers and object under rolling contact constraints.

3 Dynamics of 2D Grasping Under Rolling Contact Constraints

Modeling of dynamics of 2D grasping of a rigid object with arbitrary smooth contour curves is discussed on the basis of the following postulates on rolling contact constraints (see Figs. 2 and 5):

(1) Two contact points on each contour curve must coincide at a single common point without mutual penetration, and
(2) The two contours must have the same tangent at the common contact point.

This set of postulates reflects rolling motion of pointwise contacts when contact areas are sufficiently small.

Let us denote the fingerend contour curve by $\gamma_1(s_1) = (X_1(s_1), Y_1(s_1))^{\mathrm{T}}$ along the arclength parameter s_1 expressed in the local coordinates $O_1 - X_1Y_1$ (see Fig. 5). Also, denote the object contour by $\gamma_{01}(s_{01}) = (X_{01}(s_{01}), Y_{01}(s_{01}))^{\mathrm{T}}$ along s_{01} based on the local object coordinates $O_m - XY$ (see Fig. 5). According to the postulates 1) and 2), the two contour curves $\gamma_{01}(s_{01})$ and $\gamma_1(s_1)$ during continuous motion by rolling contacts can be described in terms of the same length parameter s_1 in such a way that $s_{01} = s_1 + c_1$ where c_1 is a constant. In other words, suppose that at some s_1 the curve $\gamma_1(s_1)$ meets $\gamma_{01}(s_{01})$ at contact point P_1 where $s_{01} = s_1 + c_1$. It is now necessary to express this object contour curve in terms of the frame coordinates $O - xy$, which is obviously given by

$$\bar{\gamma}_{01}(s_1) = \Pi_0(\theta)\gamma_{01}(s_1) \tag{3}$$

where

$$\Pi_0 = (\,\boldsymbol{r}_X,\ \boldsymbol{r}_Y), \quad \boldsymbol{r}_X = \begin{pmatrix} \cos\theta \\ -\sin\theta \end{pmatrix}, \quad \boldsymbol{r}_Y = \begin{pmatrix} \sin\theta \\ \cos\theta \end{pmatrix} \tag{4}$$

and θ denotes the inclination angle of the object as depicted in Fig. 5. Note that $\boldsymbol{r}_X$ denotes the unit vector of X-axis of the object expressed in terms of the frame coordinates $O - xy$ and $\boldsymbol{r}_Y$ does that of Y-axis of the object. Hence, $\Pi_0(\theta) \in SO(2)$. Similarly, denote by $\bar{\gamma}_1(s_1)$ the fingertip contour curve in terms of the frame coordinates such that

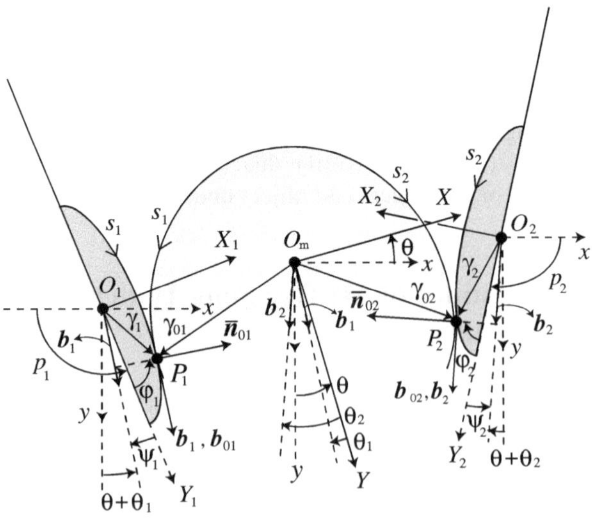

Fig. 5 Definitions of the inertial frame coordinates $O - xy$ and local coordinates $O - X_iY_i$ for $i = 1, 2$

$$\bar{\gamma}_1(s_1) = \Pi_1(p_1)\gamma_1(s_1) = (\ \boldsymbol{r}_{X1},\ \boldsymbol{r}_{Y1})\gamma_1(s_1) \tag{5}$$

$$\boldsymbol{r}_{X1} = \begin{pmatrix} \cos(p_1 - \frac{\pi}{2}) \\ -\sin(p_1 - \frac{\pi}{2}) \end{pmatrix}, \quad \boldsymbol{r}_{Y1} = \begin{pmatrix} \sin(p_1 - \frac{\pi}{2}) \\ \cos(p_1 - \frac{\pi}{2}) \end{pmatrix} \tag{6}$$

where $p_1 = q_{11} + q_{12} + q_{13}$ or $p_1 = \boldsymbol{q}_1^{\mathrm{T}}\boldsymbol{e}_1$. Similarly, it is possible to define $\bar{\gamma}_{02}(s_2)$, $\bar{\gamma}_2(s_2)$, together with orthogonal matrices $\Pi_0(\theta)$ and $\Pi_2(p_2)$, where $p_2 = q_{21} + q_{22}$. Note that $\Pi_0(\theta)$ is common to both fingertip contours. According to the postulate 2), it is possible to introduce the unit tangent vector $\boldsymbol{b}_{0i}(s_i)$ tangent to the fingerend contour curve at contact point P_i for $i = 1, 2$ expressed in terms of the object coordinates $O_m - XY$. Samely, the unit tangent vector of the fingertip at P_i is denoted by $\boldsymbol{b}_i(s_i)$ expressed in terms of the local coordinates $O_i - X_iY_i$. Similarly, it is possible to introduce unit normals $\boldsymbol{n}_{0i}(s_i)$ and $\boldsymbol{n}_i(s_i)$ for $i = 1, 2$. All these unit vectors can be expressed in the frame coordinates in such a way that

$$\begin{cases} \bar{\boldsymbol{b}}_{0i} = \Pi_0\boldsymbol{b}_{0i}, & \bar{\boldsymbol{n}}_{0i} = \Pi_0\boldsymbol{n}_{0i} \\ \bar{\boldsymbol{b}}_i = \Pi_i\boldsymbol{b}_i, & \bar{\boldsymbol{n}}_i = \Pi_i\boldsymbol{n}_i \end{cases}, \quad i = 1, 2 \tag{7}$$

Next, denote the position vector of O_i, the fingerend center of finger i, in terms of the frame coordinates $O - xy$ by $\boldsymbol{r}_i$ for $i = 1, 2$. According to the postulate 1), the contact point P_1 on the fingerend contour curve coincides with that on the object center, that is spelled out in the following equality (see Fig. 5)

$$\boldsymbol{r}_1 + \Pi_1\gamma_1 = r_0 + \Pi_0\gamma_{01} \tag{8}$$

where r_0 denotes the position vector of the object mass center expressed in terms of $O - xy$. Since the same equality holds at the contact point P_2, it follows that

$$\boldsymbol{r}_i + \Pi_i\gamma_i(s_i) = \boldsymbol{r}_0 + \Pi_0\gamma_{0i}(s_i), \quad i = 1, 2 \tag{9}$$

This equality at P_i holds along evolution of the length parameter s_i and s_i itself can be a function of time t. The time derivative of (9) reduces to

$$\dot{\boldsymbol{r}}_i + \dot{\Pi}_i\gamma_i + \Pi_i\frac{\partial\gamma_i}{\partial s_i}\frac{\mathrm{d}s_i}{\mathrm{d}t} = \dot{\boldsymbol{r}}_0 + \dot{\Pi}_0\gamma_{0i} + \Pi_0\frac{\partial\gamma_{0i}}{\partial s_i}\frac{\mathrm{d}s_i}{\mathrm{d}t}, \quad i = 1, 2 \tag{10}$$

where $\dot{\boldsymbol{r}}_i$ denotes the time derivative of $\boldsymbol{r}_i$, and $\dot{\Pi}_i$ and $\dot{\Pi}_0$ denote that of Π_i and Π_0 respectively. Note that

$$\frac{\partial\gamma_i(s_i)}{\partial s_i} = \boldsymbol{b}_i(s_i), \quad \frac{\partial\gamma_{0i}(s_i)}{\partial s_i} = \boldsymbol{b}_{0i}(s_i), \quad i = 1, 2 \tag{11}$$

and

$$\dot{\Pi}_0 = \Pi_0\Omega_0, \quad \dot{\Pi}_i = \Pi_i\Omega_i \tag{12}$$

where

$$\Omega_0 = \dot{\theta}\begin{pmatrix} 0 & 1 \\ -1 & 0 \end{pmatrix}, \quad \Omega_i = \dot{p}_i\begin{pmatrix} 0 & 1 \\ -1 & 0 \end{pmatrix} \tag{13}$$

Substituting (11) and (12) into (10) yields

$$(\dot{r}_i - \dot{r}_0) + \Pi_i \Omega_i \gamma_i - \Pi_0 \Omega_0 \gamma_{0i} = 0, \quad i = 1,2 \tag{14}$$

where the equality $\overline{\boldsymbol{b}_i} = \overline{\boldsymbol{b}}_{oi}$ is taken into account. At this stage it is important to remark that the positive sign of angles θ, θ_i, p_i, and q_{ij} is taken in the counter-clockwise direction. Hence, Ω_0 in (13) expresses the rotation of positive angle $\pi/2$ with associating the angular velocity $\dot{\theta}$. In other words, we have

$$\Omega_0 b_{01} = \dot{\theta} n_{01}, \ \Omega_0 n_{01} = -\dot{\theta} b_{01} \tag{15}$$

in $O_1 - X_1 Y_1$ and

$$\Omega_0 b_{02} = -\dot{\theta} n_{02}, \ \Omega_0 n_{02} = \dot{\theta} b_{02} \tag{16}$$

in $O_2 - X_2 Y_2$ and, from the skew symmetry of Ω_0 and Ω_i $(i = 1,2)$, we have

$$\begin{cases} b_{01}^{\mathrm{T}} \Omega_0 = -\dot{\theta} n_{01}^{\mathrm{T}}, & n_{01}^{\mathrm{T}} \Omega_0 = \dot{\theta} b_{01}^{\mathrm{T}} \\ b_{02}^{\mathrm{T}} \Omega_0 = \dot{\theta} n_{02}^{\mathrm{T}}, & n_{02}^{\mathrm{T}} \Omega_0 = -\dot{\theta} b_{02}^{\mathrm{T}} \end{cases} \tag{17}$$

It is possible to obtain similar relations concerning Ω_i, b_i, and n_i by replacing $\dot{\theta}$ with $\dot{p}_i$ for $i = 1,2$, i.e.,

$$\Omega_i b_i = (-1)^i n_i, \ \Omega_i n_i = -(-1)^i b_i, \ i = 1,2 \tag{18}$$

The relation of (14) for i may be treated as a Pfaffian constraint with two degrees-of-freedom corresponding to its integral form of (9) that may be considered to be a holonomic constraint. However, (14) includes explicitly the rotation matrices Π_0 and Π_i which are governed by the first-order differential equations of (12). Explicit appearance of these rotational matrices in (14) makes a cause of complication in derivation of the Euler–Lagrange equation of motion of the overall fingers-object system. To avoid this and gain a physical insight into the meaning of rolling contact constraints, we rewrite the zero relative-velocity form by using the moving frame coordinates at contact point P_i for $i = 1,2$. This is carried out by taking inner products of (14) and b_i, and (14) and n_i, for $i = 1,2$ in the following way:

$$\begin{cases} \overline{b}_{0i}^{\mathrm{T}} (r_i - r_0) + \overline{b}_i^{\mathrm{T}} \Pi_i \Omega_i \gamma_i - \overline{b}_{0i}^{\mathrm{T}} \Pi_0 \Omega_0 \gamma_{0i} = 0 \\ \overline{n}_{0i}^{\mathrm{T}} (r_i - r_0) - \overline{n}_i^{\mathrm{T}} \Pi_i \Omega_i \gamma_i - \overline{n}_{0i}^{\mathrm{T}} \Pi_0 \Omega_0 \gamma_{0i} = 0 \end{cases}, \quad i = 1,2 \tag{19}$$

where equalities $\overline{b}_i = \overline{b}_{0i}$ and $\overline{n}_i = -\overline{n}_{0i}$ are used. Noting that $\overline{b}_i^{\mathrm{T}} \Pi_i = b_i^{\mathrm{T}}$, $\overline{n}_i^{\mathrm{T}} \Pi_i = n_i^{\mathrm{T}}$, $\overline{b}_{0i}^{\mathrm{T}} \Pi_0 = b_{0i}^{\mathrm{T}}$, and $\overline{n}_{0i}^{\mathrm{T}} \Pi_0 = n_{0i}^{\mathrm{T}}$, and bearing in mind (15) to (17), Eq. (19) is reduced to

$$\begin{cases} R_{bi} = (r_i - r_0)^{\mathrm{T}} \, \overline{b}_{0i} - (-1)^i \left(\dot{p}_i \, n_i^{\mathrm{T}} \gamma_i + \dot{\theta} n_{0i}^{\mathrm{T}} \gamma_{0i} \right) = 0 \\ R_{ni} = (r_i - r_0)^{\mathrm{T}} \, \overline{n}_{0i} - (-1)^i \left(\dot{p}_i \, b_i^{\mathrm{T}} \gamma_i - \dot{\theta} b_{0i}^{\mathrm{T}} \gamma_{0i} \right) = 0 \end{cases}, \quad i = 1,2 \tag{20}$$

In parallel to this Pfaffian form, it is important to show that the arclength parameter $s_i(t)$ is governed by the first-order differential equation of the form

$$(\kappa_{0i} + \kappa_i)\frac{\mathrm{d}s_i}{\mathrm{d}t} = (-1)^i\left(\dot{\theta} - \dot{p}_i\right), \quad i = 1, 2 \tag{21}$$

where κ_i denotes the curvature of γ_i at point P_i and κ_{0i} does that of γ_{0i}, i.e., $\kappa_i = |\partial^2\gamma_i/\partial s_i^2|$ and $\kappa_{0i} = |\partial^2\gamma_{0i}/\partial s_i^2|$ for $i = 1, 2$. In fact, the derivative of $b_i = \overline{b}_{0i}$ in t reduces to

$$(\kappa_{0i} + \kappa_i)\ \overline{n}_{0i}\frac{\mathrm{d}s_i}{\mathrm{d}t} = \Pi_i\Omega_i\ b_i - \Pi_0\Omega_0\ b_{0i} = (-1)^i\left(\dot{\theta}\ \overline{n}_{0i} + \dot{p}_i\ \overline{n}_i\right), \quad i = 1, 2 \tag{22}$$

which apparently reduces to (21). It should be noted that each of Pfaffian forms in (20) has an integral form shown in the form

$$\begin{cases} Q_{bi} = (r_i - r_0)^{\mathrm{T}}\overline{b}_{0i} + b_i^{\mathrm{T}}\gamma_i - b_{0i}^{\mathrm{T}}\gamma_{0i} = 0 \\ Q_{ni} = (r_i - r_0)^{\mathrm{T}}\overline{n}_{0i} - n_i^{\mathrm{T}}\gamma_i - n_{0i}^{\mathrm{T}}\gamma_{0i} = 0 \end{cases}, \quad i = 1,\ 2 \tag{23}$$

which is obtained by taking inner products of (9) and b_{0i}, and (9) and $\overline{n}_{0i}$ for $i = 1, 2$. In other words, it follows that

$$\frac{\mathrm{d}}{\mathrm{d}t}Q_{bi} = R_{bi} = 0, \ \frac{\mathrm{d}}{\mathrm{d}t}Q_{ni} = R_{ni} = 0, \quad i = 1, 2 \tag{24}$$

To see this, let us define the position vector as $X = (x, y, \theta, q_1^{\mathrm{T}}, q_2^{\mathrm{T}})^{\mathrm{T}}$, and introduce the following four 8-dimensional vectors as follows:

$$\begin{cases} \Re_{b1} = \begin{bmatrix} -\overline{b}_{01} \\ n_{01}^T\gamma_{01} \\ J_1^T\ \overline{b}_{01} + (\ n_1^T\gamma_1)\ e_1 \\ 0_2 \end{bmatrix}, & \Re_{b2} = \begin{bmatrix} -\overline{b}_{02} \\ -\ n_{02}^T\gamma_{02} \\ 0_3 \\ J_2^T\ \overline{b}_{02} - (\ n_2^T\gamma_2)\ e_2 \end{bmatrix} \\ \Re_{n1} = \begin{bmatrix} -\overline{n}_{01} \\ -\ b_{01}^T\gamma_{01} \\ J_1^T\ \overline{n}_{01} + (\ b_1^T\gamma_1)\ e_1 \\ 0_2 \end{bmatrix}, & \Re_{n2} = \begin{bmatrix} -\overline{n}_{02} \\ -\ b_{02}^T\gamma_{02} \\ 0_3 \\ J_2^T\ \overline{n}_{02} - (\ b_2^T\gamma_2)\ e_2 \end{bmatrix} \end{cases} \tag{25}$$

where $J_i(q_i) = \partial\ r_i/\partial q_i^{\mathrm{T}}$ for $i = 1, 2$ and $r_0 = (x, y)^{\mathrm{T}}$ that denotes the position of the object mass center in O-xy (the frame coordinates). Then, Eq. (24) can be written into

$$R_{bi} = \Re_{bi}^{\mathrm{T}}\frac{\mathrm{d}X}{\mathrm{d}t}, \quad R_{ni} = \Re_{ni}^{\mathrm{T}}\frac{\mathrm{d}X}{\mathrm{d}t}, \quad i = 1, 2 \tag{26}$$

where it is possible to confirm that

$$\Re_{bi} = \partial Q_{bi}/\partial X, \ \Re_{ni} = \partial Q_i/\partial X, \quad i = 1, 2 \tag{27}$$

Now, we are at the stage to derive the Euler–Lagrange equation of motion of the overall fingers-object system by introducing the total kinetic energy in the following form:

$$\begin{aligned} K(X, \dot{X}) &= \frac{M}{2}\left(\dot{x}^2 + \dot{y}^2\right) + \frac{I}{2}\dot{\theta} + \sum_{i=1,2} \frac{1}{2}\dot{q}_i^{\mathrm{T}} G_i(q_i)\dot{q}_i \\ &= \frac{1}{2}\dot{X}^{\mathrm{T}} G(X)\dot{X} \end{aligned} \tag{28}$$

where $G(X)$ denotes the 8×8 matrix defined by $G = \mathrm{diag}(M, M, I, G_1, G_2)$. The Lagrangian of the system is given by introducing Lagrange multipliers f_i and λ_i for Q_{ni} and Q_{bi} respectively in the following form:

$$L(X; s_1, s_2) = K(X, \dot{X}) - \sum_{i=1,2} (f_i Q_{ni} + \lambda_i Q_{bi}) \tag{29}$$

Note that $K(X, \dot{X})$ is independent of the shape parameters s_1 and s_2 but Q_{ni} and Q_{bi} are dependent on s_i for $i = 1, 2$. From the variational principle applied to the formula

$$\int_{t_0}^{t_1} \left\{\delta L + u_1^{\mathrm{T}} \delta q_1 + u_2^{\mathrm{T}} \delta q_2\right\} \mathrm{d}t = 0 \tag{30}$$

it follows that

$$G(X)\,\ddot{X} + \left(\frac{1}{2}\dot{G}(X) + S(X, \dot{X})\right)\dot{X} + \sum_{i=1,2} (f_i \Re_{ni} + \lambda_i \Re_{bi}) = B\begin{pmatrix} u_1 \\ u_2 \end{pmatrix} \tag{31}$$

where $S(X, \dot{X})$ is a skew-symmetrix matrix, $B = (O_{3\times 5}, I_5)^{\mathrm{T}}$, $0_{3\times 5}$ signifies the 3×5 zero matrix, and I_5 the 5×5 identity matrix, u_i denotes the finger joint control input for $i = 1, 2$. Equation (31) can be spelled out in detail as

$$M\ddot{x} - \sum_{i=1,2} \left(f_i\ \bar{n}_{0i} + \lambda_i\ \bar{b}_{0i}\right) = 0 \tag{32}$$

$$I\ddot{\theta} + \sum_{i=1,2} (-1)^i \{f_i(\ b_{0i}^{\mathrm{T}} \gamma_{0i}) - \lambda_i(\ n_{0i}^{\mathrm{T}} \gamma_{0i})\} = 0 \tag{33}$$

$$\begin{aligned} G_i(q_i)\,\ddot{q}_i + \left\{\frac{1}{2}\dot{G}_i(q_i) + S_i(q_i, \dot{q}_i)\right\}\dot{q}_i + f_i\{J_i^{\mathrm{T}}(q_i)\ \bar{n}_{0i} - (-1)^i(\ b_i^{\mathrm{T}} \gamma_i)\} \\ + \lambda_i\{J_i^{\mathrm{T}}(q_i)\ \bar{b}_{0i} - (-1)^i(\ n_i^{\mathrm{T}} \gamma_i)\ e_i\} = u_i, \quad i = 1, 2 \end{aligned} \tag{34}$$

This Euler–Lagrange equation should be integrated under the constraints $Q_{ni} = 0$ and $Q_{bi} = 0$ for $i = 1, 2$ together with the update law of arclength parameters s_i described in (21) and the Frenet-Serret equations:

$$\left\{\frac{\mathrm{d}}{\mathrm{d}s_i}(\, b_i,\; n_i) = (\, b_i,\; n_i)\begin{pmatrix} 0 & -\kappa_i \\ \kappa_i & 0 \end{pmatrix},\; \frac{\mathrm{d}}{\mathrm{d}s_i}(\, b_{0i},\; n_{0i}) = (\, b_{0i},\; n_{0i})\begin{pmatrix} 0 & -\kappa_{0i} \\ \kappa_{0i} & 0 \end{pmatrix}\right\} \tag{35}$$

4 Stability of Blind Grasping

Let us consider the control signal of blind grasping defined by (2) for the pair of planar robot fingers depicted in Fig. 2. Since physical values of every component of u_i in (2) stand for the physical unit of torque [Nm], the integral of the sum of inner products of $\dot{q}_1^{\mathrm{T}}\, u_1$ and $\dot{q}_2^{\mathrm{T}}\, u_2$ over time interval $[\mathrm{t}_0, \mathrm{t}_1]$ gives rise to the work done over $[\mathrm{t}_0, \mathrm{t}_1]$. To gain a physical insight into this fact, first note that

$$\sum_{i=1,2} \dot{q}_i u_i = -\frac{\mathrm{d}}{\mathrm{d}t}\left\{\frac{k}{2}\parallel r_1 - r_2 \parallel^2 + \sum_{i=1,2}\frac{\alpha_i k}{2}(p_i - p_i(0))^2\right\} - \sum_{i=1,2} c_i \parallel \dot{q}_i \parallel^2 \tag{36}$$

that is obtained by taking inner products of $\dot{q}_1$ and u_1 and $\dot{q}_2$ and u_2. Then, it is possible to see that the closed-loop equation of motion when u_i of (2) is substituted into (31) is expressed in the form

$$\begin{aligned} G(X)\,\ddot{X} + \left\{\frac{1}{2}\dot{G}(X) + S(X,\dot{X}) + C\right\}\dot{X} + \frac{\partial W(X)}{\partial X} \\ + \sum_{i=1,2}(f_i \Re_{ni} + \lambda_i \Re_{bi}) = 0 \end{aligned} \tag{37}$$

where $C = \mathrm{diag}(0, 0, 0, c_1 I_3, c_2 I_2)$ and

$$W(X) = \frac{k}{2}\parallel r_1 - r_2 \parallel^2 + \sum_{i=1,2}\frac{\alpha_i k}{2}\hat{N}_i^2 \tag{38}$$

and for later convenience, we introduce symbol $\hat{N}_i$ by setting $\hat{N}_i = p_i - p_i(0)$ for $i = 1, 2$. Note that from (36), (28), and taking the inner product of (37) and $\dot{X}$ it follows that

$$\frac{\mathrm{d}}{\mathrm{d}t}\{K(X,\dot{X}) + W(X)\} = -\sum_{i=1,2} c_i \parallel \dot{q}_i \parallel^2 \tag{39}$$

It is interesting to note that the Euler–Lagrange equation of (37) is derived under the condition of four holonomic constraints of rolling contact and thereby four Lagrange multipliers f_i and λ_i ($i = 1, 2$) explicitly enter into (37). Nevertheless, these multipliers do neither enter explicitly into the total energy of the system defined by $E(X, \dot{X}) = K(X, \dot{X}) + W(X)$ nor the energy conservation law described in (39). Since the Lagrangian can be defined originally by $L(X, \dot{X}) = K(X, \dot{X}) - W(X)$ and $W(X)$ can be regarded as a potential function, it is important to remind of the notion of Lagrange stability for a class of mechanical systems. Generally it states:

If a critical position state $X = X^*$ satisfying $\partial W(X^*)/\partial X = 0$ attains a local minimum of $W(X)$ under the holonomic constraints then the position state X^* is stable. Further, if the Euler–Lagrange equation is fully dissipated, that is, the damping matrix C in (37) is positive definite, then the state X^* is asymptotically stable.

In the case of the concerned Euler–Lagrange equation of (38), $\bar{W} = W(X) - W(X^*)$ is not positive definite in X in a neighborhood of $X = X^*$ even if $W(X)$ attains a local minimum at $X = X^*$. We must bear in mind at this stage that the total degrees-of-freedom of the pair of robot fingers depicted in Fig. 2 is too redundant to only preserve the stability of a 2D-grasp. However, such redundancy of finger joints is necessary to some extent if any additional control objective is imposed such as control of the orientation angle of an object. Returning to the concerned problem of stability of blind grasping, first we note that $W(X)$ is only dependent on position variables p_1, p_2, and θ and hence we rewrite $W(X)$ into $W(X; z)$ by defining $z = (p_1, p_2, \theta)^{\mathrm{T}}$. Note that the vector z expresses necessarily and sufficiently each corresponding geometric composition of the fingerends and object under rotational motion with rolling contacts (see Fig. 7). Thus, we introduce the notion of "optimal composition" in the following meaning:

If z attains a geometric composition at $z = z^*$ shown in Fig. 6 such that the line connecting the two contact points P_1 and P_2 becomes parallel to the line $\overline{O_1 O_2}$ connecting the two fingertip centers, we call z^* an optimal coordinate composition.

In what follows we show that $W(X; z)$ attains a local minimum when z reaches an optimal coordinate composition. To do this, it is necessary to note that the first term of $W(X; z)$ is determined geometrically by four contact positions $\gamma_i(s_i)$ and

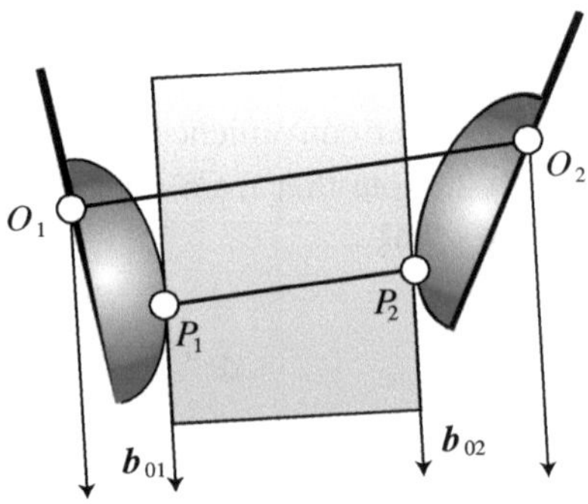

Fig. 6 Geometric composition of the steady state of a grasp stable in a dynamic sense. The stable grasp is realized when $\overline{O_1 O_2}$ is parallel to $\overline{P_1 P_2}$

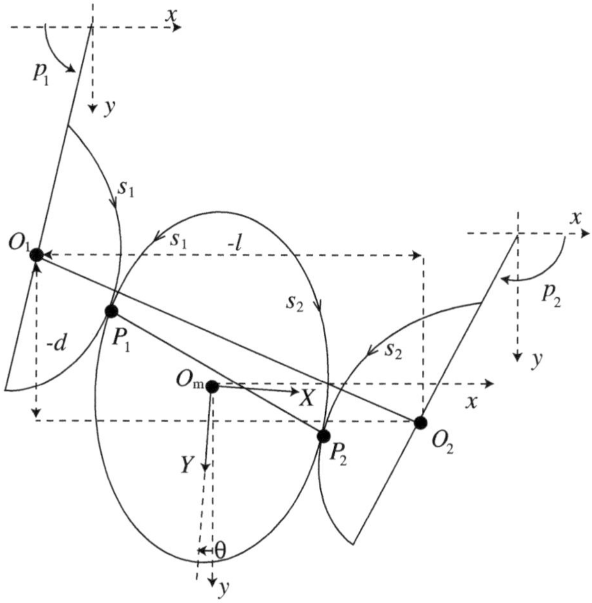

Fig. 7 Geometric composition of two fingerends contacting with a 2D object. The triplet $z = (p_1, p_2, \theta)$ determines rotational motion uniquely if both the two rolling contacts are maintained

$\gamma_{0i}(s_i)$ in each corresponding local coordinates for $i = 1, 2$. More precisely, it is evident from Fig. 7 to have

$$\frac{k}{2} \| r_1 - r_2 \|^2 = \frac{k}{2}(l^2 + d^2) \tag{40}$$

since l denotes the X-component of $r_2 - r_1$ in $O_m - XY$ and d the Y-component. In other words, we have

$$r_1 - r_2 = \Pi_\theta (l, d)^{\mathrm{T}} \tag{41}$$

where

$$\Pi_\theta = \begin{pmatrix} \cos\theta & \sin\theta \\ -\sin\theta & \cos\theta \end{pmatrix} \tag{42}$$

and Π_θ is an element of $SO(2)$ expressing the rotational operation of angle θ. Next, in order to see the gradient of $W(X; z)$ in z, it is necessary to see that l and d are written in the form

$$l = \sum_{i=1,2} (l_i + l_{0i}), d = \sum_{i=1,2} (d_i + d_{0i}) \tag{43}$$

where

$$\begin{pmatrix} l_i \\ d_i \end{pmatrix} = \Pi_{\theta_i} \begin{pmatrix} n_i^T \gamma_i \\ (-1)^i \ b_i^T \gamma_i \end{pmatrix}, \ \begin{pmatrix} l_{0i} \\ d_{0i} \end{pmatrix} = \Pi_{\theta_i} \begin{pmatrix} n_{0i}^T \gamma_{0i} \\ -(-1)^i \ b_{0i}^T \gamma_{0i} \end{pmatrix} \tag{44}$$

Note that angle θ_i $(i = 1, 2)$ are defined in Fig. 5 and Π_{θ_i} $(i = 1, 2)$ signify a similar meaning to Π_θ of (42). Then, it is possible to obtain

$$\frac{\mathrm{d}}{\mathrm{d}t}\begin{pmatrix} l_{0i} \\ d_{0i} \end{pmatrix} = -\frac{\dot{\theta} - \dot{p}_i}{\kappa_{0i} + \kappa_i}\begin{pmatrix} \sin\theta_i \\ \cos\theta_i \end{pmatrix} \tag{45}$$

$$\frac{\mathrm{d}}{\mathrm{d}t}\begin{pmatrix} l_i \\ d_i \end{pmatrix} = \left\{\begin{pmatrix} -d_i \\ l_i \end{pmatrix} + \frac{1}{\kappa_{0i} + \kappa_i}\begin{pmatrix} \sin\theta_i \\ \cos\theta_i \end{pmatrix}\right\}\left(\dot{\theta} - \dot{p}_i\right) \tag{46}$$

which, according to (44), lead to

$$\frac{\mathrm{d}}{\mathrm{d}t}\begin{pmatrix} l \\ d \end{pmatrix} = \begin{pmatrix} -d_1 - d_2 \\ l_1 + l_2 \end{pmatrix}\dot{\theta} + \begin{pmatrix} d_1 \\ -l_1 \end{pmatrix}\dot{p}_1 + \begin{pmatrix} d_2 \\ -l_2 \end{pmatrix}\dot{p}_2 \tag{47}$$

Thus, it follows that

$$\begin{aligned}\frac{\mathrm{d}}{\mathrm{d}t}W &= k(l, d)\frac{\mathrm{d}}{\mathrm{d}t}\begin{pmatrix} l \\ d \end{pmatrix} + \sum_{i=1,2}\alpha_i k\hat{N}_i\dot{p}_i \\ &= k(ld_0 - dl_0)\dot{\theta} + \sum_{i=1,2}k\left(d_i\hat{N}_i + ld_i - dl_i\right)\dot{p}_i\end{aligned} \tag{48}$$

where $l_0 = l_{01} + l_{02}$ and $d_0 = d_{01} + d_{02}$. For convenience, we denote coefficients of the right hand side of (48) by

$$S_N = ld_0 - dl_0, \quad \Delta N_i = \alpha_i\hat{N}_i + ld_i - dl_i, \ i = 1, 2 \tag{49}$$

and rewrite (48) in the form

$$\frac{\mathrm{d}}{\mathrm{d}t}W = kS_N\dot{\theta} + \sum_{i=1,2}k\Delta N_i\dot{p}_i \tag{50}$$

This means that, if $S_N = 0$ and $\Delta N_i = 0$ for $i = 1, 2$, then $\mathrm{d}W/\mathrm{d}t = 0$ and hence the gradient of W in z reduces to zero. Further, note that the vector becomes orthogonal to the vector $(d_0, -l_0)$. Bearing in mind that the vector (l_0, d_0) expresses $\overleftarrow{O_2O_1}$, we see that $S_N = 0$ if and only if the two fingertip centers O_1 and O_2 is parallel to the line connecting the two contact points P_1 and P_2, i.e., z reaches the optimal coordinate composition $z = z^*$ such that $S_N = 0$ at $z = z^*$ and p_1 and p_2 are determined to satisfy $\Delta N_i = 0$ respectively for $i = 1, 2$. It is also important to know that if we define

$$\begin{cases} \Delta f_i = f_i + k(l\cos\theta_i - d\sin\theta_i) \\ \Delta\lambda_i = \lambda_i - (-1)^i k(l\sin\theta_i - d\cos\theta_i) \end{cases} \tag{51}$$

then, by substituting these equalities into (32) $\sim$ (34), we obtain

$$M\ddot{x} - \sum_{i=1,2}\left(\Delta f_i\,\overline{n}_{0i} + \Delta\lambda_i\,\overline{b}_{0i}\right) = 0 \tag{52}$$

$$I\ddot{\theta} + \sum_{i=1,2}(-1)^i\{\Delta f_i(\ b_{0i}^{\mathrm{T}}\gamma_{0i}) - \Delta\lambda_i(\ n_{0i}^{\mathrm{T}}\gamma_{0i})\} + kS_N = 0 \tag{53}$$

$$\begin{aligned} G_i\ddot{q}_i + \left\{\frac{1}{2}\dot{G}_i + S_i\right\}\dot{q}_i + c_i\dot{q}_i + \Delta f_i\{J_i^{\mathrm{T}}\ \bar{n}_{0i} - (-1)^i(\ b_i\gamma_i)\ e_i\} \\ + \Delta\lambda_i\{J_i^{\mathrm{T}}\ \bar{b}_{0i} - (-1)^i(\ n_i^{\mathrm{T}}\gamma_i)\ e_i\} + k\Delta N_i\ e_i = 0,\ i = 1,2 \end{aligned} \tag{54}$$

This shows that at the optimal coordinate composition $z = z^*$ the position terms kS_N of (53) and $k\Delta N_i\ e_i$ of (54) can vanish. Further, it is possible to confirm that the Hessian matrix of $W(X;z)$ in z at $z = z^*$ becomes positive definite by choosing control gains $k > 0$, $\alpha_i > 0$, $c_i > 0$ for $i = 1, 2$ appropriately. Finally, owing to the Pfaffian forms of rolling contact constraints described in (20), it is possible to see that there exist positive constants c_0, $c_{\theta i}$, and c_{Xi} for $i = 1, 2$ such that

$$\begin{cases} I\dot{\theta}^2 \leq c_{\theta 1} \parallel \dot{q}_1 \parallel^2 + c_{\theta 2} \parallel \dot{q}_2 \parallel^2 \\ M(\dot{x}^2 + \dot{y}^2) \leq c_{X1} \parallel \dot{q}_1 \parallel^2 + c_{X2} \parallel \dot{q}_2 \parallel^2 + c_0\dot{\theta} \end{cases} \tag{55}$$

These inequalities imply that, regardless the closed-loop dynamics of motion of the object described by (52) and (53) is not dissipative, it can be regarded as being fully dissipated through the fingers' dynamics of (54) that is fully dissipative. Thus, we conclude that in a neighborhood of an optimal coordinate composition $z = z^*$ a solution trajectory to the closed-loop dynamics of (52) to (54) satisfies

$$\frac{\mathrm{d}}{\mathrm{d}t}\bar{E}(X,\dot{X};z,z^*) \leq -\gamma_0\bar{E}(X,\dot{X};z,z^*) \tag{56}$$

with some positive constant $\gamma_0 > 0$, where

$$\bar{E}(X,\dot{X};z,z^*) = K(X,\dot{X}) + W(X;z) - W(X;z^*) \tag{57}$$

Note that $\bar{E}$ is not positive with respect to the full state vector $(X,\dot{X})$ but it is positive definite with respect to $(z - z^*, \dot{X})$. Although the overall fingers-object system is redundant, the Lyapunov-like relation of (56) leads to the conclusion that the optimal coordinate composition is asymptotically stable in the sense of Lagrange stability.

Before closing this section, let us discuss the gravity effect that affects motion of the object during object manipulation. In this case, the equation of translational motion of the object is governed, instead of (32), by the form

$$M\ddot{x} - \sum_{i=1,2}(f_i\ \bar{n}_{0i} + \lambda_i\ \bar{b}_{0i}) - Mg\begin{pmatrix}0\\1\end{pmatrix} = 0 \tag{58}$$

if the direction of gravity force is in y-axis of the frame coordinates. In this case, we suggest a blind control signal with PI terms of p_i in such a way that

$$
\begin{aligned}
u_i =& (-1)^i k J_i^{\mathrm{T}}(q_i)(r_1 - r_2) - c_i \dot{q}_i \\
& - \alpha_i k \hat{N}_i e_i - \beta_i \int_0^t \hat{N}_i(\tau) e_i \mathrm{d}\tau, \ i = 1, 2
\end{aligned} \tag{59}
$$

where $\hat{N}_i = p_i - p_i(0)$ and β_i is a positive constant for $i = 1, 2$. Compared with (2), the last term of (59) is additional, that can compensate the gravity effect at the contact points. Another important problem is to control the orientation angle $\theta(t)$ of the object toward a desired value θ_d, provided that $\theta(t)$ can be measured by visual sensing. This problem will be treated in the next section. Generally, we remark that the total degrees-of-freedom of the fingers must be redundant to some extent for the effective use of design methodology of control signals on the basis of the principle of superpose like (59).

5 Numerical Simulation

The computational model of 2D grasping described by (31) or, in detail, (32) to (34) is a set of eight second-order differential equations that implicitly depend on the arclenght parameters s_1 and s_2. Therefore, computer simulation of the dynamics of (31) should be carried out by numerically integrating (32) to (34) and simultaneously integrating the first-order differential equation of (21) together with the Frenet-Serret equation of (35), that shows the update law of arclength parameters. Since all four rolling contact constraints expressed by (20) are integrable and correspond to integral forms of holonomic constraints shown in (24), the well-known CSM (Constraint Stabilization Method) can be used by corresponding f_i and λ_i to each critically-damped second-order system. By using two robot fingers with fingerends imitating the contours of human index-finger and thumb (see Fig. 8), we

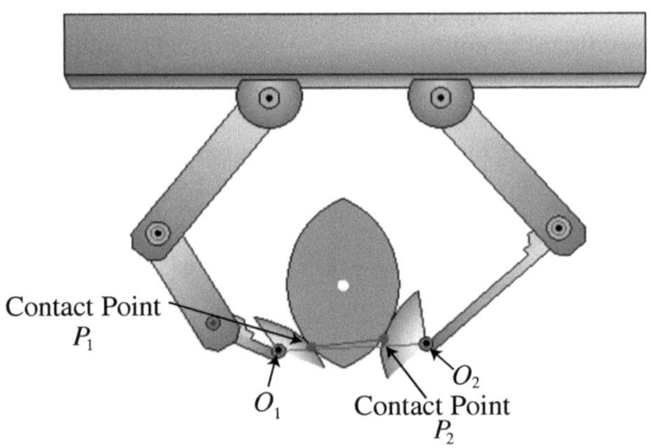

Fig. 8 Initial composition

Table 1 Physical parameters of fingers and object

$l_{11} = l_{21} = l_{22}$	Length	0.065 [m]
l_{12}	Length	0.039 [m]
l_{13}	Length	0.026 [m]
m_{11}	Weight	0.045 [kg]
m_{12}	Weight	0.025 [kg]
m_{13}	Weight	0.015 [kg]
m_{21}	Weight	0.045 [kg]
m_{22}	Weight	0.040 [kg]
L	Base length	0.063 [m]
M	Object weight	0.040 [kg]

Table 2 Parameters of control signals

k	Position feedback gain	75.000 [N/m]
c_i	Damping coefficient	0.006 [Nms]
$\bar{\alpha}_i$ $(i = 1, 2)$	Gain ($\bar{\alpha}_i = \alpha_i k$)	1.0 [s^2/kg]

carried out numerical simulation. The control input (Eq. 2) was used for stabilizing a grasped object with smooth contours as shown in Fig. 8 or 9. In the simulations, physical parameters of the fingers-object system listed in Table 1 and the control signal parameters listed in Table 2 are used. Figure 8 shows an initial composition of the system in the simulations, and Fig. 9 shows the composition after 1 s. The results in Figs. 10, 11, 12, 13, and 14 indicate that all the velocities converged to zero. Figures 15, 16, 17, and 18 show that the normal force components f_1 and f_2 were adequately large enough compared to the tangential force components λ_1 and λ_2. In Fig. 19, $\hat{N}_1$ and $\hat{N}_2$ converge to some constant values. For checking whether the equilibrium state the system converged to is an optimal composition, we define the following two vectors:

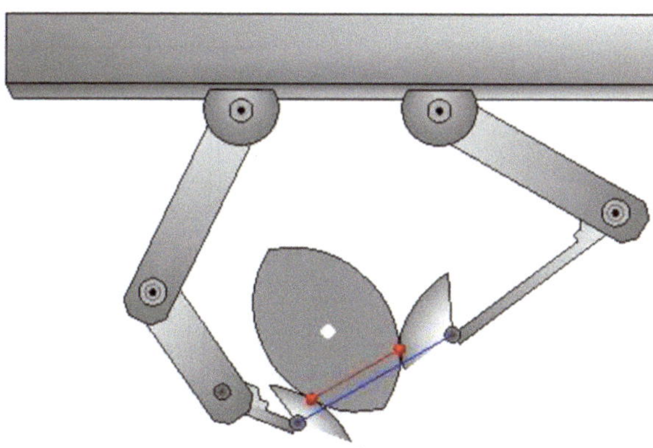

Fig. 9 Composition after 1 s

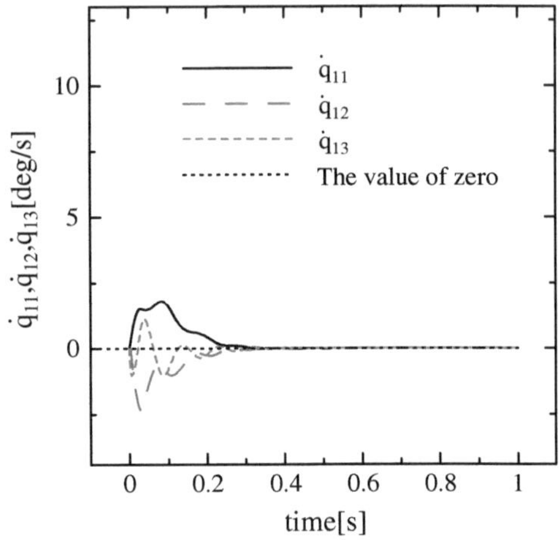

Fig. 10 Trajectories of angular velocities $\dot{q}_{11}$, $\dot{q}_{12}$, and $\dot{q}_{13}$

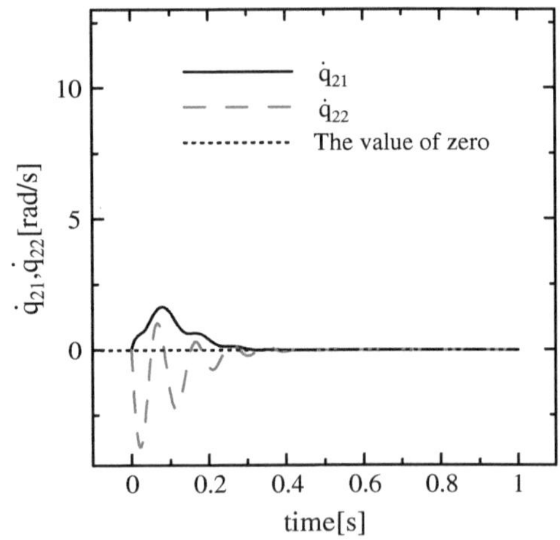

Fig. 11 Trajectories of angular velocities $\dot{q}_{21}$ and $\dot{q}_{22}$

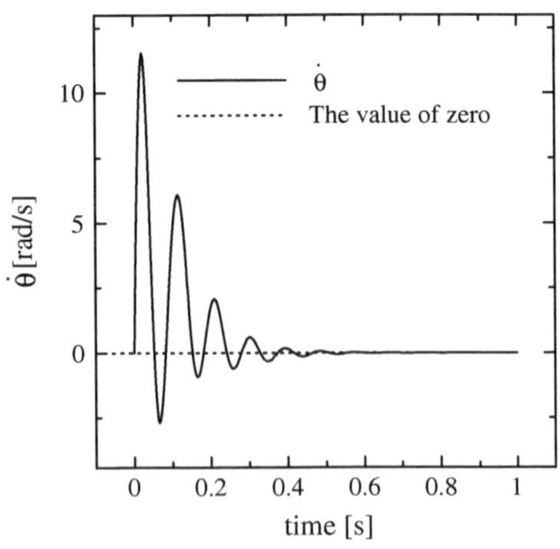

Fig. 12 Trajectory of angular velocity $\dot{\theta}$

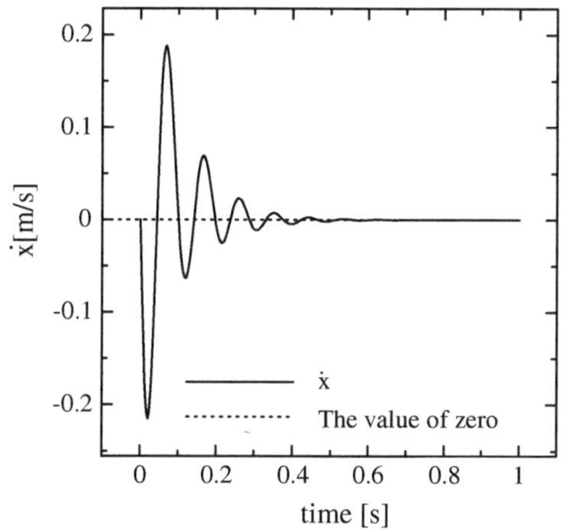

Fig. 13 Trajectory of translational velocity $\dot{x}$

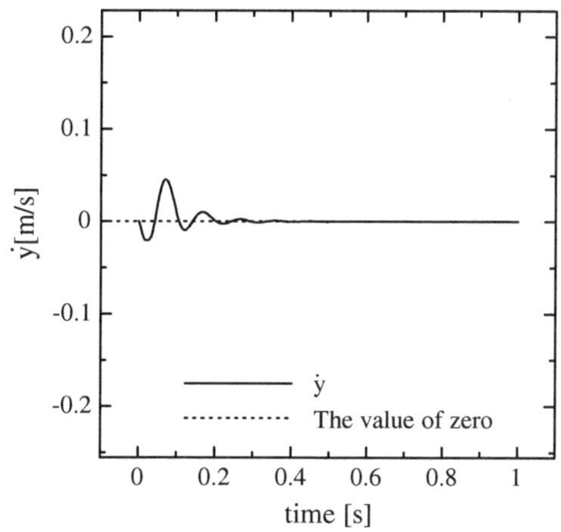

Fig. 14 Trajectory of translational velocity $\dot{y}$

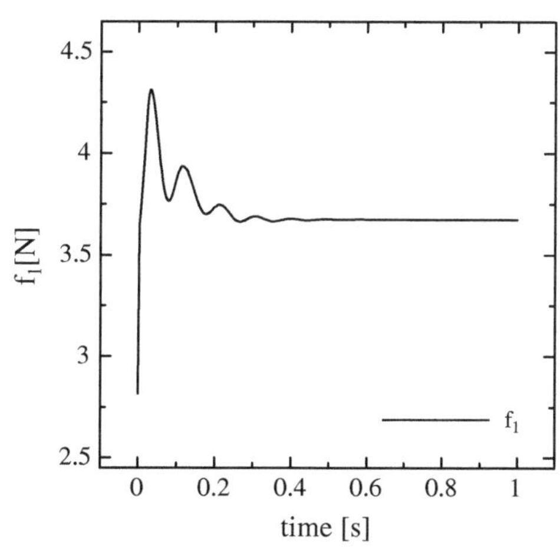

Fig. 15 Trajectory of f_1

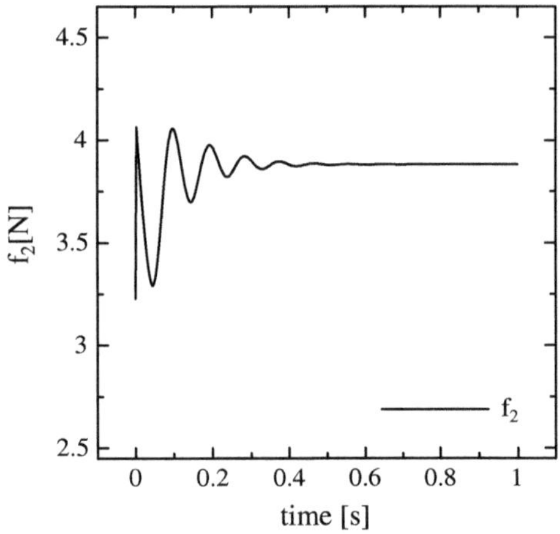

Fig. 16 Trajectory of $f2$

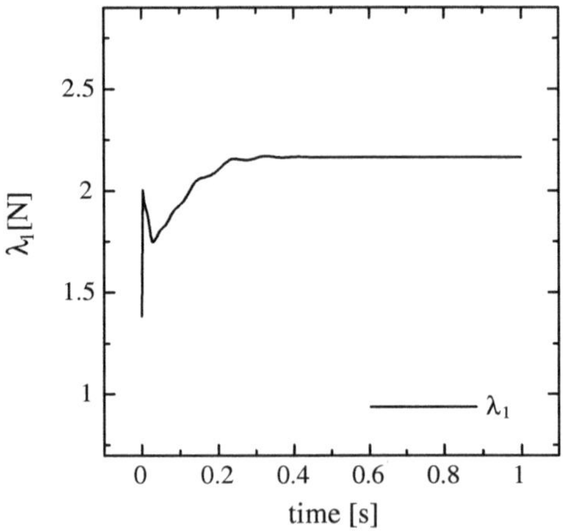

Fig. 17 Trajectory of λ_1

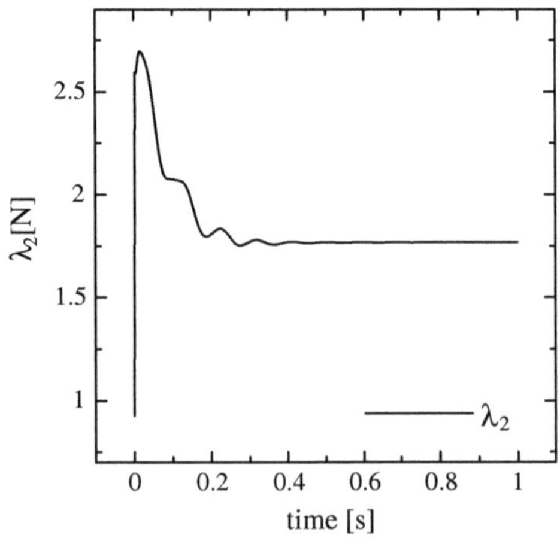

Fig. 18 Trajectory of λ_2

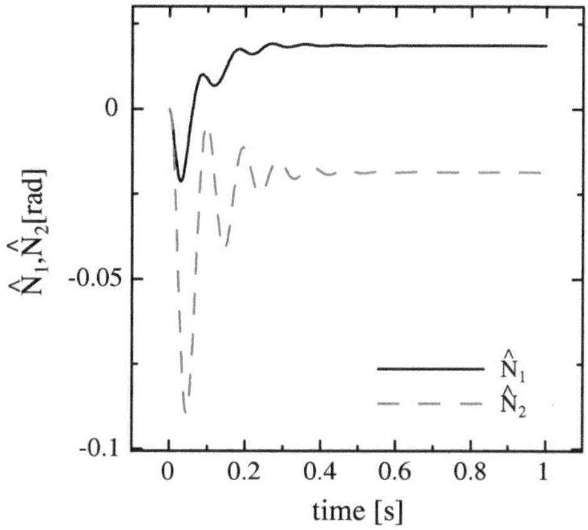

Fig. 19 Trajectories of $\hat{N}_1$ and $\hat{N}_2$

$$r_P = \begin{pmatrix} \cos\frac{\pi}{2} & \sin\frac{\pi}{2} \\ -\sin\frac{\pi}{2} & \cos\frac{\pi}{2} \end{pmatrix} \left(\frac{\overleftarrow{P_1P_2}}{\| \overleftarrow{P_1P_2} \|} \right), \quad r_O = \frac{\overleftarrow{O_1O_2}}{\| \overleftarrow{O_1O_2} \|} \tag{60}$$

where r_P is a unit vector rotating 90 degrees with respect to a unit vector connecting two contact points P_1 and P_2 between the fingerends and the object, and r_O is a unit vector connecting O_1 and O_2. Figure 20 shows that $r_P^{\mathrm{T}} r_O$ converged to zero. At the same time, Figs. 9 and 20 show that $\overline{P_1P_2}$ converged to be parallel to $\overline{O_1O_2}$ after 1 s, although $\overline{P_1P_2}$ was not parallel to $\overline{O_1O_2}$ at the initial composition shown in Fig. 8. Therefore, these results validate that the blind control scheme is effective to establish stable grasping under rolling contact constraints and smooth geometry of fingerends and objects.

In addition to stabilization of a grasped object, we try to control the object orientation angle θ toward a desired value θ_d, provided $\theta(t)$ can be measured real time. Based on the principle of superposition, we examine a control signal of the form

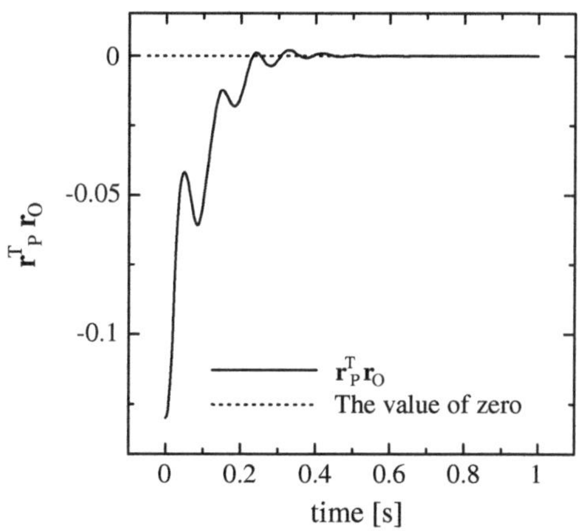

Fig. 20 Trajectory of $r_P^{\mathrm{T}} r_O$

$$u_i = (-1)^i k J_i^T(q_i)(r_1 - r_2) - c_i \dot{q}_i - \bar{\alpha}_i \{p_i(t) - p_i(0)\} e_i - k_I \int_0^t \Delta\theta d\tau, \quad i = 1, 2 \tag{61}$$

where $\Delta\theta = \theta - \theta_d$ and θ_d denotes the desired angle of an object. In the simulations to check the validity of the control input (61), the same initial pose shown in Fig. 8 was used. Figure 21 shows the pose after 7 s. Figure 22 shows that an object angle θ converged to the desired angle θ_d. Figure 23 shows that $r_P^T r_O$ converged to zero. Figures 21 and 23 show that $\overline{O_1 O_2}$ was parallel to $\overline{P_1 P_2}$ after 7 s. This shows the geometrical condition showing the equilibrium state of the system was satisfied while realizing the desired angle. It was confirmed that the proposed orientation control is effective from the viewpoint of numerical simulations (Table 3).

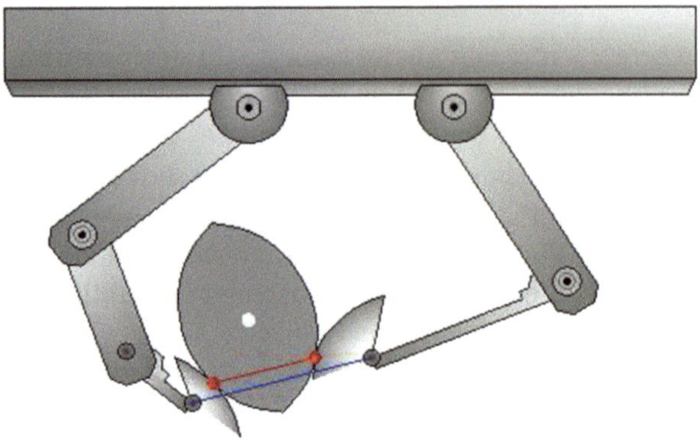

Fig. 21 Composition after 7 s

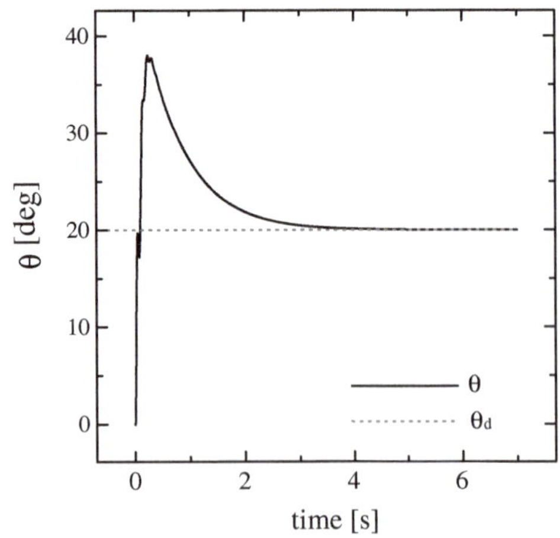

Fig. 22 Trajectory of θ

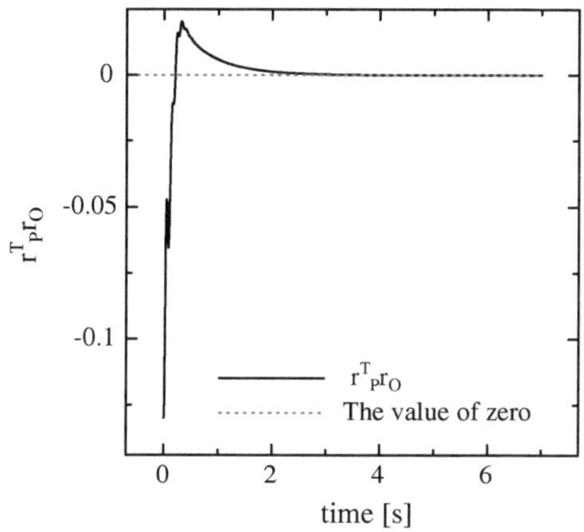

Fig. 23 Trajectories of $r_P^{\mathrm{T}} r_O$

Table 3 Parameters of control signals

k	Position feedback gain	75.000 [N/m]
k_I	Integral feedback gain	1.000 [Nm/s]
c_i	Damping coefficient	0.006 [Nms]
$\bar{\alpha}_i$ $(i = 1, 2)$	Gain	1.0 [Nm]
θ_d	Desired angle	20.0 [°]

6 Modeling of 3D Object Grasping Under Rolling Contacts

This section presents a computable modeling of 3D grasping of a rigid object with arbitrary smooth surfaces under the following set of postulates mathematically characterizing 3D rolling contact constraints:

(a1) Two contact points on each curved surface must coincide at a single common point without mutual penetration.
(a2) The two curved surfaces have the same tangent plane at the common contact point, that is, each surface has the same unit normal with mutually opposite direction at the common contact point, and
(a3) the two path lengths running on their corresponding surfaces must be coincident.

Let us consider a physical situation when a pair of multi-joint robot fingers is grasping a 3D rigid object as shown in Fig. 24. In the figure, the inertial frame denoted by $O - xyz$ is fixed in the euclidean space E^3 and local coordinates systems $O_i - X_iY_iZ_i$ for $i = 1, 2$ are introduced at each robot fingerend. The local coordinates system of the object is denoted by $O_m - XYZ$, where O_m stands for the object mass center. Next, denote each locus of points of contact between the two surfaces by a curve $\gamma_i(s_i)$ (3-dimensional vector in E^3) with length parameter s_i on its corresponding surface S_i $(i = 0, 1)$, where $i = 0$ signifies the object and $i = 1$

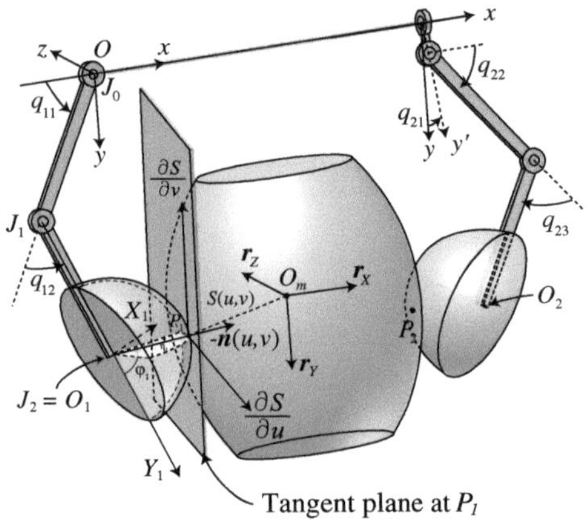

Fig. 24 A pair of robot fingers grasping a 3D rigid object with smooth surfaces

does the left hand fingertip. It is possible to assume that, given a curve $\gamma_0(s_0)$ as a locus of contact points on S_0 and another curve $\gamma_1(s_1)$ on S_1, the two curves coincide at contact point P_1 and share the same tangent plane T_1 at P_1 (see Fig. 25). During continuation of rolling contact, the two curves can be described in terms of the same length parameter s in such a way that $s_0 = s + c_0$ and $s_1 = s + c_1$, where c_0 and c_1 are constant.

Suppose that at some s of the length parameter the two curves $\gamma_0(s_0)$ and $\gamma_1(s_1)$ coincide at $P_1(s)$. Since $\gamma_0(s_0)$ is described in $O_m - XYZ$, its expression in the frame coordinates is given by

$$\bar{\gamma}_0(s_0) = \Pi_0 \gamma_0(s_0) \tag{62}$$

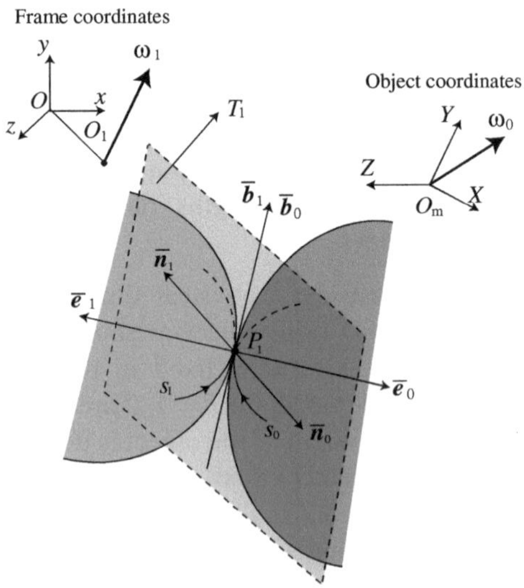

Fig. 25 Definition of the moving frame coordinates system centering at the rolling contact point

where Π_0 denotes a 3×3 rotational matrix composed of three unit orthogonal vectors r_X, r_Y, and r_Z that are expressed in the inertial frame $O - xyz$ as shown in Fig. 24, that is, $\Pi_0 = (r_X,\ r_Y,\ r_Z)$. Since $\gamma_0(s_0)$ is parameterized by length parameter, its derivative in s_0, i.e., $\gamma_0'(s_0) = \mathrm{d}\gamma_0(s_0)/ds_0$ must be expressed by the unit tangent vector $b_0(s_0)$ at $P_1(s)$ lying on the tangent plane T_1. Similarly, $b_1(s_1) = \gamma_1'(s_1)(= \mathrm{d}\gamma_1(s_1)/\mathrm{d}s_1)$. According to a1) and a2), it is possible to suppose that there exist the two unit normals $n_0(s_0)$ and $n_1(s_1)$ expressed in corresponding local coordinates $O_\mathrm{m} - XYZ$ and $O_1 - \mathrm{X_1Y_1Z_1}$ respectively (see Fig. 25). Then, it is possible to certify that

$$\overline{n}_0 = -\ \overline{n}_1 \ \text{ and } \ \overline{b}_0 = -\ \overline{b}_1 \tag{63}$$

at $s_0 = s_1 = s$, where $\overline{n}_0 = \Pi_0\ n_0$, $\overline{b}_0 = \Pi_0\ b_0$, $\overline{n}_1 = \Pi_1\ n_1$, and $\overline{b}_1 = \Pi_1\ b_1$ and $\Pi_1 = (\ r_{X1},\ r_{Y1},\ r_{Z1})$, r_{X1} denotes the unit vector of X_1-axis of $O_1 - X_1Y_1Z_1$ expressed in the inertial frame coordinates, and r_{Y1} and r_{Z1} have a similar meaning.

In what follows, let us denote the derivative Π_i in time t by $\dot{\Pi}_i$ and similarly the derivatives of vectors r_i in t by r_i, where r_1 denotes the position vector of O_m in the inertial frame coordinates. At this stage, it is important to note that

$$\dot{\Pi}_0 = \Pi_0\Omega_0,\quad \Omega_0 = \begin{pmatrix} 0 & -\omega_Z & \omega_Y \\ \omega_Z & 0 & -\omega_X \\ -\omega_Y & \omega_X & 0 \end{pmatrix} \tag{64}$$

For convenience, let us define $\omega = (\omega_X, \omega_Y, \omega_Z)^\mathrm{T}$. In the illustrative case of a spherical left hand fingertip shown in Fig. 24, we have

$$\dot{\Pi}_1 = \Pi_1\Omega_1,\quad \Omega_1 = \begin{pmatrix} 0 & \dot{p}_1 & 0 \\ -\dot{p}_1 & 0 & 0 \\ 0 & 0 & 0 \end{pmatrix} \tag{65}$$

where $\dot{p}_1 = \dot{q}_{11} + \dot{q}_{12}$ because both the rotational axes of joints J_1 and J_2 have the same direction in z-axis of $O - xyz$.

Let us now interpret the first postulate a1) in a mathematical form described by the vector form:

$$r_1 + \Pi_1\gamma_1(s_1) = \ r_0 + \Pi_0\gamma_0(s_0) \tag{66}$$

Then, the derivative of this form in t reduces to

$$(\dot{r}_1 - \ \dot{r}_0) + \dot{\Pi}_1\gamma_1 + \ \overline{b}_1\left(\frac{\mathrm{d}s_1}{\mathrm{d}t}\right) = \dot{\Pi}_0\gamma_0 + \ \overline{b}_0\left(\frac{\mathrm{d}s_0}{\mathrm{d}t}\right) \tag{67}$$

Since, during rolling contact motion, $\overline{b}_0 = \ \overline{b}_1$ and $\mathrm{d}s_1/\mathrm{d}t = \mathrm{d}s_0/\mathrm{d}t (= \mathrm{d}s/\mathrm{d}t)$, Eq. (67) reduces to

$$(\dot{r}_1 - \ \dot{r}_0) + \Pi_1\Omega_1\gamma_1 - \Pi_0\Omega_0\gamma_0 = 0 \tag{68}$$

Similarly to the 2D case, multiplication of the rotation matrix of the moving frame coordinates defined by $\Pi_0\Psi_0$ from the right yields

$$\begin{aligned}(R_{n1}, R_{b1}, R_{n1}) &\triangleq (\dot{r}_1 - \dot{r}_0)^{\mathrm{T}}\Pi_0\Psi_0 - \omega^{\mathrm{T}}(\gamma_0 \times \Psi_0) + \omega_1^{\mathrm{T}}(\gamma_1 \times \Psi_1^*) \\ &= 0\end{aligned} \tag{69}$$

where $e_0 = n_0 \times b_0$, $e_1 = n_1 \times b_1$,

$$\Psi_i = (n_i,\ b_i,\ e_i),\ \textit{for}\ i = 0, 1,\ \Psi_1^* = (-\,n_1,\ b_1, -\,e_1) \tag{70}$$

and $\gamma_0 \times \Psi_0 = (\gamma_0 \times n_0,\ \gamma_0 \times b_0,\ \gamma_0 \times e_0)$ and $\gamma_1 \times \Psi_1^*$ has a similar meaning. Equation (69) implies the three equations $R_{n1} = 0$, $R_{b1} = 0$, and $R_{e1} = 0$ that constitute the set of three Pfaffian forms expressing the rolling contact constraint of zero relative velocity. In the recent paper [2], it is shown that the Pfaffian forms of (69) are integrable with the integral forms

$$\frac{\mathrm{d}}{\mathrm{d}t}(Q_{n1},\ Q_{b1},\ Q_{e1}) = (R_{n1},\ R_{b1},\ R_{e1}) = 0 \tag{71}$$

where

$$\begin{aligned}(Q_{n1}, Q_{b1}, Q_{e1}) &= (r_1 - r_0)^{\mathrm{T}}\Pi_0\Psi_0 - \gamma_1^{\mathrm{T}}\Psi_1^* - \gamma_0^{\mathrm{T}}\Psi_0 \\ &= 0\end{aligned} \tag{72}$$

By virtue of the integrability of each Pfaffian form, the Lagrangian of the system is written into

$$L = K - Mgy - \sum_{i=1,2} P_i(q_i) - \sum_{i=1,2} (f_i Q_{ni} + \lambda_i Q_{bi} + \xi_i Q_{ei}) \tag{73}$$

where

$$K = \sum_{i=1,2} \frac{1}{2}\dot{q}_i^{\mathrm{T}} g_i(q_i)\dot{q}_i + \frac{M}{2}\parallel \dot{x} \parallel^2 + \frac{1}{2}\ \omega^{\mathrm{T}} H\ \omega \tag{74}$$

and H denotes the inertia tensor (3×3 matrix) of the object around its mass center, P_i the potential energy of finger i, and f_i, λ_i, and ξ_i express Lagrange multipliers corresponding to constraints $(Q_{n1}, Q_{b1}, Q_{e1}) = (0, 0, 0)$, and $x = \mathrm{d}r/\mathrm{d}t$. Then, by applying the variational principle to L, we obtain the following set of Euler–Lagrange equations of motion of the system:

$$M\ddot{x} - \sum_{i=1,2} (f_i\ \overline{n}_i + \lambda_i\ \overline{b}_i + \xi_i\ \overline{e}_i) - Mge_y = 0 \tag{75}$$

$$H\dot{\omega} + \omega \times H\omega - \sum_{i=1,2} \gamma_{0i} \times (f_i\ n_{0i} + \lambda_i\ b_{0i} + \xi_i\ e_{0i}) = 0 \tag{76}$$

$$
\begin{aligned}
&G_i \ddot{q}_i + \left\{ \frac{1}{2} \dot{G}_i + S_i \right\} \dot{q}_i + \frac{\partial P_i}{\partial q_i} + J_i^{\mathrm{T}} \{ f_i \, \overline{n}_i + \lambda_i \, \overline{b}_i + \xi_i \, \overline{e}_i \} \\
&\quad - W_i^{\mathrm{T}} \{ \gamma_i \times (f_i \, n_i + \lambda_i \, b_i + \xi_i \, e_i) \} = u_i, \quad i = 1, 2
\end{aligned}
\tag{77}
$$

where $e_y = (0, 1, 0)^{\mathrm{T}}$ and W_i can be given in the illustrative example of Fig. 24 as follows:

$$
W_1 = \begin{pmatrix} 0 & 0 \\ 0 & 0 \\ 1 & 1 \end{pmatrix}, \quad W_2 = \begin{pmatrix} 1, & 0 & 0 \\ 0 & \sin q_{21} & \sin q_{21} \\ 0 & \cos q_{21} & \cos q_{21} \end{pmatrix}
\tag{78}
$$

It should be noted that the 6-dimensional vectors appearing in (75) and (76)

$$
w_{fi} = \begin{pmatrix} \overline{n}_{0i} \\ \gamma_{0i} \times \, n_{0i} \end{pmatrix}, \quad w_{\lambda i} = \begin{pmatrix} \overline{b}_{0i} \\ \gamma_{0i} \times \, b_{0i} \end{pmatrix}, \quad w_{\xi i} = \begin{pmatrix} \overline{e}_{0i} \\ \gamma_{0i} \times \, e_{0i} \end{pmatrix}
\tag{79}
$$

are regarded as a wrench vector affecting motion of the object at contact point P_i for $i = 1, 2$. In fact, a steady state of the object when $x = 0$, $\omega = 0$, and $\omega = 0$ is established when the force/torque of the object is achieved in such a manner that, in accordance with (71) and (72),

$$
\sum_{i=1,2} \left(f_i \, w_{fi} + \lambda_i \, w_{\lambda i} + \xi_i \, w_{\xi i} \right) + Mg \begin{pmatrix} e_y \\ 0_3 \end{pmatrix} = 0
\tag{80}
$$

Finally it is necessary to derive the update law of path length parameter $s_1 (= s_0 + c_0)$ from the assumption of (63) and $\mathrm{d}\, n_0 / \mathrm{d}t = -\mathrm{d}\overline{n}_1 / \mathrm{d}t$. Since it follows that

$$
\begin{aligned}
\frac{\mathrm{d}}{\mathrm{d}t} \overline{n}_0 &= \dot{\Pi}_0 n_0 + \Pi_0 \dot{n}_0 = \Pi_0 \Omega_0 n_0 - \kappa_{n0} \Pi_0 b_0 \frac{\mathrm{d}s_0}{\mathrm{d}t} \\
&= \left(\omega_0^{\mathrm{T}} e_0 - \kappa_{n0} \frac{\mathrm{d}s_0}{\mathrm{d}t} \right) \overline{b}_0 - \left(\omega_0^{\mathrm{T}} b_0 \right) \overline{e}_0
\end{aligned}
\tag{81}
$$

$$
\frac{\mathrm{d}}{\mathrm{d}t} n_1 = \left(\omega_1^{\mathrm{T}} e_1 - \kappa_{n1} \frac{\mathrm{d}s_1}{\mathrm{d}t} \right) \overline{b}_1 - \left(\omega_1^{\mathrm{T}} b_1 \right) \overline{e}_1
\tag{82}
$$

we obtain

$$
(\kappa_{n0} + \kappa_{n1}) \frac{\mathrm{d}s_1}{\mathrm{d}t} = \omega^{\mathrm{T}} e_0 + \omega_1 \mathrm{T} e_1
\tag{83}
$$

provided that $\omega_0^{\mathrm{T}} b_0 = \omega_1^{\mathrm{T}} b_1$, where κ_{n0} denotes the normal curvature of the object surface at contact point P_1 in the direction of b_0 and κ_{n1} does that of the fingerend surface at P_1 in the direction of b_1. In order to update the moving frame coordinates appearing in (75) to (77), we need to take integration of the Frenet-Serret equations

$$\frac{\mathrm{d}}{\mathrm{d}t}\Psi_i = \Psi_i \begin{pmatrix} 0 & -\kappa_{ni} & 0 \\ \kappa_{ni} & 0 & \kappa_{ei} \\ 0 & -\kappa_{ei} & 0 \end{pmatrix} \frac{\mathrm{d}s_i}{\mathrm{d}t}, \quad i = 1, 2 \tag{84}$$

where κ_{e0} denotes the geodesic curvature of the object at P_1 and κ_{e1} that of the fingerend at P_1. Regardless that coefficients κ_{ni} can be determined from the surface geometry, the coefficients κ_{ei} must be determined via the instantaneous motion of rolling contact. In fact, we need the equality $\omega_0^{\mathrm{T}} b_0 = \omega_1^{\mathrm{T}} b_1$ as pointed out below (79). To assume this equality, it is possible to prove that if κ_{e0} and κ_{e1} are determined such that

$$\kappa_{e0}\frac{\mathrm{d}s_0}{\mathrm{d}t} = \frac{-\omega_1^{\mathrm{T}} e_1 \beta_1 + \xi_1}{\alpha_1}, \quad \kappa_{e1}\frac{\mathrm{d}s_1}{\mathrm{d}t} = \frac{-\omega_0^{\mathrm{T}} e_0 \beta_1 + \xi_1}{\alpha_1} \tag{85}$$

where

$$\begin{cases} \alpha_1 = \omega_1^{\mathrm{T}} e_1 + \omega_0^{\mathrm{T}} e_0, \beta_1 = \omega_0^{\mathrm{T}} n_0 + \omega_1^{\mathrm{T}} e_1 \\ \xi_1 = \left(-\kappa_{n0}\omega_0^{\mathrm{T}} n_0 + \kappa_{n1}\omega_1^{\mathrm{T}} n_1\right)\frac{\mathrm{d}s_0}{\mathrm{d}t} + \dot{\omega}_1^{\mathrm{T}} b_1 - \dot{\omega}_0^{\mathrm{T}} b_0 \end{cases} \tag{86}$$

then the equality $\omega_0^{\mathrm{T}} b_0 = \omega_1^{\mathrm{T}} b_1$ is maintained once at the initial time $\omega_0^{\mathrm{T}} b_0 = \omega_1^{\mathrm{T}} b_1$. The details of the derivation of (85) and (86) and the proof of sufficiency of (86) for keeping $\omega_0^{\mathrm{T}} b_0 = \omega_1^{\mathrm{T}} b_1$ are discussed in [2]. It should be noted that the geodesic curvatures must be determined dependently on angular accelerations $\dot{\omega}_i (i = 0, 1)$, which are subject to (76).

Thus, it is possible to conclude that the Euler–Lagrange equations of motion of the system are numerically integrable together with the update law of length parameters and the Frenet-Serret equations along running of each rolling contact point.

7 A Historical Note and Concluding Remarks

The importance of sensory feedback control for multi-finger hands from the viewpoint of everyday physics was addressed by one of the authors in an article [3] dedicated to a special memorial issue of International Journal of Robotics in memory of welcoming the third millennium. Toward the beginning of 21st century, Bicchi [4] spelled out the difficulty in unveiling secrets of dexterity of human multi-finger hands from the kinematics and control point of view. However, it is surprising to know that, far before the beginning of this century, the mathematical details of rolling contact constraints between two rigid bodies with arbitrary smooth surfaces were given by Montana [5] in a complete manner. In the well-known text book by Murray et.al. [6], the rolling contact constraint was defined rigorously by following Montana's formulation of a set of relations of velocity variables based on the condition of zero relative-velocity between the two velocity vectors at the contact point. Following Montana's velocity relations, Sarkar et al. [7] attempted to simulate dynamics of motion and control for a 3D object manipulation under rolling contacts. Since their proposed control was based on the computed torque and numerical simulation was carried out in a special case, there still remained unsolved what is a numerically computable model of motion equations of 3D grasping under rolling contacts. Limited to a simple case of 2D ball-plate grasping under rolling contacts, an Euler–Lagrange equation of motion of 2D object grasping and manipulation was first derived, that accompanies a sum of wrench vector forms explicitly exerting the objects [8]. Sensory feedback controls for stabilization of a grasp were proposed subsequently based on D'Alembert's principle [9, 10]. The idea of blind grasping was first proposed in [11] and extended to a more general case [12]. Extensions of modeling and control of 2D grasping to the case of an arbitrary shape of fingerends and objects were presented very recently in [13, 14]. An Euler–Lagrange equation of 3D grasping accompanied with wrench vector forms was presented in [15] by showing integrability of Pfaffian forms expressing zero relative-velocity relations of rolling contacts. All update laws of length parameters and moving frame coordinates are presented very recently in [2], where the nonholonomic constraint of relative twisting motion around the contact normal is resolved into determination of a geodesic curvature of each 3D curve of contact points. However, to practice computer simulation in this line, there remains a lot of difficult problems such as a penetration checking between 3D rigid bodies to find a contact point and a common tangent.

References

1. J. Napier, *Hands* (Princeton University Press, Princeton, NJ, 1993)
2. S. Arimoto, M. Yoshida, A Mathematical and numerically integrable modeling of 3D object grasping under rolling contacts between smooth surfaces. Mod. Simul. Eng. **2011**, 7 (2011). doi:10.1155/2011/684034
3. S. Arimoto, Robotics research toward explication of everyday physics. Int. J. Robot. Res. **18**(11), 1056–1063 (1999)
4. A. Bicchi, Hands for dexterous manipulation and robust grasping: a difficult road toward simplicity. IEEE Trans. Robot. Autom. **16**(6), 652–662 (2000)
5. D.J. Montana, The kinematics of contact and grasp. Int. J.of Robot. Res. **7**(3), 17–32 (1988)
6. R.M. Murray, Z. Li, S.S. Sastry, *A Mathematical Introduction to Robotic Manipulation* (CRC Press, Boca Raton, 1994)
7. N. Sarkar, X. Yun, V. Kumar, Dynamic control of 3-D rolling contacts in two-arm manipulation. IEEE Trans. Robot. Autom. **13**(3), 364–376 (1997)
8. S. Arimoto, P.T.A. Nguyen, H.Y. Han, Z. Doulgeri, Dynamics and control of a set of dual fingers with soft tips. Robotica **18**(1), 71–80 (2000)
9. S. Arimoto, K. Tahara, M. Yamaguchi, P.T.A. Nguyen, H.-Y. Han, Principle of superposition for controlling pinch motions by means of robot fingers with soft tips. Robotica **19**(1), 21–28 (2001)
10. S. Arimoto, Intelligent control of multi-fingered hands. Ann. Rev. Control **28**(1), 75–85 (2004)

11. R. Ozawa, S. Arimoto, S. Nakamura, B. Ji-Hun, Control of an object with parallel surfaces by a pair of finger robots without object sensing. IEEE Trans. Robot. **21**(5), 965–976 (2005)
12. S. Arimoto, R. Ozawa, M. Yoshida, Two-dimensional stable blind grasping under the gravity effect'', in *Proceedings of the 2005 IEEE International Conference on Robotics and Automation*, pp. 1208–1214 (2005)
13. S. Arimoto, M. Yoshida, M. Sekimoto, K. Tahara, Modeling and control of 2-D grasping of an object with arbitrary shape under rolling contact. SICE J. Control Meas.Syst. Integr. **2**(6), 379–386 (2009)
14. S. Arimoto and M. Yoshida, Modeling and control of 2D grasping under rolling contact constraints between arbitrary shapes: a riemannian-geometry approach. J. Robot. **2010,** 13 (2010) Article ID 926579. doi:10.1155/2010/926579
15. S. Arimoto, Dynamics of grasping a rigid object with arbitrary smooth surfaces under rolling contacts. SICE J. Control Meas. Syst. Integ. **3**(3), 199–205 (2010)

Chapter 16
Force Perception of Human Finger Using a Multi-Fingered Haptic Interface

Takahiro Endo and Haruhisa Kawasaki

Abstract In contrast with single-point haptic interfaces, multi-fingered haptic interfaces have great potential to dramatically increase the believability of the haptic experience. When a user uses a multi-fingered haptic interface, the user receives a force sensation through each finger holder. The force sensation is derived from the contact surfaces between the human fingertips and the finger holders. Therefore, to increase the believability of the haptic experience of the multi-fingered haptic interface, it is necessary to investigate and clarify the human perception of the force presented by the haptic interface. Here, we introduce the multi-fingered haptic interface robot HIRO III that was developed by our research group, and we then investigate the human perception of fingertip force using HIRO III.

Keywords Haptic interfaces · Multiple-fingers · Human factors · Force perception · Fingertip

1 Introduction

Haptic interfaces are devices that allow human-machine interaction through force and touch. These interfaces have been utilized in areas such as telemanipulation, interaction with micro/nanoscale phenomena, and medical training and evaluation. The multi-fingered haptic interfaces that allow multipoint contact between users

T. Endo (✉) · H. Kawasaki
Department of Mechanical Engineering, Gifu University, 1-1 Yanagido, Gifu 501-1193, Japan
e-mail: tendo@gifu-u.ac.jp

H. Kawasaki
e-mail: h_kawasa@gifu-u.ac.jp

R. Balasubramanian and V. J. Santos (eds.), *The Human Hand as an Inspiration for Robot Hand Development*, Springer Tracts in Advanced Robotics 95,
DOI: 10.1007/978-3-319-03017-3_16,

and a virtual environment have greater potential for various applications than do single-point haptic interfaces. For example, the multipoint interaction allows a user to perform natural actions such as grasping, manipulation, and exploration of virtual objects, and such interaction will dramatically increase the believability of the haptic experience [1–3]. In performing activities in our daily lives, we usually use multiple fingers and grasp and manipulate objects with dexterity. Furthermore, the exploration of an object is affected by the number of fingers used, and exploration with multiple fingers is more effective than exploration with a single finger [4, 5]. Thus, it is important to be able to exert force at multiple fingertips to generate a sensation that is highly realistic to human beings. In addition, since we are familiar with activities that involve multiple fingers, the multi-fingered haptic interface can be used naturally and without extensive familiarization.

Based on these considerations, the development of a multi-fingered haptic interface has been eagerly anticipated, and we have developed a multi-fingered haptic interface robot, named HIRO III ("Haptic Interface RObot III"), which can present three-directional forces at five human fingertips [6]. When a user uses a multi-fingered haptic interface, the user receives a force sensation through each finger holder, which connects the human fingertips and the haptic fingers of the multi-fingered haptic interface. This means that the force sensation is derived from the contact surfaces between the human fingertips and the finger holders. Therefore, to increase the believability of the haptic experience of the multi-fingered haptic interface, it is necessary to investigate and clarify the human perception of the force presented by the haptic interface. Recently, many systems based on the human perceptual ability have been developed, and their effectiveness has been reported [7–9]. We believe that the measurement of the human perception ability and its application can be effective for developing the multi-fingered haptic interface.

The force that the human being exerts using the fingers or feels from the fingers can be expressed based on the magnitude and the direction of the force. Many studies of the human perception of force magnitude have been reported. Although there have been only a few studies about the perception of the force direction, the subject has been researched aggressively in recent years. In this chapter, we investigate the human perception of fingertip force using the multi-fingered haptic interface robot HIRO III. In particular, we report on the force perception of the human finger with regard to the magnitude and the direction of the force, and we evaluate the perception of the spatial fingertip force, that is, the force in three-dimensional space.

This chapter is organized as follows. In the next section, we review the studies about the human perception of the fingertip force that uses a haptic interface, and we introduce the multi-fingered haptic interface robot HIRO III that was developed by our research group in Sect. 3. Next, we introduce the human perception of the force direction in Sect. 4, and then we investigate the human perception of the force magnitude in Sect. 5. Finally, Sect. 6 contains our conclusions.

2 Related Works: Human Perception of the Fingertip Force that Uses Haptic Interface

With the development of the haptic technology, the measurement of various perceptions, such as force, torque, force and torque, and position, has become possible. Here, we review the existing studies about the human perception ability of the fingertip force that uses a haptic interface. Research of the human perception ability with regard to force has a long history, and many studies of the human perception of force magnitude have been reported (see [7, 10–14] and the references therein). For example, Pang et al. [10] instructed participants to pinch a force display device using the thumb and the index finger. They examined a just noticeable difference (JND) of the reaction force presented at the fingertips, and they reported that the JND was in the range from 5 to 10 % and its average was 7 %. In their experiment, the examined force was one-dimensional force. In similar research, Yamakawa et al. [7] asked participants to pinch a force display device using the thumb and the index finger, and one-dimensional force was presented to the participants by the force display device, where the range of the presented force was 2–4.5 N. They reported that the difference threshold (DL) of the constant force was 0.39 N.

There are fewer studies about the perception of the fingertip force direction than about the perception of the force magnitude, but the perception of the force direction has been researched aggressively in recent years [15–17]. Barbagli et al. [15] asked participants to insert their index finger into the thimble of a Phantom, and the force was presented to the index finger. They reported that the DL of the human fingertip for the force direction was 25.6°. Tang et al. [16] investigated the perception of the force direction using the same experimental environments as were used by Barbagli et al. [15] and concluded that the DL of the human fingertip for the force direction was 33°. In both studies, the presented force was in the vertical plane, and the force was two-dimensional force. Kammerl et al. [17] investigated the perception of the force direction using force presented in the vertical plane and the horizontal plane, that is, three-dimensional force. However, in the experiment, they instructed the participants to grasp the end-effector of the Phantom, and the force was not presented at the fingertips. Thus, the measured perception ability was not for the fingertip but for the entire hand.

Based on these considerations, in this chapter we evaluated the force perception of the fingertip force magnitude and the direction using the spatial fingertip force, namely three-dimensional force.

3 Multi-Fingered Haptic Interface

A haptic interface consisting of an arm and fingertips can be used in a large workspace [18–21]. However, most of them consist of a hand-exoskeleton-and-arm or a hand-and-arm-exoskeleton system. With this system, it is difficult to present three-directional forces or the weight of virtual objects through the fingertips because the hand mechanism is mounted on the back of a human hand, and it is difficult to use the haptic interface for a long time because of its weight [3]. With these issues in mind, we have developed the multi-fingered haptic interface robot HIRO III [6], shown in Fig. 1, to meet the following points: the haptic interface must be safe, function in a wide space, and present not only three-directional forces at the user's five fingertips but also the sensations of the weights of virtual objects. In addition, it should neither cause an oppressive feeling when attached to the user's hand nor represent its own weight.

HIRO III is placed opposite a human hand and can present three-directional forces at five human fingertips. The specifications of HIRO III are shown in Table 1. This haptic interface consists of an interface arm and a five-fingered haptic hand. The interface arm is a PUMA-type robot arm consisting of an upper arm (humerus), a lower arm (forearm), and a wrist. The interface arm has 3 degrees of freedom (DOF) at the arm joint and 3 DOF at the wrist joint. The interface arm, therefore, has six joints allowing 6 DOF.

The haptic hand begins at the wrist but does not include it, and ends with the fingertips. A hand-base and five haptic fingers form the haptic hand. Each haptic finger has 3 joints, allowing 3 DOF. The first joint relative to the hand-base allows abduction/adduction, while the second and third joints allow flexion/extension.

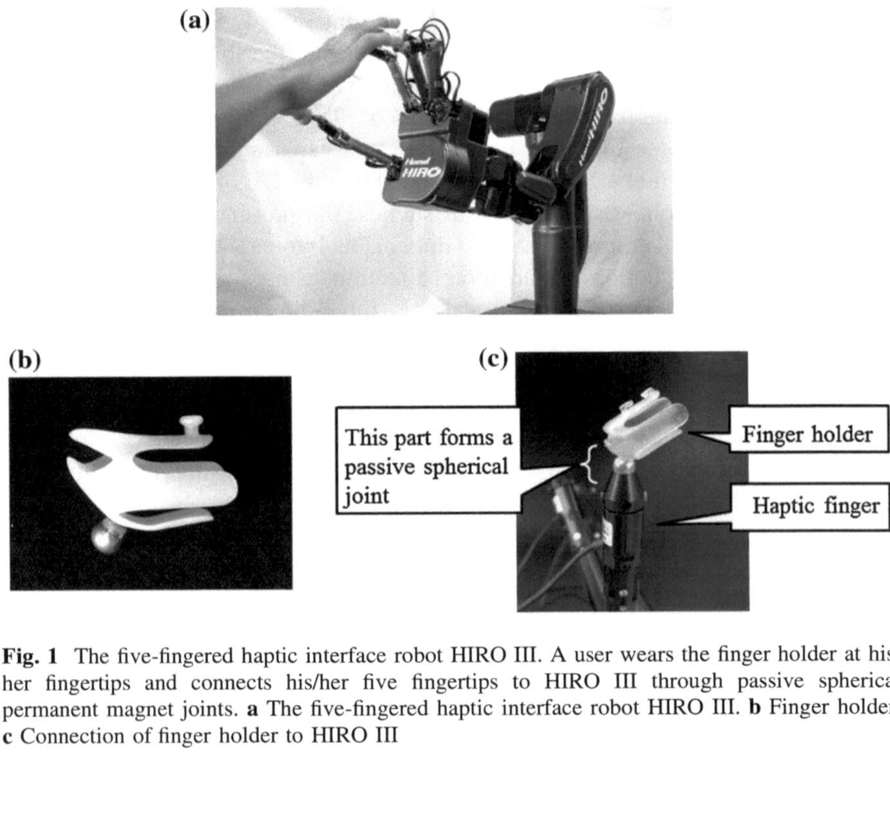

Fig. 1 The five-fingered haptic interface robot HIRO III. A user wears the finger holder at his/her fingertips and connects his/her five fingertips to HIRO III through passive spherical permanent magnet joints. **a** The five-fingered haptic interface robot HIRO III. **b** Finger holder. **c** Connection of finger holder to HIRO III

Table 1 Specifications of HIRO III

Degrees of freedom	Haptic hand	15 DOF (Number of haptic fingers: 5)
	Interface arm	6 DOF
Performance	Maximum output force of haptic finger	Over 3.6 N
	Maximum displayable stiffness	5 kN/m
	Frequency response	8 Hz
	Sampling time of control	1 kHz

Therefore, the total DOF of HIRO III is 21, and its work space covers virtual reality (VR) manipulation on the space of a desktop. A three-axis force sensor is installed at the top of each haptic finger to read the finger loading forces, and the measured forces are used in the control law of the haptic interface.

To manipulate HIRO III, the user has to wear a finger holder on each of his/her fingertips. Figure 1b and c show a finger holder and its connection to the haptic finger of HIRO III. Twelve types of finger holders were designed to adjust to various forms and sizes of an adult's fingers. The finger holder has a steel sphere which, when attached to the permanent magnet at the fingertip of the haptic finger, forms a passive spherical joint. Its role is to adjust for differences between the human and haptic finger orientations. This connection method ensures that HIRO III is safe to use and involves no oppressive feeling for the user, because the coupled part between the user and the haptic interface is centered only on the fingertips of the user. To summarize, the main features of HIRO III are as follows: It can present three-dimensional forces at the thumb and four fingertips of a human hand, and thus, it allows the grasping and manipulation of an object in a VR environment. Further, it allows for a large work space, and it minimizes discomfort in the human user at the connection. For more details, please see [6].

4 Human Perception of Fingertip Force Direction

In order to know how human beings accurately perceive fingertip force by the multi-fingered haptic interface, we consider the human perception ability with regard to the direction of fingertip force by using HIRO III. The perception results of the fingertip force direction were presented previously [22], and we summarize them briefly here.

4.1 Experimental Setup

Ten people in their twenties (nine males and one female) participated in this measurement. All of the participants were right-handed and connected their index finger to the index finger of HIRO III, as shown in Fig. 2, and HIRO III presented the force to the index finger of the participants. In this experiment, the hand part of HIRO III was fastened to a fixture, as shown in Fig. 2a, and the finger holder was fixed to the index finger of the participants using surgical tape, as shown in Fig. 2b. During the measurement, a cloth covered both HIRO III and the participant's hands, and thus sight information was not available, as shown in Fig. 2c. The participants responded to the direction of the presented force by using the measuring instrument, a goniometer, as shown in Fig. 2c, and we measured the angular error between the presented force direction by HIRO III and the answered angle by the measuring instrument. To measure the human perception of the force

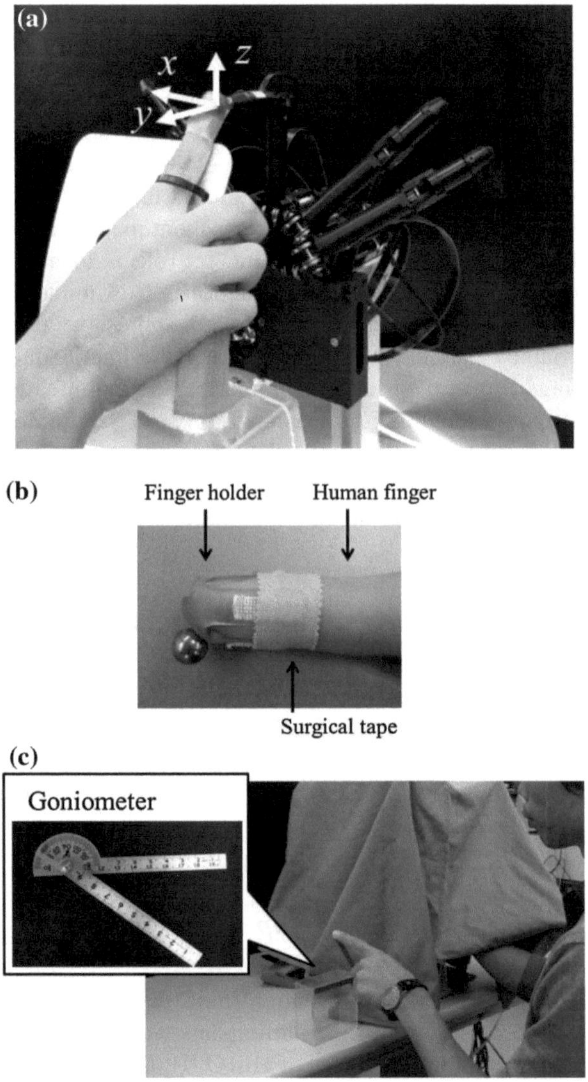

Fig. 2 Experimental environment. **a** Connection of HIRO III to the index finger **b** Installation of finger holder to the index finger. **c** Experimental setup and measuring instrument (a goniometer)

direction with regard to the spatial variation, we considered the force in the horizontal plane (the x–y plane in Fig. 2a) and in the vertical plane (the x–z plane in Fig. 2a). We divided the presented force in both plane every 15°, and 13 kinds of forces were presented to the participants in random order (Fig. 3c and d), where the magnitude of the presented force was 1.5 N. In the experiment, HIRO III presented the force to the participant after an operator of HIRO III gave the signal

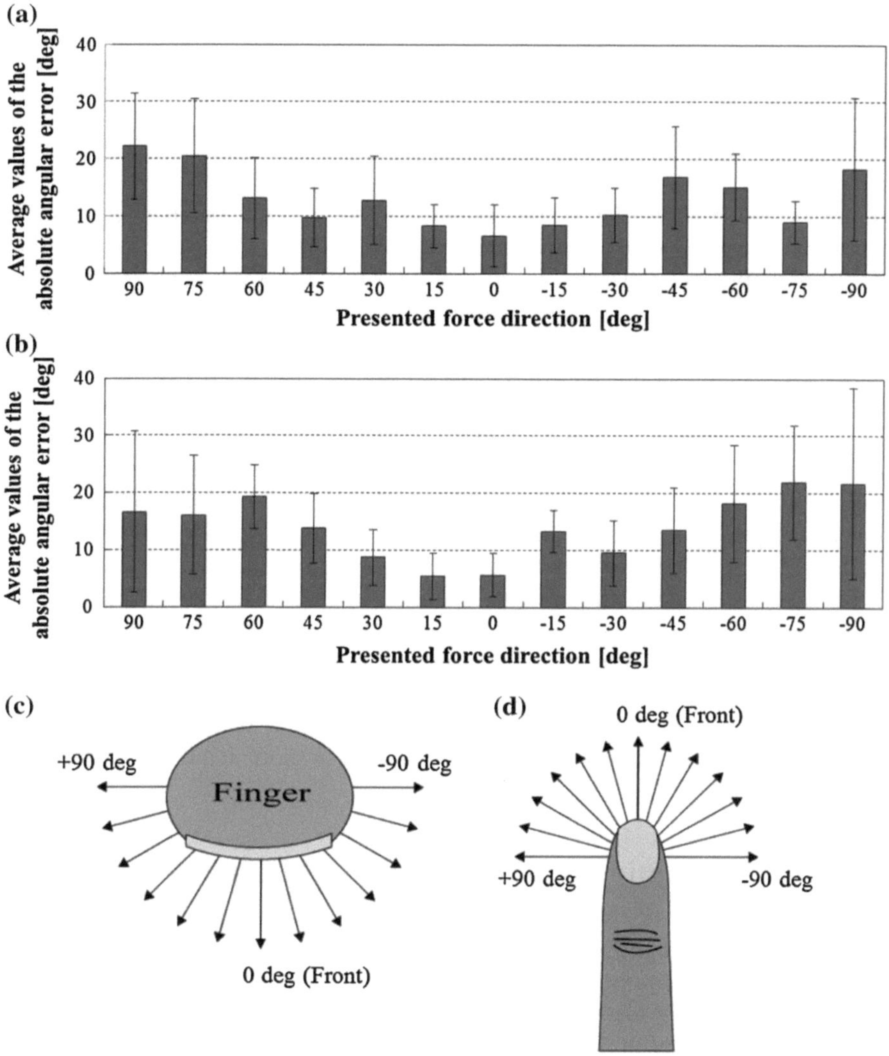

Fig. 3 Measurement results of the angular error at each presented force direction. **a** Average values of the absolute angular error in the horizontal plane. **b** Average values of the absolute angular error in the vertical plane. **c** The direction of the presented force in the horizontal plane. **d** The direction of the presented force in the vertical plane

to start, and the participant felt the force. After the participant recognized the force direction, he/she signaled the operator and the operator stopped the presentation of the force. The participant then indicated the direction of the presented force using the measuring instrument, a goniometer. As the control law of HIRO III, the manipulability-optimized control law was used. For the control, please see [6]. Here note that the settling time of HIRO III for the step response was less than 0.2 s, and the angular errors occurred in all the presented force directions (the 13 kinds of force in the horizontal and in the vertical plane) were very small, with an average value of 1.5 ± 1.3 (SD)°.

4.2 Experimental Results

Figure 3a and b show the average values of the absolute angular error in the horizontal plane and vertical plane, respectively. In the figures, the error bars show the 95 % confidence intervals. From Fig. 3a, we see that the angular error when the presented force at 0° is the smallest, and it grows as the presented angle approaches −90 and +90°. This is true in both the horizontal and the vertical planes. Here note that this tendency, anisotropy, was also described previously for the perception of the human hand (not the finger) [23]. In [23], the participants held a joystick, and the perception ability for the direction of the human hand was investigated. The anisotropy of the human hand perception regarding the direction was shown. Based on these findings, it seems that like the human hand, the human finger perception for the force direction has anisotropy. Whether the perception of the force magnitude influences this phenomenon is examined in the next section.

The average value of the absolute angular error in the horizontal plane was in the range from 6.6 to 22.2°, and its average was 13.2°. The average value of the absolute angular error in the vertical plane was in the range from 5.4 to 21.9°, and its average was 14.1°. The significant difference test with a 5 % significance level by t test indicated no significant difference between the average angular error in the horizontal plane and that in the vertical plane. For more details, please see [22].

5 Human Perception of Fingertip Force Magnitude

Regarding the magnitude and the direction, which are components of the force, we introduced the measurement of the human perception of the force direction in the previous section. We also measured the human perception of the fingertip force magnitude. In view of the influences of the force direction on the force magnitude, we measured the followings: (i) the difference threshold (DL) of the force magnitude at each direction, (ii) the point of subject equality (PSE) of the force magnitude when the presented force direction was known, and (iii) the PSE of the force magnitude when the presented force direction was unknown.

5.1 Difference Threshold of the Fingertip Force Magnitude

We evaluated the human perception of the fingertip force magnitude. In particular, based on the results in the previous section, we measure the DL of the fingertip force magnitude in both the horizontal and vertical planes.

Experimental Setup
Ten people in their twenties (nine males and one female) participated in this measurement. All of the participants were right-handed. The experimental environment was the same as that described in Sect. 4.1. To obtain the DL, we used the constant method. As in the experiment described in Sect. 4.1, the participant's index finger was connected to the index finger of HIRO III. The following three steps were carried out: (1) Using HIRO III, the standard stimulus force was presented to the participant; (2) Using HIRO III, the comparison stimulus force was presented to the participant, where the direction of the comparison stimulus force and that of the standard stimulus force were the same; (3) The participant selected one of the following three answers: (i) the comparison stimulus force was larger than the standard stimulus force, (ii) the comparison stimulus force was the same as the standard stimulus force, or was not distinguishable, (iii) the comparison stimulus force was smaller than the standard stimulus force. The experiment was carried out in the following directions: +90, +45, 0, −45, and −90° in the horizontal and vertical planes (Fig. 4e and f). Here note that the direction of ± 90° in the horizontal plane and that in the vertical plane were the same, thus only the directions of +45, 0, and −45° were measured in the vertical plane. In the measurement, the operator of HIRO III informed the participant about the direction of the presented force in advance, and we set the standard stimulus force at 2.2 N and considered the following eleven comparison stimulus forces: 1.0, 1.2, 1.4, 1.6, 1.8, 2.0, 2.2, 2.4, 2.6, 2.8, and 3.0 N. We set the standard stimulus force at 2.2 N to compare the results in Sects. 5.2 and 5.3 with the results in this subsection (see Sect. 5.2 for details). The force was presented until the participant answered. For each participant, all the comparison stimulus forces were presented, and the order of presentation was random.

Experimental Results
Figure 4 shows the experimental results. Figure 4a and c show the ratios of how often the participants answered "the comparison stimulus force was larger than the standard stimulus force" in the horizontal plane and the vertical plane, respectively. From the data in Fig. 4a and c, we obtained the psychometric function at each direction in the horizontal and the vertical plane, respectively. Here the psychometric functions were derived by fitting the sigmoid function $1/(1 + \exp(\alpha + \beta x))$, where x is the presented force magnitude, and α and β are real numbers. Then the upper difference threshold (UDL) and the lower difference threshold (LDL) were derived from the psychometric function, and the DL was derived by (UDL − LDL)/2. Figure 4b and d show the DLs in the horizontal plane

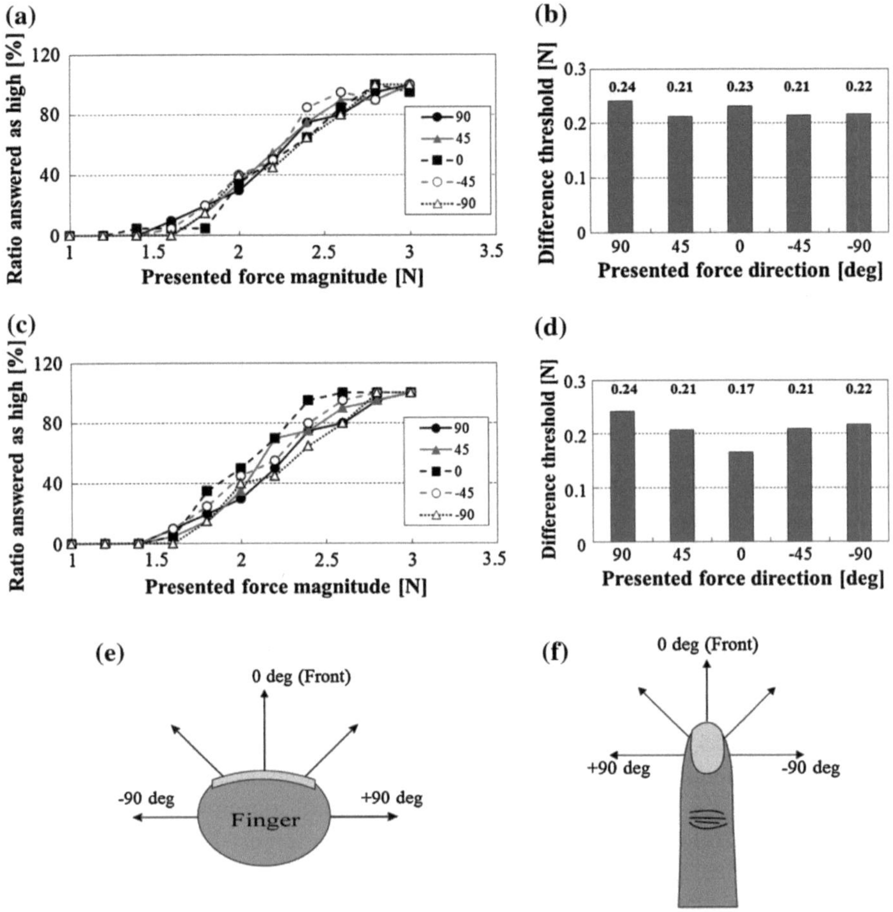

Fig. 4 Measurement results of the difference threshold of the fingertip force magnitude. **a** Ratio answered as high in the horizontal plane. **b** Difference thresholds in the horizontal plane. **c** Ratio answered as high in the vertical plane. **d** Difference thresholds in the vertical plane. **e** The direction of the presented force in the horizontal plane. **f** The direction of the presented force in the vertical plane

and the vertical plane, respectively. In the figures, the horizontal axis is the direction of the presented force, and the vertical axis is the DL.

From Fig. 4a and b, we see that the same tendency was obtained at all directions of the presented force in the horizontal plane, and all participants perceived the difference of force magnitude within the range from the standard stimulate force 2.2 N to the ± 0.8 N. In analysis of variance (ANOVA) with a 5 % significance level, there was no significant difference in the participants' answered ratios caused by the direction of the presented force. Therefore, the DL in the horizontal plane was 0.224 N, which was the average value of the difference thresholds at each direction. The standard deviation was 0.013 N, and this value was less than the resolution of the force sensor, 0.05 N, used here.

From Fig. 4c and d, we see that the results in the vertical plane differed from the results in the horizontal plane, and the DL grew slightly worse as the presented angle approached −90 and +90°. All participants perceived the difference of force magnitude within the range from the standard stimulate force of 2.2 N to the plus or minus 0.8 N. However, there was no significant difference between the ratios of how often the participants answered "the comparison stimulus force was larger than the standard stimulus force" in the vertical plane by ANOVA with a 5 % significance level. In addition, using the *t*-test with a 5 % significance level, we found no significant difference between the five DLs in the horizontal plane and the five DLs in the vertical plane. The DL in the vertical plane was 0.208 N, which was the average value of the DLs in the vertical plane, and the standard deviation was 0.027 N. This standard deviation was twice the standard deviation in the horizontal plane, but this value was less than the resolution of the force sensor, too. Here note that, in the following subsection, we consider the PSE using the obtained DL.

5.2 *PSE of the Fingertip Force Magnitude: Known Force Direction Case*

In the previous subsection, we measured the DL of the fingertip force at each direction. Here, we consider whether there is a difference in the perception of the fingertip force magnitude based on a difference of the force direction.

Experimental Setup

Ten people in their twenties (nine males and one female) participated in this measurement. The experimental environment was the same as that described in Sect. 4.1. Here we measured the PSE of the force magnitude. The results of this experiment were previously reported [22] and can be summarized as follows. As in the experiment of Sect. 4.1, the participant's index finger was connected to the haptic finger of HIRO III. Then the following three steps were carried out in the horizontal and vertical planes: (1) Using HIRO III, the standard stimulus force was presented to the participant, where the standard stimulus force was presented in the reference direction, (2) Using HIRO III, the comparison stimulus force was presented to the participant, where the comparison stimulus force was presented in a comparison direction, (3) The participant selected one of the following three answers: (i) the comparison stimulus force was larger than the standard stimulus force, (ii) the comparison stimulus force was the same as the standard stimulus force, or was not distinguishable, (iii) the comparison stimulus force was smaller than the standard stimulus force. The reference directions of the horizontal and vertical planes were 0° in Fig. 4e and f, respectively. The comparison directions with respect to the reference direction were the following five directions: +90, +45, 0, −45, and −90°. We set the standard stimulus force at 2.2 N and considered the following eleven comparison stimulus forces: 1.0, 1.2, 1.4, 1.6, 1.8, 2.0, 2.2, 2.4, 2.6, 2.8, and 3.0 N. Before the experiment, we conducted a preliminary experiment several times. From the preliminary experimental result, we found there were comparison directions for which we could not obtain the PSE when the standard stimulus force was set smaller than 2.2 N. Thus, to obtain the PSE in all comparison directions, we set the standard stimulus force to 2.2 N. For each participant, all the comparison stimulus forces were presented in the comparison directions, where the order of the presentation of the comparison stimulus force was random, and the comparison direction was selected from the above five directions before the measurement. After the measurement, we selected another comparison direction and then conducted measurements (1)–(3). Here note that the comparison direction was transmitted to the participant preliminarily, and thus the participant knew the direction of the presented force before the measurement.

Experimental Results

Figure 5 shows the measurement results. Figure 5a and c show the ratios of how often participants answered "the comparison stimulus force was larger than the standard stimulus force" in the horizontal and the vertical plane, respectively. From these data, we derived the psychometric functions at each direction, and then we obtained the PSE by (UDL + LDL)/2. Figure 5b and d show the PSE at each direction in the horizontal and the vertical planes, respectively. From Fig. 5a and b, we see that the PSEs in the horizontal plane have the maximum difference of

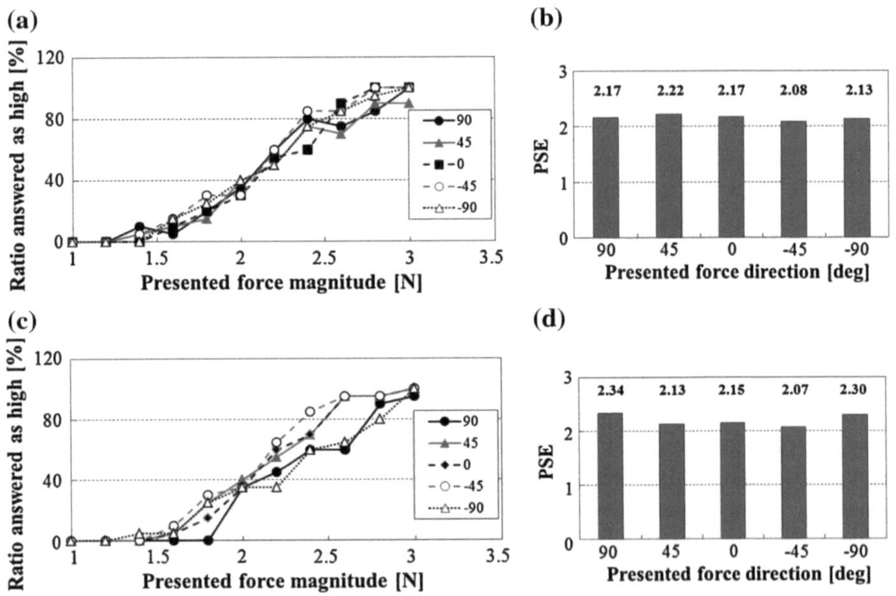

Fig. 5 Measurement results of the PSE at each direction of the presented force, where the direction was known. **a** Ratio answered as high in the horizontal plane. **b** PSE in the horizontal plane. **c** Ratio answered as high in the vertical plane. **d** PSE in the vertical plane

0.138 N, but from the results in Sect. 4.1, we see that the 0.224 N force or less in the horizontal plane is in the interval of uncertainty, and thus, 0.138 N is not perceivable. The constant error (CE) is defined as CE = PSE − POS, where POS is a point of objective equality, and CEs at each direction were −0.035 N for 90°, 0.020 N for 45°, −0.028 N for 0°, −0.119 N for −45°, and −0.068 N for −90°. The maximum of the absolute CE was 0.119 N, and this value was also within the interval of uncertainty. Thus, we found that the accuracy of the judgment from the force magnitude in directions of 0° and that in other directions is high. We concluded that PSE in the horizontal plane was not changed by the difference of the force direction.

In the results in the vertical plane (Fig. 5c and d), the maximum difference between PSEs at each direction was 0.267 N, which was the difference between PSE at −45° and PSE at 90°. This value was larger than the DL in the vertical plane, 0.208 N. However, CEs at each direction were 0.137 N for 90°, −0.069 N for 45°, −0.049 N for 0°, −0.130 N for −45°, and 0.095 N for −90°, and all CEs were within the DL in the vertical plane. Further, by the *t*-test with a 5 % significance level, there was no significant difference between five PSEs in the horizontal plane and five PSEs in the vertical plane. Thus the accuracy of the judgment from the force magnitude in directions of 0° and that in other directions is high, and PSE in the vertical plane was not changed by the difference of the force direction, like the finding with the horizontal plane.

These findings showed that, in both the horizontal and vertical planes, there was no difference in the feeling of magnitude caused by the difference of the presented force direction, and when forces were presented from a variety of directions, the participants recognized that the forces had almost the same magnitude from one to the next.

5.3 PSE of the Fingertip Force Magnitude: Unknown Force Direction Case

From the experimental results in Sects. 5.1 and 5.2, we learned that the DL and the PSE in both the horizontal and vertical planes resulted in no difference in the feeling of magnitude caused by the difference of the presented force direction. However, in Sect. 4.1, the angular error grew as the presented angle approached −90 and +90° in both the horizontal and vertical planes. It is possible that the difference was not caused by the participant knowing the direction of the presented force beforehand in the experiments described in Sects. 5.1 and 5.2. Thus, in this experiment, we measured the PSE of the force magnitude with the direction of the presented force remaining unknown, and we examined the influence of the direction on the perception of the force magnitude. The measurements in this experiment were the same as those described in Sect. 5.2, with only the following point modified: The comparison direction was not transmitted to the participant preliminarily.

Experimental Setup
The experimental environment, the number of participants, and the experimental procedure were the same what was described in Sect. 5.2, with one exception, as follows: the measurement in Sect. 5.2 was carried out per the force direction, and thus the participants knew the presented force direction before the measurement, while in the experiment of this subsection, the force direction was presented in random order, and the participants did not know the direction before the measurement. In the measurement, all of the comparison stimulus forces and the reference directions were presented in random order.

Experimental Results
The measurement results are shown in Fig. 6. Figure 6a and c show the ratios of how often participants answered "the comparison stimulus force was larger than the standard stimulus force" in the horizontal and the vertical planes, respectively, and Fig. 6b and d show the PSE in the horizontal and vertical planes, respectively. The CEs in the horizontal plane were −0.088 N for 90°, −0.129 N for 45°, −0.119 N for 0°, −0.261 N for −45°, and −0.108 N for −90°, and the CEs in the vertical plane were 0.113 N for 90°, −0.058 N for 45°, −0.089 N for 0°, −0.170 N for −45°, and −0.069 N for −90°. The absolute values of CE were within the corresponding DL, except for −45° in the horizontal plane. The results in this subsection (Figure 6a and c) had great variability compared with the results in Sect. 5.2 (Fig. 5a and c). In fact, the variability of the CE, as shown by the

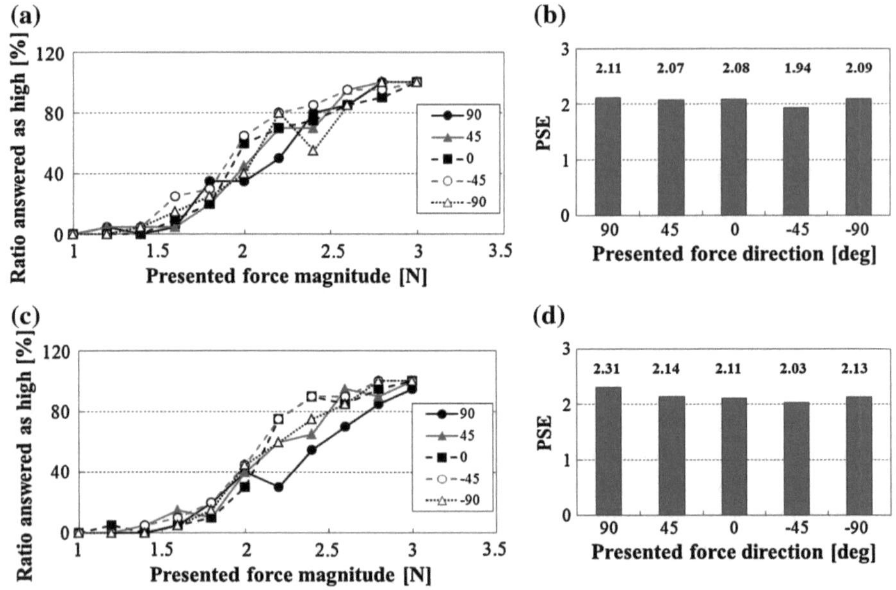

Fig. 6 Measurement results of the PSE at each direction of the presented force, where the direction was unknown. **a** Ratio answered as high in the horizontal plane. **b** PSE in the horizontal plane **c** Ratio answered as high in the vertical plane. **d** PSE in the vertical plane

maximum of CE minus the minimum of CE, was 0.267 N for the results in Sect. 5.2 and 0.374 N for the results in this subsection, and we consider that the small PSE at −45° in the horizontal plane is because of this variability. In the *t*-test with a 1 % significant level, there was a significant difference between ten PSEs in Sect. 5.2 and ten PSEs in this subsection. The only difference between the measurement method in this subsection and that in Sect. 5.2 was that the direction was unknown or known, and thus we believe that participants focused attention on the direction when the direction was known, which led to the higher accuracy of the perception in those tests. Based on these considerations, we conclude that the accuracy of the force perception worsens somewhat when the direction is unknown. In addition, since the perception of the force magnitude showed no difference between the presented force directions, the measurement results in Sect. 4.1, that is, that the angular error grew as the presented direction approached −90 and +90°, were not influenced by the perception of the force magnitude.

6 Conclusion

In this chapter, we described experiments that involved the multi-fingered haptic interface robot HIRO III. The haptic interface consisted of a five-fingered haptic hand and an interface arm. The interface was placed opposite a human hand and could present three-directional forces at the user's five fingertips. Thus, the interface allowed multipoint contact between user and a virtual environment, and it is utilized in a wide range of applications.

The user wore a finger holder while manipulating the multi-fingered haptic interface. The force sensation was derived from the contact surface between the human fingertip and the finger holder. With the goal of further developing the multi-fingered haptic interface, we investigated the human perception of the fingertips force using HIRO III. In particular, we measured the difference threshold and PSE of the fingertips force under the use of the finger holder, and we collected the fundamental data about the human perception ability of fingertip force using the multi-fingered haptic interface robot HIRO III.

The force consists of the magnitude and the direction, and thus we introduced the measurement results of the perception for the fingertip force direction. The average perception error with regard to the direction was 13.2° in the horizontal plane and 14.1° in the vertical plane, and it grew as the direction of the presented force approached −90 and +90°. Next, we examined the perception of the fingertip force magnitude. Our experimental results showed the following: (i) there was no difference in the DL caused by the difference of the presented force direction, and the average value was 0.22 N for the horizontal plane and 0.21 N for the vertical plane, (ii) there was no difference in the feeling of the magnitude caused by the difference of the presented force direction, and (iii) the human users focused their attention on the force direction when the direction was known beforehand, and this increased the accuracy of the force perception. We obtained these results of these tests of human perception by using the finger holder, and we consider this method to be an effective way to study the multi-fingered haptic interface.

The next problem to be tackled is to use these results in the skill transfer system. In the transferring of an expert skill, it takes a great deal of time and effort for beginners to obtain skills, and it is difficult to teach the skills by using only words. For these reasons, a skill transfer system that uses virtual reality and haptic interface technology is very attractive, and we have studied the skill transfer system using HIRO III [22]. By using the obtained results regarding the perception ability, we hope to improve the skill transfer system.

Acknowledgments This work was supported in part by SCOPE of the Ministry of Internal Affairs and Communications and by Grant-in-Aid for Young Scientists (B) (23700143) from the Japan Society for the Promotion of Science (JSPS).

References

1. N. Magnenat-Thalmann, U. Bonanni, Haptics in Virtual Reality and Multimedia. IEEE Multimedia **13**, 6–11 (2006)
2. F. Barbagli, K. Salisbury, R. Devengenzo, Toward virtual manipulation: from one point of contact to four. Sens. Rev. **24**, 51–59 (2004)
3. Z. Najdovski, S. Nahavandi, *Extending Haptic Device Capability for 3D Virtual Grasping*, in *Proceedings of EuroHaptics 2008*, Madrid, 2008 pp. 494–503
4. G. Jansson, M. Bergamasco, A. Frisoli, A new option for the visually impaired to experience 3D art at museums: manual exploration of virtual copies. Visual Impairment Res. **5**, 1–12 (2003)
5. S. McKnight, N. Melder, A. Barrow, W. Harwin, and J. Wann, Psychophysical *Size Discrimination using Multi-fingered Haptic Interfaces*, in *Proceedings of EuroHaptics 2004*, Munich, 2004 pp. 274–281
6. T. Endo, H. Kawasaki, T. Mouri, Y. Ishigure, H. Shimomura, M. Matsumura, K. Koketsu, Five-fingered haptic interface robot: HIRO III. IEEE Trans. Haptics **4**, 14–27 (2011)
7. S. Yamakawa, H. Fujimoto, S. Manabe, Y. Kobayashi, The necessary conditions of the scaling ratio in master-slave systems based on human difference limen of force sense. IEEE Trans. Syst. Man Cybern. Part A Syst. Hum. **35**(2), 275–282 (2005)
8. D. Botturi, M. Vincentini, M. Righele, C. Secchi, Perception-centric force scaling in bilateral teleoperation. Mechatronics **20**, 802–811 (2010)
9. P. Hinterseer, S. Hirche, S. Chaudhuri, E. Steinbach, M. Buss, Perception-based Data Reduction and Transmission of Haptic Data in Telepresence and Teleaction Systems. IEEE Trans. Signal Process. **56**(2), 588–597 (2008)
10. X.D. Pang, H.Z. Tan, N.I. Durlach, Manual discrimination of force using active finger motion. Perception and Psychophysics **49**(6), 531–540 (1991)
11. H.Z. Tan, M.A. Srinivasan, B. Eberman, B. Cheng, Human factors for the design of force-reflecting haptic interfaces, in *Proceedings of the 3rd International Symposium on Haptic Interfaces for Virtual Environment and Teleoperator Systems*, ASME Dynamic Systems and Control Division, vol. 55(1) (New York, 1994) pp. 353–359
12. S. Allin, Y. Matsuoka, and R. Klatzky, Measuring just noticeable differences for haptic force feedback: implications for rehabilitation, in *Proceedings of the 10th Symposium on Haptic Interfaces for Virtual Environment and Teleoperator Systems (HAPTICS 2002)*, 2002 pp. 299–303

13. L.A. Jones, Perception of force and weight: theory and research. Psychol. Bull. **100**(1), 29–42 (1986)
14. S.A. Cholewiak, H.Z. Tan, D.S. Ebert, Haptic identification of stiffness and force magnitude, in *Proceedings of Symposium on Haptic Interfaces for Virtual Environment and Teleoperator Systems (HAPTICS 2008)*, 2008 pp. 87–91
15. F. Barbagli, K. Salisbury, H.Z. Tan, Haptic discrimination of force direction and the influence of visual information. ACM Trans. Appl. Percept. **3**(2), 125–135 (2006)
16. H.Z. Tan, F. Barbagli, K. Salisbury, C. Ho, C. Spence, Force-direction is not influenced by reference force direction, Haptics-e, **4**(1) (2006)
17. J. Kammerl, R. Chaudhari, E. Steinbach, Exploiting directional dependencies of force perception for lossy haptic data reduction, in *Proceedings of Haptic Audio-Visual Environments and Games (HAVE)* (Phoenix, AZ, 2010)
18. Cyber Glove Systems, Cyber Force [Online]. http://www.cyberglovesystems.com/products/hardware/cyberforce.php
19. A. Frisoli, F. Rocchi, S. Marcheschi, A. Dettori, F. Salsedo, and M. Bergamasco, A new force-feedback arm exoskeleton for Haptic interaction in Virtual Environment, in *Proceedings of WorldHaptics*, 2005 pp.191–201
20. Y. Adachi, T. Kumano, A. Ikemoto, A. Hattori, N. Suzuki, "Development of a haptic device for multi fingers by macro-micro structure", (in Japanese). J. Robot. Soc. Jpn. **20**, 725–733 (2002)
21. H. Maekawa, J. Hollerbach, Haptic display for object grasping and manipulating in virtual environment, in *Proceedings of 1998 IEEE International Conference on Robotics and Automation*, 1998 pp. 2566–2573
22. T. Endo, T. Kanno, M. Kobayashi, H. Kawasaki, Human perception test of discontinuous force and a trial of skill transfer using a five-fingered haptic interface, J. Robot. **2010**, Article ID 542360, p. 14 , (2010) doi:10.1155/2010/542360
23. I. Elhaji, H. Weerasinghe, A. Dika, and R. Hansen, Human perception of haptic force direction, in *Proceedings of IEEE/RSJ International Conference on Intelligent Robots and Systems*, 2006, pp. 989–993

Chapter 17
Multi-finger Haptic Displays for Characterization of Hand Response

Blake Hannaford, Pietro Buttolo and Hawkeye King

Abstract This chapter will describe some properties of multi-finger haptic interaction and two devices which support it. Multi-finger haptic interaction can involve many contacts with the environment, but can also involve only one contact point when mediated by a tool such as a pen. As multiple fingers interact with the environment, their individual biomechanics and their sensory properties interact to form the net mechano-sensory properties of the interaction. This chapter will look at such interactions in two particular cases, spatially varying stiffness of the pen grasp, and sensory thresholds of multi-finger versus single finger interaction with haptic features. To characterize the stiffness of the pen-like grasp in various directions, we describe experiments in which force steps (randomized in amplitude and direction) were applied to subjects' pen-like tools in the plane tangential to the tip. From these, the stiffness ellipse could be identified. A dynamical model of the fingers positioned similarly to the user's grasp was used to predict the stiffness ellipsoids with similar results. The ellipsoids were shown to be a function of the squeezing force with which the subjects performed the grasps. Much of the research on sensitivity and sensory thresholds is based on measurements with a single finger. We developed a multi-finger haptic device (MFHD) to allow two high quality degrees of freedom for each of four fingers in a natural pose. With this device we could compare the sensory thresholds between single finger and multiple finger haptic exploration.

Keywords Pen-grasp · Pen-based haptic device · Multi-finger haptic device · Hand impedance/performance

B. Hannaford (✉) · H. King (✉)
Biorobotics Lab, Department of Electrical Engineering, The University of Washington, Seattle, WA, USA
e-mail: blake@u.washington.edu

H. King
e-mail: hawkeye1@uw.edu

P. Buttolo
Research & Advanced Engineering, Ford Motor Company, Dearborn, MI, USA
e-mail: pbuttolo@ford.com

R. Balasubramanian and V. J. Santos (eds.), *The Human Hand as an Inspiration for Robot Hand Development*, Springer Tracts in Advanced Robotics 95,
DOI: 10.1007/978-3-319-03017-3_17, © Springer International Publishing Switzerland 2014

1 Introduction and Literature Review

The rich behaviors of the hand derive in major part from the five flexible fingers. Only a few natural haptic interactions involve just one finger, but because of the complexity of multi-finger grasping and perception there is less research on the properties of the fingers working together or their individual differences.

In this chapter we will highlight just two aspects of multi-finger haptics: the spatial stiffness properties of pen-type multi-finger grasping, and perception of small haptic features using all four fingers. The former is of interest because of increasing focus on performance of tasks using the pen-grasp, such as surgery. The later is significant for designers of multi-finger haptic devices and algorithms whose specifications are ideally derived from human perceptual characteristics.

The material in this chapter on pen-grasp has appeared in [1]. The material on multi-finger perception has appeared in [2].

1.1 Multi-finger Haptic Displays

Many engineers have tackled the challenge of multi-finger haptic devices (see Burdea [3] for a comprehensive 1996 review). These devices tend to be mechanically very complex as structure, sensing, and actuation needs to be provided for a large number of coupled degrees of freedom (DOF) in a small space. The following review is not meant to be comprehensive, but instead to convey the common and necessary mechanical tradeoffs.

The SARCOS dexterous master [4] provided force sensing and hydraulic drive to the thumb and one finger in a 3 DOF configuration optimized for grasping and tool use. The U. Tokyo Sensing Glove II [5] was a tendon driven exoskeleton, with 20 DOF, aimed at manipulation of virtual objects. The "Tactuator" [6], was a very high bandwidth device designed and used for psychophysical threshold measurements on a single DOF to each of three fingers. With disk drive flat coil actuators, the Tactuator achieved bandwidths of over 200 Hz and up to 25 mm displacement. The motion axes drove the thumb, index finger, and middle finger in a relaxed cup-shaped posture. The Rutgers Master [3] used four custom pneumatic pistons on gimbal mounts to generate internal forces between the palm and the tips of the thumb and three fingers. The Cyberglove/Cyberforce system [7] was a multi-finger glove and wrist gimbal mounted in a haptic device. The finger actuators were removed to ground (for mass and volume reduction) by tendon drives. Kron and Schmidt [8] designed compact fingertip tactile actuators to overcome some of the bandwidth limitations of the Cyberglove's tendon drives. Gosselin et al. [9] developed a two-finger spatial device worn on the wrist which had three actuated degrees of freedom. Gillespie [10] studied a piano keyboard haptic device capable of simulating the dynamics of linkages (such as piano mechanisms). High bandwidth and multi-finger display was achieved in one degree of freedom per finger.

Casiez et al. developed a 3-finger skin-slip device [11] and two recent devices [12, 13] supplied three high-bandwidth DOF to two fingers. The Spidar-8 [14], supplied 3DOF to 8 fingers through tension cables.

The human hand gives us at least 26 DOF (including the wrist) inside a very compact space (estimate: 17.2 ml per DOF). This complexity makes it inevitable that many compromises are made by engineers of haptic devices.

All of the above devices, as well as the devices we describe below, trade away many desireable properties. High bandwidth (e.g. [6]) can be achieved with only 3 degrees of freedom while high degrees of freedom (e.g. [7]) can be achieved with high friction tendon drives which limit force feedback fidelity.

1.2 Mechanical Impedance of the Hand and Fingers

In most of the related experiments found in literature, a haptic device is used to measure the mechanical properties of the human arm/hand. The subject is constrained to the device. Then, either the device applies a signal of force, and position is measured, or the device position is displacement controlled, and the force applied by the operator onto the device is measured. A model of the arm/hand can than be identified by the data collected.

Hogan [15, 16] stated and experimentally confirmed that the human arm dynamics can be described as a modulated, lumped parameter, linear impedance varying with the amount of muscular contraction, position of the arm, and difficulty of the task. He experimentally observed that the human arm can be described as an adaptive second order system, composed of mass, damper and spring. Impedance can be defined as the dynamic relation between force exerted by the muscle and imposed stretch, or as the relation between imposed force and stretch of the muscle.

$$F \rightarrow \text{admittance} \rightarrow \Delta x$$

$$\Delta x \rightarrow \text{impedance} \rightarrow F$$

Whenever this relation can be modeled by a lumped-parameter linear system, the dynamics of the muscle-limb can be represented as a Laplace Transform in the frequency domain:

$$Z(s) = \frac{F(s)}{\Delta X(s)}$$

where $Z(s)$ is the muscle-limb impedance.

Mussa-Ivaldi [17] modeled the human arm, for tasks constrained to the horizontal plane, as a pair of springs and dampers, oriented in the plane. The notion of mechanical impedance is expressed in 2 DOF, as:

$$\begin{bmatrix} F_1(s) \\ F_2(s) \end{bmatrix} = \begin{bmatrix} Z_{11}(s) & Z_{12}(s) \\ Z_{21}(s) & Z_{22}(s) \end{bmatrix} \begin{bmatrix} \Delta X_1(s) \\ \Delta X_2(s) \end{bmatrix}$$

where $[F_1(s) \quad F_2(s)]^T$ and $[\Delta X_1(s) \quad \Delta X_2(s)]^T$ are the force and position vectors respectively. The matrix impedance is a 2nd order linear model of the impedance in the plane at the limb end point:

$$\begin{bmatrix} Z_{11}(s) & Z_{12}(s) \\ Z_{21}(s) & Z_{22}(s) \end{bmatrix} = \begin{bmatrix} M_{11} & M_{12} \\ M_{21} & M_{22} \end{bmatrix} s^2 + \begin{bmatrix} B_{11} & B_{12} \\ B_{21} & B_{22} \end{bmatrix} s + \begin{bmatrix} K_{11} & K_{12} \\ K_{21} & K_{22} \end{bmatrix}$$

where the matrix K is symmetric such that $K_{21} = K_{12}$.

If we plot the set of points F obtained from mapping a circle in X around the equilibrium,

$$\begin{bmatrix} F_1(s) \\ F_2(s) \end{bmatrix} = \begin{bmatrix} K_{11} & K_{12} \\ K_{21} & K_{22} \end{bmatrix} \begin{bmatrix} \Delta \bar{X}_1 \\ \Delta \bar{X}_2 \end{bmatrix}$$

where $||\Delta \bar{X}|| = 1$, we obtain a stiffness ellipse in the Cartesian plane, that tell us the stiffness along different directions. The ellipse is characterized by size (principal axis), shape (ratio of axis sizes) and orientation (angle that the principal axis makes with the reference frame) Mussa-Ivaldi's experimental results showed that the stiffness ellipse size and orientation are to an extent under control of the subject according to co-contraction and the posture of the arm.

Tsuji [18], qualitatively confirmed the results of Mussa-Ivaldi, but found differences in the size of the ellipses, probably due to the differences in the experimental setup. Tsuji's experimental data shows that:

- The human hand inertia characteristics can be explained from basic biomechanics of the passive effects (stiffness, damping, and inertia).
- Increasing the grip force increases the size of the stiffness and viscosity ellipses. However, the published data consist only of two different types of grasp: relaxed and tight;
- The orientation and shape characteristics of the stiffness and viscosity ellipses are mostly explained from the kinematics of the human arm.

Haijan and Howe [19] measured the impedance of the metacarpal-phalangeal joint of the index finger, constraining the distal and the proximal interphalangeal joints. They found that the joint can be reasonably modeled as a linear spring, with stiffness a function of how much the subject is trying to co-activate the muscles of the finger. Milner [20], modeled the stiffness of the full finger as a 2×2 matrix, as was done by Mussa-Ivaldi and Tsuji for the arm impedance. However, for a given posture, he measured five different ellipses that were different not only in size, as measured by Mussa-Ivaldi, but also in shape and orientation, depending on the presence and orientation of an additional constant bias of force. Stiffness maximum eigenvalues ranged from about 8–22 N/cm, for the extended finger, and from about 4–20 N/cm for the flexed finger. The ratio between maximum and minimum eigenvalues ranged from about 2.45–8 for the extended finger, and from about 1.3–2.44 for the flexed finger. Force measurements were made 70–90 ms. after displacement onset, so that measurements included a contribution from reflexes activated by the displacement.

In a manual about surgical technique, Anderson [21] describes how to handle a scalpel according to the particular task to be performed, and he suggests the best direction to execute a clean cut. This information is the result of centuries of practical evidence. At Berkeley, Tendick [22] used a model of the kinematic structure of the human hand, including the stiffness of the muscles, to see if there is a mechanical advantage, in terms of impedance, in the current surgical technique. The parallel configuration of the grasp is a very stiff structure, in particular along some specific directions, depending on the fingers' orientation. If we are to perform a clean cut, it is better to move orthogonally to the direction of maximum stiffness. This is for two reasons: first of all, the low stiffness allows force control in the direction of the cut; second, the high stiffness improves position control in the direction orthogonal to the cut, filtering mechanical noise that could cause to deviate from the straight cut trajectory. Tendick partially confirms the correctness of surgical techniques using an analytical model of the human grasp, and experimentally measured the impedance at the tip of a firmly grasped scalpel.

1.3 Psychophysical Thresholds

Psychophysical literature relevant to sensory thresholds at the fingertips has been reviewed in detail in [23]. A review from the point of view of haptic interface design is available in [24]. Physiological responses can be detected from stimuli as high as 10 kHz, and these perceptions have been linked to specific neural discharges and receptor types [25, 26]. Tan and Rabinowitz's device [6] confirmed earlier measurements of a declining vibrotactile threshold (indicating increased sensitivity) up to 200 Hz.

Other researchers have quantified the spatial acuity of human tactile perception with the bare finger [27] as well as perception of textures via a rigid probe [28, 29]. A study of Braille perception contributed adaptive thresholding algorithms to the study of tactile perception [30]. In terms of amplitude, Jones [31] measured a 6 % ability (Weber fraction $\Delta F/F$) to haptically discriminate forces applied to the extended finger. Allin et al. [32] got a Just Noticeable Difference (JND) of 9.9 % in a similar experiment.

To our knowledge, the only similar work with *multiple* fingers has used vibrotactile stimulation. Yuan et al. [33] studied ability to detect onset time differences between the thumb and index finger. They found a threshold of 34 ms below which onset order could not be distinguished. Craig [34] measured about 2 dB drop in threshold when 100 Hz vibrotactile stimuli were applied to two fingertips simultaneously. This spatial summation disappeared when the frequency of vibration was 9 Hz. A similar result, when fingers contacted a vibrating cylinder, was obtained in [35]. However, Refshauge et al. [36] found that tonic stimulation of adjacent fingers did not reduce thresholds for detection of passive movements. Physiological mechanisms for aspects of these sensations are explored by Collins et al. [37].

What these studies have in common is an input/output view of perception. One or more physical stimulus variables (e.g. texture roughness or orientation, vibration frequency or energy) is rigorously controlled, and the human central nervous system returns information. This is important knowledge in its own right, but when humans interact with haptic devices, as well as with most of the physical world, no single variable is held constant. There is a bi-directional flow of energy and information between the human and external environment. Furthermore, the haptic device is designed with different goals and constraints than psychophysical experimental apparatus. It is not straightforward to predict whether or not a haptic effect is detectable on a given device from psychophysical thresholds alone. Nor is it straightforward to use psychophysical data to predict the effect of design variations in a haptic device on feature detection performance.

West and Cutkosky [38] compared the bare finger, hand-held stylus, and stylus/haptic device in terms of users' ability to detect sinusoidal gratings along one dimension and count the number of cycles present. They found that detection performance with the haptic device was inferior to the bare finger or stylus, and depended on the stiffness parameter of the virtual surface model.

Venema and Hannaford [39] compared haptic feature detection performance with a single finger haptic device and found optimal values of stiffness and damping gains. The variable of interest in this experiment was the magnitude of C1 discontinuity between two linear segments. Thus, in most cases where general purpose haptic devices have been used in psychophysical research (e.g. [29, 38]), only the gain of the haptic rendering has been varied. Other variables, notably the number of fingers supported, have been held constant and we do not learn how performance varies with these variables.

2 Pen-Based Haptic Device

In this section we describe a haptic display designed to measure the mechanical impedance of the human hand [1, 40, 41] (Fig. 1). The design specifications were to obtain a large stiffness, low damping and low inertia, and to achieve a spatially invariant dynamic response. The operator interacts with the manipulator using the tip of a real scalpel, other pointed tools, or the fingertip.

The pen grasp, in which individual digits jointly support a rod-like tool, is used for many high precision manipulation tasks including writing, drawing, soldering and surgery. While it is clear that the multiple fingers provide stability to the pen grasp, a quantitative framework would be useful to answer questions such as ''Which direction has the maximal stiffness for a give type of pen-like grasp?''.

Fig. 1 Pen-based force display, a high-bandwidth 3 DOF device for application of force to the tip of a pen-like instrument

The device is composed of a 2 DOF, actuation redundant, planar parallel structure, rotated around a horizontal axis by an additional pair of actuators to create vertical motion. Because of the limited motion range, it is possible to consider the rotation around the up and down axis as a linear motion along the vertical z axis. The actuators are direct-drive, flat-coil actuators from hard disk drives [42]. The device is characterized by a high force output bandwidth (over 100 Hz), low friction and no backlash. The maximum torque output of the 3 actuators on the planar structure is about 0.01 Nm, while the maximum torque

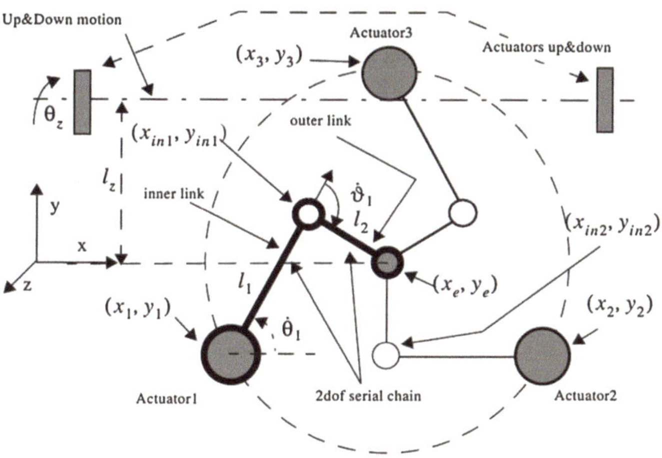

Fig. 2 Schematic representation of the pen-based force display (reprinted with permission from [1]). The *shaded circles* represent actuated joints. The *non-shaded circles* represents non-actuated joints. The *small filled circle* at the center represents the end effector

output of the other 2 actuators, connected to the up and down axis is about 0.05 Nm. The kinematic structure of the device is shown in Fig. 2. We can consider the 2 DOF parallel structure as three 2 DOF serial Cartesian manipulators connected together at the end-effector. Each serial chain is composed of an inner and an outer link.

3 Mechanical Impedance of the Pen-Grasp

3.1 Theory

We focus here on the relationship between displacement and force provided instantaneously by the biomechanics of a particular grasp:

$$F = K_0 \Delta X_0 \tag{1}$$

To determine the stiffness matrix K_0, we will use the grip transform analysis. This method has been extensively analyzed in [22, 43, 44]. It consists of three steps, during which we will determine:

- the relation between force at the tip, and torques at the joints of each single finger;
- the relation between forces at the tip of the finger and net force applied to the pen;
- the relation between finger joints displacement and pen tip displacement.

However, the data obtained in these three steps alone it is not enough to obtain the stiffness matrix in the Cartesian space, K_0. We also need to know the stiffness matrix in the joint space, that relates the displacement induced in the angular position of the joint due to applied torque. Unfortunately, there is not much data in the literature about finger joint stiffness. Most of the literature reports experiments which were performed to obtain the stiffness of a single joint, since inter joint stiffness is very difficult to obtain [19, 20]. Since we will use incomplete data for the stiffness matrix in the joint space, the model in this section is intended to be qualitative, to get a better overall idea of the pen grasp. The experimental setup described later will be used to determine quantitative data.

Before the detailed grip transform analysis, we will briefly describe the hand kinematic model and introduce some notation [45, 46] (Fig. 3). The model permits grips consisting of one to four fingers, with points of contact only at the fingertips. In reality, a one finger grip is impossible. However, in our model we make the assumption that the tool is the distal phalanx of the finger, or it is rigidly connected to it, such as in the Phantom haptic interface [47]. We rely on the Jacobian Matrix, J_i:

$$\tau_i = J_i^T C_i \tag{2}$$

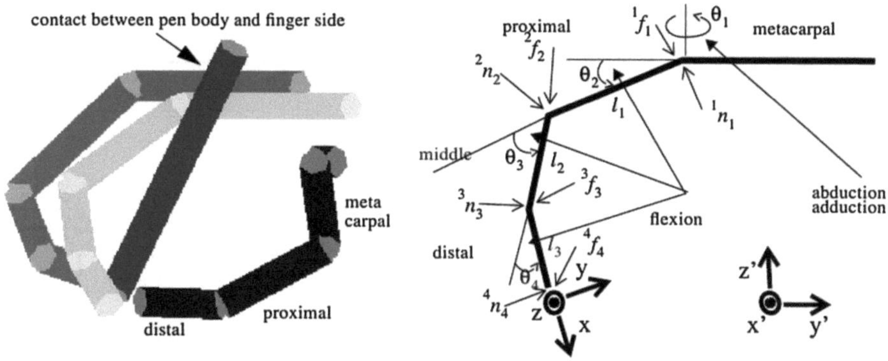

Fig. 3 3-finger pencil-grasp model (reprinted with permission from [1]). *Left* screen capture of simulation software package. Each finger is modeled as three moving links plus a fixed link. Each finger has three degrees of motion. Shown is the index finger (*light color*), the middle finger (*intermediate color*), the thumb (*dark color*), and the pencil, held at the tip (*black*). *Right*, kinematic representation of single finger: reference frames, joint angles, dimensions and rotation axis. Thumb phalange names are on the plot on the *left*

where C_i is the generalized contact force and τ_i is the joint torque vector for the ith finger. J_i can be determined using Craig's approach as

$$J_i^T = \begin{bmatrix} 0 & 0 & (s_{34}+c_{34})s_2 l_3 - c_{23} l_2 - c_2 l_1 \\ l_2 s_4 + l_1 s_{34} & l_3 + l_2 c_4 + l_1 c_{34} & 0 \\ l_2 s_4 & l_3 + l_2 c_4 & 0 \\ 0 & l_3 & 0 \end{bmatrix} \quad (3)$$

considering all n fingers together,

$$\tau = \begin{bmatrix} J_1^T & 0 & \dots & 0 \\ 0 & J_2^T & \dots & 0 \\ \dots & \dots & \dots & \dots \\ 0 & 0 & \dots & J_n^T \end{bmatrix} \begin{bmatrix} C_1 \\ C_2 \\ \dots \\ C_n \end{bmatrix} = J^T C \quad (4)$$

The contact forces produce a net force, F, on the object related by

$$F = WC$$

where W is the grasp matrix which depends on grasp configuration.

Figure 4 shows the relation between a fingertip frame and the object (pen) frame. On the lower right is the sequence of rotations $(\theta_{xi}, \theta_{yi}, \theta_{zi})$ that defines the orientation of the finger frame with respect to the pen. All rotations are around the fixed pen frame, with the fingertip lying on the x axis. The sequence of rotation is *ZYX*. θ_{pi} is used to define the point of contact of the finger on the pen and is the angle between the y axis and the point of contact. Δcp_i is the offset of the contact position from the tip of the pen.

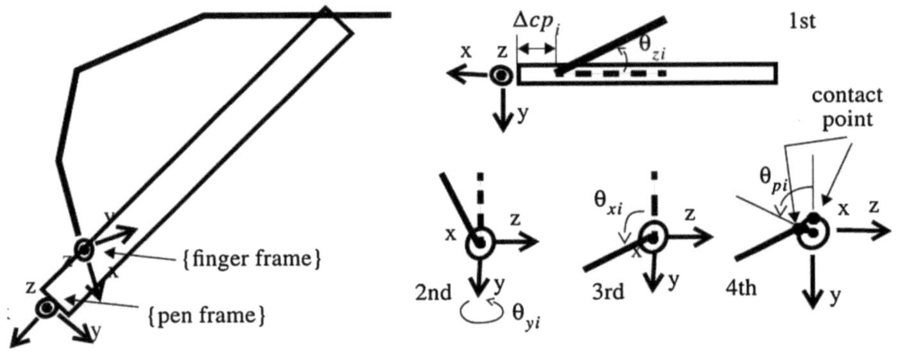

Fig. 4 Fingertip-Pen Orientation. *Left* representation of the fingertip and pen (object) frames. *Right* Definition of pen offset and rotations

The four angles $\theta_{xi}, \theta_{yi}, \theta_{zi}, \theta_{pi}$ and the offset Δcp_i completely define the position and orientation of the finger relative to the pen. However, since the effect of the offsets is a torque applied onto the pen, they will not be used on our model which only studies forces in a 3 DOF space. Alternatively, it can be assumed that the *i* forces and offsets always produce zero torque in combination or that the pen is at rotational equilibrium.

For each finger, the relation between the contact force expressed in the fingertip frame and the same force expressed in the pen frame is given by

$$F_i = W_iC_i = rot(\hat{x}, \theta_{xi})rot(\hat{y}, \theta_{yi})rot(\hat{z}, \theta_{zi})C_i$$

Multiplying the rotations together we get:

$$W_i = \begin{bmatrix} c_zc_y & -s_zc_y & s_y \\ s_zc_x + c_zs_ys_x & c_zc_x - s_zs_ys_x & -c_ys_x \\ s_zs_x - c_zs_yc_x & c_zs_x + s_zs_yc_x & c_yc_x \end{bmatrix} \tag{5}$$

The total force on the object becomes:

$$F = WC = \sum_i W_iC_i = [W_1W_2 \ldots W_n] \begin{bmatrix} C_1 \\ C_2 \\ \ldots \\ C_n \end{bmatrix} \tag{6}$$

By the principle of virtual work, infinitesimal displacements of the finger joints, fingertips, and grasped object, $(\Delta\Phi, \Delta X_C, \Delta X_0)$ are related by the following equations.

$$\Delta X_C = J\Delta\Phi$$

$$\Delta X_C = W^T \Delta X_0$$

Linear stiffness relationships between generalized displacements and forces can be expressed either in joint or pen spaces:

$$\tau = K_\theta \Delta\Phi$$

$$F = K_0 \Delta X_0$$

where K_θ and K_0 are stiffness matrices. Using the previous equations, the stiffness in the two spaces is related by

$$K_0 = WJ^{-T} K_\theta J^{-1} W^T$$

To simplify the analysis, we will consider joint stiffness matrices with non-zero elements only on the diagonal.

Note that J_i^T are 3×4 matrices, and therefore J^T is a 12×16 non-square matrix. Therefore, the notation J^{-1} and J^{-T} in the above equation are not formally correct. We will assume these two matrices to be the pseudo inverses of J^T and J. An interesting consequence of the kinematic redundancy of the hand is that, in theory, there could be infinite stiffness matrices in the pen frame K_0, for a given grasp configuration and joint stiffness matrix K_θ, depending on how the pseudo inverses are computed [48].

Once it is identified, K_0 is the stiffness matrix in the 3 DOF Cartesian space. In our study we are interested in determining the stiffness at the tip of the pencil when the motion is constrained to be on a specific plane, such as, for example, the surface of a table during the action of writing. We choose the table frame so that X is perpendicular to the surface and points straight down, Y lies on the surface and points toward the subject, and Z lies on the surface and points to the right of the subject. We will call this plane the "horizontal plane". Because the motion is constrained to be on a plane, we determine the 2 DOF stiffness ellipse by computing the section obtained by cutting the 3 DOF stiffness ellipsoid by the $\{Y, Z\}$ plane.

3.2 Results: Experimental Measurements

In this section we will experimentally compare the pen-grasp and the single finger configurations. We will first describe the experimental protocol, and the analysis method. Then we will discuss the results.

Experimental setup and protocol The pen-based force display was used to apply a series of 32 force steps ΔF, of random intensity and orientation. The two components of ΔF were uniformly distributed in the interval [0.1N, 0.5N]:

$$\Delta F_y = U(0.1N, 0.5N)$$

$$\Delta F_z = U(0.1N, 0.5N)$$

$$\vec{\Delta F} = \left[\Delta F_y, \Delta F_z\right]^T$$

The subjects interacted with the device first using only the index finger, and then grasping a stylus. In both configurations the subjects were asked to ground the heel of their hand on the flat surface in front of the haptic display to eliminate their arm and wrist dynamics from the measurements. Each force step lasted 100 ms. and was followed by 1 s during which no force was applied, to allow the pen tip to go back to its at rest position. The force applied and the pen tip displacement induced were recorded 30 ms. after the beginning of the step. This should guarantee that the measured response is not influenced by spinal reflexes, and is only the mechanical impedance of the pen grasp. Because the pen-based force display has a very linear force response [49], and almost no friction and moving mass, we can assume for the purposes of this experiment that the force applied to the pencil tip is equal to that commanded by the controller. The position of the end-effector is measured with a resolution of about 20 μm. Due to the particular design, when force is applied on the end-effector there is no significant backlash. The subjects were asked to keep their grasp strength as constant as possible throughout each single experiment. For this reason a 4 mm force sensing resistor[1] was interposed between the thumb and the pencil. The pressure sensor signal was displayed as a yellow bar on top of the computer display. The pressure sensor has a non-linear response, so calibration was necessary. Each experiment was repeated twice. The subjects were first asked to start trying to resist force as much as possible, and then stay relaxed.

Analysis Data from three subjects were acquired and used to identify the parameters of the following 2 DOF linear system, representing the stiffness of the pen grasp at the tip of the pencil

$$\begin{bmatrix} \Delta F_y \\ \Delta F_z \end{bmatrix} = \begin{bmatrix} K_{yy} & K_{yz} \\ K_{zy} & K_{zz} \end{bmatrix} \begin{bmatrix} \Delta y \\ \Delta z \end{bmatrix}$$

where $[\Delta y, \Delta z]^T$ is the position displacement induced by the force step.

An estimate, $\hat{K}$ of the K matrix was computed from the force step response data. We estimated the symmetric component of $\hat{K}$ by singular value decomposition. The singular value decomposition of $\hat{K}$ yields two singular values and a rotation matrix which encode the maximum and minimum directions of stiffness in the plane perpendicular to the fingertip/pen-tip: K_y, K_z, α. These parameters are conveniently plotted as a ellipse whose size and direction convey the spatial stiffness of the grip.

In data from the three subjects (Fig. 5) stiffness magnitude varied between 1.15 and 27.99 N/mm for the various conditions and multiple regression coefficients varied from 0.74 to 0.98. The angle of peak stiffness was an average of 27.5° for the index finger conditions and 61.2° for the pen grasp conditions.

Sources of error in this computation include asymmetric stiffness component (non-conservative forces, which were no more than 10 % of the symmetric stiffness) and variance which was about 10 % of the peak stiffness.

---

[1] INTERLINK Electronics, Santa Barbara, CA.

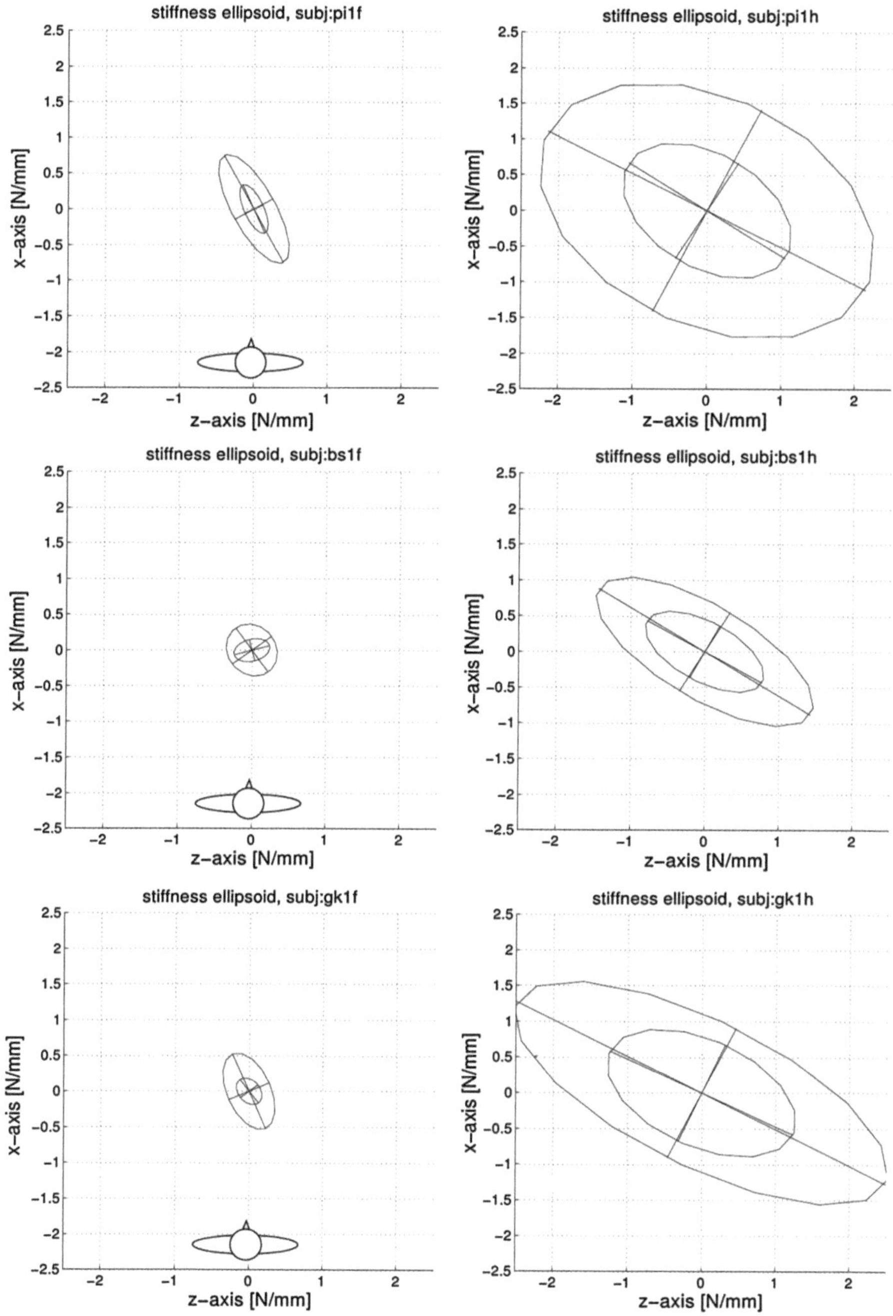

Fig. 5 Stiffness elipsoids measured from three subjects for single finger (*left*) and pen-grasp (*right*) (reprinted with permission from [1]). In each figure, the smaller ellipse was measured when subjects were instructed to use a relaxed pose/grip, and the larger ellipse resulted from a firm pose/grip

3.3 Results: Different Grasp Types

In this section we will characterize different types of grasp, to see if there are any obvious advantages or disadvantages in particular kinematic configurations. We are particularly interested in grasps in which the index finger can be freely detached from the pen without a significant loss of stability. These grasps would allow the user, for example, to press a micro-switch during surgery to control an additional function of an advanced tool. We are also interested in grasps whose stiffness ellipse is almost a circle in which

$$\frac{\lambda_{\max}}{\lambda_{\min}} \approx 1$$

These grasps would guarantee similar user capabilities, such as force control and position control, in all directions. We will consider four different grasps (Fig. 6). We will first analyze the differences in shape, size and orientation of the stiffness ellipses using the theoretical model. Then we will experimentally measure the stiffness ellipses, as subjects assumed different grasp types.

3.3.1 Four Grasp Types

Surgeon Precision Grasp The first grasp is the surgeon precision grasp [22]. The three fingers (index, middle, thumb) contact the pen at approximately the same distance from the pen-tip. The contact points are equally spaced around the pen. Figure 6 shows views from the CAD model (left), and a photograph (right) taken looking at the hand from the side. The stiffness ellipse was computed using the model described in Sect. 3.1.

Four-Finger Grasp The middle and ring fingers and thumb contact the pen at approximately the same distance from the pen-tip, while the index is about 1 cm further up. The contact points for the middle and ring fingers and thumb are equally spaced around the pen. The index finger contact point is in between the middle finger and thumb. We observed that this second grasp is routinely used for writing and drawing by some of the subjects. The stiffness ellipse, obtained again from the theoretical model, has a more elongated shape, and it is rotated counter-clockwise, compared to the previous grasps.

This grasp is particularly interesting because it is possible to lift the index finger, and keep a stable grip with the remaining three fingers. The index finger is therefore free to move around and touch optional sensors and micro-switches. When the index finger is not touching the pen, the stiffness ellipse slightly shrinks in size, but the orientation does not change significantly.

Modified Four Finger Grasp This is a four-finger configuration similar to the four-finger grasp, but with the index finger moved further away from the tip of the pen. We observed that this grasp is routinely used for writing and drawing by one

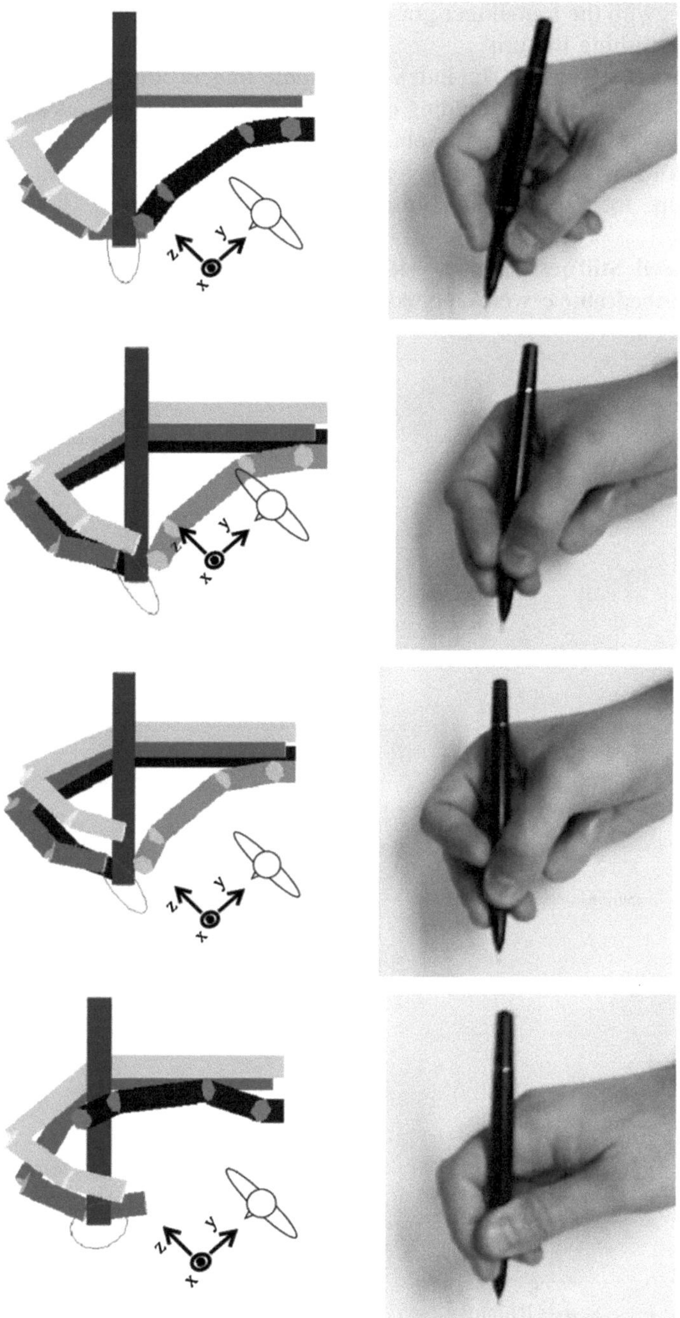

Fig. 6 Four grasp types studied for relative stiffness (reprinted with permission from [1]) *Top* to *Bottom* Surgeon Precision Grasp, Four-Finger Grasp, Modified Four Finger Grasp, "Chopsticks" Grasp (see text). *Left* Kinematic model (looking down on plane of table-top) *Right* Photograph

subject. As with the four-finger grasp, the index finger can be freely moved around without perturbing the grip.

"**Chopsticks Grasp**" The index and middle fingers strongly grip the pen from both sides, as if they were holding chop-sticks. The thumb presses against the pen further away from the tip. We obtained this particular three finger grasp trying different kinematic configurations on the model, until we obtained a more "circular" shape for the stiffness ellipse. As can be seen in Fig. 7 (dotted lines), the ratio $\lambda_{\max}/\lambda_{\min} = 1.2$ is very close to one.

Measured Stiffness Elipses with Different Grasp Types The step perturbations described above were applied to pen tips held by four subjects in the four grasp types above.

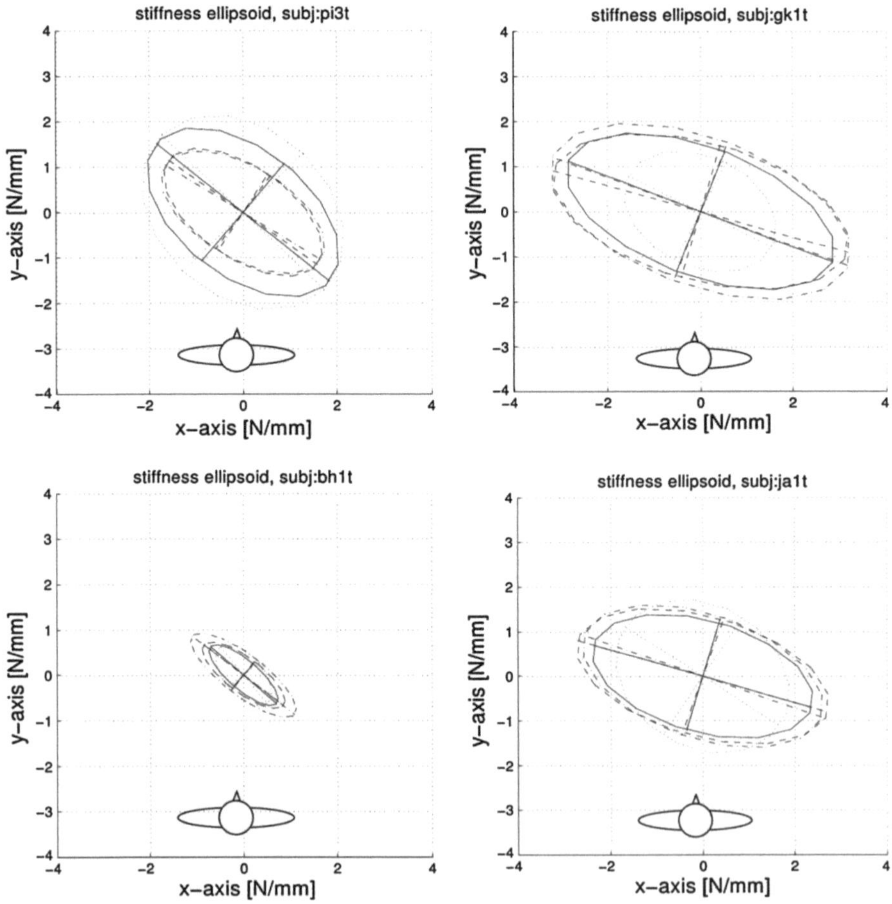

Fig. 7 Experimental stiffness ellipses measured from four subjects using four different grasps (reprinted with permission from [1]): Surgeon Precision grasp (*continuous line*), Four Finger grasp (*dashed line*), Modified Four Finger grasp (*dashed-dotted line*), "Chopsticks" grasp (*dotted line*)

Because of the variability of the experimental results (Fig. 7), we cannot draw general conclusions for all subjects about typical stiffness ellipse characteristics for each grasp type. From these preliminary results it seems like there is not a universally advantageous grasp, valid for every human. Each person, depending on their own hand structure, will perform better, in terms of stiffness, with a particular grasp. The experimental setup can be effectively used to assess the characteristics of different types of grasp for a single subject. As an example, this could be used for the training of surgeons, allowing trainees to check among different type of grasps, measuring the stiffness ellipses. It would be possible to choose the type of grasp that best fit the specific individual or, an individual could choose different grasps for different tasks.

4 Multi-finger Haptic Device

The University of Washington Biorobotics Lab built a four finger, eight DOF haptic device, the MultiFinger Haptic Device (MFHD) [50] by making four copies of its 1997 single-finger device [39, 51] (Fig. 8). The device support planar motion of the four fingertips. Each finger contains two custom wound flat-coil actuators driven by Neodymium-Iron-Boron permanent magnets and having 90° of motion range. Thermal modeling enables peak torques of up to 0.6 Nm—equivalent to about 6 N finger tip force. The lightweight linkages were computer synthesized to match human finger anthropometric data. In the MFHD, miniature interferometric optical encoders from Micro-E Inc. were integrated inside to allow the fingers close proximity and to increase position sensing resolution. We installed two grades of this encoder, the M1500 in three fingers and the M2000 in one finger, to allow testing at extremely high resolutions (see Table 1).

The user's fingers are attached to the end of each mechanism using a fitted glove with plastic clips on each fingertip [52].

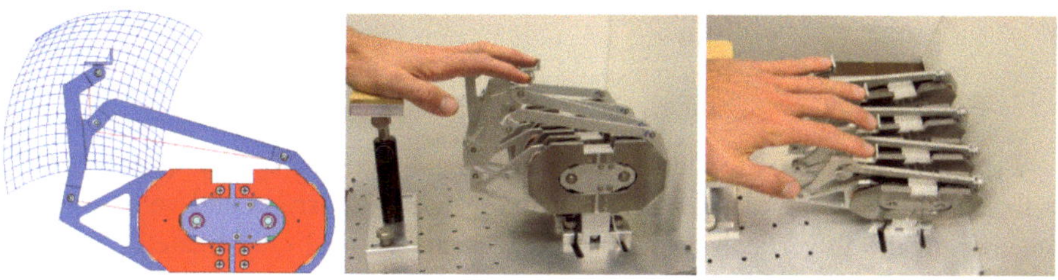

Fig. 8 Multi-finger Haptic Device completed in 2005. Each finger is a two-DOF planar mechanism, computer optimized to cover the flexion–extension workspace of human fingers. Each finger is driven by two hand-built low friction, low inertia actuators. The base contains all electronics and interfaces to the computer through a single USB 2.0 cable

Table 1 Specifications of position sensors from Micro-E Inc. used in the Multi-finger Device (Fig. 8). In the earlier single finger version reported in [39, 52], angular resolution was 48 μrad

Encoder type	M1500	M2000
	3 fingers	1 finger
Counts per revolution (interpolated)	163840	1048576
Bits per revolution	17.3	20
Angular resolution	38.3 μrad	6 μrad
Approximate XY resolution	3.83 μm	0.6 μm

5 Multi-finger Haptic Thresholds

5.1 Related Previous Work

A successful haptic interaction involves a haptic device in contact with the human operator in a bi-directional exchange of information and energy. Thus it is difficult to predict haptic feature detection ability from psychophysical thresholds. Here we find the point at which a human can perceive meaningful information from a realistic haptic device, and additionally explore how that performance changes as we vary one or more engineering parameters. Our measurement depends on the specific haptic device as well as on the user and so is not a threshold in the classical psychophysical sense. Our prior work in this area involved only a single finger. This section looks at how results change as stimuli are presented to multiple fingers (Fig. 9).

5.2 Haptic Target Rendering

For this experiment haptic icons are presented to the user on a horizontal plane. The horizontal plane lies perpendicular to the fingertips' workspace, and rendering of the plane is described by:

$$F_y = \begin{cases} k_p\left(y_{plane} - y\right) - k_d v_y & y \le y_{plane} \\ 0 & y > y_{plane} \end{cases} \tag{7}$$

$$F_x = 0; \tag{8}$$

F_x and F_y are forces rendered in the x and y directions. k_p and k_d describe the stiffness and damping of the surface, y and v_y are the position and velocity of the fingertip in the y, vertical, direction. y_{plane} is the vertical coordinate of the horizontal plane.

The haptic plane forms lines of intersection with the vertical plane at each of the fingertips' workspaces. Haptic icons are presented to the subject on these lines. Along the line for each finger two potential locations for a 5 mm sawtooth icon are

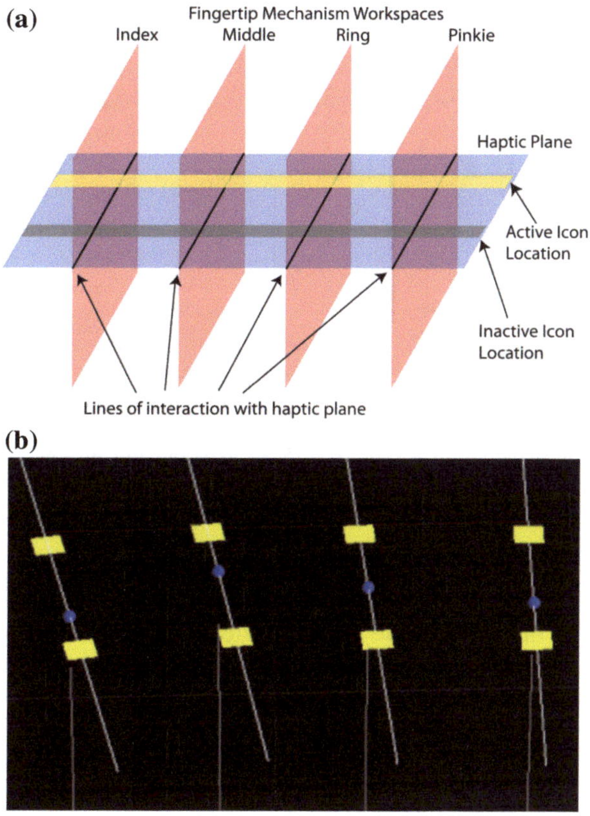

Fig. 9 **a** Illustration of the rendering environment shows planar workspace of the fingertip mechanisms, and icon locations on a *horizontal plane* in the workspace. **b** Screenshot of a multi-finger trial. A *blue sphere* indicates the subject's position in the environment, and the icon turns from *yellow* to *pink* when in contact (used with permission from [2])

spaced 30 mm apart, center-to-center. One of the two locations has a sawtooth force icon, while the other location has no force. The sawtooth is described by:

$$F_x = \begin{cases} 0 & x - x_0 < -\dfrac{\Delta}{2} \\ F_{\max}(x - x_0) & -\dfrac{\Delta}{2} \leq x - x_0 \leq \dfrac{\Delta}{2} \\ 0 & \dfrac{\Delta}{2} < x - x_0 \end{cases} \tag{9}$$

This is illustrated in Fig. 10 (left). For all active fingers, the active icon was in the same location in the workspace.

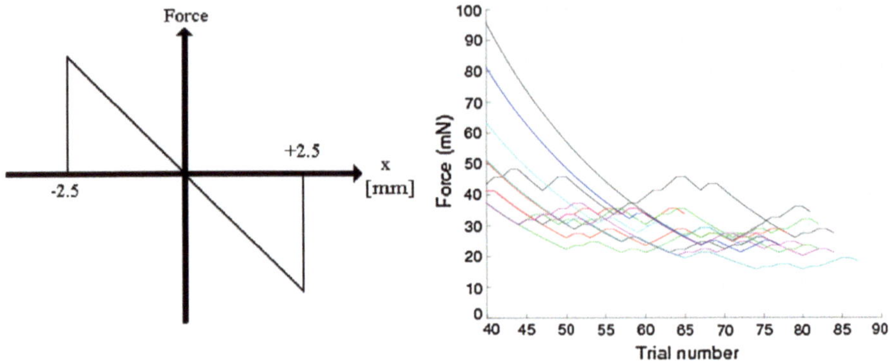

Fig. 10 *Left* Sawtooth profile of tangential force inside the icon (5 mm width). Amplitude of sawtooth was adaptively converged to subject's threshold. *Right* Convergence path of adaptive thresholding for 8 subjects—75–85 iterations to convergence. Average threshold was 28.6 mN (about the weight of a dime) [54]. (used with permission from [2])

5.3 Detecting Small Haptic Effects

In this section we briefly review our related work (reported in more detail in [53, 54]) using the FHD with a single finger to study limiting factors for perception and usefulness of small haptic effects. Pairs of targets (icons) were presented to the subjects in which one randomly selected target had force feedback and the other had none. The targets consisted of two 5 mm regions, 30 mm apart. The subject's fingertip was supported by a virtual plane approximately mid-workspace, $y \geq 100$ mm. The attractive force profile (Fig. 10, left) was a linear function of the end effector's distance from the center of the target. Subjects were allowed to sample both targets indefinitely before indicating which target they perceived to contain a force and indicated their choice by pressing a button held in their opposite hand and wore noise reducing ear protection with masking music to eliminate possible sound queues.

An adaptive thresholding method [30] converged the force output to a level at which approximately 71 % of responses observed from the subject were correct. Two or more correct responses decreased the force by 5 %, while a single incorrect response increased the force back to its previous value. Each deviation from the current trend in force was termed a ''force reversal''. The experiment started at 500 mN of force (a very easily detected amplitude) and continued with initially declining force amplitudes until twelve force reversals occurred.

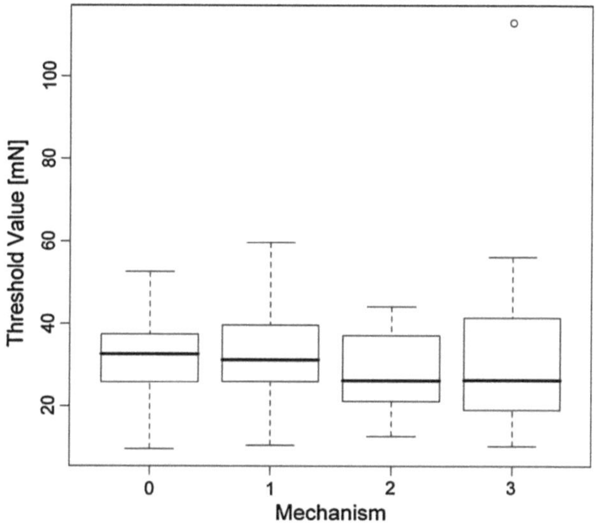

Fig. 11 Detection thresholds in which the subjects used their index finger singly on each mechanism. There is no statistically significant difference between the devices thresholds. (used with permission from [2])

5.4 Results: Active Exploration

Figure 10, right, shows force-adaptation paths taken by seven test subjects [54]. Each line represents the peak magnitude of force present for each trial during the experiment (the first 40 trials are not plotted). Subjects took 75–85 trials per experiment before incurring the necessary twelve force-path reversals. The mean of all converged values was 28.6 mN.

Device Friction It is interesting to compare this value with the static friction of the device. By slowly increasing the applied force and watching for end effector motion (without finger contact), we measured the static friction level at 100 mN. However, observing the convergence pathways, we see that most of the force reversals occur with $F < 100$ mN. Because the subject kept the device in nearly constant motion, substantially smaller force levels could be detected at the 71 % reliability level. This illustrates how the haptic effect detection threshold is in fact a combined human–machine property so neither machine or human properties should be considered alone.

It is possible that differences between the mechanisms could account for some observed effects. To assess this possibility, subjects performed a single-finger experiment using their index finger on each of the four mechanisms (Fig. 11. The differences in thresholds (means between 30 and 35 mn) were not statistically significant.

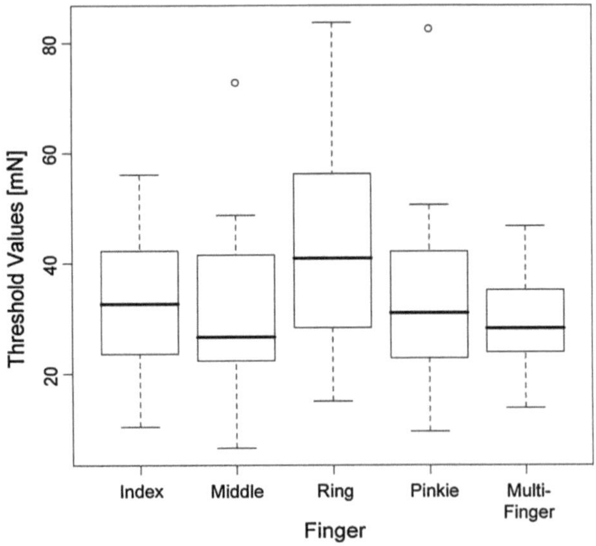

Fig. 12 71 % detection threshold statistics for each individual finger (index, middle, ring, pinkie) on one fingertip mechanism, with the multi-finger result. (used with permission from [2])

Table 2 The mean 71 % detection threshold, standard deviation, and number of subjects in each of the eight trial conditions. (used with permission from [2])

Finger	MFHD mechanism			
	0	1	2	3
Index	31.8	32.9	27.8	33.5
	10.3	12.2	9.8	24.7
	16	16	16	16
Middle		40.7		23.4
		16		9
		8		8
Ring			43.5	43.7
			21.7	20.5
			8	8
Pinkie				34.0
				17.2
				16
Multi-	28.9/9.9/16			

5.5 Results: Finger/Device Variations

Figure 11 shows the mean and distribution of 71 % detection thresholds of the index finger on each mechanism. A one-way ANOVA for repeated measures was used to compare results. Analysis showed no significant difference between fingertip mechanisms ($F = 0.83$, $P = 0.48$).

Single finger thresholds were compared with the multi-finger threshold. The distribution is shown in Fig. 12, and mean values for each trial are shown in Table 2. The overall mean and standard deviation for the middle and ring finger are 32.1/15.5 and 43.6/20.4 respectively. For the index finger, the force threshold mean and standard deviation from mechanism 0 (31.8/10.3) are used. We see in the figure that all fingers and the multi-finger case appear to have comparable mean value, except for the ring finger which appears to be higher. A one-way ANOVA for repeated-measures demonstrated statistical significance with $F = 3.8$, $P = 0.008$.

Four paired t-tests compared each individual finger's threshold to the multi-finger threshold using the Bonferroni correction over four trials. This correction requires a $p < 0.0125$ for rejection of the null hypothesis at the 0.05 confidence level. The t-values showed no significant difference between the multi-finger and the index, middle and pinkie fingers ($p = 0.12, 0.27, 0.18$ respectively). The ring finger showed a statistically significant difference compared to the multi-finger with $p = 0.006$.

6 Conclusion

In this chapter we have demonstrated (1) some aspects of how the multi-finger pen-like grasp affects the spatial stiffness which is achieved at the instrument tip, and (2) variations in sensitivity to detect haptic features as various combinations of fingers are used.

Our pen-based haptic device was a suitable experimental platform to determine mechanical properties of pen-like grasps. Using a theoretical model and experimental data, we found that pen grasp manipulation is superior to single finger manipulation, both in terms of mechanical impedance and accuracy of motion control. Moreover, because of its parallel structure, the pen grasp stiffness ellipse is relatively insensitive to changes in the kinematic configuration of the fingers.

We analyzed four different types of grasp, with particular attention to four-fingered grasps, and found similar stiffness ellipses for all of them. Some of these grasps were chosen by observing different people interacting with a pen, others by trying different configuration on a custom computer animated model.

A new Multi-finger haptic device, specialized for supplying two high bandwidth DOF to each finger's flexion–extension plane, allowed us to measure sensitivities of the different fingers and compare them to all four fingers working together to detect a line stimulus. Note that the multi-finger stimulus presented here occured in a straight line, yet the fingertips more naturally form a curve. If the subjects used a more natural finger pose, each finger would in general encounter the stimulus at a different time. Our ongoing work addresses the possible affects of time synchronization versus spatial alignment on perception thresholds.

Many challenges remain in scientific understanding and technological innovation in multi-finger haptics. New mechanisms, actuators, sensing, and control techniques must be developed to convey kinesthetic sensations effectively to the whole hand. Teleoperation systems, especially those which support precision and highly dexterous interactive tasks like surgery could take advantage of the operator's grasp characteristics to accomplish tasks more accurately and robustly.

References

1. P. Buttolo, Characterization of human pen grasp with haptic displays. Ph.D. Dissertation, June 1996
2. H.H. King, R. Donlin, B. Hannaford, Perceptual thresholds for single vs. multi-finger haptic interaction, in *Haptics Symposium, 2010 IEEE*, pp. 95–99. IEEE (2010)
3. G.C. Burdea, *Force and Touch Feedback for Virtual Reality* (Wiley Interscience, New York, 1996)
4. S. Jacobsen, F. Smith, D. Backman, E. Iversen, High performance, high dexterity, force reflective teleoperator II, in *Proceedings, ANS Topical Meeting on Robotics and Remote Systems* (ANSI, New York, 1991)
5. H. Hashimoto, M. Boss, Y. Kuni, F. Harashima, Intelligent cooperative manipulation system using dynamic force simulator, in *Proceeding of IEEE International Conference on Robotics and Automation*, pp. 2598–2603 (1994)
6. H.Z. Tan, W.M. Rabinowitz, A new multi-finger tactual display. Proc. Haptics Symp. ASME Dyn. Syst. Control Div. **DSC-58**, 515–522 (1996)
7. M.L. Turner, D.H. Gomez, M.R. Tremblay, M.R. Cutkosky, Preliminary tests of an arm-grounded haptic feedback device in telemanipulation. Proc. Haptics Symp. ASME Dyn. Syst. Control Div. **DSC-64**, 145–149 (1998)
8. A. Kron, G. Schmidt, Multi-fingered tactile feedback from virtual and remote environments, in *Haptic Interfaces for Virtual Environment and Teleoperator Systems, 2003. HAPTICS 2003 Proceedings*, 22–23 March 2003, pp. 16–23
9. F. Gosselin, T. Jouan, J. Brisset, C. Andriot, Design of a wearable haptic interface for precise finger interactions in large virtual environments, in *Haptic Interfaces for Virtual Environment and Teleoperator Systems, 2005. WHC 2005. First World Haptics Congress*, 18–20 March 2005, pp. 202–207
10. B. Gillespie, L. Rosenberg, Design of high-fidelity haptic display for one-dimensional force reflection applications. in *Telemanipulator and Telepresence Technology, Proceedings of the SPIE East Coast Conference*, pp. 44–54 (1994)
11. G. Casiez, P. Plénacoste, C. Chaillou, B. Semail, The digihaptic, a new three degrees of freedom multi-finger haptic device, in *Proceedings of Virtual Reality International Conference*, pp. 35–39 (2003)
12. M. Monroy, M. Oyarzabal, M. Ferre, A. Campos, J. Barrio, Masterfinger: Multi-finger haptic interface for collaborative environments, in *Haptics: Perception, Devices and Scenarios*, pp. 411–419. Springer (2008)
13. C.A. Avizzano, S. Marcheschi, M. Angerilli, M. Fontana, M. Bergamasco, T. Gutierrez, M. Mannegeis, A multi-finger haptic interface for visually impaired people, in *Proceedings. ROMAN 2003. The 12th IEEE International Workshop on Robot and Human Interactive Communication, 2003*, pp. 165–170. IEEE (2003)
14. Y. Kohno, S. Walairacht, S. Hasegawa, Y. Koike, M. Sato, Evaluation of two-handed multi-finger haptic device spidar-8, in *ICAT2001*, pp. 135–140 (2001)
15. Neville Hogan, Adaptive control of mechanical impedance by coactivation of antagonist muscles. Trans Autom Control IEEE **29**(8), 681–690 (1984)
16. N. Hogan, Controlling impedance at the man/machine interface. In *Proceedings.of the 1989 IEEE International Conference on Robotics and Automation, 1989*, pp. 1626–1631. IEEE (1989)

17. F.A. Mussa-Ivaldi, N. Hogan, E. Bizzi, Neural, mechanical, and geometric factors subserving arm posture in humans. J. Neurosci. **5**(10), 2732–2743 (1985)
18. T. Tsuji, K. Goto, M. Moritani, M. Kaneko, P. Morasso, Spatial characteristics of human hand impedance in multi-joint arm movements, in *Intelligent Robots and Systems' 94. 'Advanced Robotic Systems and the Real World', IROS'94. Proceedings of the IEEE/RSJ/GI International Conference on*, vol. 1, pp. 423–430. IEEE (1994)
19. R.D. Howe, A.Z. Hajian, Identification of the mechanical impedance at the human finger tip. J. Biomech. Eng. **119**, 109–114 (1997)
20. T.E. Milner, D.W. Franklin, Two dimensional endpoint stiffness of human fingers for flexor and extensor loads, in *Proceedings of the 1995 ASME Winter Annual Meeting, Dynamic Systems and Control Divison*, San Francisco, DSC-Vol 57, 649–656 (1995)
21. R.M. Anderson, R.F. Romfh, S.L Wangensteen, *Technique in the Use of Surgical Tools* (Appleton-Century-Crofts, New York, 1980)
22. F. Tendick, R. Jennings, G. Tharp, and L.W. Stark, Sensing and manipulation problems in endoscopic surgery: experiment, analysis, and observation. *Presence* **2**(1), 66–81 (1993) (Winter)
23. S.C. Venema, Experiments in surface perception using a haptic display. Ph.D. Thesis, April 1999
24. K.B. Shimoga, A survey of perceptual feedback issues in dexterous telemanipulation. I. finger force feedback, in *Proceedings IEEE VRAIS-93*, pages 263–270, Seattle, WA, Sept 1993
25. M.A. Srinivasan, R.H. LaMotte, Tactile discrimination of shape: responses of slowly and rapidly adapting mechanoreceptive afferents to a step indented into the monkey fingerpad. J. Neurosci., **7**, 1682–1697 (1987)
26. R.H. LaMotte, R.F. Friedman, C. Lu, P.S. Khalsa, M.A. Srinivasan, Raised object on a planar surface stroked across the fingerpad: responses of cutaneous mechanoreceptors to shape and orientation. J. Neurophysiol. **80**, 2446–2466 (1998)
27. G. Moy, U. Singh, E. Tan, R.S. Fearing, Human psychophysics for teletaction system design. Haptics-e Electron. J. Haptics Res. **1**(3), 18 (2000)
28. S.J. Lederman, R.L. Klatzky, Feeling through a probe. Proc. ASME Haptics Symp. Dyn. Syst. Control **DSC-64**,127–131 (1998)
29. J.M. Weisenberger, M.J. Krier, M.A. Rinker, Judging the orientation of sinusoidal and square-wave virtual gratings presented via 2-dof and 3-dof haptic interfaces. Haptics-e Electron. J. Haptics Res. **1**(3), 1–20 (2000)
30. J.C. Stevens, E. Foulke, M.Q. Patterson, Tactile acuity, aging, and braille readings in long-term blindness. J. Exp. Psychol. Appl. **2**(2), 91–106 (1996)
31. L.A. Jones, Perception and control of finger forces. Proc. Haptics Symp. ASME Dyn. Syst. Control Div. **DSC-64**, 133–137 (1998)
32. S. Allin, Y. Matsuoka, R. Klatzky, Measuring just noticeable differences for haptic force feedback: implications for rehabilitation, in *Haptic Interfaces for Virtual Environment and Teleoperator Systems, 2002. HAPTICS 2002. Proceedings*, 24,25 March 2002, pp. 299–302
33. H.F. Yuan, C.M. Reed, N.I. Durlach, Temporal onset-order discrimination through tactual sense. J. Acoust. Soc. Am. **117**(5), 3139–3148 (2005)
34. J.C. Craig, Vibrotactile spatial summation. Percept. Psychophys. **4**, 351–354 (1968)
35. A.J. Brisben, S.S. Hsiao, K.O. Johnson, Detection of vibration transmitted through an object grasped in the hand. J. Neurophysiol. **81**, 1548–1558 (1999)
36. K.M. Refshauge, D.F. Collins, S.C. Gandevia, The detection of human finger movement is not facilitated by input from receptors in adjacent digits. J. Physiol. **551**, 371–377 (2003)
37. D.F. Collins, K.M. Refshauge, S.C. Gandevia, Sensory integration in the perception of movements at the human metacarpophalangeal joint. J. Physiol. **529**(2), 505–515 (2000)
38. A.M. West, M. R. Cutkosky, Detection of real and virtual fine surface features with a haptic interface and stylus. Proc. ASME Haptics Symp. Dyn. Syst. Control Div. **DSC-61**,159–165 (1997)

39. S.C. Venema, B. Hannaford, Experiments in fingertip perception of surface discontinuities. Int. J. Robot. Res. **19**(7), 684–696 (2000)
40. P. Buttolo, B. Hannaford, Pen based force display for precision manipulation of virtual environments, in *Proceedings VRAIS-95*, pp. 217–225, Raleigh, NC, March 1995
41. P. Buttolo, B. Hannaford, Advantages of actuation redundancy for the design of haptic displays, in *Proceedings, ASME Fourth Annual Symposium on Haptic Interfaces for Virtual Environment and Teleoperator Systems*, San Francisco, vol. DSC-57-2, pp. 623–630, Nov 1995
42. P. Buttolo, D.Y. Hwang, B. Hannaford, Hard disk actuators for mini-teleoperation, in *Proceeding SPIE Telemanipulator and Telepresence Technologies Symposium*, Boston, 31 Oct 1994, pp. 55–61
43. Jeffrey Kerr, Bernard Roth, Analysis of multifingered hands. Int J Robot Res **4**(4), 3–17 (1986)
44. M.T Mason, J. Kenneth Salisbury Jr, *Robot Hands and the Mechanics of Manipulation* (The MIT Press, Cambridge, 1985)
45. Robert N Rohling and John M Hollerbach. Optimized fingertip mapping for teleoperation of dextrous robot hands. In *Robotics and Automation, 1993. Proceedings., 1993 IEEE International Conference on*, pages 769–775. IEEE, 1993
46. Thomas Speeter, Transforming human hand motion for telemanipulation. Presence Teleoper Virtual Environ **1**(1), 63–79 (1992)
47. T.H Massie, J. Kenneth Salisbury, The phantom haptic interface: a device for probing virtual objects, in *Proceedings of the ASME Winter Annual Meeting, Symposium on Haptic Interfaces for Virtual Environment and Teleoperator Systems*, vol. 55, pp. 295–300. Kluwer (1994)
48. Y. Nakamura, *Advanced Robotics: Redundancy and Optimization* (Addison-Wesley Longman Publishing Co. Inc., Reading, 1990)
49. P. Buttolo, B. Hannaford, Direct drive manipulator for pen-based force display. U.S. Patent #5,642,469, 24 June 1997
50. R. Leuschke, E.K.T. Kurihara, J. Dosher, B. Hannaford, High fidelity multi finger haptic display in *Proceedings World Haptics Congress 2005*, pp. 606–608, March 2005
51. S.C. Venema, E. Matthes, B. Hannaford, Flat coil actuator having coil embedded in linkage. U.S. Patent #6,437,770, 20 Aug 2002
52. R. Donlin, R. Leuschke, B. Hannaford, Experimental evaluation of attachment methods for a multifinger haptic device, in *Proceedings of World Haptics 2007*, Japan, 22 March 2007
53. J. Dosher, G. Lee, B. Hannaford, How low can you go? Detection thresholds for small haptic effects. in *Touch in Virtual Environments, Proceedings USC Workshop on Haptic Interfaces*, ed. by M. McLaughlin (Prentice Hall, New York, 2001)
54. J. Dosher, B. Hannaford, Detection of small haptic effects, in *Proceedings, SPIE Teleoperator and Telemanipulator Workshop, Boston MA, 29 Oct 29* 2001

Chapter 18
Dynamic Tactile Sensing

Mark R. Cutkosky and John Ulmen

Abstract Dynamic tactile sensing is an important capability for interacting with the world to identify textures and identify contact events such as objects making and breaking contact with the skin and rolling or slipping on the fingers. It is also used for identifying friction between the fingers and a grasped object and regulating the grasp force accordingly. Humans are endowed with multiple types of mechanoreceptors capable of detecting dynamic events with frequencies in the tens or hundreds of Hz. Increasingly, robots are also being equipped with tactile sensors capable of detecting dynamic phenomena, using a variety of different transducers depending on application-specific design considerations. Advances in electronics have made it possible to do the requisite amplification, signal processing and communication within the hand, with improved performance and greatly reduced wiring in comparison to early efforts.

Keywords Active sensors · Ampliers · Capacitive sensors · Contact motion · Design principles · Dynamic range · Frequency response · Human sensing · Instrumentation · Mechanoreceptors · Optical sensors · Piezoelectric · Psychophysical tests · PVDF · Skin · Skin acceleration · Sliding · Slip · Strain rate sensor · Stress rate sensor · Vibration

1 Introduction

If you press your finger against the corner of a table, you can feel the corner for as long as you hold your finger in place. In contrast, if you rest your finger gently on the table surface you feel relatively little, until you start moving it gently back and

M. R. Cutkosky (✉) · J. Ulmen
Department of Mechanical Engineering, Stanford University, Stanford, CA 94305, USA
e-mail: cutkosky@stanford.edu

J. Ulmen
e-mail: ulmenj@stanford.edu

R. Balasubramanian and V. J. Santos (eds.), *The Human Hand as an Inspiration for Robot Hand Development*, Springer Tracts in Advanced Robotics 95,
DOI: 10.1007/978-3-319-03017-3_18,

forth. Suddenly, the texture of the tabletop is apparent, including how dusty it is and whether it has any fine scratches or slippery patches. These sensations are provided by dynamic, or "fast acting" tactile sensors including the Pacinian corpuscles and Meissner endings [1, 2].

Normally, we integrate the signals from fast- and slow-acting mechanoreceptors to give us a comprehensive picture of surfaces, contact events and other conditions important for exploring and manipulating objects. However, even without local contact sensing—for example when we work with tools like scissors or pliers—we can use the vibrations sensed by our Pacinian corpuscles to tell us about events at the tool tip and the textures of surfaces that the tools are interacting with [3, 4].

Dynamic tactile sensing is also useful for event detection. When we grasp an object, it takes a short while for the forces to ramp up to a perceptible level. In contrast, the skin immediately experiences a large deceleration at the instant of contact. Even gently brushing against a soft surface, such as a silk scarf or the fur of an animal, elicits ample dynamic sensation, although the forces and pressures are very small. Another indication of the importance of vibration sensing for event detection is the ubiquity of small vibrating motors in cell phones, pagers and gaming consoles. We are "wired" to respond to vibrational event cues.

One kind of event that is particularly important for dexterous manipulation is the onset of sliding. When humans grasp and manipulate objects, they maintain a consistent margin of safety with respect to the minimum grasp force [2]. The regulation of the grasp force is informed by small vibrations that accompany incipient slippage. If a person cannot sense these vibrations, for example due to the application of an anesthetic, or peripheral neuropathy, or simply because the fingers are getting numb due to cold, the grasp force will gradually relax and a grasped object may be dropped.

In summary, dynamic tactile sensing is an indispensable part of human grasping and manipulation, and of our tactile interaction with the world around us. In robotics, it has received comparatively little attention relative to pressure and force/torque sensing. However it has the potential to be equally useful as robots become more adept at controlling forces on grasped objects and more responsive to changes in sensed forces and vibrations when interacting with the environment.

The following section briefly reviews the current understanding of dynamic tactile sensing in humans and the insights that can be drawn for dynamic tactile sensing in robotics and teleoperated devices. Section 3 reviews progress in developing dynamic tactile sensors for robots, examining the different operating principles employed and their advantages and disadvantages for responding to different types of phenomena. Section 4 then examines the basic characteristics of dynamic tactile sensing in more detail, using an example to illustrate some of the issues regarding mechanical design, transduction principles and signal processing. Section 5 concludes with some observations about the nature and utility of dynamic tactile sensing and challenges for future widespread use in robotics.

2 Dynamic Tactile Sensing in Humans

The human skin is endowed with fast- and slow-acting mechanoreceptors (Fig. 1). The fast-acting (FA) mechanoreceptors respond strongly to high-frequency signals and events characterized by accelerations, vibrations or rapid changes of the strain in the skin. However, if a force is imposed and then remains constant, they produce an initial flurry of spikes that quickly subside. In contrast, the slow acting (SA) mechanoreceptors continue to respond to steady skin deformation.

From the perspective of robotic dynamic tactile sensing, the FA mechanoreceptors are particularly of interest. The relatively superficial FA-I mechanoreceptors, which include the Meissner corpuscles, have an approximate frequency range of 5–50 Hz and a density of over $100/cm^2$ in the fingertips [1]. They are most responsive to changing contact conditions, as when a finger makes or breaks contact, or when an object slips over the finger surface. Indeed, slippery surfaces produce more excitation than rough ones [2]. Because of neural branching, and how FA-I receptors are attached to the skin and subcutaneous tissue, contact events affect not only receptors immediately at the contact site but also those nearby. Timing differences in the first spikes produced by these neighboring FA-I receptors can provide information about the shapes and type of contacts even as contacts are still being made, providing information of immediate use in controlling grasp forces. In addition, because each mechanoreceptor responds most strongly to a particular direction, it is possible to ascertain something of the shear stress distribution and the overall direction of force at the contact [5].

In comparison, the FA-II mechanoreceptors, or Pacinian corpuscles, have a broad receptive field and respond strongly to vibrations in the range of 50–500 Hz anywhere in the hand. They are excited, for example, by the vibrations of a tool or object held in the hand as it contacts or drags over surfaces in the environment [2].

Many events will excite multiple types of mechanoreceptors. For example, the act of grasping an object will initially excite the FA-I mechanoreceptors as the skin deforms. Vibrations arising from the contact will also be picked up by the FA-II

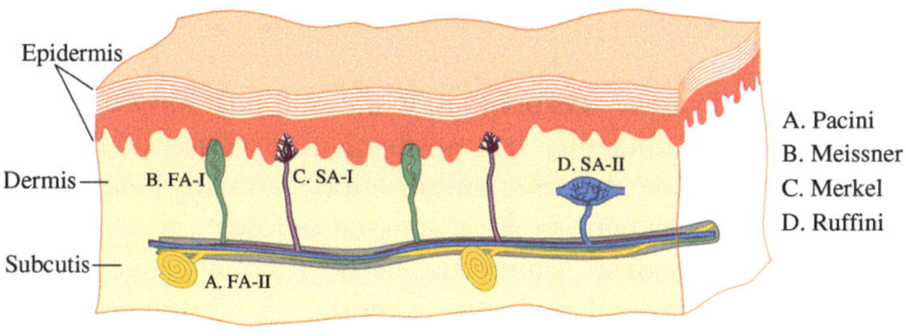

Fig. 1 Cross section of human skin showing slow-acting (SA) and fast-acting (FA) mechanoreceptors, with small (Type I) and large (Type II) receptive fields. Receptors near the surface have smaller receptive fields while those deeper respond to stimuli over a broader area. The superficial FA-I and deeper FA-II mechanoreceptors each have approximate analogs in robotic dynamic tactile sensing

mechanoreceptors. The SA-I mechanoreceptors will also respond to the continued deformation of the skin.

Given that friction is essential for manipulating objects, it is not surprising that humans are quite good at quickly determining the friction conditions associated with handling an object. Reacting to dynamic tactile cues, humans maintain a consistent margin of safety (10–40 % above the minimum grasp force) when handling objects.

Another commonly noted feature of human mechanoreception is its wide dynamic range, defined here as the ratio between the largest and smallest resolvable stimulus. The neural response typically follows a power law: $R \propto S^b$ where R is the response, S is the stimulus magnitude and b is an exponent that averages around 0.7 in many cases [6]. However the variation in intensity reported by subjects in psychophysical tests is often nearly linear, suggesting that the neural and cortical systems compensate for the nonlinear response. One potential advantage of nonlinear response with $b<1$ is a greater ability to discriminate between low stimulus levels (e.g. a light touch) without quickly saturating for large stimulus levels. At the lowest levels, FA-I mechanoreceptors can detect dynamic skin displacement amplitudes as small as 10 μm.

In summary, tactile sensing in humans is multi-modal, with different kinds of mechanoreceptors specialized for responding to different kinds of phenomena, but also highly integrated. Further, the lessons taught by human mechanoreception can provide insight into the design of robotic dynamic tactile sensing systems.

3 Developments in Robotic Dynamic Tactile Sensing

The advantages of dynamic tactile sensing have been noted in robotics as well as in biology, although dynamic sensors remain a relatively small part of the overall literature on robotic tactile sensing. Broadly speaking, dynamic tactile sensing includes several categories of sensors that are either meant to *detect* motion or incipient motion (slippage), or that *utilize* motion of the fingertips to produce results. A few other sensors are dynamic in the sense that they are actively stimulated, and monitor a change in impedance as they contact objects or surfaces. Finally, there are tactile array sensors that, while not inherently designed to detect or utilize motion, have sufficiently fast mechanical response, and can be sampled rapidly enough, to provide dynamic information as contact conditions change. Examples from of these categories are discussed below.

Motion detection sensors: Among the earliest dynamic tactile sensors were small rollers attached to encoders to detect motion at the fingertips [7, 8]. Other motion-detecting sensors include whiskers, akin to the vibrissae of animals [9, 10]. Still other contact motion-detecting sensors use a transduction technique that is particularly sensitive to motion—for example, a fabric with conductive yarns, for which large changes in resistance can occur in responses to contact movement [11].

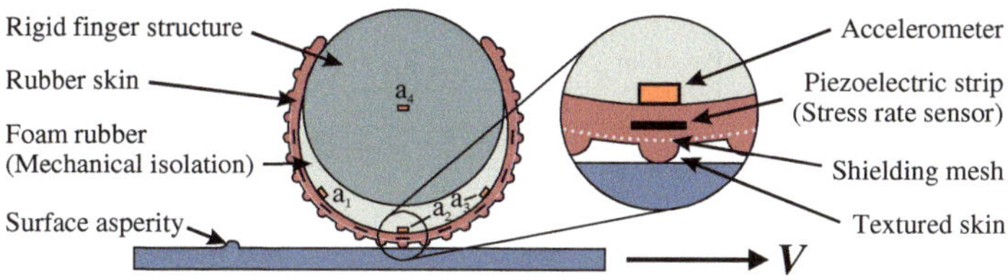

Fig. 2 Cross section of robotic skin, adapted from [12–14], showing accelerometers mounted to the skin and partially isolated from the finger structure by a compliant layer. Small piezoelectric strips embedded in the skin provide localized dynamic response to changes in skin stress. A flexible grounded layer of conductive fabric is added for shielding from electromagnetic interference

Accelerometers: A rough analog to human FA-II mechanoreceptors can be obtained by attaching small accelerometers to the skin (Fig. 2). These can provide large signals when the fingers make or break contact with an object or when an object starts to slip. When part of a suite of tactile sensors, they can be used to detect important events in manipulation tasks (e.g. fingers make or break contact with an object, an object is lifted or replaced on a table top, or slippage occurs) [12, 15].

One potential drawback to accelerometers is that they are excited by all kinds of vibrations, including those emanating from the drivetrain in the robot hand and arm. One way to mitigate this effect is to mount the accelerometers directly to an outer skin, separated from the finger structure by a layer of soft foam or similar material, so they are partially isolated from the finger structure and respond primarily to vibrations or accelerations of the skin itself during manipulation. It can also be useful to compare the signals from accelerometers located immediately at the contact site and nearby on the fingertip, as shown in Fig. 2; the neighboring accelerometers will respond to microslips at the periphery of the contact patch before gross sliding occurs [16].

strain rate sensors: While accelerometers respond broadly to all kinds of vibrations in the hand, a more selective response can be obtained by embedding piezoelectric or other strain rate transducers (also called stress rate transducers, because the strains are produced by stresses in the skin), producing a directional response to changing stresses. PVDF, a flexible piezoelectric film, is particularly useful in this context. Figure 2 shows PVDF strips embedded in an elastomeric skin.

A useful circuit for employing PVDF as a stress rate sensor is depicted in Fig. 3. Two thin strips of metallized, poled PVDF are laminated with opposite polarity so that the outer metallic surfaces are grounded for electromagnetic shielding. A large (approximately 1 MΩ) feedback resistor R_f, and amplifier servo the voltage across the film to zero, minimizing leakage effects. The film has different piezoelectric constants for each direction: $\mathbf{d} = [d_1\, d_2\, d_3]$ where the effect in the d_3 direction, perpendicular to the film, is strongest, the effect in the d_1 direction is also strong, and opposite to d_3, and the constant in the d_2 direction is much weaker. Thus, for triaxial stress, the charge, q, is

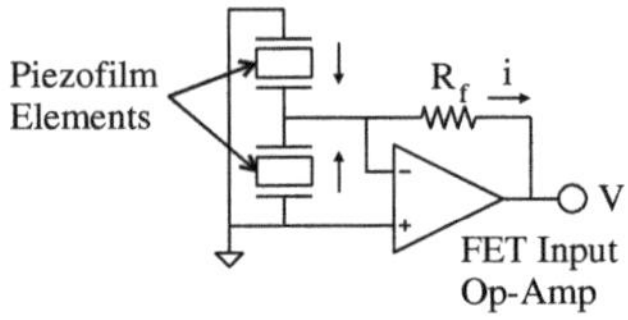

Fig. 3 Circuit for utilizing piezoelectric PVDF film as a stress rate sensor. Two pieces of film are laminated with opposite polarity and outer conductive surfaces grounded, to reduce noise (adapted from [13])

$$q = A\mathbf{d} \cdot \sigma \tag{1}$$

where $\sigma = [\sigma_1 \sigma_2 \sigma_3]^t$ and A is the area of the film, typically on the order of 0.5–1 cm^2. Since $i = d\sigma/dt$ the output of the circuit is proportional to the rate of change of stress:

$$v = AR_f\mathbf{d} \cdot \frac{d\sigma}{dt}. \tag{2}$$

As the fingertip drags over small surface features, the PVDF sensors provide large signals in proportion to the local rate of change of stress in the skin [13, 14]. In this manner, stress-rate sensors exploit motion. If the finger slides along a surface with a velocity $v = dx/dt$, the signal will be a function both of stress gradients within the skin, which depend on the sharpness of features on the object surface, and the sliding velocity: $d\sigma/dt = (d\sigma/dx)(dx/dt)$. This effect allows a finger with a single stress rate sensor to scan across an object, detecting small features such as ridges or bumps, with profiles as small as a few micrometers high [13].

It is also possible to instrument piezoelectric sensors with charge amplifiers so that they produce an output proportional to the local stress rather than the stress rate, although their high impedance makes them particularly suited for measuring dynamic forces [17–25]. However, with care they can even be used to measure static loads [26]. There is also the possibility to integrate PVDF directly with MOS circuitry for miniaturization and high signal/noise ratio [27, 28]. Local integration of amplifying elements could greatly reduce the complexity of wiring and readout electronics in this class of sensor.

Other transducers can also be used as stress or strain-rate sensors. For example, Kikkwe et al. [29] demonstrate a device using a viscous fluid and baffles that responds only to transient changes in loading.

Actively stimulated sensors: Capacitive and piezoelectric transducers can also function as actuators, leading to the possibility of actively stimulated tactile arrays. Variations include ultrasonic arrays that measure changes in thickness as a soft skin makes contact with surfaces [30], stimulated piezoelectric probes [31] or resonant cavities [32, 33], piezoelectric array sensors [34] and pneumatically driven cells with piezoresistive elements [35].

Other sensors with fast response: There are a number of sensor designs that, while not designed specifically to detect or utilize motion, can nonetheless produce dynamic information by virtue of having a frequency response on the order of tens or hundreds of Hz. The performance of these sensors depends both on the available sampling or addressing rate for individual elements and, especially, on their mechanical construction. A particular challenge is to achieve a useful frequency response of 50 Hz or greater with soft materials. The sensors need to have low hysteresis, which in turn requires that they use materials without significant damping or viscoelasticity, and that they be constructed as a single bonded unit, instead of assemblies with internal contacts having friction or adhesion. Examples of such sensors include capacitive arrays that utilize a silicone rubber foam or molded pattern, bonded to a conductive outer skin for shielding [36–41]. Other sensors that employ an incompressible low-viscosity fluid can also provide dynamic signals [42].

With sufficient response, small arrays of piezoresistive or capacitive sensors can also be used in a scanning mode, to discriminate among different textures [43–46] or to detect incipient object slippage based on the ratios of strains as an elastic fingertip is pressed against a surface and loaded in shear [47]. Optical tactile sensors can also detect changing textures and slip, depending on the frame rate of the associated optical imaging device [48]. In many cases, mechanical design features in the skin and how the sensors are connected to the skin can be used to enhance the sensor response. For example ridges can produce a "plucking" action that enhances response as a sensor slides over small features [49, 50]. More generally, texturing the robot skin leads to more predictable sliding behavior, often with a characteristic vibration frequency [13].

4 Dynamic Tactile Sensing Design Considerations

As the previous section reveals, there are various ways to achieve dynamic tactile sensing. However, they share a number of common design principles:

- low hysteresis for good dynamic response
- mechanical isolation to measure forces or motions at the location of interest only
- sensitivity aligned with force direction of interest, i.e. normal forces may be important for contact, whereas normal and shear forces may be important for friction estimation
- strong event correlation, e.g. fingerprint ridges vibrating at a characteristic frequency during slipping.

In addition, dynamic tactile sensors share a number of basic considerations with all tactile sensors:

- coverage and density
- repeatability
- minimum resolvable force or acceleration
- maximum force or acceleration before saturation

- packaging and robustness
- provisions for sampling and signal routing.

Unlike a CMOS imaging chip, a tactile sensor must be curved, compliant, flexible (especially around joints), tough enough to survive repeated impacts and scrapes, and distributed over large surface areas. As a result, wiring becomes a challenge. These factors have conspired to make progress in tactile sensing slower than in computer vision.

Achieving a suitable combination of the (sometimes competing) objectives listed above typically requires a design that is customized to a particular application. Thus the sensors and skin for a fingertip are different from those on an arm. To illustrate some of the design considerations in more detail, the following section considers a capacitive tactile array designed to provide both steady, or slow-acting, and dynamic, or fast-acting, performance.

4.1 Dynamic Sensing Example

4.1.1 Sensor Design

A capacitive transducer can be constructed as in Fig. 4 with two conductive plates separated by a compressible dielectric medium. As forces are applied normal to the surface, the gap reduces, producing a change in capacitance governed by the well known parallel plate capacitor equation:

$$C = \frac{\epsilon A}{d} \tag{3}$$

where C, ϵ, A, and d are the capacitance, dielectric constant, plate area, and separation, respectively. With this simple structure, it can be seen that the sensitivity to small changes in force, F, depends on the initial plate separation, d_0, and the dielectric stiffness. Let $F = k(d_0 - d)$ for small deflections, where the

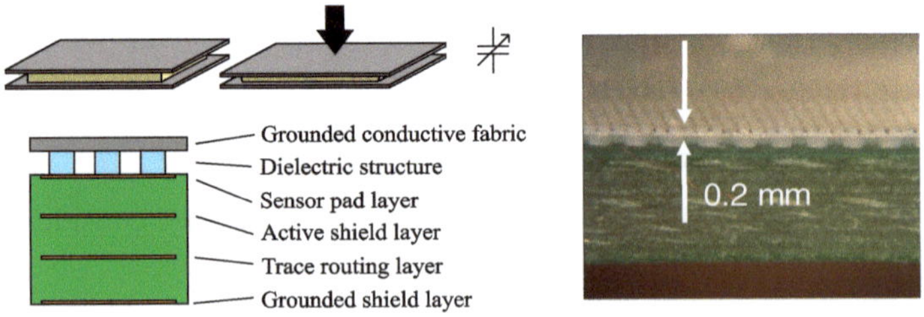

Fig. 4 **a** Two conductive plates are separated by an elastic dielectric. As force is applied, the gap between the plates reduces, changing the capacitance. **b** In practice, additional shielding layers are usually required for immunity to noise and to prevent the sensor from responding in part as a proximity sensor. **c** In construction, the dielectric and upper conductive layer can be quite thin

$k = AE/d$ and E is the effective modulus (i.e. the average modulus for initial compression, given that the dielectric may not be homogeneous). Then for initial contact:

$$\lim_{d \to d_0} \frac{\partial C}{\partial F} = \frac{\epsilon}{E d_0} \tag{4}$$

For sensitivity to small forces, E should be low. A low density open cell foam can be quite compressible and indeed, yields a high sensitivity. Unfortunately, such a foam also tends to have significant hysteresis, which reduces the dynamic response. A better solution for dynamic response is to use a pattern molded from an elastomer such as silicone rubber, which has low viscoelasticity and low hysteresis in comparison to other polymers (e.g. urethanes) of similar stiffness [51, 52]. The other way to increase sensitivity is to make the sensor quite thin (small d_0), as shown in Fig. 4 at right. In this example, the dielectric consists of a dense array of short elastic posts, bonded at the base and top, to create a monolithic structure with low hysteresis and moderate stiffness (approximately 1 MPa per mm of compression) for initial deflections. However the stiffness increases for large deflections leading to a nonlinear response, which can be useful for delaying saturation at high loads.

The thin sensor is intended to be located beneath a thicker artificial skin, which provides the desired compliance and has the added advantage of "blurring" the pressure distributions associated with sharp contacts. This well known effect can lead to higher accuracy in resolving the locations of isolated contacts for a given taxel size [53, 54]. However, if one is interested in dynamic response, the skin material, like the dielectric, should be chosen to avoid significant hysteresis and viscoelasticity. At higher frequencies (e.g. above 100 Hz for the sensor shown in Fig. 4), the mass of the skin should also be considered.

4.1.2 Instrumentation and Signal Processing

Many options are available for measuring the change of capacitance in a sensor like that shown in Fig. 4. With the advent of smart cell phones with capacitive touch screens, an attractive solution is to use dedicated integrated circuits (ICs) that provide active shielding, high sampling and analog/digital conversion rates, filtering and communication over a high-speed bus. In particular, a solution used by the authors, and by others for sensors on the iCub robot [39, 40], is the Analog Devices AD7147-1. With this IC, small arrays of n sensors can be sampled at $1200/n$ Hz, well into the range of human FA-I and FA-II sensitivity. Multiple arrays can be controlled by small microcrontrollers located adjacent to the sensors [41].

Figure 5 shows a three fingered robot hand with tactile sensors as described above, pulling a thin object out of a slot. It is of interest to see whether object/hand slips can be distinguished from object/world slips, something that humans do easily using their suite of mechanoreceptors, as discussed in Sect. 2. In the present case, it is difficult to distinguish between the two cases using accelerometers or individual piezoelectric sensors, as both types of slippage excite numerous vibrations in the skin. However, by comparing the average power per taxel with the maximum power, a distinction emerges.

Let $p_n^{(i)}$ be the nth sample from the ith sensor on the finger surface. The sum of signals over the surface is

$$s_n = \sum_i p_n^{(i)} \tag{5}$$

and the power in a given frequency band is given by

$$S(f, w) = \sum_{k=f-w}^{f+w} S_k \tag{6}$$

$$M(f, w) = \max_i \sum_{k=f-w}^{f+w} P_k^{(i)} \tag{7}$$

where S_k and $P_k^{(i)}$ are the power in the kth frequency bin from the Discrete Fourier Transform of s and $p^{(i)}$, respectively, and f and w are the center and half-width, respectively, of the frequency band to consider. The sensor with the most power in the band of interest is used to calculate M.

Figure 5 shows the results of taking the ratio $S(f, w)/M(f, w)$ for $w = 7.5$ Hz and a range of frequencies, for a group of taxels at the fingertips. The taxels were sampled at 300 Hz and filtered using a discrete 1st order Butterworth high-pass filter with a cutoff frequency of 5 Hz, and power spectrum was computed using the FFT. The power spectra were averaged across 10 trials to give an estimate of a "typical" power spectrum for the manipulation trials. What stands out in Fig. 5 is that over a range of frequencies, the ratios of the sum and the maximum power are typically different for object/hand and object/world slips. The insensitivity to frequency suggests that such a metric should hold for a range of range of speeds and textures [41].

4.1.3 Instrumenting as a Dynamic Sensor

While a single sensor package providing broad frequency response is a highly desirable solution, there may remain advantages to designing a purpose built dynamic sensor for high frequency signals. In such a sensor, the steady state or "DC" component of the signal can be ignored. In the case of the capacitive tactile sensor, the direct analog output from the sensor may be high pass filtered prior to amplification, potentially allowing for much higher amplification levels due to the lack of a large bias offset that would cause premature saturation. This is especially true if there is a mismatch in the required force range at high and low frequencies (i.e. it may be desirable to sense large static loads and light contact events). Furthermore, some mechanical/material properties such as thermal drift will no

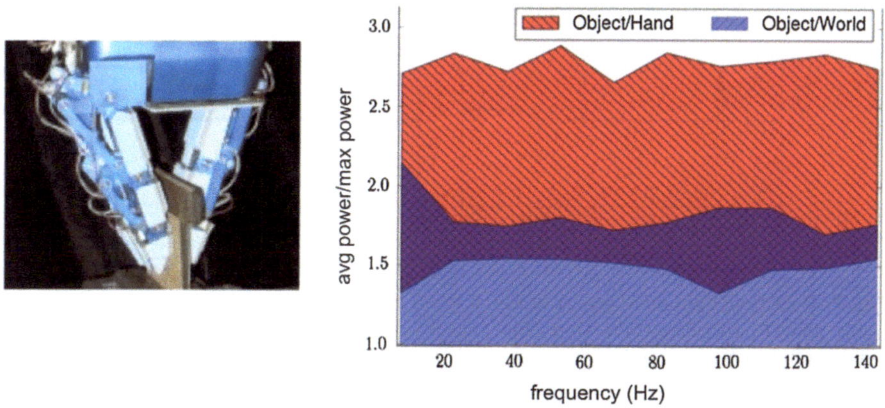

Fig. 5 A robot hand grasps a thin object and pulls it out of a slot; object/hand and object/world slips can occur (*left*). The ratios of average to maximum power for a collection of taxels help to distinguish object/hand from object/word slips, for a wide range of frequencies and a collection of 10 trials (*right*)

longer affect response and a wider group of materials can be considered for mechanical transduction.

An example dynamic sensor circuit for a capacitive transducer is shown in Fig. 6. The schematic is a typical amplifier for a condenser microphone. The circuit provides high dynamic sensitivity due to its high pass filtered input stage and allows correspondingly high gain. The preponderance of microphones in devices such as cell phones means that small and inexpensive microphone amplifier IC's are readily available. The filtering and amplifier circuits can be located adjacent to the transducer allowing for minimal noise coupling from cabling. As seen in Fig. 7, for the same transducer, the response and signal/noise ratio from the microphone circuit are much greater for light dynamic contact events such as dropping or placing a small weight on the sensor surface.

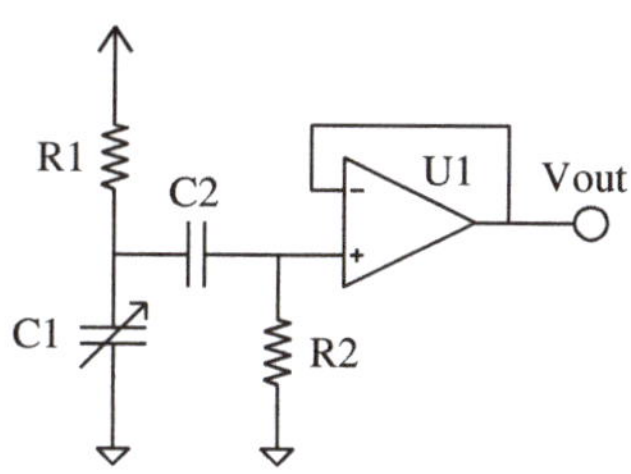

Fig. 6 Condensor microphone circuit that can be used with capacitive sensor. Resistor R1 allows a static bias charge to accumulate on sensor C1 and, in dynamic operation, C1 can be considered to operate in a constant charge condition. C2 and R2 (R2 is typically the input impedance of the amplifier) high pass filter the voltage across C1 prior to amplification. High amplification is possible because the filter removes bias and the amplifier can be located adjacent to transducer to minimize noise

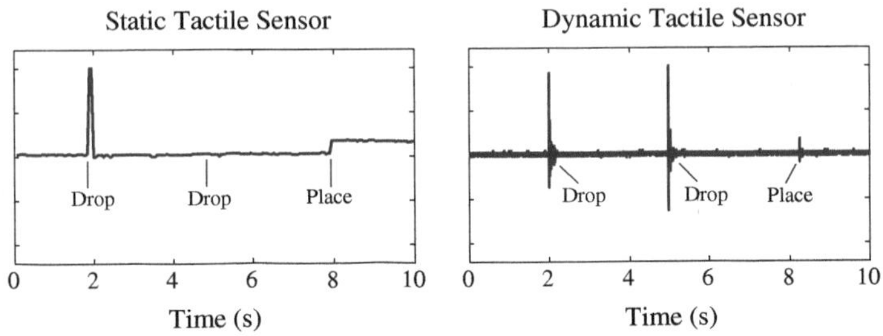

Fig. 7 Data from capacitive sensor in Fig. 4 as 10 g weight is dropped from a few cm and then gently placed on surface (*left*). Data from sensor instrumented using circuit from Fig. 6 for the same loading sequence (*right*)

5 Summary

As the example in this section shows, a common transducer can, with attention to mechanical construction and materials, provide data at frequencies high enough to detect dynamic events such as object slippage. As noted in [15], comparisons of low-pass and high-pass filtered data from a commercially available tactile array, and data from an accelerometer, can be combined to detect various dynamic events in manipulation. As Fig. 5 reveals, the ability to compare correlated and uncorrelated signals from adjacent taxels is also helpful in distinguishing between different kinds of dynamic events in a manner somewhat analogous to the use of groups of FA-I sensors in humans. Finally, the distinction between a tactile array with good frequency response and a dedicated dynamic tactile sensor often comes down to the circuits used to filter and amplify signals. Constructing a circuit that only amplifies high-frequency signals, as in Figs. 3 and 6, allows much greater sensitivity to transient phenomena.

6 Conclusions and Future Challenges

Dynamic tactile sensing is extremely important in human manipulation: it allows us to detect events like contact and slippage, and to distinguish textures and perceive tiny surface features as we slide our fingers over surfaces. Humans use combinations of mechanoreceptors, with different inherent properties, both at the site of a contact and nearby, to distinguish among events like objects slipping in the hand and grasped objects slipping against a surface in the world. As robots emerge from structured and predictable environments like manufacturing they too will increasingly require dynamic tactile sensing to be informed of contact events and changes in texture, friction conditions, etc.

Many approaches are available for dynamic tactile sensing including accelerometers and high-frequency piezoelectric transducers. It is also possible to obtain dynamic information from conventional capacitive, piezoresistive, optical or other tactile arrays provided that care has been exercised in their design and materials choices. However, for the greatest sensitivity to phenomena like a light grazing touch, it is ideal to configure at least some of the transducing elements explicitly as high frequency sensors, permitting high amplification and a high signal/noise ratio for transient forces or vibrations.

References

1. R.S. Johansson, A.B. Vallbo, Tactile sensibility in the human hand: relative and absolute densities of four types of mechanoreceptive units in glabrous skin. J. Physiol. **286**(1), 283 (1979)
2. R.S. Johansson, J.R. Flanagan, Coding and use of tactile signals from the fingertips in object manipulation tasks. Nat. Rev. Neurosci. **10**(5), 345–359 (2009)
3. S.J. Lederman, R.L. Klatzky, C.L. Hamilton, G.I. Ramsay, Perceiving surface roughness via a rigid probe: effects of exploration speed and mode of touch. Haptics-e Electron. J. Haptic Res. **1**(1), 1–20 (1999)
4. T. Yoshioka, J. Zhou, Factors involved in tactile texture perception through probes. Adv. Robot. **23**(6), 747–766 (2009)
5. I. Birznieks, P. Jenmalm, A.W. Goodwin, R.S. Johansson, Encoding of direction of fingertip forces by human tactile afferents. J. Neurosci. **21**(20), 8222–8237 (2001)
6. A.B. Vallbo, R.S. Johansson, Properties of cutaneous mechanoreceptors in the human hand related to touch sensation. Hum. Neurobiol. **3**, 3–14 (1984)
7. M. Ueda, K. Iwata, H. Shingu, Tactile sensors for an industrial robot to detect a slip, in *Proceedings of the 2nd International Symposium on Industrial Robots*, pp. 63–70, 1972
8. R. Masuda, K. Hasegawa, Slip sensor of industrial robot and its application. Electr. Eng. **96**(5), 219–226 (1976)
9. P.A. Schmidt, E Mael, R Wurtz, E. Maël, R.P. Würtz, A sensor for dynamic tactile information with applications in human robot interaction and object exploration. Robot. Auton. Syst. **54**(12), 1005–1014 (2006)
10. R.A. Russell, Using tactile whiskers to measure surface contours, in *Proceedings 1992 IEEE International Conference on Robotics and Automation* (IEEE Computer Society Press, 1992), pp. 1295–1299
11. Van Anh Ho, D. Kondo, S. Okada, T. Araki, E. Fujita, M. Makikawa, S. Hirai, Development of a low-profile sensor using electro-conductive yarns in recognition of slippage, in *2011 IEEE/RSJ International Conference on Intelligent Robots and Systems* (IEEE, 2011), pp. 1946–1953
12. R.D. Howe, N. Popp, P. Akella, I. Kao, M.R. Cutkosky. Grasping, manipulation, and control with tactile sensing. In *Proceedings of IEEE International Conference on Robotics and Automation* (IEEE Computer Society Press, 1990), pp. 1258–1263
13. R.D. Howe, M.R. Cutkosky, Dynamic tactile sensing: perception of fine surface features with stress rate sensing. IEEE Trans. Robot. Autom. **9**(2), 140–151 (1993)
14. J.S. Son, E.A. Monteverde, R.D. Howe, A tactile sensor for localizing transient events in manipulation. In *Proceedings of the 1994 IEEE International Conference on Robotics and Automation* (IEEE Computer Society Press, 1994), pp. 471–476
15. J.M. Romano, K. Hsiao, G. Niemeyer, S. Chitta, K.J. Kuchenbecker, Human-inspired robotic grasp control with tactile sensing. IEEE Trans. Rob. **27**(6), 1067–1079 (2011)

16. M.R. Tremblay, M.R. Cutkosky, Estimating friction using incipient slip sensing during a manipulation task, in *Proceedings of IEEE International Conference on Robotics and Automation,* vol. 1, pp. 429–434, 1993
17. P. Dario, D. De Rossi, C. Domenici, R. Francesconi, Ferroelectric polymer tactile sensors with anthropomorphic features, in *Proceedings of IEEE International Conference on Robotics and Automation,* vol. 1 (Institute of Electrical and Electronics Engineers, 1984), pp. 332–340
18. R.W. Patterson, G.E. Nevill, The induced vibration touch sensor—a new dynamic touch sensing concept. Robotica **4**(01), 27–31 (1986)
19. E.S. Kolesar, C.S. Dyson, R.R. Reston, R.C. Fitch, D.G. Ford, S.D. Nelms, Tactile integrated circuit sensor realized with a piezoelectric polymer, in *Proceedings of Eighth Annual IEEE International Conference on Innovative Systems in Silicon* (IEEE, 1996), pp. 372–381
20. S. Ando, H. Shinoda, Ultrasonic emission tactile sensing. IEEE Control Syst. Mag. **15**(1), 61–69 (1995)
21. G. Canepa, R. Petrigliano, M. Campanella, D. De Rossi, Detection of incipient object slippage by skin-like sensing and neural network processing. IEEE Trans. Syst. Man Cybern. Part B **28**(3), 348–356 (1998)
22. J.T. Dennerlein, R.D Howe, P.A. Millman, Vibrotactile feedback for industrial telemanipulators, in *Proceedings of the Sixth Annual Symposium on Haptic Interfaces* (ASME IMECE, Dallas, 1997), pp. 1–7
23. B. Choi, H.R. Choi, S. Kang, Development of tactile sensor for detecting contact force and slip, in *2005 IEEE/RSJ International Conference on Intelligent Robots and Systems* (IEEE, 2005), pp. 2638–2643
24. A. Mingrino, P. Dario, A. Sabatini. A hand prosthesis with slippage control by tactile sensors, in *Proceedings of the 15th Annual International Conference of the IEEE Engineering in Medicine and Biology Society* (IEEE, 1993), pp. 1276–1277
25. A.M. Dollar, L.P. Jentoft, J.H. Gao, R.D. Howe, Contact sensing and grasping performance of compliant hands. Auton. Robot. **28**(1), 65–75 (2009)
26. M.A. Qasaimeh, S. Sokhanvar, J. Dargahi, M. Kahrizi, PVDF-based microfabricated tactile sensor for minimally invasive surgery. J. Microelectromech. Syst. **18**(1), 195–207 (2009)
27. R.S. Dahiya, M. Valle, G. Metta, L. Lorenzelli, A. Adami, Design and fabrication of posfet devices for tactile sensing, in *Transducers 2009, International Conference on Solid-State Sensors, Actuators and Microsystems* (IEEE, 2009), pp 1881–1884
28. R.S Dahiya, L. Lorenzelli, G. Metta, M. Valle, POSFET devices based tactile sensing arrays. Ferroelectrics 893–896 (2010)
29. R. Kikuuwe, A. Sano, H. Mochiyama, N. Takesue, H. Fujimoto, A tactile sensor capable of mechanical adaptation and its use as a surface deflection detector. Strain 256–259 (2004)
30. B.L. Hutchings, A.R. Grahn, R.J. Petersen, Multiple-layer cross-field ultrasonic tactile sensor, in *Proceedings of the 1994 IEEE International Conference on Robotics and Automation* (IEEE Computer Society Press, 1994), pp. 2522–2528
31. S. Omata, Real time robotic tactile sensor system for the determination of the physical properties of biomaterials. Sens. Actuators A **112**(2–3), 278–285 (2004)
32. H. Shinoda, K. Matsumoto, S. Ando, Acoustic resonant tensor cell for tactile sensing, in *Proceedings of International conference on Robotics* and *Automation,* vol. 4, pp. 3087–3092, 1997
33. H. Shinoda, S. Sasaki, K. Nakamura, Instantaneous evaluation of friction based on ARTC tactile sensor, in *Proceedings ICRA '00 IEEE International Conference on Robotics and Automation,* vol. 3, pp. 2173–2178, 2000
34. V. Todorova, D. Kolev, Piezoelectric resonance sensor array, in *Proceedings of 27th International Spring Seminar on Electronics Technology: Meeting the Challenges of Electronics Technology Progress,* vol 2 (IEEE, 2004), pp. 247–251
35. M. Shikida, T. Shimizu, K. Sato, K. Itoigawa, Active tactile sensor for detecting contact force and hardness of an object. Sens. Actuators A **103**(1–2), 213–218 (2003)
36. R.D. Howe, W.J. Peine, D.A. Kantarinis, J.S. Son, Remote palpation technology. IEEE Eng. Med. Biol. Mag. **14**(3), 318–323 (1995)

37. D.T.V. Pawluk, J.S. Son, P.S. Wellman, W.J. Peine, R.D. Howe, A distributed pressure sensor for biomechanical measurements. J. Biomech. Eng. **120**(2), 302–305 (1998)
38. J. Ulmen, M. Cutkosky, A robust, low-cost and low-noise artificial skin for human-friendly robots, in *IEEE International Conference on Robotics and Automation (ICRA)*, pp. 4836–4841, 2010
39. G. Cannata, M. Maggiali, G. Metta, G. Sandini, An embedded artificial skin for humanoid robots, in *IEEE International Conference on Multisensor Fusion and Integration for Intelligent Systems* (IEEE, 2008), pp. 434–438
40. A. Schmitz, M. Maggiali, L. Natale, B. Bonino, G. Metta, A tactile sensor for the fingertips of the humanoid robot iCub. Inform. pp. 2212–2217 (2010)
41. B. Heyneman, J. Ulmen, M.R. Cutkosky, Scalable and responsive articial skin for mobile manipulation, in *7th ACM/IEEE International Conference on Human-Robot Interaction* ed. by G. Cannata, F. Mastrogiovanni, G. Metta, L. Natale, *Advances in tactile sensing and touch based human-robot interaction*, Boston, MA, 2012
42. N. Wettels, V.J. Santos, R.S. Johansson, G.E. Loeb, Biomimetic tactile sensor array. Adv. Robot. **22**(8), 829–849 (2008)
43. H.B. Muhammad, C. Recchiuto, C.M. Oddo, L. Beccai, C.J. Anthony, M.J. Adams, M.C. Carrozza, M.C.L. Ward, A capacitive tactile sensor array for surface texture discrimination. Microelectron. Eng. **88**(8), 1811–1813 (2011)
44. Florian De Boissieu, C. Godin, B. Guilhamat, D. David, C. Serviere, D. Baudois, Tactile texture recognition with a 3-axial force MEMS integrated articial finger. Houille Blanche Revue Internationale De L'Eau, 2010
45. C.M.Oddo, L.Beccai, M.Felder, F.Giovacchini, M.C.Carrozza, Artificial roughness encoding with a bio-inspired MEMS-based tactile sensor array. Sensors **9**(5), 3161–3183 (2009)
46. C. Schürmann, R. Haschke, H. Ritter, Modular high speed tactile sensor system with video interface, in *Tactile sensing in Humanoids—Tactile Sensors and beyond IEEE-RAS Conference on Humanoid Robots*, no. 12. IEEE Paris, France, 2009
47. T. Maeno, T. Kawamura, S.-C. Cheng, Friction estimation by pressing an elastic finger-shaped sensor against a surface. IEEE Trans. Robot. **20**(2), 222–228 (2004)
48. V. Maheshwari, R.F. Saraf, High-resolution thin-film device to sense texture by touch. Science **312**(5779), 1501–1504 (2006)
49. D. Yamada, T. Maeno, Y. Yamada, Artificial finger skin having ridges and distributed tactile sensors used for grasp force control, in *Proceedings 2001 IEEE/RSJ International Conference on Intelligent Robots and Systems. Expanding the societal role of robotics in the next millennium* (Cat. No.01CH37180), vol. 2 (IEEE, 2001), pp. 686–691
50. A.M. Mazid, R.A. Russell. A robotic opto-tactile sensor for assessing object surface texture, In *2006 IEEE Conference on Robotics, Automation and Mechatronics* (IEEE, 2006), pp. 1–5
51. S. O'Sullivan, R. Nagle, J.A. McEwen, V. Casey. Elastomer rubbers as deflection elements in pressure sensors: investigation of properties using a custom designed programmable elastomer test rig. J.Phys. D Appl. Phys. **36**(15), 1910–1916 (2003)
52. J.C. Lötters, W. Olthuis, P.H. Veltink, P. Bergveld, The mechanical properties of the rubber elastic polymer polydimethylsiloxane for sensor applications. J. Micromech. Microeng. **7**(3), 145–147 (1997)
53. R.S. Fearing, J.M. Hollerbach, Basic solid mechanics for tactile sensing. Int. J.Robot. Res. **4**(3), 40–54 (1985)
54. M.R. Cutkosky, R.D. Howe, W.R. Provancher, *Force and Tactile Sensors* ed. by B. Siciliano, O. Khatib. Springer Handbook of Robotics, chapter 20 (Springer, Heidelberg, 2008), pp. 455–476

Chapter 19
Multimodal Tactile Sensor

Nicholas Wettels, Jeremy A. Fishel and Gerald E. Loeb

Abstract We have developed a finger-shaped sensor array (BioTac®) that provides simultaneous information about contact forces, microvibrations and thermal fluxes, mimicking the full cutaneous sensory capabilities of the human finger. For many tasks, such as identifying objects or maintaining stable grasp, these sensory modalities are synergistic. For example, information about the material composition of an object can be inferred from the rate of heat transfer from a heated finger to the object, but only if the location and force of contact are well controlled. In this chapter we introduce the three sensing modalities of our sensor and consider how they can be used synergistically. Tactile sensing and signal processing is necessary for human dexterity and is likely to be required in mechatronic systems such as robotic and prosthetic limbs if they are to achieve similar dexterity.

Keywords Force and tactile sensing · Biomimetics · Dexterous manipulation · Thermal sensing · Texture

N. Wettels (✉) · J. A. Fishel · G. E. Loeb
Syntouch LLC, 2222S Figueroa St. PH2, Los Angeles, CA 90007, USA
e-mail: nick.wettels@gmail.com

J. A. Fishel
e-mail: jeremy.fishel@syntouchllc.com

G. E. Loeb
e-mail: gloeb@usc.edu

N. Wettels · J. A. Fishel · G. E. Loeb
Department of Biomedical Engineering, University of Southern California,
Los Angeles, CA, USA

R. Balasubramanian and V. J. Santos (eds.), *The Human Hand as an Inspiration for Robot Hand Development*, Springer Tracts in Advanced Robotics 95,
DOI: 10.1007/978-3-319-03017-3_19,

1 Introduction

The performance of robotic and prosthetic hands in unstructured environments is substantially limited by their having little or no tactile information compared to the multi-modal sensory feedback of the human hand. The necessity of tactile information is evidenced in clinical cases where patients who suffer peripheral nerve damage to their hands are able to initiate, but not maintain stable grasp due to lack of sensory feedback from cutaneous receptors [1]. Rapid reflexive adjustment of grip is essential for handling objects and depends on tactile feedback via the spinal cord [2]. Autonomous robots can deal only with rigid objects in known orientations, specifically because they lack tactile feedback. Overcoming this limitation would enable many commercial applications, including anthropomorphic robotic assistants, teleoperated dexterous manipulators, autonomous robots, quantitative palpation for medical diagnostics, and prosthetic hands.

Tactile sensing in robotic end-effectors must meet two types of haptic requirements: those geared toward object identification and characterization (e.g. determination of compliance, thermal and textural properties) and those designed for manipulation (e.g. closed loop control of grip force). Most applications will involve both: first exploratory movements are made to gain information about the object and its properties in order to select and implement intelligent strategies to handle it. For example, characterization of friction coefficients will influence the required grip force applied in manipulation strategies.

One of the limiting factors in robotics has been the absence of sensitive yet robust sensors that can be incorporated into anthropomorphic mechatronic fingers and used in the often hostile environments in which hands function. A wide variety of tactile sensing technologies have been attempted. Transduction mechanisms such as optics, capacitance, piezoelectric, ultrasound, conductive polymers, etc. provide some useful sensing but only for limited environments or applications. Most require large numbers of delicate transducers and connections in deformable media that will be in harm's way. Table 1 summarizes various types of sensors based on transduction mechanism. In-depth surveys of tactile sensing technologies can be found in [3–6]; additional information can be found in the dynamic tactile sensing chapter in this book.

While a wide array of technologies have been developed, the great majority of sensors have focused on individual sensing modalities rather than the multimodal combination of sensory capabilities found in human skin. We have developed a finger-shaped sensor array (BioTac®) that provides simultaneous information about the contact forces [7], microvibrations [8], and thermal fluxes induced by contact with external objects, thus mimicking the full cutaneous sensory capabilities of the biological finger (Fig. 1).

The biomimetic BioTac array has an elastomeric skin inflated by a conductive liquid over a bone-like core, resulting in mechanical properties similar to a human fingertip [9]. Contact force deforms the skin and underlying fluid, resulting in changes in the electrical impedance of an array of electrodes arranged on the

Table 1 Summary of tactile sensing technologies

Transduction Method	Advantages	Disadvantages	Examples
Capacitive	Can be flexible, wide dynamic range, sensitive	Hysteresis, noise, limited resolution	Pressure profile systems robotouch [37, 38]
Inductive	High sensitivity, repeatability	Complex construction, electronics in workspace, low spatial resolution	[39]
Resistive: deformable contact area	Flexible, thin	Hysteresis	Inaba: Inastomer [40]
Resistive: conductive fabric	Flexible, robust, simple	Unable to resolve more than one contact point	[41]
Resistive: quantum tunneling composite	Sensitive, wide dynamic range	Hysteresis, gas absorption	Peratech QTC [42]
Resistive: strain gauge	Sensitive, wide dynamic range	Bulky, expensive	ATI: Nano 17 load cell [43]
Resistive: piezoresistive conductive polymer	Thin, low cost, simple	Hysteresis, stiff	Tekscan flexiforce [44]
Resistive: piezo-MEMS	Small, multi-element	Large number of wires in workspace	[45]
Resistive/Capacitive: liquid channels embedded in elastomer	Small, flexible, multidimensional force sensing capability	Large number of wires in workspace	[46]
Polymer-MEMs (multi-modal)	Measures 6-DOF force, heat-flow, temperature, roughness	Large number of wires in workspace, wiring complexity	[47]
Piezoelectric	Detects dynamics for slip and texture	Only detects dynamic events, thermal sensitivity	[48, 49]
Optical: video processing	Very high resolution, sensitive	Computationally intensive, sensitive to ambient light	[50, 51]
Optical: resistive	Flexible, low hysteresis	Complex fabrication	[27]
Magneto-elastic	Very sensitive, low hysteresis	Sensitive to external magnetic fields	[52]
Magneto-resistive	Robust, sensitive, low hysteresis	Noisy	[53]
Ultrasound	Can resolve static and dynamic information	High voltage, complex electronics	[54, 55]

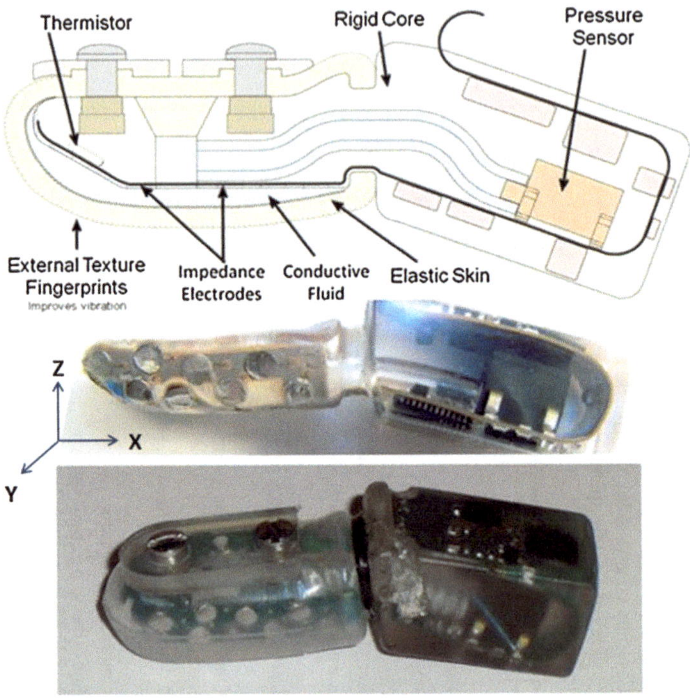

Fig. 1 (*Top*) Cross-sectional drawing of BioTac showing various components used for each sensing modality. (*Middle*) Coordinate frame with lateral picture of device with skin removed; large disks on the distal portion are platinum electrodes. (*Bottom*) Angular view of sensor with skin and nail installed; device is inflated with 200 μL of fluid

surface of the core. Artificial neural networks (ANN) and Gaussian mixture model regression (GMMR) can be used to extract three dimensional force vectors from a moderate number of nonlinear impedance sensing channels. Sliding the skin over textured surfaces results in microvibrations that propagate as sound waves through the fluid to a pressure transducer (hydrophone) that is high-pass filtered and amplified to improve sensitivity. Because the conductivity of the fluid increases with temperature, a commercial thermistor is located on the distal, flat portion of the core for temperature compensation of the fluid. The BioTac is heated above ambient (like a biological finger), so the slope of temperature changes (thermal flux) provides an indicator of the thermal properties of contacted objects.

This chapter will discuss the prior art and problems of using thermal and vibration sensing modalities to discriminate objects and detect slip. We demonstrate that thermal characterization of objects obtained from principal component analysis of temperature derivatives provides information about their respective thermal effusivities. We also show that information about texture and slip can be derived from vibrations of skin ridges sliding over a surface. In both cases it is necessary to calibrate for the force vectors applied to the sensor, as well as the point of application of this force. By merging these three modalities into a single device, robotic and prosthetic hands can benefit from a sensing package that provides the data required to identify and manipulate objects.

2 Methods

2.1 BioTac Design: Electronic Architecture

The BioTac contains integrated electronic circuitry to condition and digitize signals for all three sensing modalities. The complete signal processing chain for the integrated electronics is realized as a miniature 3-layer flex-circuit that carries all sensing electrodes, transducers and electronic components for placement in the mold that forms the rigid core [10]. The impedance sensing electrodes are switched by multiplexer (Analog Devices, Inc., #ADG732) and connected in series to a load resistor and in turn to the internal analog-to-digital (ADC) converter in the PIC microcontroller (Microchip Technology Inc., #dsPIC33FJ128GP802). This circuitry measures the voltage produced by the current passing through the fluid path from the excitation electrodes—four similar contacts distributed around the fingertip and driven by an AC-coupled, 4 kHz clock that is synchronous with the multiplexer and ADC operations. The conventional MEMS pressure sensor (Honeywell, #26PC15SMT) for the fluid is amplified by operational amplifiers (Analog Devices, Inc., #AD8630) to produce both DC (pressure) and AC (vibration, 1 kHz bandwidth) signals. The conventional thermistor (GE, #EC95) is similarly amplified to provide both DC (absolute temperature of the heated core) and AC (thermal flux into contacted objects) signals. All sensory data are acquired by the PIC microprocessor and sent out in a digital SPI data stream to the host computer.

2.2 Force Sensing

In detecting thermal and microvibration transients, it is important to control and know the amount of force applied. In the instance of thermal events, normal and tangential forces dictate the surface area between the contacted object and the compliant sensor and thus the heat-flux. Sensing the location of contact force is also critical because thermal characterization requires the object to be contacted in a precise location with respect to the thermistor. Because the sensor has a displaceable fluid layer, these forces will control the sensor's geometry and how the heat-flux is conducted to the thermistor. In the instance of the texture and slip sensing experiments, normal forces directly affected sensed signals. Sensing the normal and tangential forces when slip occurs can also be used to infer the friction coefficient between the sensor and the object. In order to account for these phenomena, it is important that the sensor also be able to encode tri-axial forces. For forces normal to an electrode, there is a monotonic increase in electrode impedance; the slopes of the curves depend complexly on probe curvature. As the skin deforms above a given working electrode, it constricts the conductive fluid path between the electrode and ground, increasing impedance [7]. Because the skin can slide laterally, tangential forces produce impedance changes at electrodes away from the contacting surface. However, this behavior is non-linear and also depends on the point of application of force as well as radius of curvature of the incident object [7].

2.2.1 Data Collection and Preprocessing

In order to extract the richest data set possible, the sensor was exposed to a wide variety of radius of curvature objects (flat, 20, 7.2 and 2 mm diameter), points of contact and force vectors/movements. The tactile sensor was secured facing upwards in a vise positioned on a 6-axis Advanced Mechanical Technology Inc. HE6x6-16 force-plate (Fig. 2).

Force directions were defined in a global reference frame with respect to the force-plate: Fz is vertical (downward = positive), Fy is lateral (left = positive), and Fx is along the longitudinal axis (towards the base of the vise = positive); see Figs. 1 and 2. The digital protractor ensured the vise was level. The data were

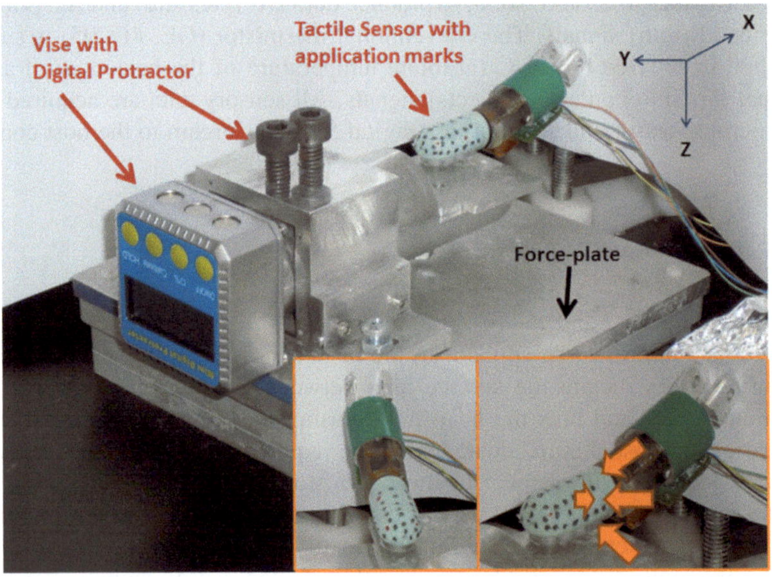

Fig. 2 Data collection assembly with BioTac mounted to vise. *Inset Lower Right*: *Orange arrows* highlight force vectors applied (4 of 5 showing); *Inset Lower Center*: Alternate view of sensor (figure reproduced from [35] with permission)

collected at 100 samples/sec using a NI USB-SPI/I2C-8451 data acquisition block in conjunction with LabVIEW. All data collected were digitally filtered by a 3-pole, 5 Hz low-pass filter.

Probes were manually held with the hand steadied atop a solid block that was level with the vise. Contact was made normal to the surface of the skin and force increased to 30–40 N. For example, probe forces applied to the palmar, center-line dots would be normal to the skin and parallel with the Z-axis. Probe forces applied on the lateral dots would again be normal to the skin, but parallel with the Y-axis. The force was relaxed, contact retained and the probe tilted approximately ±30° away from the axis of stimulation (in both planes, see Fig. 2, inset), then the motion repeated for a total of 5 stimulations. Stimulation motions were sufficiently slow relative to sample rate (100 Hz) to assume quasi-static loading and care was taken to observe that no dynamic events occurred (e.g. probe slippage along the sensor surface). Earlier attempts to use a stepping motor under computer control to generate the force vectors were abandoned because of a tendency to produce excessive force and potential damage to the BioTac skin. The goal was to demonstrate that the sensor provides information that encodes force over a three-dimensional workspace. It is not clear yet how such information will be used, so precise, position-controlled experiments were not performed.

We compared various methods for extracting these force vectors using machine learning, in particular Gaussian mixture modeling regression and artificial neural networks, as well as Support Vector Machines. These preliminary results show that ANNs and GMMR have the best performance in estimating force vectors. To extract force vectors using ANNs, a three-layer back-propagation perceptron was used because it is capable of approximating any given nonlinear relation when a sufficient number of neurons are provided in the hidden layer [11]. MATLAB's Neural Network Toolbox 6.0.4 was used; data for each voltage channel were preprocessed by subtracting the mean and dividing by the variance.

Prior to machine learning, the primary data sets were randomly divided into three sets: (1) a working set for training (70 %), (2) a test set to measure the ability to generalize after training (15 %), (3) a validation set to determine when training should be stopped (15 %). Training and subsequent use of the machine learning algorithms (MLA) are segregated by prior knowledge and extraction of probe radius of curvature and contact location. This process is beyond the scope of this discussion; further information can be found in [12].

2.2.2 Adaptive Neural Network Training

The basic mathematical structure of the ANN is a series of functional transformations. First we construct a linear combination of D input variables (Eq. 1).

$$Aj = \sum_{i=1}^{D} WjiXi + Wj0 \tag{1}$$

where parameters Wji refer to the biases and Wj0 refers to the weights of the activation function Aj. Each of these activation functions are transformed using a differentiable, non-linear function:

$$Zj = H(Aj) \tag{2}$$

These M basis function outputs, referred to as the hidden units, are linearly combined to form the K outputs on which the system was trained on.

$$Ak = \sum_{j=1}^{M} WkjZj + Wk0 \tag{3}$$

This software employed the Levenberg–Marquardt backwards propagation algorithm [13] to tune the weights and biases of the ANN to maximize the correlation between the model predictions and the recorded data. Hidden units used hyperbolic tangent activation functions; outer units used linear functions. Hidden layer size was such that an adequate number of units existed relative to the inputs (2x inputs, [14]).

To prevent the ANN from overfitting and to improve generalization, we used Bayesian regularization and early stopping. Bayesian regularization maximizes the posterior probability of the weights and biases over an error function based on the training data. In this framework, performance function parameters are iteratively updated with the weights and biases while a search for the minimum of the sum-of-squared error function occurs via the Hessian [15]. Early stopping examines the performance of the ANN during training by examining its performance on the validation set. If the network's performance on the test set is no longer improved over six iterations, then training is stopped.

2.2.3 Gaussian Mixture Model Regression

With regard to GMMR, a training dataset with N samples and D dimensionality (number of electrodes), $\{\xi^j = \xi_{(V,j)}, \xi_{(F,j)}\}(j = 1)^N$, ($\xi_{(V,j)}$: impedance signals, $\xi_{(F,j)}$: force vectors) can be modeled by a mixture of Gaussian distributions [16]; K components are determined by system error minimization:

$$p(\xi_j) = \sum_{k=1}^{K} \pi_k N(\xi_j; \mu_k, \Sigma_k) = \frac{1}{\sqrt{(2\pi)^D |\Sigma_k|}} e^{-\frac{1}{2}((\xi_j-\mu_k)^T \Sigma_k^{-1} (\xi_j-\mu_k))} \tag{4}$$

where $\{\pi_k, \mu_k, \Sigma_k\}$ are the prior probability, mean, and covariance matrix of the Gaussian mixture component k. Expectation–Maximization (EM) algorithm is applied to estimate the $\{\pi_k, \mu_k, \Sigma_k\}$ by optimizing the maximum likelihood. K-means clustering is used to set the initial estimation of $\{\pi_k, \mu_k, \Sigma_k\}$.

Thus for each Gaussian component k, there is a separate mean and covariance matrix:

$$\mu_k = \{\mu_{V,k}, \mu_{F,k}\}, \Sigma_k = \begin{pmatrix} \Sigma_{V,k} & \Sigma_{VF,k} \\ \Sigma_{FV,k} & \Sigma_{F,k} \end{pmatrix} \tag{5}$$

Given a novel dataset of $\xi_{V^{\wedge\prime}}$, Gaussian Mixture Regression is used to estimate $\xi_{F^{\wedge\prime}}$ [17]. In this case $\{\xi_{V^{\wedge\prime}}, \xi_{F^{\wedge\prime}}\}$ are the novel distributed impedance signals and estimated three force vectors, respectively.

$$\xi'_{F,k} = \mu_{F,k} + \Sigma_{FV,k}(\Sigma_{V,k})^{-1}(\xi'_V - \mu_{V,k}) \tag{6}$$

$$\omega_k = \frac{\pi_k N\left(\xi'_V; \mu_{V,k}, \Sigma_{V,k}\right)}{\sum_{k=1}^{K} \pi_k N\left(\xi'_V; \mu_{V,k}, \Sigma_{V,k}\right)} \tag{7}$$

$$\xi'_F = \sum_{k=1}^{K} \omega_k \xi'_{F,k} \tag{8}$$

where ξ_(F,k)^′ is estimated force vectors for each Gaussian component k, ω_k is the corresponding weight, ξ_F^′ is the estimated force vector. Estimated force vectors are presented based on novel data; R-squared, standardized mean square error (SMSE) and percentage Error (Eq. 9) were reported where appropriate (integral evaluated numerically using the trapezoidal rule).

$$Error = 100\,\% \times \int \frac{\sqrt{(Xpred - Xactual)^2}}{Xactual} \tag{9}$$

2.3 Vibration Sensing

Sensitivity of the receptors is a limiting factor both for slip detection and texture discrimination. Fine control of grip force for the human hand is made possible largely by the wealth of tactile sensory information delivered to the central nervous system. When making a precision pinch, the muscles deliver just enough grip force so that an object does not slip out of grasp [2]. This desirable behavior requires finely tuned sensory neurons capable of detecting microslips between the skin and the object when the ratio of gripping to lateral forces at the fingertip approaches a critical threshold [2]. In the biological hand, Pacinian corpuscles with frequency responses of 60–500 Hz [18] are capable of measuring such vibrations associated with slipas small as a micrometer in amplitude with center frequency of 200 Hz [19]. There have been no microvibration sensors robust enough to be suitable for practical application in mechatronic hands. Various mechano-electrical transducers have been described [20–22] but they had to be located on or near the surface of the skin to achieve the requisite sensitivity, where they could easily be damaged.

To measure microvibrations, our sensor takes advantage of the excellent transmission properties and long wavelengths of low-frequency sound waves in incompressible liquids ($\lambda = 3$ m at 500 Hz in water), allowing for the fragile pressure transducer to be located away from the region of contact. This design was validated to detect microvibrations correlated with slip against a smooth skin [8]. More recent experiments have focused on the effects of fingerprints and exploratory strategies (contact force and sliding velocity) on the amplitude and spectra of measured vibrations and the ability to sense slip [21].

2.3.1 Texture Discrimination

A prototype tactile sensor was molded with a fluid pressure sensor (Honeywell 26PC15). Two sets of 1/16" thick skins were molded from silicone (Silastic E, Dow Corning), one smooth and one with a fingerprint-like ridged pattern (0.0075" deep, 0.0150" spacing). The assembled sensor was mounted on a bracket that allowed controlled forces to be applied to interchangeable textures that could then be slid past the sensor while recording the hydro-acoustic pressure fluctuations. To investigate the role of fingerprints, data were collected for both the smooth skin as well as the fingerprinted skin over a range of various surface textures. Contact forces and slip velocities used while exploring these textures were controlled manually by the operator, but were typical to those that human use when exploring textures (forces between 0.2–2.0 N and velocities between 1–10 cm/s). Acoustic pressure signals were analyzed with a Short Time Fourier Transform and presented as spectrogram versus time for various trials to indicate intensity. A systematic analysis comparing the precise contributions of forces and velocities has been presented more recently [21].

2.3.2 Slip Detection Algorithm

Previous studies with our BioTac indicated that the spectral power of slip-related signals resided in the frequency band of 100–500 Hz, permitting them to be discriminated from other contact events with lower frequency components [22]. Therefore the overall strategy to sense the occurrence of slip is to sense rapidly the onset of power within this band. If we desire to measure accurately the power of the signal within a certain frequency band, we must first determine an appropriate response delay. Human grip reflexes have been observed to be roughly 60–80 ms [2], largely due to the transmission delays of axons. If an artificial slip signal is required to produce a similar grip reflex to control an actuator, it will require similar response times.

Considering that sensing and transmitting the slip signal is only part of the reflex loop time (the other part is the transmission to and response of the actuator), we have chosen a signal analysis time of no more than 30 ms, which represents half of the total desired response time. To accomplish this, a simple 44-tap band-pass finite impulse response filter (100–250 Hz) was used to eliminate low frequency interference and high frequency sampling noise. The AC pressure signal is sampled at 2200 Hz, so this represents a total of 20 ms window. The signal is then rectified and smoothed with an additional 22-tap averaging filter, which represents an additional 10 ms window and coincides with the 100 Hz lower limit of the first filter. This results in an output that represents slip power within the filtered bandwidth and a total window length of 30 ms. The resulting effect is that the slip detection algorithm only requires the previous 30 ms of data; fainter signals will be expected to require the whole window length before detection, resulting in a delay of 30 ms, while larger signals would be detected in less time after their origination. To produce controlled slip, the BioTac was mounted onto a bracket in contact with an ultrasonic motor (UPM46, DTI Industries). This motor was chosen because its high stepping frequency (39 kHz) avoids interference with the lower frequency vibrations being sensed in the BioTac. An oscillator drove the stepping frequency of the motor; varying the resistive element in this circuit could modulate the overall speed. Due to the high power of the motor, the change in velocity from rest to full speed was virtually instantaneous.

2.4 Thermal Sensing

When a heated end-effector contacts an object at room temperature, there will be heat transfer between the two objects that is dependent on their geometry and thermal properties. If the thermal properties of the BioTac are known, the thermal properties of the unknown object can be inferred from the changes in temperature when the two come in contact. In the case of the BioTac, as the amount of force between them increases, the skin deforms around the object, increasing the contact area between them. This increase in contact area will cause an increase in heat transfer between the objects, so careful regulation of contact force is required for repeatability.

Several exotic technologies have been applied to the thermal sensing problem in haptics, such as carbon nanotubes [23] and MEMS embodiments yielding skin-like configurations [24, 25]. Others groups use more traditional devices such as thermistors, but also use them in conjunction with force sensors and heating elements in the sensors [26] when applied to grippers. The strategy to replicate heat-flow sensing for object discrimination is not new; more recent strategies by Engel et al. and Takamuku et al. used strain sensors, heating elements and temperature detectors to track thermal and force profiles of contacted objects [26, 27]. Many of these analyses do not extend beyond features of the DC temperature profile of the sensor. We speculate that there may be many discriminable features in the first and second derivatives of temperature. Our goal is to produce a device that can discriminate objects thermally as well as or better than human subjects [28].

Early prototypes of the BioTac were heated above ambient temperature so that contact with an external object facilitated heat flow dependent on the thermal properties and size of the object. This is analogous to biological mechanisms for thermal object discrimination—the human body is typically a few degrees warmer than the surrounding environment. All other things being equal, for a given temperature differential the thermal flux at the interface between two contacted objects will depend mutually on the magnitudes of their thermal effusivities ε as calculated in Eq. (10) below [29].

$$\varepsilon = \sqrt{\kappa \rho c} \tag{10}$$

where κ equals thermal conductivity, ρ equals density and c equals specific heat capacity. We posit that this larger thermal flux will be reflected as greater variance in the data of temperature, its first and second derivatives. By using principal components analysis (PCA), we can extract and compare the relevant variant features against a known data set to identify objects thermally. We also hypothesized that each of the thermal features (ε, the product of ρ and c (relative heat capacity) and κ) can be associated with features in the heat-flow signals. For example, consider two massive objects with equal thermal effusivities, but one has a large thermal conductivity and low relative heat capacity and vice versa. The two thermal signals will look different: the first object will likely have a higher heat-flow initially but it will rapidly taper off, while the second will not peak as high but the heat-flow will be sustained for a longer period of time. Therefore the net effect of thermal processes over time would be necessary to identify objects. To further discriminate based on thermal conductivity and relative heat capacity, one must examine landmark features in the heat-flow signals. This method is under development and is elaborated further in the Discussion section.

2.4.1 Heat-Flow Sensing

The BioTac sensor is heated by on-board electronics in the proximal section and power resistors in the distal section. The combined electronic circuitry generates 250 mW of power, resulting in approximately 10 °C increase in temperature from ambient after the requisite 15 min warm-up cycle from rest. While the sensor is touching an object, force (impedance) sensors gather information to analyze the contact forces and shape of the object, which is necessary to characterize the contact properties that affect heat flow. At the same time, the PIC microcontroller gathers thermistor information.

A prototype approximating the thermal generation and sensing components of the BioTac was used to demonstrate the feasibility of using heat-flow sensing via a thermistor to discriminate objects. The sensor consisted of two 40-Ohm resistors (heaters) in the back section and three thermistors (GE EC95, Type F): two monitoring the temperature of the heaters and a third in the tip for thermal characterization of contacted objects. These electronic components were placed into a mold and cast with an epoxy-based encapsulant (Stycast 1264) to generate a core with the size and shape of the BioTac. The core was covered with a molded silicone elastomer skin and inflated with approximately ~1 cc of propylene glycol as used in earlier versions of the BioTac. The two heaters were powered with a 5 V source providing 1.3 W of power. After powering and reaching equilibrium, the sensor's temperature was approximately 85 °C at the heaters and 31 °C in the tip (ambient 25 °C). Voltage was measured via voltage divider, analog low-pass filtered (single-pole, cut-off frequency of 1.6 Hz) and recorded by a custom LabVIEW program. The signal was also passed through a differentiator to record AC transients; the digitized voltages were converted to their corresponding temperatures and temperature derivatives using MATLAB at 500 samples/sec. The sensor was mounted to a pivot on the stepper motor's base, allowing the finger to contact the sample material consistently (Fig. 3).

Six materials: copper, aluminum, 316-stainless steel, potter's ceramic, glass and Teflon, were placed on an insulating piece of foam mounted on the force plate (Advanced Mechanical Technology, Inc., Model HE6X6-16). The samples consisted of a 101 mm diameter by 25 mm thick puck of material; these samples were large enough to be considered infinite heat sinks relative to the mass of the sensor. A stepper motor (Nippon Pulse America, Inc., PFL35T-48Q4C (120), NPAD10BF chopper drive) was used to press the tip of the finger against the sample material with a force of 10 N (Fig. 3) for approximately 5 min.

To identify materials thermally, the temperature and dT/dt data were first smoothed with a 2001-point moving average filter. This 2 s filter period was substantially shorter than the relatively slow time constant of thermal flux. The smoothed dT/dt signal was then numerically differentiated and smoothed by a

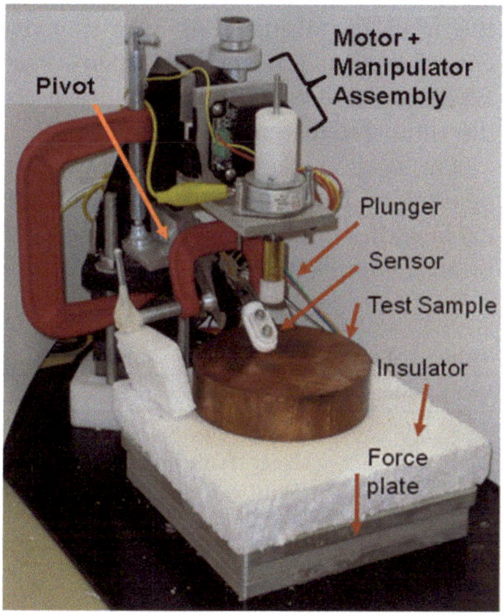

Fig. 3 Thermal test assembly (figure modified from [10] with permission)

2001-point Savitzky-Golay second-order differentiator [28]. A 3-row matrix M was formed with the first row consisting of temperature, the second dT/dt and the third d^2T/dt^2. The data were preprocessed for PCA by subtracting the mean and dividing by the variance. To get the principal-component-space (PCS) for a particular material, PCA was performed in MATLAB on the preprocessed matrix M resulting in the principal components for that material. To derive the PCS, the data for that material were multiplied by the respective principal components. To explore if a new material has the same thermal effusivity as a known material we compare the results of:

$$PCS_{KM} \times Data_{KM} = PCS_{KM} \times Data_{UM} \quad (11)$$

where KM refers to known material and UM refers to unknown material. This will produce a mapping of the 2 materials in PCS; if they have the same variance, the data points will overlap precisely.

2.4.2 Object Contact Period

It should be possible to identify the material nature of contacted objects by comparison with a database of known objects. Contact pose can be controlled by the robotic manipulator and we posit that deterministic machine learning techniques like artificial neural networks (ANNs) or Gaussian mixture modeling and regression (GMMR) can provide force feedback (See Sect. 3.1. below). In our initial experiments, object contact times were 5 min but most of the distinguishing features appeared in the early parts of the response (<30 s). For practical use, it will be important to determine the shortest period of contact that can be used to discriminate materials reliably. To evaluate the effect of contact time, each of the five trials for all materials were split into six "windows" based on features discussed in Sect. 3.3. (the last window encompassing the full time period; Table 2). Euclidean cluster analysis was then used to determine if the materials were discriminable as time progressed.

For each material's set and window, a mean and standard deviation was determined based on their mapping onto copper principal component space (as explained in Sect. 2.4.1; the choice of the mapped material is arbitrary). The 2 principal component mean values became the center of an ellipse and 2 times the

Table 2 Time contact window summary

Window marker	Reason for marker	% of test period/time
1	Near inflection point	~1.5 %/5 s
2	Near dT/dt minimum #2	~7 %/20 s
3	Half-way between inflection point and 50 % marker	34 %/1.7 min
4	50 % of test period	50 %/2.5 min
5	75 % of data	75 %/3.75 min
6	End of data	100 %/5.0 min

standard deviation in each principal component became the major and minor axis of the ellipse. The degree to which two ellipses overlap indicates the similarity in material. It is expected that as more data are used in the comparison (i.e. the time window gets larger), the means of ellipses will diverge if two materials have largely different effusivities or they will remain close if the effusivities are similar.

3 Results

3.1 Force Sensing

Table 3 shows a summary of 5-cross fold generalization error for all forces, ANNs and GMMR (K = 9). The ANN set performed better and a typical generalization set across three forces is presented below for visualization (Fig. 4).

Table 3 A and B: Statistical metrics for MLA tests sets; ± refers to one standard deviation

A: Artificial Neural Network (ANN)			
	Force (X)	Force (Y)	Force (Z)
R^2	0.829 ± 0.0341	0.943 ± 0.00434	0.876 ± 0.0218
Error	40.7 ± 2.19	18.6 ± 0.808	23.5 ± 2.56
SMSE	0.183 ± 0.0188	0.0571 ± 0.00451	0.127 ± 0.00456
B: Gaussian Mixture Modeling Regression (GMMR)			
	Force (X)	Force (Y)	Force (Z)
R^2	0.408 ± 0.0598	0.851 ± 0.00521	0.445 ± 0.0311
Error	67.9 ± 2.99	32.6 ± 1.67	59.3 ± 3.23
SMSE	0.616 ± 0.0198	0.149 ± 0.00897	0.737 ± 0.00587

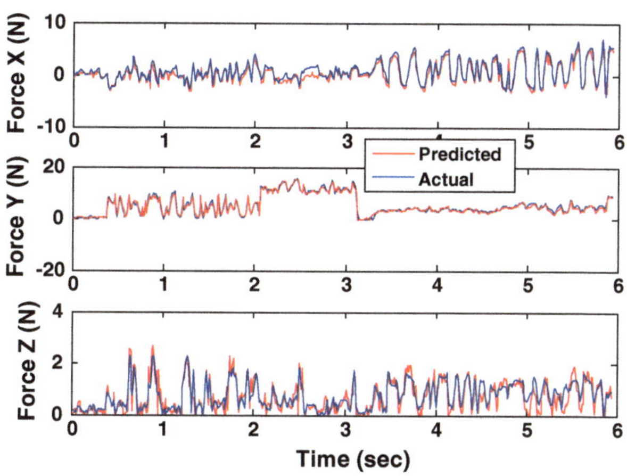

Fig. 4 Tri-graph of forces versus ANN predictions over a representative generalization test of novel data Tri-axial force extraction showing $R^2 > 0.9$ and errors 10–30 % (figure reproduced from [35] with permission)

3.2 Vibration Sensing

Vibrations sensed by the BioTac originate from the power in the sliding friction between the object and the BioTac. Total friction power is proportional to the sliding velocity and the contact force. It would be intuitive to assume that increasing either of these parameters would result in larger signals but this was, in fact, not the case as demonstrated in Fig. 5. Instead some of the largest signals were found at the lower force levels. The authors speculate that this may be due to a damping mechanism in the BioTac at larger forces. Clearly it will be important to carefully regulate the contact forces used during exploratory strategies. The addition of fingerprints had a profound effect on sensed vibration amplitude and spectral content for all possible combinations of surface texture, contact force, and sliding velocities. Spectrograms demonstrating this contrast can be seen in Fig. 5.

In general, fingerprints increased signal amplitudes roughly 10–30 times and produced more complex spectra than the smooth skins. In all cases the spectral content was richer with the addition of fingerprints and the signal-to-noise was greatly improved, which would be desirable for texture discrimination.

The ability to detect slip by sensing power within the 100–250 Hz power range proved to be reliable (Fig. 6) and reflected the slight delay as expected, but further studies over a wider range of forces and velocities are still required.

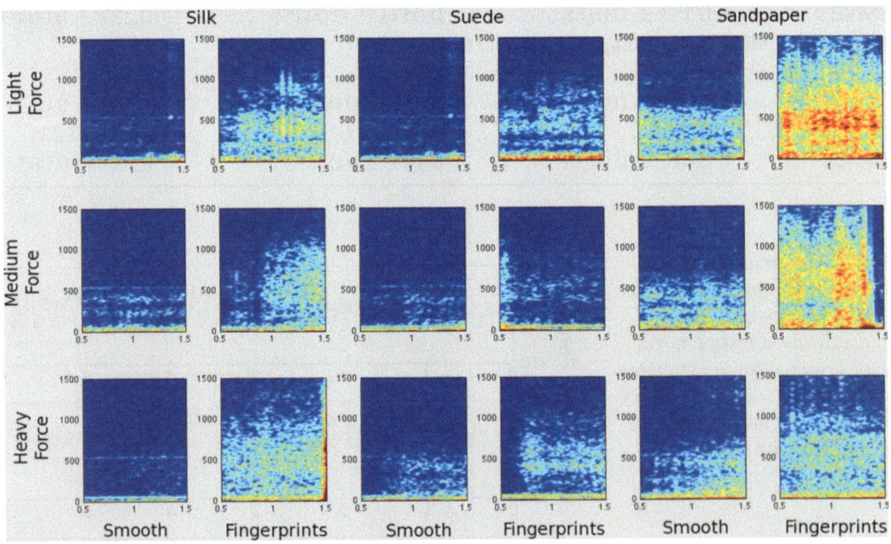

Fig. 5 Spectrograms for light (~0.1 N), medium (~1.0 N) and heavy (~10 N) forces across three materials: Silk (*left*), Suede (*center*) and 100 grit Sandpaper (*right*), sliding velocities were relatively slow (~1 cm/s). Each abscissa spans 1 s and ordinate spans 0–1500 Hz (Figure reproduced from [36] with permission)

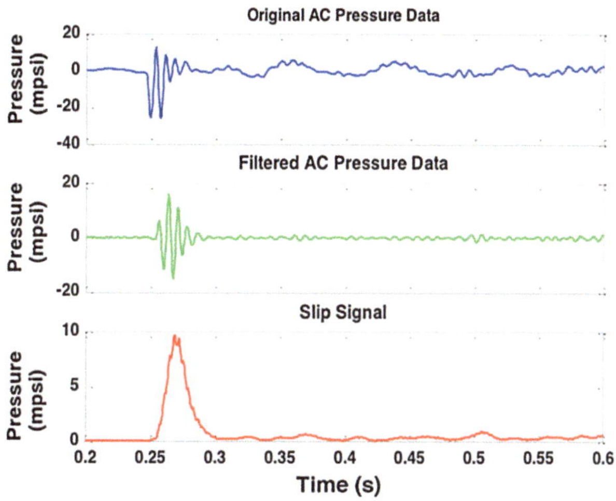

Fig. 6 Signal processing to detect slip that began at 0.24 s; response confirms a small time delay due to the filter design, but otherwise a fairly rapid response and detection of slip. X-Axis time (seconds), y-axis pressure (mpsi)

3.3 Thermal Sensing

Upon contact with a test object, the derivative of temperature (T_{AC}) has several reproducible features. The initial negative peak rate of change is similar for all materials because this cooling effect is due to the cooler skin displacing the fluid and contacting the core near the thermistor, typically 5 s in these experiments (Fig. 7). The skin surrounding the sensor is closer to ambient temperature than the core because of the intervening fluid and proximity to ambient air.

After the initial transient, features of the contacted object emerge. The next notable feature is an inflection point in the rate of temperature change. Following contact with plastic, T_{AC} reverts to a gradually decreasing rate of cooling. During contact with copper, cooling is faster and exhibits a second negative peak. Similar behavior can be seen for the other metal samples. Materials with low thermal

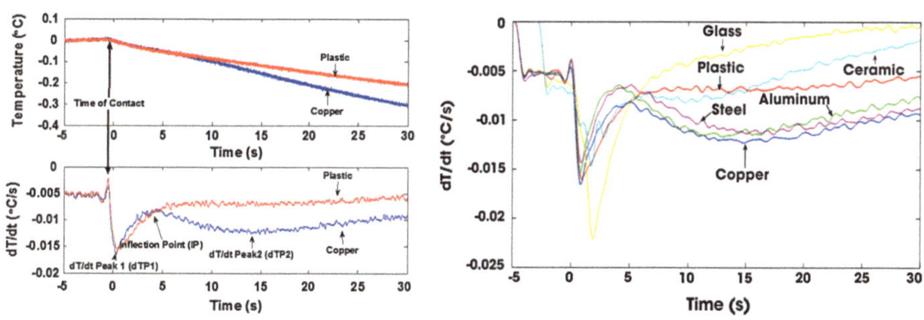

Fig. 7 (*Left*) TDC (*top*) and TAC (*bottom*) following contact (*vertical arrows*) with large plastic and copper test pucks (figure reproduced from [36] with permission). (*Right*) AC transient responses for large test samples: digitally filtered with 1 Hz Butterworth LPF for clarity

conductivity do not have this second negative peak but exhibit other material-specific behaviors that remain to be examined in more detail. The prominent initial negative peak for glass probably reflects the specular surface of this sample and a high level of tackiness to the silicone skin; all other materials were polished to a satin finish, resulting in lower contact surface area at the microscopic scale.

3.3.1 Thermal Object Identification: Principle Components Analysis

To form the component set for PCA, we used the thermistor temperature signal and its first two derivatives. The principal component breakdown was very similar for all materials and trials and percent energy variance was: 88.7 % (1st Eigenvalue), 9.87 % (2nd) and 1.52 % (3rd). To determine which features to map onto the principal components, we examine absolute variance to mean ratio. It was found that d^2T/dt^2 exhibited the highest average absolute variance/mean ratio of 1.72, followed by dT/dt 0.948, and Temperature 0.00916; unsurprisingly, the derivatives appear to provide the most discriminative information.

The proximity between trials in PC space (red dots to blue dots in Fig. 8) is an indication of the likelihood that the materials tested have the same thermal effusivities. Two copper runs and aluminum (close in effusivity) have co-located data points, whereas copper versus ceramic and plastic are much further apart,

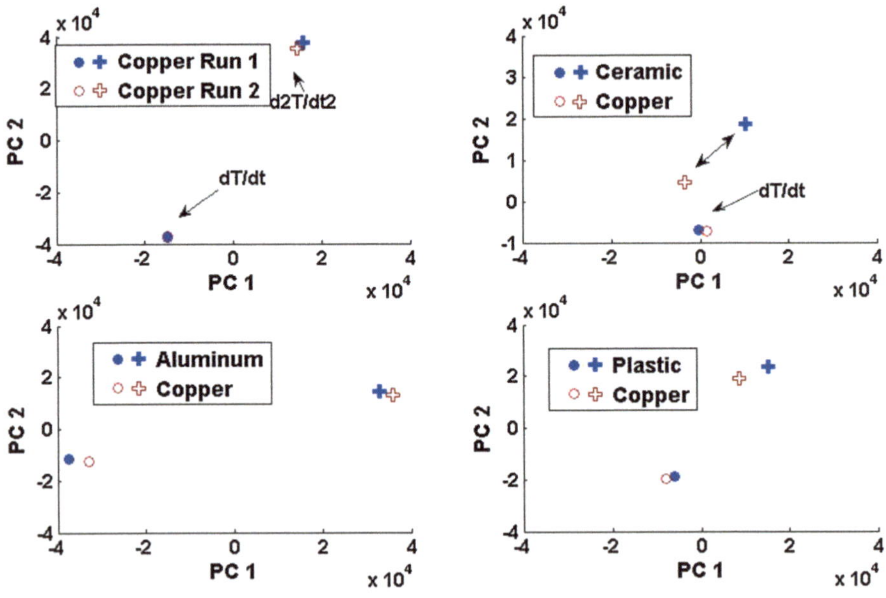

Fig. 8 One copper trial plotted in principal component space with a different copper trial (*upper left*), aluminum (*lower left*), ceramic (*upper right*) and Teflon (*lower right*); second derivative pairs occupy the upper section of the graph, first derivative pairs the lower

indicating discriminability. Figure 9 shows various trials of thermal data (1st and 2nd derivatives) mapped on to the 1st and 2nd principal components of a representative trial of copper. The data start after the initial skin contact transient for all trials.

We were surprised by the relatively larger distance between ceramic and copper versus plastic and copper because ceramic and copper are closer to each other in thermal effusivity than plastic and copper. We also wanted to confirm that plastic could be discriminated from ceramic. This was confirmed in plots of plastic versus ceramic for a PCS constructed from the plastic data.

3.3.2 Object Contact Period

Our goal was to discover the minimal amount of time required to reliably discriminate two objects from one another. Several contact time windows were determined as indicated in Table 3. Window markers were established based on thermal signal landmarks observed in Fig. 7 and by evenly dividing data at later intervals. We used thermal data collected at 10 N normal force, which compresses the skin against the core, and a pose of 45°, which causes the surface of the core in the vicinity of the thermistor to be parallel to the surface of the puck. The two cases examined are copper versus aluminum (the two materials with the most similar thermal effusivities) and copper versus Teflon plastic (the most different effusivities). In the case of aluminum, it appears that the first time window is the most reliable (Fig. 9, Left). While the centers of the ellipses are never co-located, as the time window increases, the size of the ellipse grows, decreasing discriminability as heat systems equalize (Fig. 9, Right). As time goes on, variability increases as the ellipses enlarge and only the second derivative component maintains discriminability. To test that the means of the ellipses diverge over time between two materials of different effusivity, copper and plastic means were

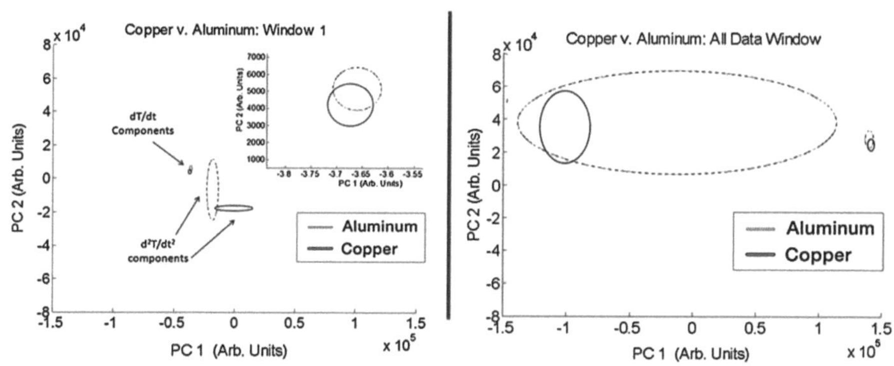

Fig. 9 (*Left*) Window 1: Euclidean cluster analysis ellipses for copper and aluminum. (*Inset*) *Zoom view* of dT/dt ellipses. (*Right*) Window 6: Ellipses for Cu and Al

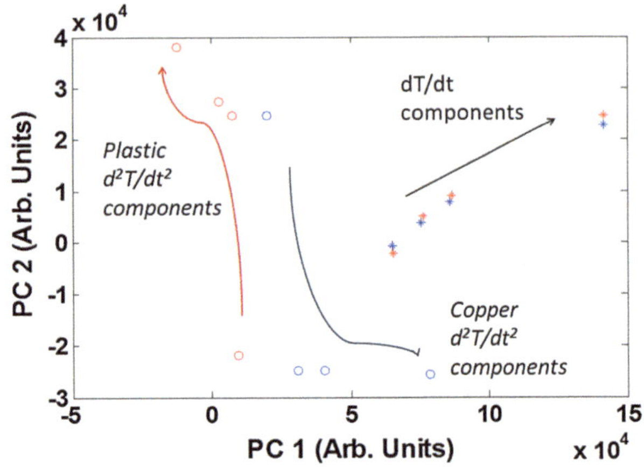

Fig. 10 Means of ellipses for copper (*blue*) and plastic (*red*) (windows 2–4 only shown for clarity). Circles indicate d2T/dt2 components and stars indicate dT/dt components as they deviate over time

plotted for progressive time-windows (Fig. 10). The dT/dt components maintain relative close proximity, although slight divergence can be seen as the plot progresses (black straight arrow). The d^2T/dt^2 components show the most spreading as indicated by the red curved arrow (plastic) and dark blue curved arrow (copper).

4 Discussion

4.1 Force Sensing

Both MLAs were able to represent both the magnitude and dynamics of the applied force with reasonable accuracy. ANNs were able to represent direct force calculations better; this probably reflects the larger number of free parameters and robustness of the software version (Mathworks). However, the goal was not high specificity in either MLA, but rather to demonstrate that information about such forces is actually embedded in the impedance data. What is useful to note is that representations for Y forces are much better than Z or X forces; this is most likely due to the higher number of electrodes respectively facing the ±Y versus singular Z or X planes of action. The ANN produces force errors (18–40 %) that are higher, but not unreasonable given typical human force discrimination errors (6–9 %) [30, 31]. This error could be further reduced by more elegant tuning of ANN parameters and more accurate calculation of error (e.g. higher sample rate and use of Simpson's rule versus trapezoidal rule). Thermal compensation of the impedance signals is also required, but not addressed in this work. Empirical temperature compensation curves have been developed for each electrode using thermal data from the fingertip thermistor and are available in product packaging at www.syntouchllc.com.

It is unlikely that humans explicitly extract orthogonal force vectors from their interactions with objects and it may prove to be unnecessary for biomimetic control strategies in robots to do so, either. The biomimetic design of the BioTac core provides opportunities to use the BioTac and its individual sensor signals in simple control algorithms that may be similar to those employed by humans using their natural fingertips. For example, the flat bevel near the tip has four electrodes whose impedance signals have been used directly to control the attitude of a robotic finger while exploring tilted surfaces [9]. The indentation of the BioTac into the more compliant surfaces produced distinctive changes in the impedances of individual surrounding electrodes.

4.2 Vibration Sensing

Fingerprints have long been hypothesized to play an important role in the transduction of sliding motion to vibrations in biological skin. The actual mechanism remains unclear, however. Recently Scheibert et al. have proposed a one-to-one relationship between sliding velocity and vibration frequency that is dependent on fingerprint spacing [32]. Their data were collected at extremely low scanning velocities (0.01–0.05 cm/s), however, much slower than typical human exploratory movements (1–10 cm/s). Our own findings with smooth skins suggested that spectral frequencies of microvibrations were relatively little affected by velocities within biological scanning ranges for textures [8]. The addition of fingerprints clearly changes the mechanism of transduction in ways that are advantageous to perception but need to be better understood to permit systematic design of biomimetic sensors and feature extractors; further studies over a wider range of forces and velocities are underway [21].

We hypothesize that fingerprints result in a coherent pattern of stick-slip behavior that amplifies their individual vibrations. Tangential force deforms the elastic skin ridges until they reach the limits of static friction, whereupon they release abruptly. Because the ridges are coupled together in the elastic substrate of the dermis, abrupt release of one ridge alters the stress on adjacent ridges, changing their probability of release in a spatially coherent way. We speculate that this gives rise to coherent summation of their released energy, similar to that seen in phased-array radar. The timing and extent of these releases seems likely to depend on mechanical beating between the regular spacing of the fingerprint ridges and any repeating texture and friction of the surface being scanned. The effects of normal force, tangential force and scanning velocity are not intuitive and remain to be determined.

4.3 Thermal Sensing

In the case of Euclidean cluster analysis, we see that the centers of the ellipses gradually diverge over time as more data is accounted for, indicating improved discriminability between two samples of varying effusivity. However, the size of the ellipses also grows with time. This indicates increased variability from trial to trial as time goes on. One possible explanation is the relatively high thermal fluxes present during time-windows one and two. In the latter windows, temperature is closer to equilibrium and the signal-to-noise ratio of the temperature signals has decreased. This is especially true in the second derivative signals, which are relatively noisy. In order to choose an optimal time window, one must determine a threshold of overlapping area between ellipses based upon the relevant task—that is gathering just enough data for identification for the task at hand. Future experiments to validate proper material identification are planned for time windows 1 through 3, as well as developing further granularity within the windows themselves.

The thermal time constants were also relatively long compared to the 5–10 s required by humans to thermally identify objects [33]. Further improvements can be implemented by choosing a thermistor with a faster response and increasing the thermal conductivity of the skin (e.g. adding inert fillers prior to elastomer curing). Selecting a suitable elastomer for the skin requires careful consideration of factors such as wear resistance and absorption and diffusion of the components of the liquid filling of the BioTac [9].

Further distinction amongst materials with similar effusivities and knowledge about thermal conductivity and relative heat capacity must be considered as well. Determining features over time based on the thermal circuit models and using statistical correlation or machine learning classifiers to identify material properties are possible strategies. For example, aluminum, glass and ceramic have relatively close relative heat capacities, but aluminum has a much higher thermal conductivity. In Fig. 7 one can see this results in a sustained transient for aluminum (high thermal conductivity and large mass), but glass and ceramic show decreasing heatflow much earlier. This feature could be used to distinguish thermal conductivity from the effusivity to which it contributes, for example.

For actual use in the field, the thermal data must be deconvolved with other information about the geometry of the object, the location of the point of contact with respect to the thermistor, and the deformation of the skin around the object, which determines the contact area and thus heat flux. A smaller radius of curvature and/or a smaller thermal mass and/or a smaller temperature differential will reduce the thermal excursions. Information about all of the mechanical aspects can be extracted from the array of impedance sensing electrodes. Algorithms are being developed to account for such variables. For example, by controlling robot exploratory pose, the sensor contact area can be estimated for a given force when contacting a flat surface. One can take advantage of the fact that contact area asymptotes as force increases due to the compliant nature of the BioTac [34]. For examples, forces of 10–15 N have little change in contact area compared to 0–5 N. It is not clear, however, whether an analytical solution is possible or even necessary. In order to match the relatively coarse discriminability achieved by humans, it may be sufficient to compare the extracted features of the thermal data to a reference table of contact events collected empirically from known objects.

5 Conclusion

Each of the three sensing modalities of the BioTac appears to provide a reasonable approximation of the information available from the corresponding modality of biological sensors. Similar to biological sensors, their signals depend on their conditions of use and the exploratory movements that are made to acquire tactile information. Presumably the biological nervous system integrates information across tactile and proprioceptive modalities plus information about commands sent to actuators (efference copy) to arrive at accurate assessments of the underlying properties of objects encountered. We are just beginning to develop the corresponding algorithms that will be required to make full use of multimodal tactile sensors.

Acknowledgments This material is based upon work supported by the National Science Foundation under Grant No. 0912260 Any opinions, findings, and conclusions or recommendations expressed in this material are those of the author(s) and do not necessarily reflect the views of the National Science Foundation.

References

1. J.C. Rothwell, M.M. Traub, B.L. Day, J.A. Obesko, P.K. Thomas, C.D. Marsden, Manual motor performance in a deafferenated man. Brain **105**, 515–542 (1982)
2. G. Westling, R.S. Johansson, Factors influencing the force control during precision grip. Exp. Brain Res. **53**(2), 277–284 (1984)
3. R.D. Howe, Tactile sensing and control of robotic manipulation. J. Adv. Robot. **8**(3), 245–261 (1994)
4. M.H. Lee, H.R. Nichols, Tactile sensing for mechatronics—a state of the art survey. Mechatronics **9**, 1–31 (1999)
5. C. Melchiorri, Tactile sensing for robotic manipulation. Ramsete: Lecture Notes in Control and Information Sciences, vol. 270 (Springer, Berlin, 2001)
6. R.S. Dahiya, G. Metta, M. Valle, G. Sandini, Tactile sensing—from humans to humanoids. IEEE Trans. Rob. **26**(1), 1–20 (2010)
7. N. Wettels, V. J. Santos, R. S. Johansson, G. E. Loeb, Biomimetic tactile sensor array. Adv. Rob. 22(7), 829–849 (2008)
8. J. Fishel, V. J. Santos, G. E. Loeb, A robust microvibration sensor for biomimetic fingertips, in *Proceedings IEEE International Conference on Biomedical Robotics and Biomechatronics*, Scottsdale, AZ, pp. 659–663, 2008
9. D. Roy, N. Wettels, G. E. Loeb, Elastomeric skin selection for a fluid-filled artificial fingertip. J. Appl. Polym. Sci., in press (online 7 June 2012)
10. C. H. Lin, T. W. Erickson, J. A. Fishel, N. Wettels, G. E. Loeb, Signal processing and fabrication of a biomimetic tactile sensor array with thermal, force and microvibration modalities, in *IEEE International Conference on Robotics and Biomimetics*, 129–134, 2009
11. C.M. Bishop, *Neural Networks for Pattern Recognition* (University Press, Oxford, 1995)
12. N. Wettels, Biomimetic tactile sensor for object identification and grip control, Dissertation, University of Southern California, May 2011
13. M.T. Hagan, M. Menhaj, Training multilayer networks with the Marquardt algorithm. IEEE Trans. Neural Networks **5**, 989–993 (1994)

14. W. S. Sarle, Stopped training and other remedies for overfitting, in *Proceedings of the 27th Symposium on the Interface of Computing Science and Statistics*, pp. 352–360, 1995
15. K.T. Lau, W. Guo, B.M. Kiernan, C. Slater, D. Diamond, Non-linear carbon dioxide determination using infrared gas sensors and neural networks with Bayesian regularization. Sens. Actuators B **1**(2), 242–247 (2009)
16. G. McLachlan, D. Peel, *Finite Mixture Models* (Wiley, New York, 2000)
17. S. Calinon, *Robot Programming by Demonstration: A Probabilistic Approach* (EPFL/CRC Press, Lausanne, 2009)
18. V.B. Mountcastle, R.H. LaMotte, G. Carli, Detection thresholds for stimuli in humans and monkeys: comparison with threshold events in mechanoreceptive afferent nerve fibers innervating the monkey hand. J. Neurophysiol. **35**, 122–136 (1972)
19. R.S. Johansson, U. Landstrom, R. Lundstrom, Responses of mechanoreceptive afferent units in glabrous skin of the human hand to sinusoidal skin displacements. Brain Res. **244**(1), 17–25 (1982)
20. A. Pruski, B. Mutel, Direct contact sensors based on carbon fibre, in *IEEE international conference on Robotics and Factories of the Future*, pp. 409–415, 1984
21. J.A. Fishel, G.E. Loeb, Bayesian exploration for intelligent identification of textures. Front. Neurorobotics **6**(4), 1–20 (2012)
22. R. D. Howe, M. R. Cutkosky, Sensing skin acceleration for texture and slip perception, in *Proceedings IEEE International Conference on Robotics and Automation*, Scottsdale, vol. 1, pp. 145–150, 1989
23. S. Selvarasah et al., A Three-dimensional thermal sensor based on single walled carbon nanotubes, in *14th International Conference on Solid-State Sensors, Actuators and Microsystems,* pp. 1023–1026, Lyon, June 2007
24. Y. J. Yang et al., A wireless flexible temperature and tactile sensing array for robot applications in *Proceedings Of Fourth International Symposium on Precision Mechanical Measurements*, Dec 2008
25. T. Someya et al., Conformable, flexible, large-area networks of pressure and thermal sensors with organic transistor active matrixes. PNAS **102**(35), 2321–12325 (2005)
26. K. Shida, J. Yuji, Thermal-type tactile sensor for material discrimination and contact pressure sensing, in *Proceedings of 41st Annual SICE Conference,* vol. 1, pp. 588–589, Aug 2002
27. A. Persichetti, F. Vecchi, M. C. Carrozza, Optoelectronic-based flexible contact sensor for prosthetic hand application, in *IEEE Conference on Rehabilitation Robotics,* Netherlands, pp. 415–420, 2007
28. D.C. Spray, Cutaneous temperature receptors. Ann. Rev. Physiol. **48**, 625–638 (1986)
29. E. Marin, The Role of Thermal Effusivity. Phys. Teach. **44**, 432–434 (2006)
30. L. A. Jones, Perception and control of finger forces, in *Proceedings of ASME Dynamic Systems and Control Division*, pp. 133–137, 1998
31. S. Allin, Y. Matsuoka, R. Klatzky, Measuring just noticeable differences for haptic force feedback: Implications for rehabilitation, in *Proceedings 10th Symposium on Haptic Interfaces for Virtual Environments and Teleoperator Systems*, pp. 299–302, 2002
32. J. Scheibert, S. Leurent, A. Prevost, The role of fingerprints in the coding of tactile information probed with a biomimetic sensor. Science, **323**(5920), 1503–1506 (2009)
33. L. A. Jones, M. Berris, Material discrimination and thermal perception, in *Proceedings of the 11th Symposium on Haptic Interfaces for Virtual Environment and Teleoperator Systems,* pp. 137–142, 2003
34. Z. Su, J. A. Fishel, T. Yamamoto, G. E. Loeb, Use of tactile feedback to control exploratory movements to characterize object compliance. Front. Neurorobotics **6,** 7 (2012). (doi: 10.3389/fnbot.2012.00007)
35. N. Wettels, G. E. Loeb, Haptic feature extraction from a biomimetic tactile sensor: force, contact location and curvature, in *Proceedings of IEEE/RAS International Conference on Robotics and Biomimetics,* Phuket Thailand, 2011

36. N. Wettels, J. A. Fishel, Z. Su, C. H. Lin, G. E. Loeb, Multi-modal synergistic tactile sensing. tactile sensing in humanoids—tactile sensors and beyond workshop, in *9th IEEE/RAS International Conference on Humanoid Robots,* Paris, 2009
37. http://www.pressureprofile.com/products-robotouch
38. R.S. Fearing, Tactile sensing mechanisms. Int. J. Robot. Res. **9**(3), 3–23 (1990)
39. N. Futai, K. Matsumoto, I. Shimoyama, A flexible micromachined planar spiral inductor for use as an artificial tactile mechanoreceptor. Sens. Actuators, A **111**(2–3), 293–303 (2004)
40. http://www.inaba-rubber.co.jp/en/b_products/inastomer/index.html
41. Z. Pan, Z. Zhu, Flexible full-body tactile sensor of low cost and minimal output connections for service robot. Ind. Robot Int. J. **32**(6), 485–491 (2005)
42. http://www.peratech.com/qtctechnology.php
43. http://www.ati-ia.com/products/ft/sensors.aspx
44. http://www.tekscan.com/flexiforce.html
45. L. Beccai et al., Design and fabrication of a hybrid silicon three-axial force sensor for biomechanical applications. Sens. Actuators A **120**, 370–382 (2005)
46. Y.-L. Park, C. Majidi, R. Kramer, P. Berard, R. J. Wood, Hyperelastic pressure sensing with a liquid-embedded elastomer. J. Micromech. Microeng. **20**(12), 125029 (2010)
47. J. Engel et al, Flexible multimodal tactile sensing system for object identification, in *Proceedings of IEEE EXCO SENSORS*, Daegu, Oct 2006
48. P. Dario, D. De Rossi, C. Domenici, R. Francesconi, Ferroelectric polymer tactile sensors with anthropomorphic features, in *Proceedings IEEE International Conference on Robotics and Automation*. Washington DC, vol. 1, pp. 332–340, 1984
49. R.D. Howe, M.R. Cutkosky, Dynamic tactile sensing: perception of fine surface features with stress rate sensing. IEEE Transact. Robot. Autom. **9**(2), 140–151 (1993)
50. D. Hristu, N. Ferrier, R. W. Brockett, The performance of a deformable-membrane tactile sensor: basic results on geometrically-defined tasks, in *Proceedings IEEE International Conference on Robotics and Automation,* San Francisco, vol. 1, pp. 508–513, 2000
51. M. Ohka, Optical Three-axis tactile sensor, Mobile Robots: Perception & Navigation, InTech, 111–136 (2007)
52. E. E. Mitchell, R. DeMoyer, J. Vranish, A new MetGlas sensor. IEEE Trans. Industr. Electron. **IE-33**(2), 166–170 (1986)
53. J. M. Vranish, in *Magnetoresistive Skin for Robots,* ed. by A. Pugh. Robot Sensors, vol. 2: Tactile and Non-Vision (IFS Publications/Springer, New York, 1986), pp. 99–111
54. B. L. Hutchings, A. R. Grahn, R. J. Petersen, Multiple-layer cross field ultrasonic tactile sensor, in *Proceedings IEEE International Conference on Robotics and Automation,* vol. 3, pp. 2522–2528, 1996
55. A. R. Grahn, L. Astle, *Robotic Ultrasonic Force Sensor Arrays,* ed. by A. Pugh. Robot Sensors, vol. 2: Tactile and Non-Vision (IFS Publications/Springer, New York, 1986), pp. 297–315

Chapter 20
Biomimicry and the Design of Multigrasp Transradial Prostheses

H. Atakan Varol, Skyler A. Dalley, Tuomas E. Wiste and Michael Goldfarb

Abstract This chapter discusses some important design objectives regarding the design of multigrasp prosthetic hands, and describes two approaches toward the design of such hands. The first approach is highly biomimetic in nature, particularly with regard to the location of actuators within the prostheses, and the nature of the mapping between the neural command (e.g., electromyogram, or EMG) and movement. The second approach represents a compromised degree of biomimicry, in which some aspects of the biological system are retained, while other aspects are discarded in recognition of the spatial and sensory design constraints associated with upper extremity amputees.

Keywords Artificial limbs · Transradial prosthesis · Biomimetic design · Mechatronics

1 Introduction

The human hand contains approximately 20 degrees of freedom (DoF). In contrast, a traditional prosthetic hand (either body-powered or myoelectric) contains one. In an effort to provide upper extremity amputees with a more functional approximation of

H. A. Varol (✉)
Department of Robotics, Nazarbayev University, Astana 010000, Kazakhstan
e-mail: ahvarol@nu.edu.kz

S. A. Dalley · M. Goldfarb
Department of Mechanical Engineering, Vanderbilt University, Nashville, TN, 37235, USA
e-mail: skyler.a.dalley@vanderbilt.edu

M. Goldfarb
e-mail: michael.goldfarb@vanderbilt.edu

T. E. Wiste
Interdepartmental Research Center "E. Piaggio", University of Pisa, 56126 Pisa, Italy
e-mail: tuomas.wiste@iit.it

R. Balasubramanian and V. J. Santos (eds.), *The Human Hand as an Inspiration for Robot Hand Development*, Springer Tracts in Advanced Robotics 95,
DOI: 10.1007/978-3-319-03017-3_20,

the human hand, several multi-degree-of-freedom prosthetic hands have been developed. Because control inputs from the amputee are limited, and size and weight should be minimized, these hands have been designed with fewer degrees of freedom and/or a greater degree of coupling than contained in the native hand. Further, prosthetic hands typically include all actuation within the hand (i.e., they are intrinsically actuated), such that they accommodate a larger population of transradial amputees (i.e., low and high level amputations). Some notable examples of such multi-degree-of-freedom prosthetic hands include those described in [1–9]. Among these devices, the most highly underactuated is the hand described in [1], which utilizes a single ultrasonic motor to drive fifteen joints (three in each digit) through a fifteen-way differential coupling. The hands described in [2] and [3] each incorporate two independent actuators. The hand described in [2] (disregarding the wrist joint) contains nine powered joints (three in the thumb and three in each of the second and third digits). One actuator drives the six joints in the second and third digits through a pulley and tendon arrangement that provides kinematic coupling between the joints, while the second actuator controls the three joints in the thumb via a Geneva wheel mechanism that switches control between two thumb movements based on input range. The fourth and fifth digits are not actuated, but rather can be passively positioned. The hand described in [3] incorporates sixteen powered joints, wherein the four fingers (twelve joints) are driven by one motor through a compliant coupling mechanism, and the thumb (four joints) is driven by a second motor through a Geneva wheel mechanism in a conceptually similar manner to the hand described in [2]. The prosthetic hand described in [4, 5] contains sixteen powered joints (three in each finger and four in the thumb), which are driven by four independent actuators. Specifically, two motors are utilized to actuate the thumb, one to actuate the index finger, and one to actuate the remaining three fingers through a compliant differential coupling. The prosthetic hand described in [6] incorporates sixteen powered joints (three in each finger and four in the thumb), which are driven by six independent actuators through a linkage-based kinematic coupling mechanism. In particular, each finger is actuated by a separate motor, while the thumb is actuated by two. Like the hand in [6], the hand described in [7] utilizes six motor actuators allotted in a similar fashion, although the hand described in [7] incorporates fused distal-interphalangeal (DIP) joints in each digit, and thus incorporates eleven rather than sixteen powered joints. The underactuation in [7] is accommodated by combined tendon/linkage kinematic mechanisms. Finally, the fluidic hand described by [8, 9] utilizes a single electric motor to drive a miniature hydraulic pump, and incorporates five miniature electrohydraulic valves to control eight powered joints via miniature hydraulic actuators (two valves control the thumb, two control the index finger, and one controls the remaining digits).

In addition to the hand prostheses reported in the engineering literature, some multi-grasp hand prostheses have recently been introduced, or are currently emerging on the commercial market. These include the "i-LIMB" hand (Touch Bionics), the "Bebionic" hand (RSL Steeper), and the "Michelangelo" hand (Otto Bock). The i-LIMB and Bebionic hands, both of which have a similar configuration, each contain ten powered joints (two in each finger, two in the thumb) driven by five motors (one for each digit) via belts and linkages, respectively, in addition to one passive joint (palmar abduction/adduction of the thumb), which can be manually manipulated. The Michelangelo hand contains six joints (one in each finger and two in the thumb) which are positioned by two actuators—one that switches the thumb between palmar abduction and adduction, and one that powers a compliantly coupled flexion of all the remaining joints.

This chapter discusses design considerations for the design of a multigrasp hand prosthesis, and presents two designs that follow from these considerations. One design is highly anthropomorphic, while the other foregoes certain aspects of biomimetic design in order to better accommodate some of the spatial and sensory limitations characteristic of transradial amputees. Specifically, the first approach is highly biomimetic in nature, particularly with regard to the location of actuators within the prostheses, and the nature of the mapping between the neural command (e.g., electromyogram, or EMG) and movement. The second approach represents a compromised degree of biomimicry, in which some aspects of the biological system are retained, while other aspects are discarded in recognition of the space and sensory limitations characteristic of upper extremity amputees (e.g. the actuators and sensing are primarily located in the palm of the prosthesis instead of the forearm).

2 Design Considerations for a Multigrasp Prosthetic Hand

Ideally, a multigrasp hand prosthesis should better enable an amputee to perform activities of daily living (ADLs), relative to a single-degree-of-freedom (or single grasp) prosthesis, primarily by enabling a set of hand grasps and postures which better span the grasp taxonomies commonly used in such activities. As described by [10] and categorized by [11], these grasps include the tip, lateral, tripod, cylindrical, spherical, and hook grasps. In fact, as studied in [12] and reported in [13], this set of grasps constitutes approximately 85 % of the grasps used in activities of daily living, and as such, constitute a reasonable set of grasp objectives for the design of a multi-degree-of-freedom hand prosthesis. In addition to achieving this set of grasps, two additional postures, a pointing posture and a platform posture, are useful for activities of daily living. Specifically, a pointing posture provides the user with the ability to push buttons or keys (e.g., on a telephone or keyboard), and a platform posture enables a user to reach inside a pocket, or cradle a book or dinner plate.

In order to make such grasps (and postures) useful, a hand prosthesis must provide fingertip forces and speeds commensurate with typical activities of daily living. The fingertip forces exerted during activities of daily living (in healthy individuals) have been measured in several studies [14–20]. Pylatiuk et al. [14] conducted a study that measured the hand forces involved in three tasks representative of typical activities of daily living. One task involved (emulated) pouring from a bottle; one involved twisting the lid off of a cylindrical can; and one involved using a tip grasp to zip a large zippers (such as would open or close a backpack). In these studies, Pylatiuk et al. measured maximum fingertip forces (excluding impact) of approximately 10 N (which occurred during the zipper task). Kargov et al. [15] conducted a similar study, although in this study they considered only tasks involving the cylindrical grasping of a glass bottle, and found that healthy subjects exerted a maximum contact force of approximately 4 N while grasping the bottle. Smaby et al. [16] measured the lateral pinch (also called key grasp) forces involved in 12 insertion tasks, including inserting a key into a keyhole, a fork into putty, a plug into an electrical outlet, and pressing a button on a remote entry key. Of these tasks, Smaby et al. state that 9 of the 12 required pinch forces of less than 10.5 N. Of the remaining three tasks, zipping close a large horizontal zipper (e.g., on a suitcase) required pinch forces of up to 15 N for an average subject, pulling an electrical plug out of a socket required on average approximately 18 N, and inserting an electrical plug into an outlet approximately 25 N. Radwin et al. [17] conducted a study measuring the fingertip forces exerted when vertically grasping objects of varying size (between 45 and 65 mm span) and varying mass (between 1 and 2 kg). In their study, they report subject averaged maximum index fingertip forces of approximately 8 N, and subject averaged maximum middle, ring, and little fingertip forces of approximately 6, 4 and 4 N, respectively. In a different study, Fowler and Nicol [18] measured the index fingertip forces exerted during a number of different ADLs, including opening a jar, turning a water tap, turning a key, and pouring a jug. The study reports subject averaged maximum values of index finger (normal force) between 16 N (jug pouring) and 26 N (jar opening), where the maximum forces for the other tasks are bounded by these. Finally, Redmond et al. [20] conducted a study that measured the forces involved in texting with a typical cellular telephone, and report maximum (fingertip or thumbtip) forces of approximately 10 N.

Data characterizing typical finger speeds during ADLs is sparse. It is estimated in [21] that the joints of the hand reach angular velocities of approximately 3–4 rad/s (170–230 °/s) during typical ADLs. These estimates correlate well with the specifications of the Otto Bock Sensorhand Speed prosthetic hand (which is perhaps the fastest commercially available myoelectric prosthetic hand), which according to the manufacturer provides a fingertip speed of 300 mm/s, which assuming a finger length of 100 mm, would correspond to a maximum angular velocity (of the metacarpophalangeal joint) of 3 rad/s. For a sinusoidal motion with a peak-to-peak amplitude of 45° (i.e., approximately 50 % joint range of motion (RoM), assuming typical finger joint range of motion of 90°), these angular velocities would correspond to a frequency of movement between 1.2 and 1.6 Hz. Alternatively, assuming a minimum jerk trajectory and a nominal joint range of motion of 90°, a maximum joint angular velocity of 4 rad/s would correspond to a "time-to-close" (i.e., the time required from full open to full close) of 740 ms.

Finally, consistent with surveys of upper extremity amputees (e.g., [9, 22–24]), the prosthetic hand must be of a mass that is acceptable to the amputee user. A precise specification in this regard is difficult to obtain. According to [25], the mass of the typical human hand is approximately 400 g, which provides at least an approximate design target for a multigrasp hand prosthesis. A recent survey on the performance of anthropomorphic prosthetic hands also provides boundaries of the achievable performance with the present technology [26]. Thus, based on a collection of studies, a reasonable set of design objectives for a multigrasp prosthesis (presumably capable of conformal grasping) is as follows:

- Achieve six grasp types (tip, lateral, tripod, cylindrical, spherical, and hook).
- Achieve point and platform hand postures.
- Provide fingertip forces commensurate with typical ADLs; namely, maximum index fingertip forces of approximately 25 N, similar thumbtip force capability, and maximum combined fingertip forces of approximately 12 N for the remaining digits.
- Provide joint angular velocities of at least 4 rad/s, which correspond to a half-RoM bandwidth of 1.5 Hz.
- Total hand mass not significantly greater than 400 g (the approximate mass of a typical native hand).

These design objectives are consistent with four of the top five priorities of amputees as presented by [27], where the other priority not specifically addressed by these design objectives is the user's desire to "feel the grasping force." Also the objective of providing sensory feedback is acknowledged here as significant, but is not considered in the aforementioned design objectives. Finally, single grasp (myoelectric) hand prostheses typically provide considerably greater fingertip forces than described above, presumably because the lack of conformal grasping decreases significantly the efficiency of manipulation and contact area with the grasped object, and thus requires much greater grasping forces to achieve a given grasp stability.

3 A Biomimetic Design Approach

The design objectives did not consider the design issues of actuator location, or issues of mapping neuromuscular effort (e.g., EMG) to device output. The native hand incorporates the majority of the musculature (in terms of volume and mass) for hand movement in the forearm. As such, a strictly biomimetic multigrasp prosthesis design would similarly utilize space in the distal forearm for the location of actuators. Further, in the native limb, the neuromuscular control system is generally characterized by a low output impedance (i.e., a relaxed postural state) in the absence of neuromuscular effort, while increases in neuromuscular effort generally correspond to increases in grasping force and output impedance. As such, a strictly biomimetic design approach would map neural effort (presumably provided by either EMG or neural recordings) to grasping force through a low output impedance mechanism. That is, in the absence of neural effort, the hand would

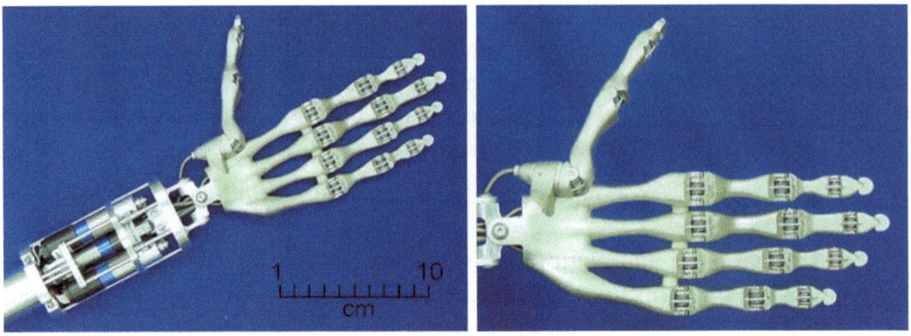

Fig. 1 The 16-joint biomimetic hand

assume a relaxed posture, while in the presence of neural effort, the grasping force exerted by the hand would relate directly to the degree of effort. The remainder of this section describes a multigrasp hand prosthesis, the design of which was guided by the objectives described in the previous section, and which was designed to comply with the biomimetic spatial and neuromuscular models described here.

The biomimetic hand, shown in Fig. 1, has 16 joints which are driven by five independent actuators. The tendon-based actuation units for the hand reside in the forearm, similar to the native human anatomy. The actuation units use brushed DC motors coupled with low-ratio gearheads and small diameter pulleys to pull hand tendons. Each joint in the hand includes embedded torsional springs in parallel with the hand tendons as discussed subsequently. The five actuators are allotted to the 16 joints in the hand as described in Table 1. In all cases, the underactuation is governed by moment isotropy (i.e., differential coupling), rather than by kinematic constraints. In other words, the hand will reach a configurational equilibrium when all joint moments are (essentially) equal. This is achieved by a combination of having the tendon span multiple joints and using a pulley differential to split the force of the actuator output equally into two tendons, as briefly described in Table 1.

Each joint in the hand incorporates embedded torsional springs, as shown in Fig. 2a. The use of torsional springs in the joints serves several purposes. First, and perhaps most obviously, the compliance provides a return force for finger

Table 1 Distribution of actuation and mechanism of coupling in the hand

Actuator	DoF's	Coupling	Mechanism of coupling
Forefinger flexion	3	Same moment across all joints	Tendon spans multiple joints
Index finger flexion	3	Same moment across all joints	Tendon Spans Multiple Joints
Ring and little finger flexion	6	Same moment across all joints	Two tendons coupled by pulley, each tendon spans multiple joints
Thumb flexion	3	Same moment across all joints	Tendon spans multiple joints
Thumb opposition	1	Direct drive	Direct drive

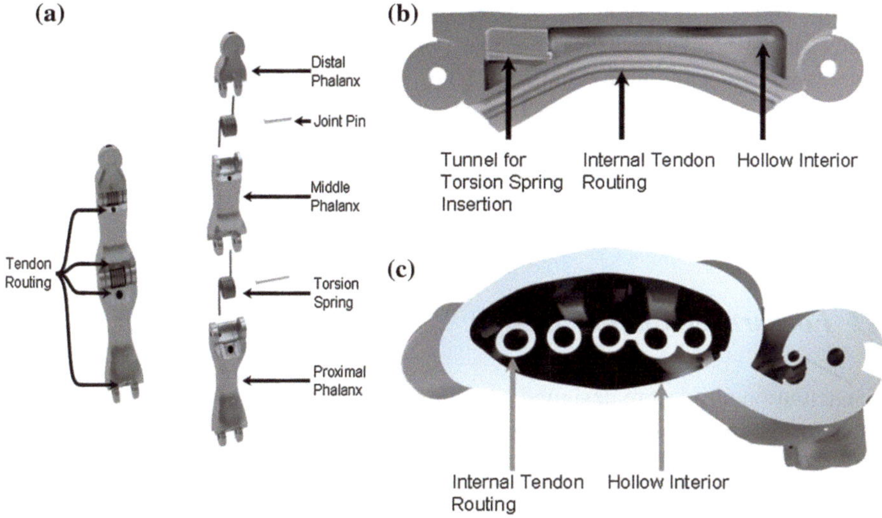

Fig. 2 Detailed views of the intrinsic hand prosthesis. **a** Unexploded and exploded views of a finger showing the phalanges, torsion springs of the joints, and joint pins. **b** Sagittal section of a proximal phalanx, showing the tunnel for torsion spring insertion, internal tendon routing, and the hollow interior. **c** Transverse section of the base of the palm, showing internal tendon routing, and the hollow interior. The base of the thumb is visible on the right

extension. This methodology simplifies tendon actuation of the joints as additional tendons for finger extension are not necessary. Second, the compliant joints map joint motion to tendon force in free space, thus eliminating the need for position sensing in the hand, and eliminating the need to switch between motion control and force control. Specifically, the hand frequently engages in both motion control (e.g., when gesturing or reaching) and force control (e.g., when grasping or squeezing). In the hand, the tendons are always under (open-loop) force control. When the fingers are not in contact with an object, the springs in the fingers map the tendon force to finger position, such that the fingers are in effect under position control. When the fingers come in contact with a rigid object, the force-controlled tendons map directly to force control, and thus the switching between position and force control is natural and seamless.

A unique aspect of the prosthetic hand is the fundamental use of additive manufacturing methods in its conception and design. The structural elements of the hand (i.e., the palm and all digits) are fabricated with a stereolithography (SLA) process from a thermoplastic resin, after which the plastic parts are strengthened with a nickel coating. The additive process enables a realization of hollow, monocoque components with complex features, which are neither possible nor practical with conventional fabrication techniques (See Fig. 2b, c). Some of these features include integrated electrical wire routing in the palm and integrated tendon routing in the palm and all digits. All tendon and wire routing paths incorporate complex curvatures to increase compactness and decrease tendon friction, and which cannot be realized with conventional fabrication techniques. The resulting hand structure has a mass of approximately 80 g, with springs and connecting pins adding an additional 20 g (i.e., 100 g hand mass, not considering the mass of the structure and actuation in the forearm). The hand structure was designed to nominal skeletal dimensions, in order to accommodate a biomimetic soft tissue covering (or cosmesis), which is a critical component of effective and efficient grasping and manipulation.

3.1 Forearm/Actuation Design

The actuation units were designed to achieve the force and speed characteristics. In order to achieve these specifications, the actuation units shown in Fig. 3a were developed. Each unit is composed of a brushed DC servomotor (Faulhaber 1724 SR12) with integrated magnetic encoder (Faulhaber 1E2 516) and planetary gearhead (Faulhaber 16/7 43:1), coupled to a bearing housing across the structural plates of the extrinsic actuation housing. Within the bearing housing are two miniature, high precision (ABEC 7) radial bearings (New Hampshire Ball Bearing SSR3 and SSRI614). A coupling shaft cantilevered within the bores of these bearings transmits torque from the motor to the pulleys about which the tendon material is wrapped. Each unit is fully backdrivable (the return force being provided by the torsion springs embedded in the joints of the digits, Fig. 2a) as per the design objectives. The actuation units transfer force to the digits via braided spectra cable (the "tendon") which runs from the motor pulleys, around idler pulleys for redirection, and through the distal-most plate of the forearm. From here, the tendons pass through Bowden cables between the forearm and hand (allowing for positioning of the wrist) and then through routing tunnels in the hand (shown in Fig. 2a, b) to terminate in the distal phalanx. These tendon paths are lined with Teflon tubing to reduce friction. Braided spectra cable of 0.75 mm diameter with 668 N rated strength was chosen for tendon material due to its high strength, high tensile fatigue resistance, low creep and low stretch characteristics. Figure 3b shows the five actuation units situated within the forearm.

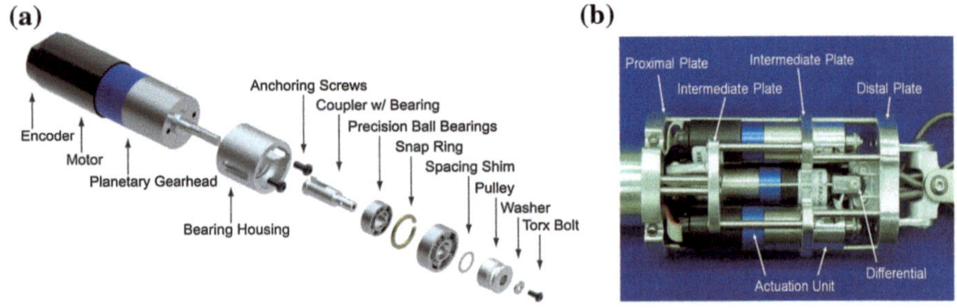

Fig. 3 Exploded view of the actuation unit (**a**) and anterior view of the actuation unit housing (**b**)

4 Experimental Performance Characterization of the Biomimetic Hand

This section provides either an experimental demonstration or characterization, as appropriate, for each of the objectives discussed in the "Design Objectives" section of this chapter. The hand was designed to achieve a set of six grasps and two postures. The ability of the hand to achieve these grasps and postures is demonstrated in Fig. 4a.

4.1 Fingertip Force

The maximum fingertip normal force was measured as a function of percent-tendon-excursion for the index finger using an Extech Instruments 475044 tension and compression force gage rigidly mounted in a variety of positions relative to the palm. The index finger was flexed and held in place for each position to be measured. The force gage was then mounted with its sensing tip placed adjacent and perpendicular to the fingertip face. With the finger in position, a current of 1 A

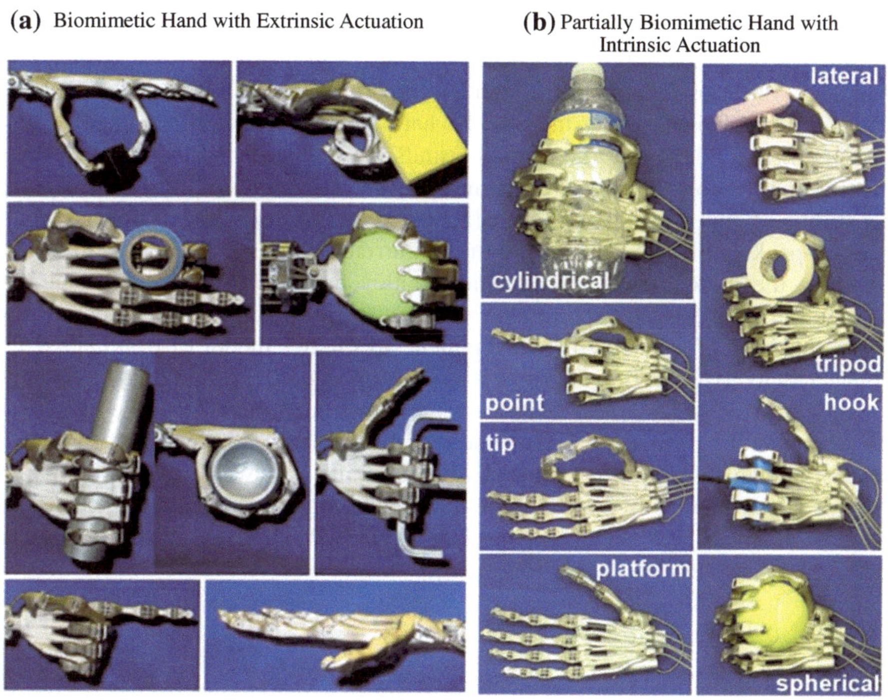

Fig. 4 Six hand grasps and two hand postures, which constitute one of the primary design objectives of the hand prototypes

(the thermal limit for short term operation) was passed through the motor so that the maximum force could be recorded with a minimum amount of initial travel. This was repeated three times and an average of the forces was taken. These values were then normalized by the initial maximum force at the fully open position, $F_o = 11.7$ N, the results of which are shown in Fig. 5a.

The curvature in the data points of Fig. 5a may be explained as follows. At small tendon excursions there is no considerable counteracting force from the springs in the joints of the finger. As tendon displacement increases, the springs provide a (linearly) increasing opposing force, causing the fingertip force to initially drop. Simultaneously, as the finger traverses its range of motion, the tendons act through increasingly greater moment arms (across the finger joints) and by 20 % of tendon excursion the maximum fingertip force begins to increase again. The drop in maximum force due to spring resistance is therefore eventually overcome by greater mechanical advantage as tendon excursion, and finger position, increases. The maximum force is found by multiplying the initial force by the normalized force at full excursion. This results in a maximum force of 19.9 N, indicating that the hand can attain the fingertip normal force of 20 N (at full flexion) specified in the actuation unit design.

4.2 Force Bandwidth

Although the closed-loop position tracking provides a characterization of the positional bandwidth of the fingers, the hand was designed to be operated in a force control mode. Since the actuation units are highly backdrivable, such force control simply constitutes current control in the DC motors. The force tracking capability of the prosthesis was characterized by commanding a positive 0.5A peak-to-peak amplitude sinusoidal current (current being linearly related to tendon force by the motor torque constant and pulley diameter) through the motor at various frequencies using open-loop force (i.e., current) control. The experimental setup consisted of a load cell (Measurement Specialties model ELFM-T2E-025L) connected to MATLAB software via an analog signal conditioning circuit and Humusoft MF624 DAQ card. Force normal to the index fingertip at the fully open (0 % excursion) position was then measured using the calibrated load cell. The results, shown in Fig. 5b, indicate a force tracking bandwidth of approximately 36 Hz. These data show the DC force to current gain at the open (0 % excursion) position as being 17.5 Newtons/Amp in dB (7.5 N/A). These forces are averaged over several cycles and do not represent peak values.

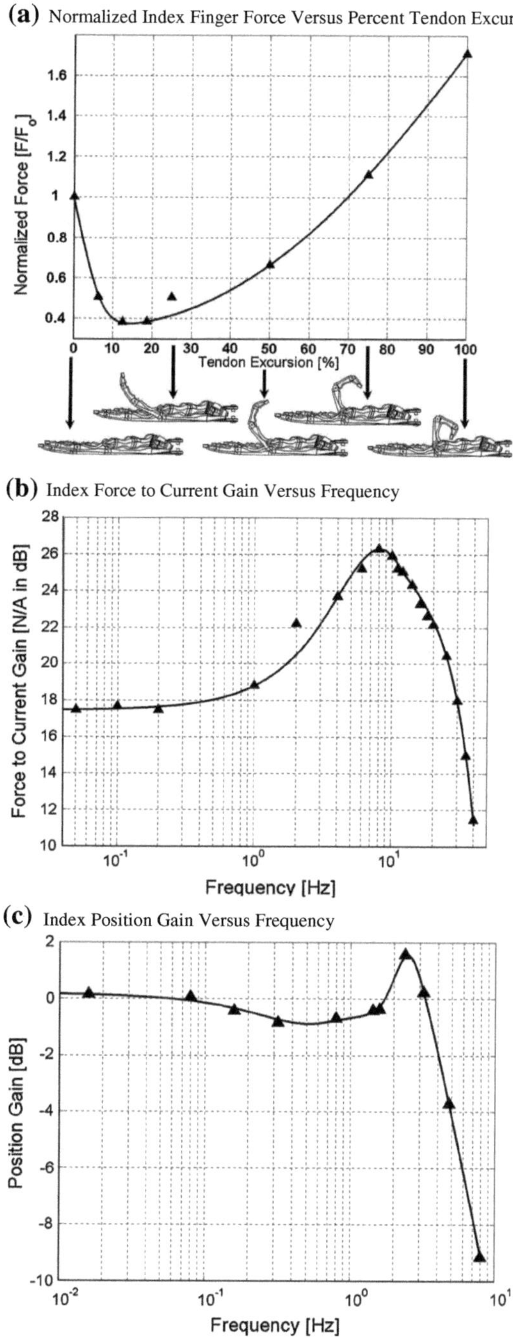

Fig. 5 Index finger performance characterization of the biomimetic hand prosthesis with extrinsic actuation

4.3 Speed Characterization

The position tracking capability of the prosthesis, shown in Fig. 5c, was investigated by commanding the position of the index finger around the mid-flexion point (corresponding to 50 % of total tendon excursion). Specifically, the index finger tracked a sinusoidal position signal which varied around the mid-flexion point by ±25 % of total tendon excursion (i.e., 50 % of the total finger motion) for various frequencies. The bandwidth was determined using the integrated encoders to send position information to MATLAB software by way of a data acquisition card (Humusoft MF624). The results indicate a bandwidth of 4.5 Hz in position tracking. This bandwidth also provides some degree of characterization of the degree of backdrivability, since the extension portion of the movement is driven entirely by the springs backdriving the motors. In comparison with the native hand, recall that joint angular velocities reach approximately 4 rad/s in typical ADL's, although the hand is capable of up to 40 rad/s [21]. The bandwidth shown in Fig. 5c represents a joint angular velocity of 12.9 rad/s, thus indicating joint speeds more than adequate for most ADL's. A summary of the technical specifications of the hand is given in Table 2.

5 A Partially Biomimetic Design Approach

Although biomimicry is a sensible objective when considering the functional replacement of the native hand, the upper extremity amputee may be better served by compromise in this regard, particularly with regard to two issues. Firstly, in order to accommodate the greatest proportion of upper extremity amputees, and in particular to accommodate transradial amputees with relatively long residual forearms, actuation for the hand should ideally be located within the hand, rather than in the forearm. Thus, the hand should be intrinsically rather than extrinsically actuated. Secondly, in the absence of kinesthetic sensory feedback to the user, the paradigm of mapping user neural effort to grasping force through a low output impedance mechanism does not provide a viable means of sustaining grasps over a

Table 2 Technical specifications of the described prosthetic hands

Specification	Biomimetic hand with extrinsic actuation	Partially biomimetic hand with intrinsic actuation
Number of actuators	5	4
Number of joints	16	16
Weight	580 g	320 g
Grasp patterns	8	8
Grasp speed (time to close)	400 ms	280 ms
Noise (dBA @ 1 m)	52.1	52.1

long time period (i.e., several seconds). Specifically, sustaining a grasp requires a continuous muscle contraction at an EMG level of sufficient strength to prevent dropping the held object. Efficiently and effectively maintaining such a muscle contraction in the absence of kinesthetic (e.g., force or slip) feedback is a challenging task, and in the absence of continuous visual feedback, a user would likely drop a held object. Thus, in the absence of kinesthetic feedback, a user is likely better served by a non-biomimetic model in which neural effort is used to change the configuration of the prosthesis, rather than to maintain it. Such a model has been incorporated for a number of years in conventional myoelectric prostheses, which effectively map neural effort to the speed of motion, such that the prosthesis maintains its current configuration in the absence of EMG input (i.e., in the circumstance of full relaxation). Although such a paradigm is non-biomimetic, it enables the user to grasp an object by using neural effort (e.g., EMG), then relax the neural effort and not be concerned about maintaining the grasp. If such an approach is combined with a non-backdrivable mechanism (NBM) between the fingers and the actuators and some compliance between the NBM and object, it also enables a grasp to be maintained for an indefinite period, without requiring any electrical power consumption to do so. Thus, although clearly non-biomimetic, a user is arguably better served by mapping neural effort to changes in hand configuration, and coupling such a mapping with non-backdrivable mechanisms between the actuators and output. As such, the remainder of this section describes a multigrasp hand prosthesis, the design of which was guided by the objectives described in Sect. 2 (Design Considerations for a Multigrasp Prosthetic Hand), but unlike the design described in Sects. 3 (A Biomimetic Design Approach) and 4 (Experimental Performance Characterization of the Biomimetic Hand), the hand described below incorporates all intrinsic actuation (i.e., actuation located within the hand), and incorporates non-backdrivable mechanisms (two-way clutches) between the actuators and the fingers.

5.1 Basic Configuration

The intrinsically actuated, non-backdrivable hand can be seen in Fig. 4b. The hand prosthesis incorporates 16 degrees of freedom, which are single-DoF metacarpophalangeal (MCP), proximal interphalangeal (PIP), and distal interphalangeal (DIP) joints in each finger, and a two-DoF carpometacarpal (CMC) and single-DoF MCP and interphalangeal (IP) joints in the thumb. That is, each finger has three DoFs, which enable finger flexion/extension, and the thumb has four DoFs, which enable thumb flexion/extension and palmar abduction/adduction. These 16 degrees of freedom enable the hand to fully achieve the six grasps and two postures (although the significance of the finger DIP and thumb IP joints is arguable). Independently achieving the aforementioned grasps and postures requires four motors, wherein one actuates thumb (digit I) flexion, one provides for palmar abduction of the thumb, one actuates index finger (digit II) flexion, and one actuates flexion of the remaining fingers (digits III-V). Relative to the extrinsically actuated hand, the independence between the middle finger (digit III) and the two smaller fingers (digits IV and V) is eliminated, such that all move together, and the full differential between the smaller fingers (digits IV and V) is replaced with a limited movement compliant differential (provided by series elastic elements, as subsequently described), which acts between the three digits (III-V). Also, only six of the aforementioned grasps and postures can be achieved independently of an object being grasped. That is, the distinction between the tripod, cylindrical, and spherical grasps are dependent on the shape and position of an object being grasped, in addition to compliance in the drive couplings. Such a dependency, however, is reasonable, since such grasps are only meaningful (with regard to ADLs) in the context of grasping an object.

Like the biomimetic design, tendon-based actuation was utilized to transmit motion between the motors and the joints of the hand to minimize the mass and size. Since the fingertip forces required for typical ADLs are in the direction of finger/thumb flexion (see, for example, [14–18, 20]), a unidirectional tendon configuration was selected, such that the tendon drives provide torques required for finger/thumb flexion, while the hand relies on torsional springs embedded in each joint to provide the torques required for finger/thumb extension. That is, closing of the hand is driven by the tendons, while opening is driven by the torsional springs. Similarly, palmar abduction of the thumb is driven by tendon actuation, while palmar adduction is provided by a torsional spring.

Each tendon is driven by a motor unit, which consists of a brushless motor and gearhead, a two-way clutch that prevents backdriving the motor, and a pulley upon which each tendon is wrapped. The construction of these motor units is described in the following section. The movement associated with each motor unit is referred to as a degree-of-actuation (DoA). The hand prosthesis therefore has four DoAs, which are thumb flexion, thumb abduction, index finger flexion, and digits III-V flexion. The thumb abduction (digit I abduction, or DI-A) motor unit drives only the palmar abduction axis of the CMC joint. The thumb flexion (digit I flexion, or DI-F) motor unit therefore drives three joints (one axis of the CMC joint, the thumb MCP joint, and the thumb IP joint), all via a single tendon. Similarly, the index finger (digit II flexion, or DII-F) motor unit drives the MCP, PIP, and DIP of the index finger, also via a single tendon. Finally, the digit III-V flexion (DIII-V-F) motor unit drives the MCP, PIP, and DIP joints of the remaining three fingers via three tendons, one for each finger. As such, the DI-A, DI-F, and DII-F motor units each drive a single tendon pulley, while the DIII-V-F motor unit drives three tendon pulleys (stacked in series along the motor unit output shaft), one for each of the DIII-V fingers. All tendons are constructed from polyethylene fiber (Spectra) rated for a load of 580 N. The DIII-V tendons all contain series elastic elements, which enable a compliant differential coupling between these fingers (i.e., they are not strictly kinematically coupled). Figure 6a highlights the layout of each of these DoAs, indicating the layout of each motor unit, and the routing of each associated tendon.

Figure 6b shows a cross-section through a representative digit (in this case, the thumb). This cross-sectional view highlights the nature of the tendon paths within each digit, the location of the series elastic springs associated with each tendon, and the location of the torsional springs (also referred to as parallel springs) that provide extension torques for each joint. As can be seen in the figure, each fingertip contains a urethane compression spring through which a tendon transmits its actuator force to the finger. The series elastic springs enable improved control of grasping force through the two-way clutches, particularly when grasping a rigid object, and also provide a compliant coupling between the DIII-V fingers. All torsional springs have a natural position that is extended 20° relative to the joint range of motion, such that all springs are effectively preloaded by a load corresponding to 20° of deflection. Finally, the thumb IP and finger DIP joints have approximately twice the stiffness of the other joints. As such, the IP and DIP joints are not fused, but generally do not flex significantly until relatively high tendon loads are imposed.

5.2 Motor Units

Each of the four motor units consists of a brushless motor, planetary gearhead, two-way clutch, and tendon pulley, and additionally includes angular position sensing via Hall effect sensors integrated into the motor. The two-way clutches prevent the load from driving the motor units, without introducing significant friction or loss in the drive train. In this manner, the hand can grasp an object and maintain that grasp, without expending electrical power to do so. Based on the force and speed design objectives, each motor unit incorporates a brushless DC servomotor (Faulhaber 1226B), which incorporate integrated Hall effect sensors for measurement of tendon excursion, coupled to a planetary gearhead (Faulhaber 12/4 64:1). The gearhead output drives a 2.5 mm diameter pulley (around which the Spectra tendon is wrapped) through a two-way clutch, which prevents back-driving without introducing significant loss. The output assembly (consisting of the two-way clutch, tendon pulley, and housing) is shown in an exploded view in Fig. 6c. The two-way clutch, which is a custom design, is shown in a cross-sectional view in Fig. 6d. In the forward drive mode, the input shaft drives the pulley through the clutch bearings, without interference from the output assembly housing. In the backward drive mode, the pulley wedges the clutch bearings against the output assembly housing, thus effectively locking the pulley against the housing and preventing the pulley from driving the input shaft. The efficacy of these clutches is characterized in Sect. 6.

5.3 Sensing

The implementation of series elasticity provides non-kinematic coupling (i.e., enables conformal grasping) between the DIII-V fingers. Of equal importance, the series elastic elements enable control of grasping force, despite the presence of the two-way clutches, by leveraging the position control loop (enabled by the Hall

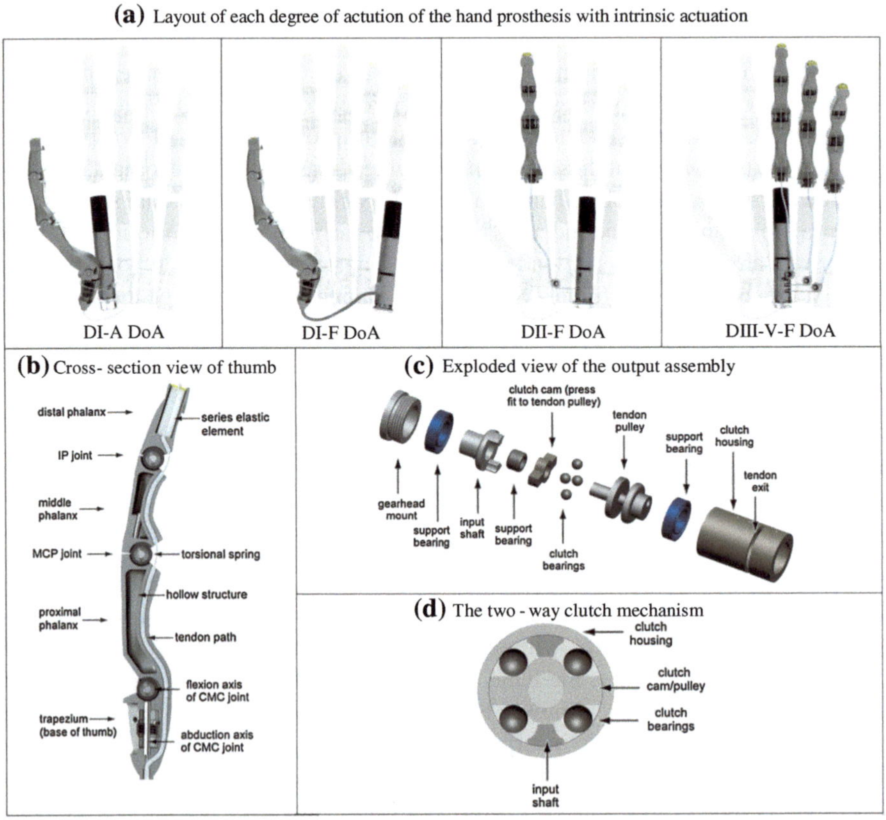

Fig. 6 Mechanical design aspects of the hand prosthesis with intrinsic actuation

Effect sensors in the motor units) that is already in place around the motor units. The hybrid position/force control aspect of the hand is designed to function as follows. When the fingertip is not in contact with an object, the tendon force is a function of the tendon displacement (based on the combination of series and parallel springs). By monitoring this relationship (i.e., via the measured motor displacement and the actuator current), one can ascertain when each finger has come into contact with an object. Once in contact with an object, the measurement and control of the tendon displacement provides measurement and control of the grasping force, where the change in applied force is related to the change in tendon displacement by the stiffness of the series elastic elements. As such, a controller can be constructed to control both displacement of the fingers (when gesturing) and control of grasp force (when grasping), both based on the measurement and control of tendon displacement (which is measured by integrated Hall effect sensing in the brushless motor unit), and by monitoring the commanded motor currents (which requires no additional instrumentation). Further, since the series elastic elements are located between the object load and the two-way clutches, presence of the clutches does not impair the ability of the hand to impose stable, force-controlled grasping. Finally, just as the clutches effectively "lock in" a given finger position under pure position control, the same clutches with the series elastic elements and described hybrid controller provide the dual function of locking in position when gesturing, or locking in force when grasping.

6 Experimental Performance Characterization of Hand with Intrinsic Actuation

This section provides either an experimental demonstration or characterization, as appropriate, for each of the design objectives discussed in the "Design Objectives" section of this chapter. The hand was designed to achieve a set of six grasps and two postures. The ability of the hand to achieve these grasps and postures is demonstrated in Fig. 4b.

6.1 Fingertip Forces

The maximum fingertip normal force was measured as a function of percent-tendon-excursion for each DoA (see Fig. 6a) using a force gage rigidly mounted orthogonally to the tip of the finger or thumb. In order to measure the combined force of the DIII-V fingers, the three fingers were coupled via extension springs to the same force gage. The maximum force is a function of the maximum allowable current in the motor, which in turn is thermally limited by the maximum allowable winding temperature. Since grasping is typically characterized by a constant force, and since the prosthesis described herein incorporates series elastic elements in the fingers and two-way clutches in the motor units, the motors need only provide a short period of active force in order to grasp an object and deform the series elastic elements, after which that force will be maintained passively by the two-way clutches (assuming the object does not continue to deform). Based on thermal simulation of the motors, if each motor is initially at room temperature, each can withstand a current of 1.8 A for approximately 5 s before reaching its recommended thermal limit. Since the fingers can fully close in approximately 300 ms, a full grasp can be obtained and held (by the clutches) in considerably less time than 5 s. Therefore, based on the same thermal modeling of the motors, each motor can continuously sustain a current of 1.8 A for approximately 800 ms every 10 s (i.e., a new object can be grasped every 10 s, with each grasp held continuously). In order to provide a margin of safety, a 400 ms current pulse of 1.8 A was used to characterize the maximum (continuous) force capability of each DoA as a function of percent tendon excursion, the results of which are shown in Fig. 7a. As shown in the figure, the index finger is capable of exerting forces between approximately 15 and 35 N in flexion; the thumb is capable of between 10 and 30 N in flexion

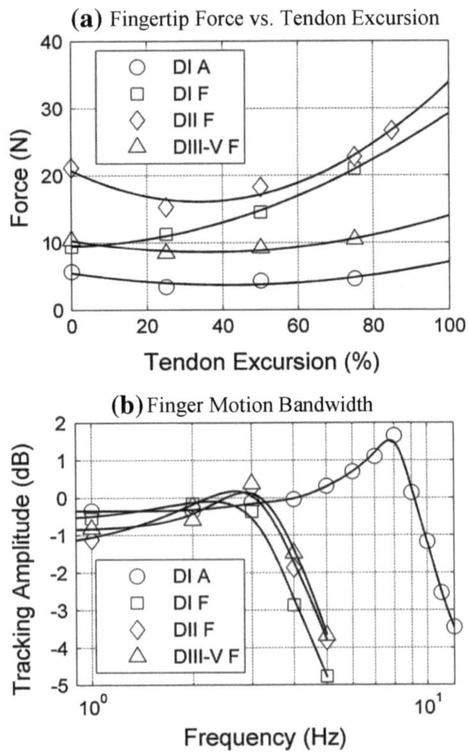

Fig. 7 Performance characterization of the hand prosthesis with intrinsic actuation

and approximately 5 N in palmar abduction; and the remaining three fingers are collectively capable of exerting between 10 and 15 N in flexion. Since most grasps are characterized by large tendon excursions (e.g., the spherical grasp shown in Fig. 4b is characterized by approximately 70 % tendon excursions), maximum usable grasp forces for most ADLs will be approximately 20–25 N for index finger flexion, approximately 18–23 N for thumb flexion, approximately 10–12 N for DIII-V finger flexion, and approximately 5 N for thumb abduction. Recall that the design objectives, based on measured forces during ADLs, were approximately 25 N in index finger and thumb flexion and approximately 12 N for flexion of the remaining digits. Based on the data presented in Fig. 7a, these objectives were fully achieved in all digits, and largely (although not fully) achieved in the range of motion in which such ADLs are likely to occur.

6.2 Clutch Holding Capacity

The clutches were tested to verify that their torque holding capacity exceeded the breaking strength of the Spectra tendons. Specifically, their torque holding capacity was measured by loading a test tendon from the clutch output pulley with a lever arm, and measuring the resultant load with a force gage. In these tests, the clutches were loaded to a maximum continuous torque of 790 mNm, which corresponds to a 630 N load in the test tendon. The clutch was not loaded beyond this point, since the actual tendon used in the hand prosthesis is rated at 580 N, and thus the failure would most likely occur in the tendon prior to failure of the two-way clutch.

6.3 Speed Characterization

The velocity-based design objective requires a half-RoM bandwidth of 1.5 Hz. As such, the half-RoM position tracking capability of each degree-of-actuation was characterized by commanding the position of the finger (or fingers) around the mid-flexion point (corresponding to 50 % of total tendon excursion). In these assessments, each tendon tracked a sinusoidal position signal which varied by ±25 % of total tendon excursion around the mid-flexion point for various frequencies. The bandwidth was determined using the integrated Hall effect sensors to measure tendon pulley position via a Humusoft 624 DAQ card and the real-time interface provided by MATLAB Real-Time Workshop. Figure 7b shows the resulting tracking bandwidth for each DoA, indicating that the index finger flexion, thumb flexion, and DIII-V finger flexion axes each have (−3 dB) position tracking bandwidths of between 4 and 5 Hz, while the thumb abduction axis has a position tracking bandwidth of approximately 11.5 Hz. Further, in terms of full tendon excursion, all fingers were able to fully close (i.e., move from the fully extended position to the fully flexed position) in approximately 280 ms. Thus, the maximum velocity capability of the hand prostheses fully satisfies the design objective for speed of movement.

6.4 Mass Projections

The total mass of the hand prosthesis as shown (i.e., without cosmesis, battery, or electronics) is 320 g. Although the prosthesis prototype described does not currently include embedded electronics, it is estimated that such electronics would have a mass less than 100 g. As such, it is estimated that the total mass of the prosthesis prototype, once embedded electronics are included, would be approximately 420 g. A summary of the technical specifications of the hand, including the force and speed characteristics, is given in Table 2.

6.5 Power Projections

Although one cannot fully characterize power consumption without embedded and battery-powered electronics, a characterization of the electrical power required for actuation is useful in understanding the first-order power requirements of the prosthesis. Based on motor current and voltage measurement, approximately 5 J of electrical energy are required at the leads of the brushless motors to fully close the hand prosthesis (recall that, once achieved, the two-way clutches will maintain the grasp without continued power consumption). Assuming 80 % efficient servo-amplifiers, 6.25 J would be required of a battery for each grasp. As such, a 740 mAh, 11.1 V lithium polymer battery (e.g., Thunderpower TP730-3SJPL2), which has a mass of 57 g, would provide sufficient energy for approximately 4700 grasps. Of course, one cannot accurately predict the expected battery life without accurate estimates of the power requirements of the embedded system, including the power required for computation, sensing, communication, and power management. Despite this, the power required at the actuators of the prosthesis prototype appears reasonable.

7 Conclusion

Two transradial prosthesis designs were described in this chapter—one based on a fully biomimetic design approach, including extrinsic location of actuation, and mapping of neural effort to grasping force, and the other based on a practical compromise in biomimicry, which compromises independence in finger control slightly, in order to incorporate intrinsic actuation, and compromises also biomimetic neural control, in order to facilitate sustained grasping and to decrease electrical power consumption. Both prostheses achieved six grasps and two postures collectively constituting over 85 % of the grasps and postures used in activities of daily living. The fingertip forces and speeds of the hands were also characterized, and shown to be comparable to the respective forces and speeds characteristic of many activities of daily living.

References

1. Y. Kamikawa, T. Maeno, Underactuated five-finger prosthetic hand inspired by grasping force distribution of humans, in *IEEE/RSJ International Conference on Intelligent Robots and Systems*, pp. 717–722, Sep 2008
2. J.L. Pons, E. Rocon, R. Ceres, D. Reynaerts, B. Saro, S. Levin, W. Van Moorleghem, The MANUS-HAND dextrous robotics upper limb prosthesis mechanical and manipulation aspects. Auton. Robots **16**(2), 143–163 (2004)
3. J.U. Chu, D.H. Jung, Y.J. Lee, Design and control of a multifunction myoelectric hand with new adaptive grasping and self-locking mechanisms, in *Proceedings of 2008 IEEE Conference on Robotics and Automation*, pp. 743–748, May 2008
4. C. Cipriani, M. Controzzi, M.C. Carrozza, Progress towards the development of the SmartHand transradial prosthesis, in *Proceedings of 2009 IEEE Conference on Rehabilitation Robotics*, pp. 682–687, June 2009

5. C. Cipriani, M. Controzzi, M.C. Carrozza, Objectives, criteria and methods for the design of the SmartHand transradial prosthesis. Robotica **28**, 919–927 (2010)
6. C.M. Light, P.H. Chappel, Development of a lightweight and adaptable multiple-axis hand prosthesis. Med. Eng. Phys. **22**, 679–684 (2000)
7. S. Jung, I. Moon, Grip force modeling of a tendon-driven prosthetic hand, in *International Conference on Control, Automation, and Systems*, pp. 2006–2009, 2008
8. C. Pylatiuk, S. Mounier, A. Kargov, S. Schulz, G. Bretthauer, Progress in the development of a multifunctional hand prosthesis, in *Proceedings of the 2004 IEEE Engineering in Medicine and Biology Society*, vol. 2, pp. 4260–4263, Sep 2004
9. A. Kargov, C. Pylatiuk, R. Oberle, H. Klosek, T. Werner, W. Roessler, S. Schulz, Development of a multifunctional cosmetic prosthetic hand, in *Proceedings of the 2007 IEEE 10th International Conference on Rehabilitation Robotics*, pp. 550–553, 2007
10. C.L. Taylor, R.J. Schwarz, The anatomy and mechanics of the human hand. Artif. Limbs A Rev. Curr. Dev. **2**(2), 22–35 (1955)
11. M.R. Cutkosky, On grasp choice, grasp models, and the design of hands for manufacturing tasks, in *IEEE Transactions on Robotics and Automation*, vol. 5, no. 3, 1989
12. C. Jacobson-Sollerman, L. Sperling, Grip function of the healthy hand in a standardized hand function test. Scand. J. Rehabil. Med. **9**, 123–129 (1977)
13. C. Sollerman, V. Ejeskar, Sollerman hand function test: a standardized method and its use in tetraplegic patients. Scand. J. Plast. Reconstr. Surg. Hand Surg. **29**(2), 167–176 (1995)
14. C. Pylatuik, A. Kargov, S. Schulz, L. Doderlein, Distribution of grip force in three different functional prehension patterns. J. Med. Eng. Technol. **30**(3), 176–182 (2006)
15. A. Kargov, C. Pylatuik, J. Martin, S. Schulz, L. Doderlein, A comparison of the grip force distribution in natural hands and in prosthetic hands. Disabil. Rehabil. **26**(12), 705–711 (2004)
16. N. Smaby, M.E. Johanson, B. Baker, D.E. Kenney, W.M. Murray, V.R. Hentz, Identification of key pinch forces required to complete functional tasks. J. Rehabil. Res. Dev. **41**(2), 215–224 (2004)
17. R.G Radwin, S. Oh, T.R. Jensen, J.G. Webster, External finger forces in submaximal five-finger static pinch prehension. *Ergonomics*, vol. 35, no. 3, pp. 275–288, 1992
18. N.K. Fowler, A.C. Nicol, Measurement of external three-dimensional interphalangeal loads applied during activities of daily living. Clin. Biomech. **14**, 646–652 (1999)
19. W.K. Purves, N. Berme, Resultant finger joint loads in selected activities. J. Biomed. Eng. **2**, 285–289 (1980)
20. B. Redmond, R. Aina, T. Gorti, B. Hannaford, Haptic characteristics of some activities of daily living, in *North American Haptics Symposium 2010*, March 2010
21. R.F. Weir, *Design of artificial arms and hands for prosthetic applications*, ed. by K.P. McCombs. Standard Handbook of Biomedical Engineering and Design, (McGraw Hill, New York, 2003), pp. 32.1–32.61
22. D.H. Silcox, M.D. Rooks, R.R. Vogel, L.L. Fleming, Myoelectric prostheses. A long-term follow-up and a study of the use of alternate prostheses. J. Bone Joint Surg. Am. **75**, 1781 (1993)
23. D.J. Atkins, D.C.J. Heard, W.H. Donovan, Epidemiologic overview of individuals with upper-limb loss and their reported research priorities. J. Prosthet. Orthot. **8**(1), 85–92 (1996)
24. P.J. Kyberd, J.J. Davey, J.D. Morrison, A survey of upper-limb prosthesis users in Oxfordshire. J. Prosthet. Orthot. **10**(4), 85–92 (1998)
25. C.E. Clauser, J.T. McConville, J.M. Young, Weight, volume and center of mass of segments of the human body. AMRL-TR-69-70, Wright Patterson Airforce Base, Dayton, Ohio, 1969
26. J.T. Belter, A.M. Dollar, Performance characteristics of anthropomorphic prosthetic hands, in *Proceedings of the 2007 IEEE International Conference on Rehabilitation Robotics*, pp. 1–7, 2011
27. C. Pylatiuk, S. Schulz, Using the internet for an anonymous survey of myoelectrical prosthesis wearers, in *Proceedings of the Myoelectric Controls Symposium*, pp. 255–257, Aug 2005

Chapter 21
Development of an Anatomically Correct Testbed (ACT) Hand

Ashish D. Deshpande and Yoky Matsuoka

Abstract We have built an Anatomically Correct Testbed (ACT) hand with the purpose of understanding the intrinsic biomechanical and control features of the human hands that are critical for achieving robust, versatile, and dexterous movements as well as object and world exploration. By mimicking the underlying mechanics and controls of the human hand in a hardware platform, our goal is to achieve previously unmatched grasping and manipulation capabilities. In this chapter we present distinguishing design philosophy and features of the ACT Hand compared to the existing robotic hands, and the details of the design and assembly of the finger bones, joints, tendons and actuators. We derive and analyze the unique muscle-to-joint relationships, called the moment arms, in the ACT Hand index finger, and present a software architecture for the control of the hand movement and forces by controlling the numerous muscle actuators. We also illustrate the grasping and manipulation abilities of the ACT Hand. The fully functional ACT Hand platform allows us to experiment with novel control algorithms to develop a deeper understanding of human dexterity.

keywords Robotic hand · Hand biomechanics · Human-inspired design · Hand control software · Moment arm variations

A. D. Deshpande (✉)
Department of Mechanical Engineering, The University of Texas, 204, E. Dean Keeton Street, ETC 1.210, Austin, TX 78756, USA
e-mail: ashish@austin.utexas.edu

Y. Matsuoka
Department of Computer Science and Engineering, The University of Washington, 185, Stevens Way, Seattle, WA 98195, USA
e-mail: yoky@cs.washington.edu

R. Balasubramanian and V. J. Santos (eds.), *The Human Hand as an Inspiration for Robot Hand Development*, Springer Tracts in Advanced Robotics 95,
DOI: 10.1007/978-3-319-03017-3_21,

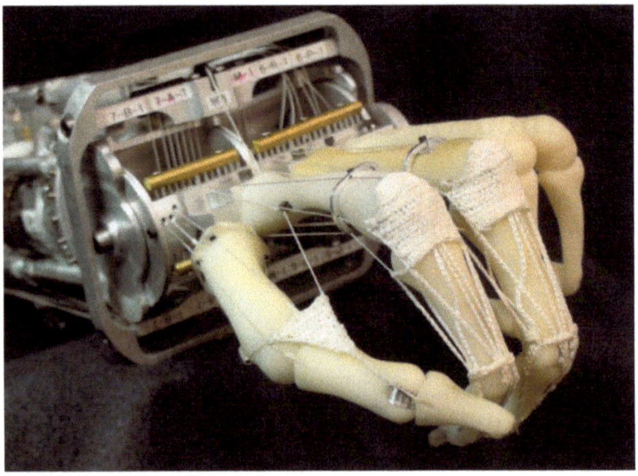

Fig. 1 The Anatomically Correct Testbed (ACT) Hand

1 Introduction

A robotic hand with capabilities comparable to the human hand has long been desired. A dexterous robotic hand has industrial applications such as manipulation tasks that are repetitive or hazardous, and medical applications such as hand prosthesis or hand rehabilitation device. For several decades, researchers have designed and developed a variety of robotic hands with focus ranging from mechanism design to sensors, from control algorithms to increasing degrees of freedom.

A number of robotic grippers have been successfully implemented in industrial applications and new designs of prosthetic hands have improved the range abilities for users. Still, the capabilities of current robotic hands are limited compared to the versatile ability of the human hand in grasping, manipulation and exploration of the world. Our approach toward the development of a dexterous robotic hand is to look closely at the form and the working mechanisms of the human hand. In the human hand, dexterity (?) is achieved by a well-evolved interaction of the intrinsic biomechanics (hardware) and the neuromuscular control (software) of the hand. To develop a robotic hand that achieves human-like dexterity, we have embarked upon a distinct approach toward robotic hand design. We have constructed the Anatomically Correct Testbed (ACT) Hand, as shown in Fig. 1, in which the mechanical elements are designed to mimic the intricate features of hand biomechanics, including bone structures and tendon arrangements, and the software control is inspired by the human hand neuromuscular control system.

While our understanding of the biomechanics of the human hand is quite advanced, there are a number open questions at the intersection of biomechanics and neuromuscular control of the hand. With its anatomical features, the ACT Hand allows us to address questions such as: What role do the biomechanics play in control? Do the intricate biomechanical features, defined by the muscle, tendons, bones and joints, complicate or simplify control? How are the seemingly redundant sets of muscles recruited to achieve posture and movement control?

How does the central nervous system (CNS) control the muscle actuation levels to achieve a certain task? Is there any hierarchy in the control architecture?

In this chapter, we present the design philosophy and details of the mechanical elements of the ACT Hand. The mechanical design leads to human-like muscle-to-joint relationships in the ACT Hand which are critical in hand control. We have developed a data-driven method to determine a mathematical model of these relationships. Since the ACT Hand has a large number of actuators and sensors, and since we want to implement human-like hierarchical neuromuscular control, we have developed a unique software architecture which is presented here. Finally, we present results from early experiments demonstrating grasping and movement control with the ACT Hand.

2 Related Work

Design of a versatile and robust robotic hand that demonstrates human-like grasping and manipulation capacity is a challenging task. The design decisions include the number of fingers, number of joints, degrees of freedom (DOFs), range of motion for the joints, speed of movements and force generation capacity. These design choices have to be made under tight space and weight constraints. A number of robotic hands have been developed to date and Chap. 20 provides detailed background on the existing robotic hand technologies.

Despite the desire for dexterity in prosthetics, the most commonly used prosthetic hand is a mechanically controlled hook prosthesis [19] that was designed over a century ago. Several researchers [7, 11, 26] and companies [20, 27] have designed robotic hands specifically for prosthetic purposes with attention toward minimizing weight, simplifying controls, and aesthetics. Some of the current commercial prostheses are controlled by means of electromyographic (EMG) signals recorded using surface electrodes, which detect electrical activity related to the patient's arm muscles [20, 27]. Because of the difficulty in translating the user intent into useful controls signals, current prosthetic hands have only one or two DOFs; therefore, they are not dexterous. Ideas of control of robotic hands using neural signals have also been explored. Recent studies enable monkeys to control the 3D movement of a robotic arm to achieve self-feeding tasks [29]. Over thirty human arm/hand amputees have received nerve reinnervation surgery to rewire the peripheral nerves that used to go into the hand/arm to the chest muscle instead [24]. The signals, amplified by the natural muscle, can then be tapped into with surface EMG for prosthetic arm/hand control.

To improve the performance and capabilities of robotic hands, new ideas have to be introduced for both the robot hardware and controls. Robotics researchers can greatly benefit from a better understanding of the biomechanics and neuromuscular control of the human hand. Although many robotic hands have been designed to be anthropomorphic, the intrinsic mechanisms of actuation and controls, in most cases, have not been anatomical. In this context, the ACT Hand is designed to be a platform to investigate human dexterity. By mimicking the biomechanical features of the human hand the ACT Hand allows for the identification of the critical factors that lead to dexterity in the human hand.

2.1 Background on Human Hands

The human hand is characterized by its unique biomechanics and neural controls. In the hand, each of the four fingers has three degrees of freedom in the flexion/extension direction and also another degree of freedom in the abduction/adduction direction. The opposable thumb also has degrees of freedom in the flexion/extension and the abduction/adduction directions. There are two types of muscles that control hand movements: (a) those located in the palm, called the intrinsic muscles, and (b) those located in the forearm, called the extrinsic muscles. The muscles are connected to the bones by long tendons that pass over the joints to the bone insertion points on the fingers. Contractions of the muscles lead to hand movement and force generation. The muscles are multi-articulate in that each muscle affects more than one degree of freedom. The moment arm, or the mechanical advantage that a muscle has on a joint motion, is defined by the tendon path around the joint. During hand movement, the tendons slide over the bones leading to variations in the moment arms. Understanding the variations in these moment arm relationships and their role in hand movement controls are topics of ongoing research [12, 21]. The physiological properties of the tendons, joints and muscles lead to the passive visco-elasticity of the hands which stabilizes the joints. The role of passive visco-elasticity during hand movement control is also being investigated [13].

Current research in neuromuscular control suggests that the CNS develops internal models of limb biomechanics, which play a fundamental role in estimation, prediction, controls, and learning [33]. In case of the human hands, it is claimed that to achieve complex manipulation, the CNS must be able to predict the motor commands [16], thus suggesting that the CNS has an internal model of the hand biomechanics, including passive dynamics. It has been shown that the internal models for grasping are physically based and contain information on biomechanics and object properties [32]. Given the current state of knowledge in the fields of hand biomechanics and neuroscience, and the current state of development of robotic hands, to take the next steps toward the development of novel robotic hands, we have determined the mathematical models of muscle-to-joint mappings, and have developed a software architecture that allows us to implement human-like control of the ACT Hand.

Table 1 ACT hand phalange lengths

Finger	Phalange	Length (cm)
Index	MCP to PIP	5.10
	PIP to DIP	2.69
	Distal phalange	1.55
Middle	MCP to PIP	5.38
	PIP to DIP	3.58
	Distal phalange	1.80
Thumb	CMC Flex to CMC Ab-Ad	2.31
	CMC Ab-Ad to MCP	4.31
	MCP to PIP	3.65
	Distal phalange	2.00

3 ACT Hand Mechanisms

In the design of the ACT Hand, our focus is on closely matching the intrinsic biomechanics, actuation and control behavior of the human hand to achieve human-like dynamic movements. The following subsections describe the mechanical components in the ACT Hand, including its finger bones, joints, tendons and actuators [15].

3.1 ACT Hand Finger Bones

We designed the finger bones by accurately matching the size, shape and mass properties of human bones. We used Stratasys Corporation's existing laser-scan model of left hand bones supplied in STL format, imported the tessellated facets into Pro/Engineer, and created solid models for each bone by fitting new surfaces to the scan geometry.

The composition of the finger bones was designed with two primary goals in mind: ease of manufacturing the complex surface shapes and high strength at the joints and tendon attachment points. The current version of the hand employs an innovative design with two separate components. The core of the bones is comprised of a set of steel beams, which offers superior strength and durability, and although not easy to machine, are much more straightforward to fabricate than the complex surface shapes in the previous design. Attached to the beams are outer plastic shells fabricated using stereolithography, which achieve the complex surface geometry. Table 1 gives the hand phalange lengths.

3.2 ACT Hand Finger Joints

The design of the finger joints plays a critical role in matching ACT Hand kinematics with human kinematics. In the ACT Hand we have mimicked the degrees of freedom (DOFs) and ranges of motions (ROMs) for the fingers by following the models

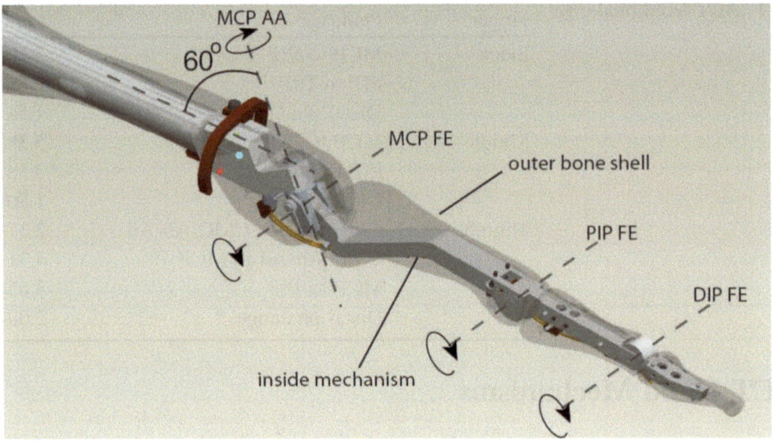

Fig. 2 ACT Hand index finger bones are made of two materials. The *outer* shell, made of plastic, matches the human shape and size; the *inner* steel beam structure allows for anatomical joints

accepted in the biomechanics community [6, 18]. The index and middle fingers have four DOFs, while the opposable thumb has five non-orthogonal, non-intersecting DOFs. We chose to defer the completion of actuation for the last two fingers until we investigate the performance of multi-finger object manipulations with three fingers.

The fingers are actuated by tendons connected to muscle-equivalent actuators. All fingers can hyper-extend, similar to the human fingers. The wrist has two DOFs, and all finger tendons are routed with moment arms preserved, i.e., wrist movements influence finger movements, similar to the human arrangement. In the current design, we implemented the joints as machined pin joints and sought to align the joint axes to best approximate the complex motion of each human finger joint. In some cases, we discovered that joints which at first might appear to be 3-DOF ball-and-socket joints were actually better represented by two carefully aligned pin joints. Figures 2 and 3 show the CAD models for the index finger and thumb.

3.2.1 Index and Middle Finger Joints

There are three joints in the index and middle fingers, namely, the metacarpophalangeal (MCP), proximal interphalangeal (PIP) and distal interphalangeal (DIP) [30]. These are modeled by a novel design involving pin joints in the ACT Hand. The PIP joint is located at the distal end of the proximal phalangeal bone, and the DIP joint is located at the distal end of the middle phalangeal bone. The MCP joint has two DOFs: one to achieve flexion–extension and another to realize abduction–adduction finger motion. These two DOFs are realized by a gimbal mechanism at the distal end of the metacarpophalangeal bone. To match the anatomical joint properties of the human index finger, as described in [5], the abduction–adduction joint axis is oriented at 60° with respect to the metacarpophalangeal bone as shown in Fig. 2.

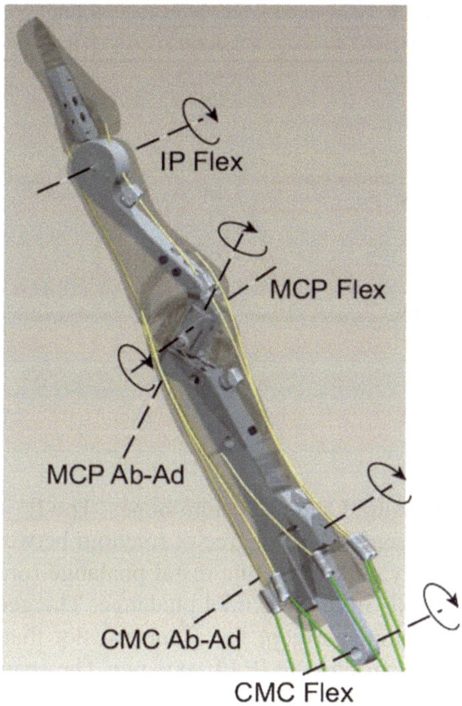

Fig. 3 Thumb bones and joints. The ACT thumb has three joints (i.e., carpometacarpal (CMC), metacarpophalangeal (MCP), and interphalangeal (IP)) and five DOFs

3.2.2 Thumb Joints

The three thumb joints are the carpometacarpal (CMC), metacarpophalangeal (MCP), and interphalangeal (IP) joints [8]. The IP joint possesses one DOF in the flexion–extension direction. The two DOFs at the CMC joint are realized by two non-perpendicular, non-intersecting pin joints. The two DOFs at the MCP joint are realized by a gimbal mechanism, as shown in Fig. 3. Supported by a pair of miniature ball bearings, the gimbal piece rotates around the MCP abduction–adduction (AA) axis fixed within the metacarpal bone. A small pin joint in the gimbal piece represents the MCP flexion–extension (FE) axis, which is fixed relative to the proximal phalange via a link arm. The sweep of the joint cavity restricts the movement of the gimbal assembly to the appropriate MCP joint range of motion.

The CMC joint involves two pin joints at the ends of a single link arm to realize the AA and FE degrees of freedom. Though the CMC and MCP joints are conceptually similar in that they both have FE and AA degrees of freedom, a gimbal design is not suitable for the CMC joint because its two rotational axes are located in separate bones. One pin joint coincides with the CMC AA axis in the proximal end of the metacarpal, while the other pin joint represents the CMC FE axis, which intersects the trapezium carpal bone (which is at the wrist of the hand). Joint range of motion for each of the two axes is constrained by narrow slot cuts in the

Table 2 ACT hand finger joint motion limits

Finger	Joint	Minimum	Maximum
Index	MCP	30° extension	90° flexion
		35° abduction	35° adduction
	PIP	0° extension	110° flexion
	DIP	0° extension	70° flexion
Middle	MCP	30° extension	90° flexion
		35° abduction	35° adduction
	PIP	0° extension	110° flexion
	DIP	0° extension	70° flexion
Thumb	CMC	40° extension	40° flexion
		40° abduction	40° adduction
	MCP	60° extension	60° flexion
		15° abduction	15° adduction
	IP	20° extension	80° flexion

metacarpal and trapezium bones. The IP joint design consists of single pin joint to represent the FE degree of freedom between the two phalangeal bones. A link arm rigidly attached to the distal phalange rotates about an axle coinciding with the IP FE axis in the proximal phalange. The geometry of the articulating bone ends was maintained except for a narrow slot that allows the small diameter link arm to rotate around the IP FE axis pin. The span of the cavity enforces the joint range of motion.

3.2.3 Joint Range of Motion

Joint limits for the flexion joints are imposed by creating internal beam features that interfere with one another at the limits. The ranges of motion were chosen to match those of the human finger, and are shown in Table 2. Because the gimbals used for the MCP joints are free to rotate by ±180°, we designed the bone shells to limit the range of motion by contacting each other at the joint limits. For example, Fig. 2 shows the CAD model of the MCP joint of the ACT Hand's index finger with the bone shell around the gimbal joint to ensure the correct range of motion.

3.3 ACT Hand Tendons

The muscles are connected to the bones by long tendons that pass over the joints, terminating at the insertion points on the finger bones. Muscle contractions lead to hand movement and force generation. In the ACT Hand, we used brushless DC motors as muscles and fabricated our tendons using 0.46 mm Spectra(R) strings. The string was chosen because of its strength (200 N breaking strength), stiffness (4,800 N/strain), and ability to slide smoothly over the bones.

Fig. 4 Wrist joint mechanism with two degrees of freedom

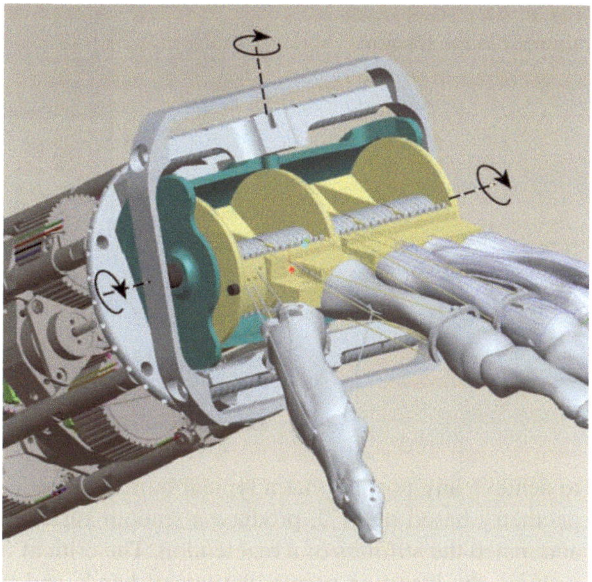

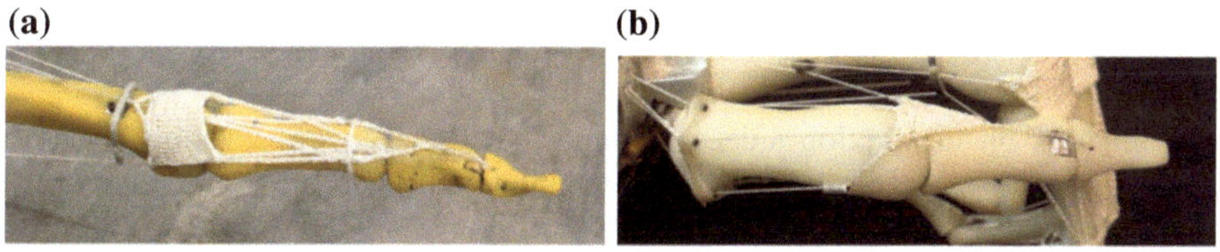

Fig. 5 Tendon hood structure for the index finger (a) and thumb (b)

3.3.1 Tendon Routing over the Wrist Joint

As shown in Fig. 4, each tendon crosses the wrist joint via a pair of sheaves: a large central sheave near the flexion/extension axis and a smaller outer sheave. The size and placement of the tendon sheaves have been chosen so that during wrist flexion/extension, the tendons unwind from one sheave as they are being wound onto the other, minimizing any change in path length.

3.3.2 Extensor Tendon Hood

On the dorsal side, an elaborate extensor tendon web mimicking the human tendon structure connects the tendons between the bone insertion points and the actuators [31]. The extensor mechanism is fabricated by crocheting the spectra strings to emulate the geometry and functionality of the human counterpart as closely as possible, as shown in Fig. 5. The primary goals of the tendon structure design are

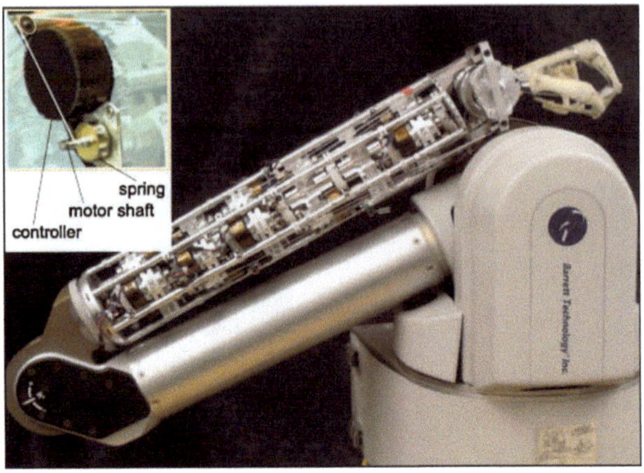

Fig. 6 ACT Hand actuators arranged in the forearm

to achieve any posture that a typical human finger can achieve, mimic the overall geometry based on [17], produce a smooth surface to facilitate a sliding motion, and match the stiffness of a real tendon. The critical features of the extensor tendon web are the insertion points, the lateral bands and the hood, as shown in Fig. 5.

3.3.3 Flexor Side

The tendons on the flexion side are connected directly from the bone insertion points (i.e., the point atop the IP flexion joint in Fig. 3) to the actuators. The flexion tendons pass through guiding "rockers" that allow for smooth travel of the tendon while holding it close to the bone forms in order to achieve accurate moment arm lengths about the finger joints. The tendons in the human hand have elastic properties [34] which play a critical role in the hand dexterity and controls. Mimicking the stiffness properties of the human tendons, either through hardware or software controls is part of our future work.

3.4 ACT Hand Actuators

The ACT Hand possesses the same number of muscles as the human hand except that one extensor muscle that controls all the finger together (EDC) is not implemented in the index and middle fingers. So the index and middle fingers each have six muscles, namely, EI, RI, PI, LUM, FDS and FDP, and the thumb has eight muscles, namely, EPL, EPB, APB, APL, OP, ADP, FPB, and FPL. The wrist is actuated by four muscles. All muscles are realized by brushless DC motors [1] located in the forearm. The tendon strings are wound on the threaded motor shaft, and the loose end is secured by a pin and knot arrangement, as shown in Fig. 6. The string-motor arrangement leads to matching the musculotendon property of one-dimensional actuation, that is, the muscles can only actuate by contraction.

3.4.1 Motor Controller-Puck

Each motor is connected to a miniature controller, called a Puck, which is the smallest motor controller currently available [3]. Each puck has an embedded photo-sensor and an encoder wheel (with 114 ticks/deg) that allows for a high precision position sensing of the motor rotation to mimic muscle spindles (see Fig. 6). The controllers are connected to an RTAI Linux machine, which provides motor position readings at high frequency (>500 Hz).

3.4.2 Motor Housing in the Forearm

All motors are located in the forearm, as shown in Fig. 6. The motors are arranged in clusters of six units, and four clusters are connected end-to-end. The modular design of the clusters and motor housings allows any motor to be easily replaced and entire clusters to be added or removed to match the number of actuators required for a particular setup. Additionally, the clusters incorporate guiding sheaves to route the tendon strings from their radial orientation after leaving the motor shaft to an axial orientation to reach the wrist.

3.5 Mounting the Hand on to an Arm

The wrist has two DoFs: a "yaw" joint ($\pm 15°$) attached to the end of the arm, and a "pitch" joint ($\pm 30°$) connected to the hand. Three banks of pulleys mounted on a gimbal structure that joins the two DOFs guide the tendons from the arm-mounted actuators to the finger joints. These pulleys minimize friction in the tendon paths and route the tendons near the pitch axis to mimic the kinematic coupling present in a human hand. The forearm and hand are mounted on a Whole Arm Manipulator (WAM) developed by Barrett Technologies, Inc. [3]. The WAM is a 4-DOF, cable-driven back-drivable manipulator. Its DC motor control allows position and force control modes. The ACT Hand forearm was connected to the WAM at the elbow joint using the same physical connection that connects the WAM forearm. Figure 6 shows the entire assembly.

4 Muscle-Joint Mappings: Moment Arms

An important characteristic of the human hand is the mechanical advantage, called the *moment arm*, that each muscle–tendon combination has on each joint. The muscle moment arms in the human hand are configuration dependent and play a critical role in hand movement control. However, the exact properties of the moment arm variations are not known. Because the ACT Hand mimics hand biomechanics, through bone shapes and tendon hood structure, its muscle moment arms are also configuration dependent, thus determining the exact moment arm relationships is critical for ACT hand controls. Also, determination of the moment arm relationships in the ACT Hand can potentially lead to a better understanding of human hand biomechanics. We developed a method to acquire the moment arm relationships for the ACT Hand that is based on an analysis of motion capture data for finger and muscle-motor movements. To determine the moment arms, we moved the index finger through its joint range of motion and recorded the joint angles and changes in muscle lengths. We then found a functional mapping, f_i, among all four joint angles and each muscle excursion using a Gaussian process (GP) based regression model.

The finger DOFs and the muscle excursions can be defined to be related by functions f_i as follows:

$$l_i = f_i(\overline{\theta}) \quad i = 1, \ldots, 6, \tag{1}$$

where l_i is the excursion length for the muscle i, which is a member of the vector of muscle excursions $\left(\overline{l} = [l_1, l_2, l_3, l_4, l_5, l_6]^T\right)$ and, $\overline{\theta}$ is a vector of finger joint angles $\left(\overline{\theta} = [\theta_1, \theta_2, \theta_3, \theta_4]^T\right)$. We define θ_1-MCP Ab-Ad, θ_2-MCP Flex, θ_3-PIP Flex and θ_4-DIP Flex, with abduction and flexion as the positive directions. The moment arm relationships were determined by taking the partial derivatives of the muscle excursion functional mappings with respect to the joint angles. In the case of the ACT Hand index finger, the moment arm is defined by a matrix R of dimension 6×4.

$$\dot{\overline{l}} = R(\overline{\theta})\dot{\overline{\theta}} \tag{2}$$

where,

$$R_{ij}(\overline{\theta}) = \frac{\partial l_i}{\partial \theta_j} = \frac{\partial f_i}{\partial \theta_j} \quad i = 1, \ldots, 6 \quad \text{and} \quad j = 1, \ldots, 4. \tag{3}$$

4.1 Method for Determination of Moment Arm Models

To determine the moment arms, the index finger of the ACT Hand was moved in the entire range of its joints. The muscle excursion data was collected by measuring the angular rotation of the motors using encoders on the motor shafts. The joint angle data was collected using a Vicon 360 motion capture system (Vicon Inc.). Six Vicon M2 cameras were set up around the finger to record motions involving all four joint angles. Thirteen markers, each 3 mm in diameter, were placed on the ACT finger and the distribution of the markers was as follows: five on the MCP bone, three on the proximal and middle phalange, and two on the distal phalange. The locations of the markers on the bones were chosen to avoid marker occlusion during finger movement. The XYZ positions of the markers were recorded at 120 Hz, and finger joint angles were determined by using an angle determination algorithm built into the Vicon software (Vicon iQ 2.5). Each data point for our experiments consisted of the time stamp, the muscle positions and the joint angles. We collected approximately 200,000 data points that cover the whole range of four joint angles which is: MCP Ab-Ad: $\in [-15^\circ \quad 15^\circ]$, MCP Flex-Extend: $\in [-30^\circ \quad 90^\circ]$, PIP Flex-Extend: $\in [0^\circ \quad 110^\circ]$ and DIP Flex-Extend: $\in [0^\circ \quad 70^\circ]$. On average, for every four dimensional bin defined by a 5° interval in the joint angles, we have 820 (sd 98.1) data points which shows satisfactory coverage of the data in the finger workspace.

The determination of the functional mapping given by Eq. 1 in a high dimensional space (six muscles and four joints) with a large data set is challenging. To address this, we implemented a machine learning technique called the Gaussian processes (GP). A GP based regression model is a probabilistic kernel method [23]. GPs are advantageous to use with high-dimensional data since it is completely data-driven and can model highly non-linear functions. Taking the derivative of the joint angle-muscle length mappings gives the moment arm. GPs have the unique property allowing for the determination of the gradient of the non-parametric functional mapping given by the GP model.

4.2 Moment Arm Variations

Table 3 gives the mean, maximum and minimum values (in mm) of the moment arm for all six muscles. The moment arms for the muscles with respect to all four joint angles show substantial variations. Figure 7a shows the variations in the excursions of EI muscles as two finger angles vary. The dots in the figure are the data points, and the surfaces show the fitted mapping functions. Figure 7c shows the moment arms derived from the muscle excursion mappings.

4.3 Physical Interpretation of Moment Arm Variations

The moment arms for all of the muscles of the index finger depend significantly on all of the joint positions of the finger and, for some of the muscles, the moment arms change sign. A negative value for the moment arm of a muscle with respect to a joint indicates that the muscle is contracting for positive change in joint angle (meaning that the muscle actively contributes to joint movement). Conversely, a positive value for the moment arm means that the muscle is stretched for positive change in joint angle. The moment arms for the muscle with respect to the primary angles show large variations and maintain the sign of the moment arms.

Determination of the variable moment arm matrix allows for quantification of the contributions of the muscle forces and velocities to the finger motions in various finger configurations.

Table 3 Moment arm (in mm/rad) mean, max and min for all six muscles of the ACT Hand due to changes in finger angles

	MCP Ab-Ad			MCP flex-extend			PIP flex-extend			DIP flex-extend		
	Mean	Max	Min	Mean	Max	Min	Mean	Max	Min	Mean	Max	Min
EI	0.11	9.35	−13.97	12.27	25.70	8.68	3.25	8.56	−1.78	−0.20	1.89	−4.07
PI	−8.24	−1.50	−13.00	−0.90	3.95	−6.04	0.11	3.32	−2.61	1.12	3.26	−1.23
FDP	−0.46	7.60	−8.83	−10.82	−9.39	−15.76	−6.35	−3.32	−8.99	4.40	3.32	−6.89
LUM	8.31	14.75	0.77	0.26	10.27	−9.01	1.26	10.21	−5.97	0.03	2.51	−3.14
FDS	0.10	8.07	−13.74	−11.26	−9.23	−15.77	−6.35	−4.25	−8.33	0.03	1.31	−1.58
RI	8.42	16.31	−7.94	−5.95	5.39	−21.57	−0.11	10.93	−6.48	3.74	6.33	1.36

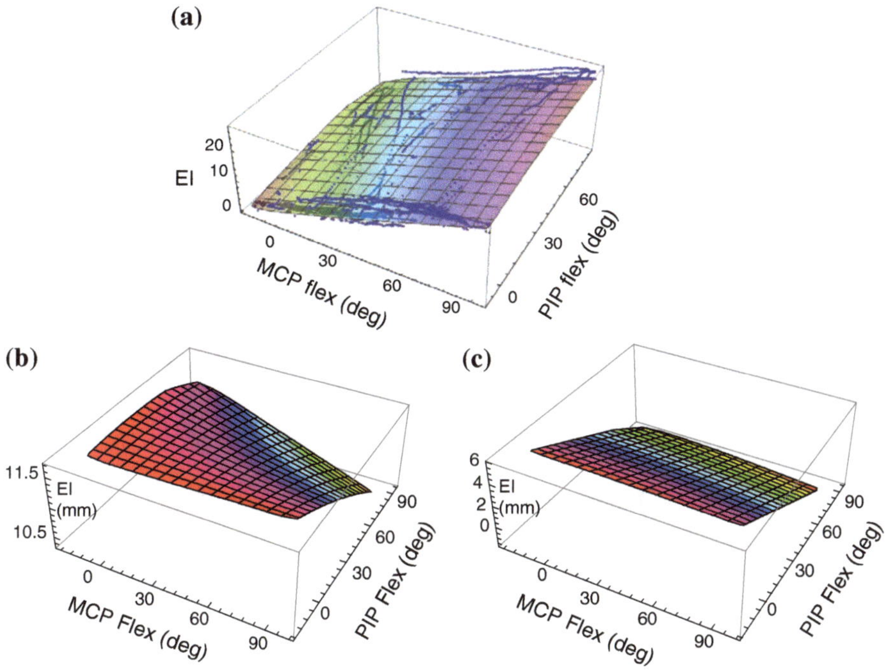

Fig. 7 Muscle excursions and moment arm variations for the EI muscle with respect to the MCP and PIP flexion–extension. **a** EI muscle excursion. **b** EI moment arm. **c** EI moment arm

Figure 8 shows variations in the MCP flexion torque for various finger configurations when all six muscles are fully activated. This surface was created using the relationship between the muscle forces and joint torque given by Eq. 4.

$$\tau_X = \sum_{i=1}^{6} -\left(R^T(2,i) \times f_m(i)_{\max}\right), \tag{4}$$

where τ_X is the flexion torque, R is the variable moment arm matrix, and $f_m(i)_{\max}$ is the maximum pull force generated by muscle i. The maximum force values used were from the literature [28]. This figure illustrates that for the same forces applied by the muscles, the joint torques vary depending on the finger configurations. The figure also shows that a constant moment arm does not capture the variations in joint torques. With the variable moment arm matrix, the flexion torque is maximum when the MCP joint is adducted and flexed. Physiologically, this allows the finger to produce the maximum palmar force when the finger is flexed and adducted.

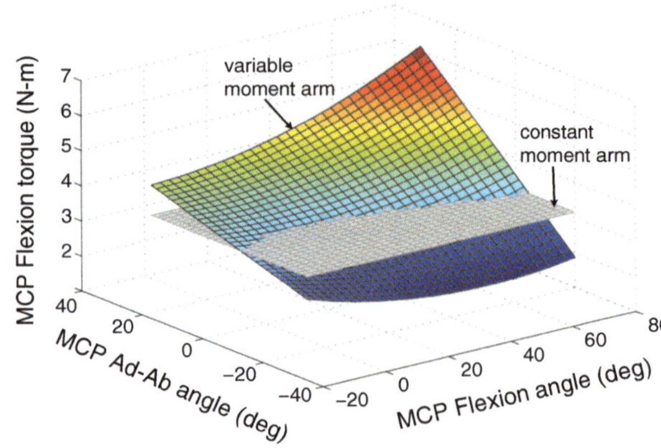

Fig. 8 MCP joint torque in the flexion (X) direction for various finger configuration as all six muscles are fully activated. The flexion torque is maximum when the finger is adducted and flexed. The figure shows the variation with a variable moment arm and with a constant moment arm

5 ACT Software Framework

We have built a software architecture for the ACT Hand control based on human neuromuscular control properties [4]. Our ultimate goal is to build a software platform that reduces the load of programming DC motor control and allows the user to interface with the ACT Hand in a biologically intuitive format. There are four main objectives of our software framework for the ACT Hand.

- *Fast communication with sensors and actuators:* To control the hand movements we need to control each motor torque based on the current torque and position values. To achieve reliable position and force control we need to communicate with all the sensors and actuators at a very high frequency.
- *Implementation of muscle dynamics:* The characteristics of the human muscles play a critical role in the control of the hand. Each muscle is activated by an electrochemical activation signal, $\alpha \in [0, 1]$, which generates a contraction force. The muscles also possess passive stiffness and damping properties, and the tension generated by a muscle depends not only on the activation signal but also on the muscle length and the muscle length change velocity. The tendons that connect muscles to the bones also possess nonlinear stiffness. In the biomechanics literature, the musculotendon system is modeled as a mechanical system with springs and dampers, along with a component that represents contraction force that results from an activation signal [34]. The generation of the contraction force for a given activation value α is modeled by a temporal first order system [34]. This leads to a length-velocity-force curve for each muscle for a given activation value [22]. We want to achieve these musculotendon characteristics using DC motor actuators.
- *Implementation of neuromuscular controls:* Each muscle is controlled by neural activity generated by a complex neural circuit. At the highest level of control the signals are generated in the motor cortex and are transmitted through the central

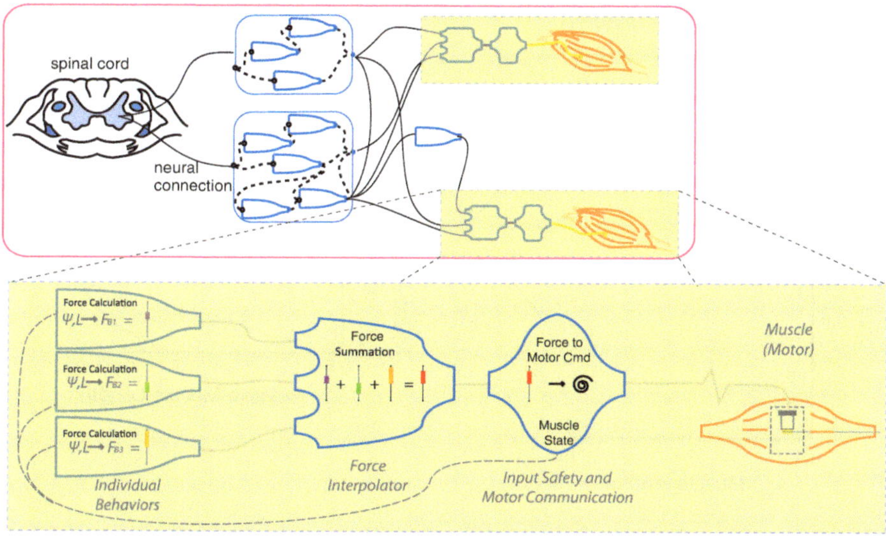

Fig. 9 A software framework is developed for the control of the ACT Hand by implementing human neuromuscular control behaviors. Figure shows different levels of hierarchy with neural connections from spinal cord to individual muscles [4]

nervous system and routed down the spinal cord towards the intended muscle targets. The command signals are transmitted by motor neurons [22]. A single motor neuron is typically attached to multiple muscles and a single muscle receives signals from multiple neurons leading to complex neuromuscular dynamics. The neurons also have local connections with the neurons in the spinal column and the neurons attached to the muscles, providing local reflex circuits that allow quick responses without the temporal delay of communicating the information to and from the cortex [4]. We want to implement the neuromuscular dynamics using a software platform.

- *Safety:* A safety net for software and hardware failures.

To achieve these objectives, we have designed a software system with three levels of hierarchy as shown in Fig. 9. At the highest, we have a muscle behavior model that is exposed to the user. The muscle physical properties such as rest length, current length, velocity, activation level etc. can be set by appropriate parameters at this level. Also, the models for calculation of the muscle passive and active forces are programmed in this level. A single muscle/motor can be commanded by more than one muscle behavior model, for example, a passive force model and an active force model. An aggregation client module, which is the next level in the software hierarchy, sums up the commands from all of the behavior models and sends the commands to the motor communication host. The motor communication host is at the lowest level. It controls the DC motor current and voltages based on the commands from the aggregation module and the current position and torque values. The motor

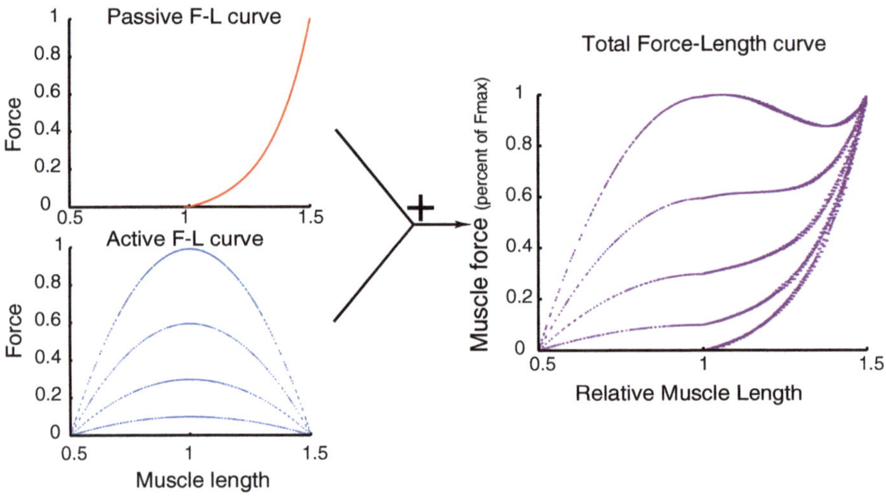

Fig. 10 Experimental results of force–length relationship, passive and active, in each muscle-motor achieved in the ACT Hand by control implemented through the software architecture [4]

communication host also ensures that the commands sent to the motors are within the safety bounds of the motors and ACT Hand motions. The framework is designed to avoid overall crashes/freezes even if part of the framework fails. The actual implementation is in C++ programming language on a Linux operating system a RTAI (Real-Time Application Interface) kernel that allows each module to run at a specific frequency with microsecond resolution.

This modular set-up allows for flexibility in implementing muscle biomechanics and neuromuscular control strategies. A number of behavior modules can be set-up or a single muscle and can run at different rates. For example, a passive behavior can run a very high rate to match physical impedance, while an active behavior can run at a rate that matches the latency of the spinal cord and cortex activities. Figure 10 shows the results of implementing passive and active behavior models on a single muscle. The muscle parameters and the behavior models are based on the physiological data and the resultant DC motor behavior simulates a physiological muscle-length curve. The behavior module provides the ability to simulate the neuromuscular control strategies. A complex neural circuit that matches the cortical signals and spinal cord loops can be implemented with software connections between parts of various muscle behaviors.

6 Grasping and Hand Movement Illustration

Taking advantage of the DOFs and ranges of motion in the ACT Hand, which are copied from the human hand, the ACT Hand is able to grasp and manipulate a number of objects, which are part of daily activities. Figure 11 shows nine

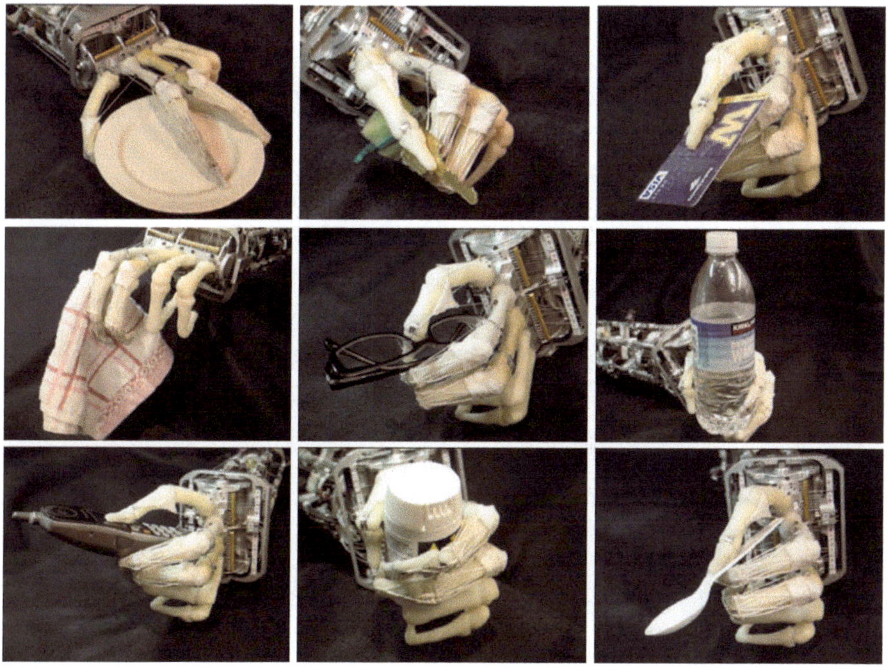

Fig. 11 The ACT Hand can grasp everyday objects with human-like finger postures. Figures show grasping of nine objects, namely, a coffee plate, key, credit card, hand towel, glasses, water bottle, cordless phone, medicine bottle, and spoon, which are identified as important for ALS patients [9]

examples of object grasping by the ACT Hand. These pictures demonstrate the DOFs and range of motion of the finger joints, and overall anthropomorphic grasping abilities. These nine objects were chosen to demonstrate the wide variety of grasps that can be achieved with the ACT Hand, such as a power grasp in the case of the water bottle and a pinch grasp in the case of the spoon. Grasping performance is greatly affected by the object interactions through the skin contact. We are currently developing a tactile skin for the ACT Hand. These grasps were achieved through a direct muscle control scheme [14]. For each grasp we started from a neutral position (finger open) of thumb, index finger and middle finger and pre-calculated the joint angles for the specific grasp. We calculated the muscle excursions (contraction and stretch) necessary to achieve the desired joint angles and implemented proportional-integral? (PI) control on the DC motor driving the muscle. Figure 12 shows the changes in lengths for the thumb and index finger muscles during the key grasp maneuver shown in Fig. 11. The top figure shows the muscle excursions for the six index finger muscles and the bottom figure shows the muscle excursions for the eight thumb muscles. For the starting neutral position the muscle excursions start at close to zero. Negative excursion means the muscle is contracting and positive excursions means that the muscle is stretching. As the

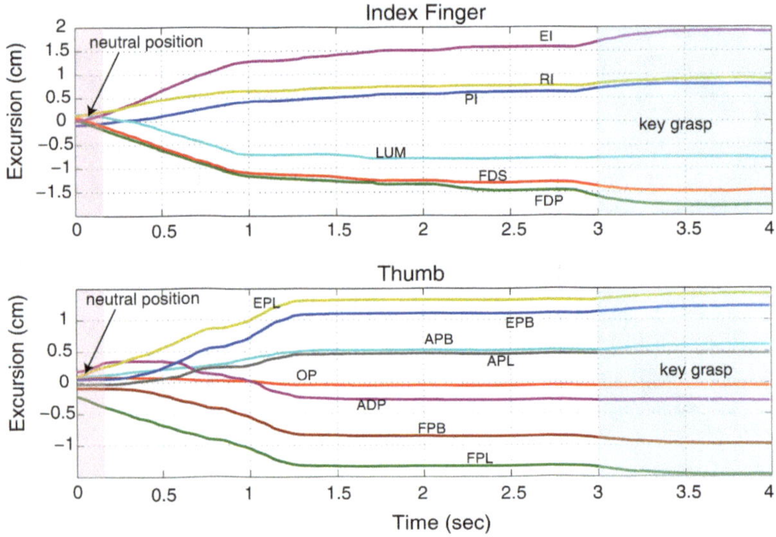

Fig. 12 Length changes in thumb and index finger muscles during key grasp maneuver starting from neutral position (fingers open)

finger joints flex, the flexors are contracted and extensors are stretched for both the thumb and index finger. Notice that the muscle lengths change smoothly over the maneuver leading to smooth motions of the fingers. The mean tracking error over all muscles is 1:67 (sd 0:34) mm.

Considering that thumb movements account for more than 50 % of hand function [10], we also implemented a thumb movement of rubbing against the index finger, which is part of many daily hand movements such as counting money and opening a plastic bag. Because human thumb biomechanics are mimicked in the ACT Hand, including axes locations, number of DOF, and muscle arrangements, we were able to generate a motion that closely matches the human thumb motion. Figure 13 shows the snapshots of the thumb motion and corresponding length changes of the eight muscles during this motion.

7 Summary and Future Work

This chapter presents the constituting mechanisms, moment arm relationships, software architecture and movement demonstrations of the thumb, index finger, middle finger, and wrist of the ACT Hand. The ACT Hand is designed to further our understanding of human hand mechanisms and control and to provide guidelines for building versatile prosthetic and dexterous hands. We have presented a number of novel mechanical elements which are designed and built to mimic human hand biomechanics including the bone structures, joints, tendon

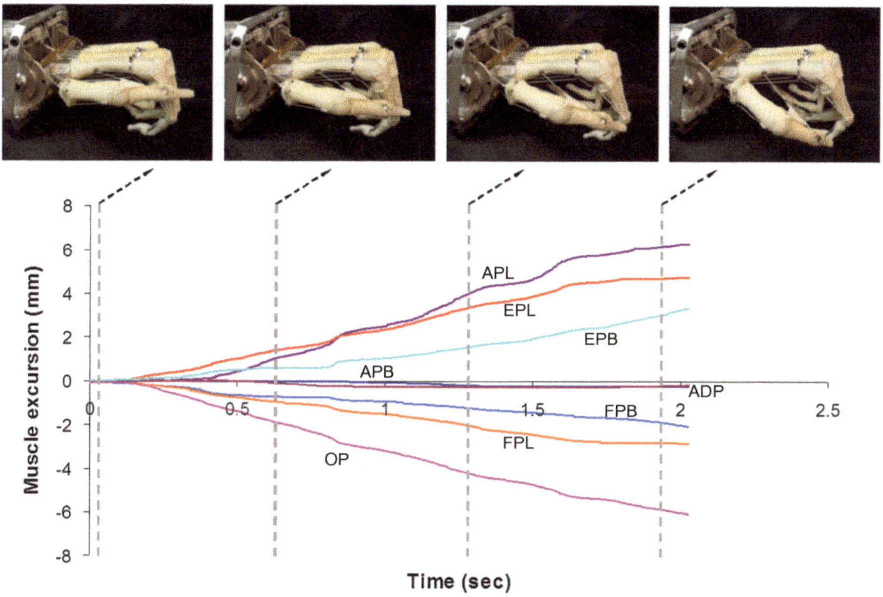

Fig. 13 With the anatomical DOFs the thumb in the ACT Hand can be moved to rub the side of the index finger which is a common and useful human motion. The *top row* shows a sequence of photos taken at different times (*dotted lines*) during the movement execution. And the *bottom row* shows the plots of length changes (in mm) in the eight muscles of the thumb during this movement

arrangements and actuators in the fingers and thumb, and the tendon routing and actuation of the wrist. Our analysis of the moment arm relationships between the ACT Hand muscles and joints shows that muscle moment arms are non-constant and depend greatly on the joint angles. Variable moment arms affect hand control strategies and can be critical in achieving stable movement and force. Using the ACT Hand, we have determined the previously unknown moment arm relationship in the human index finger, thus demonstrating that the ACT Hand can play a vital role in achieving a better understanding of human hand controls. For the control of the ACT Hand, we have developed a hierarchical software framework based on the neuromuscular behaviors of the human control system. Development of tactile skin and passive joint properties in the ACT Hand are part of ongoing research

Completion of the ACT Hand mechanisms and software platform allows us to conduct many experiments, which were hitherto impossible, in order to study human hand properties and translation of important properties into robotic forms. For example, now we can implement novel control algorithms to develop a deeper understanding of human dexterity. Our group has investigated the existence and importance of muscle synergies during hand movement and force control [1, 2] and we are implementing these mechanisms in the ACT Hand. We are also investigating control strategies for achieving neuromuscular control of ACT Hand muscles [14]. In parallel, we also plan to explore and implement existing biological neural control strategies which are investigated in the literature. For example, we are investigating the use of stiffness modulation for humans [1]. The stiffness control has been shown to utilize the intrinsic muscles in a special way [25]. We will mimic such control strategies to understand the advantage of the biological control strategies. Ultimately, we plan to compare biological and engineering control strategies side by side to compare the relative advantages and disadvantages. While we currently believe that approaches based on biological control might be superior to standard robotics approaches, some engineering solutions may allow for augmentation of prosthetic hand control and hand rehabilitation techniques. We are planning to develop replicas of the ACT Hand and make them available for other researchers. This will allow for simultaneous experimentation and growth in understanding of human hand complexities.

Acknowledgments The authors would like to thank Zhe Xu, Michael J. Vande Weghe, Benjamin H. Brown, Timothy Blakeley, Lillian Y. Chang, David D. Wilkinson, Sean M. Bidic for their role in the design and building of the ACT Hand. Also the authors would like to thank Jonathan Ko, Ravi Balasubramanian and Brian Dellon for their role in data collection and modeling of the moment arm relationships.

References

1. R. Balasubramanian, Y. Matsuoka, Biological stiffness control strategies for the anatomically correct testbed (ACT) hand, in *IEEE International Conference on Robotics and Automation*, 2008
2. R. Balasubramanian, Y. Matsuoka, The role of small redundant actuators in precise manipulation, in *IEEE/RSJ International Conference on Robotics and Automation*, 2009
3. Barrett Technology, Inc., WAM specifications, http://www.barrett.com/robot/products-arm-specifications.htm. Accessed Sept 2009
4. T. Blakely, Y. Matsuka, Software framework for human neuromuscular behavior, in *IEEE International Conference on Robotics and Automation*, 2009
5. P.W. Brand, A.M. Hollister, *Clinical Mechanics of the Hand*, 3rd edn. (Mosby, St. Louis, 1999)
6. P.W. Brand, M.H. Anne, *Clinical Mechanics of the Hand* (Mosby-Year Book Inc., St. Louis, 1993)
7. M.C. Carrozza, B. Massa, S. Micera, R. Lazzarini, M. Zecca, P. Dario, The development of a novel prosthetic hand- ongoing research and preliminary results. IEEE/ASME Trans. Mechatron. **7**, 108–114 (2002)
8. L.Y. Chang, Y. Matsuka, A kinematic thumb model for the act hand, in *Proceedings of the IEEE International Conference on Robotics and Automation*, 2006
9. Y.S. Choi, T. Deyle, C.C. Kemp, Benchmarking assistive mobile manipulation: a list of household objects for robotic retrieval prioritized by people with ALS, in *International Conference on Rehabilitation Robotics*, 2009
10. J.C. Colditz, Anatomic considerations for splinting the thumb, in *Rehabilitation of the Hand: Surgery and Therapy*, ed. by J. Hunter, E.J. Mackin, A.D. Callahan (C. V. Mosby Company, Philadelphia, 1900)
11. S.A. Dalley, T.E. Wiste, T.J. Withrow, M. Goldfarb, Design of a multifunctional anthropomorphic prosthetic hand with extrinsic actuation. IEEE/ASME Trans. Mechatron. **14**, 699–706 (2009)

12. A.D. Deshpande, R. Balasubramanian, J. Ko, Y. Matsuoka, Acquiring variable moment arms for index finger using a robotic testbed. IEEE Trans. Biol. Eng. **57**, 2034–2044 (2010)
13. A.D. Deshpande, N. Gialis, Y. Matsuoka, Contributions of the visco-elastic forces during the index finger and wrist movements, in *Press Transactions on Biomedical Engineering*, 2011
14. A.D. Deshpande, J. Ko, D. Fox, Y. Matsuoka, ACT hand finger control: muscle and joint torque control strategies, in *IEEE International Conference on Robotics and Automation*, 2009
15. A. D. Deshpande, Z. Xu, M. J. Vande Weghe, L. Y. Chang, B. H. Brown, D. D. Wilkinson, S. M. Bidic, Y. Matsuoka, Mechanisms of the anatomically correct testbed (ACT) hand. ASME/IEEE Trans. Mechatron. **10**(1), 2013
16. R.J. Flanagan, M.C. Bowman, R.S. Johansson, Control strategies in object manipulation tasks. Curr. Opin. Neurobiol. **16**, 650–659 (2006)
17. M. Garcia-Elias, K. An, L. Berglund, R.L. Linscheid, W.P. Cooney, E. Chao, Extensor mechanism of the fingers. i: a quantitative geometric study. J. Hand. Surg. **16A**, 1130–1136 (1991)
18. A. Hollister, D.J. Giurintano, W.L. Buford, L.M. Myers, A. Novick, The axes of rotation of the thumb interphalangeal and metacarpophalangeal joints. Clin. Orthop. Relat. Res. **320**, 188–193 (1995)
19. Hosmer Dorrance Corporation, Body-powered prosthetic hand, http://www.hosmer.com/products/hooks/index.html.Accessed Sept 2009
20. iLimb: Touch Bionics, ilimb, http://www.touchbionics.com/i-LIMB. (As of Jan 8, 2010)
21. D.G. Kamper, H.C. Fischer, E.G. Cruz, Impact of finger posture on mapping from muscle activation to joint torque. Clin. Biomech. **21**, 361–369 (2006)
22. E.R. Kandel, J.H. Schwartz, T.M. Jessell, *Principles of Neural Engineering*, 4th edn. (McGraw Hill, New York, 2000)
23. J. Ko, D. Klein, D. Fox, D. Hähnel, Gaussian processes and reinforcement learning for identification and control of an autonomous blimp, in *IEEE International Conference on Robotics and Automation*, 2007
24. T. Kuiken, G. Dumanian, R. Lipschutz, L. Miller, K. Stubblefield, The use of targeted muscle reinnervation for improved myoelectric prosthesis control in a bilateral shoulder disarticulation amputee. Prosthet. Orthot. Int. **28**(3), 245–253, 2004
25. T.E. Milner, D.W. Franklin, Impedance control and internal model use during the initial stage of adaptation to novel dynamics in humans. J. Physiol. **567**, 651–664 (2005)
26. D. Nishikawa, Y. Ishikawa, W. Yu, M. Maruishi, I. Watanabe, H. Yokoi, Y. Mano, Y. Kakazu, On-line learning based emg prosthetic hand, in *Electrophysiology and Kinesiology*, 2000
27. OttoBock HealthCare, Inc., Cable-controlled arm prostheses, http://www.ottobock.com. Accessed Sept 2009
28. N.S. Pollard, R.C. Gilbert, Tendon arrangement and muscle force requirements for humanlike force capabilities in a robotic finger, in *Proceedings of IEEE International Conference on Robotics and Automation*, 2002
29. M. Velliste, S. Perel, M.C. Spalding, A.S. Whitford, A.B. Schwartz, Cortical control of a prosthetic arm for self-feeding. Nature **453**(7198), 1098–1101 (2008)
30. M.V. Weghe, M. Rogers, M. Weissert, Y. Matsuoka, The ACT hand: design of the skeletal structure, in *Proceedings of the 2004 IEEE International Conference on Robotics and Automation*, 2004
31. D.D. Wilkinson, M.V. Weghe, Y. Matsuoka, An extensor mechanism for an anatomical robotic hand, in *IEEE International Conference on Robotics and Automation*, 2003
32. A. Witney, D. Wolpert, The effect of externally generated loading on predictive grip force modulation. Neurosci. Lett. **414**(1), 10–15 (2007)
33. D.M. Wolpert, Z. Ghahramani, Computational principles of movement neuroscience. Nat. Neurosci. **3**, 1212–1217 (2000)
34. F.E. Zajac, Muscle coordination of movement: a perspective. J. Biomech. **26**(no. Supplement 1), 109–124 (1993), (Proceedings of the 13th Congress of the International Society of Biomechanics)

Chapter 22
Physical Human Interactive Guidance: Identifying Grasping Principles from Human-Planned Grasps

Ravi Balasubramanian, Ling Xu, Peter D. Brook, Joshua R. Smith and Yoky Matsuoka

Abstract We present a novel and simple experimental method called Physical Human Interactive Guidance to study human-planned grasping. Instead of studying how the human uses his/her own biological hand or how a human teleoperates a robot hand in a grasping task, the method involves a human interacting physically with a robot arm and hand, carefully moving and guiding the robot into the grasping pose while the robot's configuration is recorded. Analysis of the grasps from this simple method has produced two interesting results. First, the grasps produced by this method perform better than grasps generated through a state-of-the-art automated grasp planner. Second, this method when combined with a detailed statistical analysis using a variety of grasp measures (physics-based heuristics considered critical for a good grasp) offered insights into how the human grasping method is similar or different from automated grasping synthesis techniques. Specifically, data from the Physical Human Interactive Guidance method showed that the human-planned grasping method provides grasps that are similar to grasps from a state-of-the-art automated grasp planner, but differed in one key aspect. The robot wrists were aligned with the object's principal axes in the human-planned grasps (termed low skewness in this work), while the automated grasps used arbitrary wrist orientation. Preliminary tests shows that grasps with low skewness were significantly more robust than grasps with high skewness (77–93 %). We conclude with a

Work was done when the authors were all at the University of Washington and Intel Labs Seattle.

R. Balasubramanian (✉)
School of Mechanical, Industrial, and Manufacturing Engineering,
Oregon State University, Corvallis, OR, USA
e-mail: ravi.balasubramanian@oregonstate.edu

L. Xu
The Robotics Institute, Carnegie Mellon University, Pittsburgh, PA, USA

P. D. Brook · J. R. Smith · Y. Matsuoka
Department of Computer Science, The University of Washington, Seattle, WA, USA

R. Balasubramanian and V. J. Santos (eds.), *The Human Hand as an Inspiration for Robot Hand Development*, Springer Tracts in Advanced Robotics 95,
DOI: 10.1007/978-3-319-03017-3_22,

detailed discussion of how the Physical Human Interactive Guidance method relates to existing methods for extracting the human principles for physical interaction.

Keywords Grasping • Haptic interface • Human-robot interaction • Manipulators • Telerobotics

1 Introduction

For a roboticist, the way a human grasps or manipulates an object is of great interest for at least two reasons. First, automated grasp planning is still not robust enough when implemented on a physical robot, in stark contrast to human grasps which rarely fail. Second, a personal robotic assistant that uses human-like grasps may perform better in situations when the human and the robot co-manipulate an object. For example, when a robot is handing an object to a human, it would be better if the robot grasped the object proximally rather than distally so that the person can grasp it. However, there is still much to learn about the heuristics that make human grasping or even human-planned grasping (a grasp that is planned by a human, say, for a robot) so robust. This chapter presents a novel experimental method that enables a direct comparison of human-planned grasping with the performance of a state-of-the-art automated grasp planning algorithm and simultaneously identify the heuristics humans use in performing grasps. In particular, this paper presents a previously unnoticed grasping heuristic called skewness, which significantly improves robotic grasping performance as well.

Apart from direct observation of humans using their hands [12], two primary approaches have been popular in the literature to identify how humans perform grasps: (1) Performing motion capture of the human hand itself performing grasps using vicon cameras [30, 33], data gloves [16, 27, 38], force sensors [47], or video footage [4, 11]; (2) Studying the grasps that humans plan for the robot using teleoperation either through direct sight [10, 36], using real-time video [14, 26, 29], or using a virtual environment [24].

However, there are significant challenges with these approaches. First, the human hand's complex geometry makes a direct study of its posture in grasping experiments very challenging. While the exact numbers are debated, the human hand has over twenty one degrees of freedom, including joints in the fingers, thumb, palm, and wrist [22]. Also, the joints's rotational axes are typically non-orthogonal and non-intersecting and usually differ between human subjects [5, 6, 17, 43]. Finally, the high compliance of the palm and skin and feedback control loops [20, 44] in the human body make grasp contact analysis difficult. Consequently, the large parameter space and the approximations made in describing hand kinematics and contact complicates the identification of the heuristics behind finger and wrist posture in a human grasp [11, 30]. Furthermore, if we want to use the human grasping heuristics to improve the performance of robotic grasping, then the difference in kinematics between the human hand and the robotic hand poses a further challenge. For example, the popular BarrettHand[1] [42] has only four joints compared to the many joints in the human hand. There is no straightforward procedure to map the human hand posture to the robot hand and consequently human hand grasps to robotic grasps (see [19] for an example).

Similarly, there are challenges in extracting the principles of human grasping from the human-planned grasps obtained through teleoperation. The artificial (and usually) two-dimensional visual or haptic feedback that the human is provided may limit the human subject's choice of grasps. Thus, the human may provide suboptimal grasps arising from poor access. In addition, the more physically removed the human is from the task, the role of practice and training becomes more significant to achieve optimal performance. Thus, there may be strong variability in the grasps that the human subjects provide, depending on the variability in experience operating the device. As a result of these challenges, these works have primarily yielded qualitative information about human grasping, such as a grasp taxonomy [12] and the postural synergies in hand grasping movements [38], from which it is difficult to infer which aspects of human grasping lead to its high robustness.

To achieve our goal of identifying human grasping principles, we wanted a data collection process that allowed the human to express their grasping intentions naturally with minimal training. Simultaneously, we wanted an easy and straightforward method to interpret the kinematics of the human grasp.

Our approach, in contrast to existing techniques, allowed the human subjects to plan a grasp for the robot by physically guiding the robot hand (wrist and fingers) into a grasping posture for a given task (see Fig. 1). This procedure, called Physical Human Interactive Guidance, allowed the human subject to be intimately involved in the task—arguably as involved and simple as placing a pair of tongs on an object for grasping. The advantage of the human subject using the robot hand rather than his/her own hand is that the geometry of robot wrist and finger placement are straightforward to measure through the joint encoders in the robot. Another unique aspect of our method is that instead of a stand-alone analysis of human-planned grasping, since the human plans grasps for a robotic hand, we can compare the human-planned grasps with grasps generated for the same robot by an automated grasp planner [41].

We first verified that the grasps collected using the Physical Human Interactive Guidance method performed better than grasps generated for the robot by a state-of-the-art automated grasp planning software, even though both the human and the automated planner provided the same information to the robot, namely wrist orientation and finger posture (and no dynamic information such as contact force). Second, we showed that even though the subjects were not using their biological hands to perform grasps, we could still identify critical heuristics that humans use in grasping by comparing the human-planned grasps with the grasps from the

---

[1] http://www.barrett.com/robot/index.htm

Fig. 1 The physical human interactive grasping environment: the human physically guides the robot wrist and fingers into a grasping posture

automated planner, including a new human grasping principle that to our knowledge has not been noticed before and that significantly improves grasping performance on a robot when used to filter automated grasp planning results.

After describing our method for collecting human-planned grasping data in Sect. 2 and the method for testing the grasps in Sect. 3, we then present a method for analyzing the human-planned grasps in Sect. 4. Section 5 provides the results of the experiments conducted, in terms of the success rate of the human-planned grasps and the key parameters optimized by the human-planned grasps. We also show how the Physical Human Interactive Guidance method identified a new grasp measure. In Sect. 6, we discuss how this novel method for grasp acquisition relates to previous methods in the context of human–robot interaction. Portions of this work were briefly reported earlier in [1], but that paper did not focus on the novelty of the data collection method. Also, this paper provides additional insights into human-planned grasping heuristics and the effectiveness of human involvement in teaching robots.

2 Physical Human Interactive Guidance

Our approach to acquiring grasping examples from humans allowed a human subject to teach a robot different grasps by being in the robot's workspace and physically interacting with the robot. This interaction method required the person to guide the robot to specific wrist configurations and finger postures. This experimental set-up was called the Physical Human Interactive Guidance environment (see Fig. 1) and the grasps collected "human-planned grasps".

Through the Physical Human Interactive Guidance method, the human subject had an opportunity to understand the motion capabilities of the robot arm and hand, the object's inertial and geometrical properties, and how the robot and object would interact during the grasp (including the type of contact). These aspects of the grasping process are critical since grasping is a physically interactive task where the "last few centimeters" could make the difference between a successful and unsuccessful grasp. That is, however carefully the hand's path was planned to reach the grasping posture, the grasp could still fail if the finger placement was not good. Note also that the human subject could move freely around the robot to use different views of the object–hand interaction to decide on the best grasp posture (in contrast to other work that has explored how limited visual feedback can affect human grasping [8] and reaching [46] strategies). The Physical Human Interactive Guidance method was possibly the most intimate way for human subjects to build an internal model of the grasping process using a robot hand.

Such interactive robotic grasping with a human in the loop has been explored before by the GraspIt! group [9], but their goal and approach was different from the work in this paper. Their goal was to demonstrate how GraspIt! goes through search iterations to generate a grasp for a given wrist position. Also, only wrist posture was controlled by the human and finger posture was controlled by GraspIt!. The purpose of our experiment was to identify what was unique about human-planned grasping strategies, which might not be expressed properly in other methods. Also, in our method, the human had full control over the wrist and finger posture, both of which are critical to grasp quality.

2.1 Robotic Equipment

The Physical Human Interactive Guidance environment used a robotic platform consisting of a seven degree-of-freedom Barrett Whole Arm Manipulator robotic arm and a three-fingered four degree-of-freedom BarrettHand[2] [42]. The robotic system was run on Willow Garage's ROS software,[3] and the robotic hand was equipped with electric field sensors [45] which enabled the fingers to detect their proximity to objects. The electric field sensors were used primarily to close the fingers on the object simultaneously. Note that the choice of robotic equipment used in this paper is only incidental to available resources, and other robotic arms and hands could be used to recreate the experimental set-up.

2.2 Grasp Guidance and Acquisition

Grasp data acquisition was kept as simple as possible. The object to be grasped was placed by the experimenter at a known location and orientation in the robot's workspace. The robot arm was placed in a "gravity compensation" mode, where the robot arm had negligible weight and could be easily moved by a human subject. The robot arm was reset to a neutral position in the workspace and the

[2] http://www.barrett.com/robot/index.htm

[3] http://www.ros.org/

robot hand's fingers were kept open. The grasp guidance procedure included the following four steps.

Step 1: The human subject physically guided the robot arm to an initial wrist pose at which the object could be grasped (see Fig. 2a and b). The human subject was free to move around the robot and use as many views as necessary to position the robot wrist with or without an offset with respect to the object axes.

Step 2: Using electric-field sensing, the finger motors were commanded to close on the object so that each fingerpad was approximately 5 mm from object surface. At this point, the BarettHand motors were turned off to allow the human subject to physically adjust the spread angle of the fingers, depending on whether a parallel gripper-like grasp or a three-finger triangular grasp was desired. Additionally, the subject could adjust wrist pose again to better align the fingers with the object (see Fig. 2c). Importantly, the human subject was again given ample freedom, time, and space to move around the robot in order to choose what he/she believed was the best grasp for that task.

Step 3: When the subject was satisfied with this grasp pose, the robotic fingers were commanded to close on the object, completing the grasp guidance procedure. The final closure step was guided by the electric-field sensors so that all fingers contacted at the same time as to not perturb the object (see Fig. 2d).

Step 4: Subjects were then allowed to lift and shake the robotic arm to determine if they liked the grasp. Note that this light shaking performed by the subject is different from the vigorous programmed shaking that was performed during the

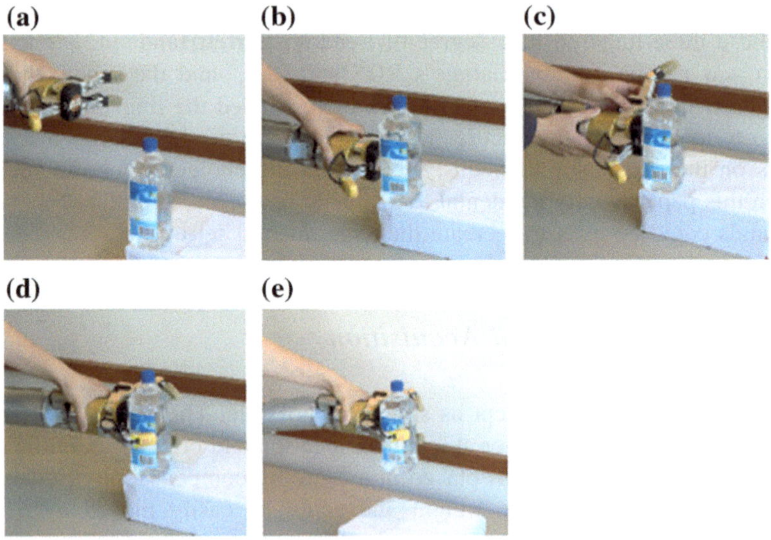

Fig. 2 The experimental procedure of a human subject guiding the robot to grasp an object: **a**, **b** approach the object, **c** adjust wrist orientation and finger spread, **d** fingers close in on the object, and **e** lift object. Note that the subject was free to move around the workspace to view the physical interaction from multiple angles

grasp testing phase (see Sect. 3). If the subject did not like the grasp or if the object slipped out, the grasp was disregarded (see Fig. 2e). We eliminated such grasps because the goal was to collect the best grasps that humans could provide.

Since the subjects had less than 5 min of practice with the system before experiment data was collected, the grasp guidance procedure provided an opportunity for the subjects to review the grasps. This allowed the subjects to understand the grasping process with the robotic hand and build an internal model based on their grasping experiences with their own hands. It turned out in the experiment described in the next section that less than five percent of all the human guidance grasps were eliminated because the subject was not satisfied with the grasp. Therefore, the grasp review process did not significantly affect the set of grasps collected.

Each grasp was represented simply as the kinematic configuration of the robot arm and hand relative to the object reference frame. Thus, a grasp was an 11 dimensional vector containing the seven degree-of-freedom robot arm joint angles and the four degree-of-freedom hand joint angles (one spread and three flexion) relative to the object's reference frame.

2.3 *Human-Subject Experiment Paradigm*

Seven human subjects participated in a study approved by the University of Washington Human Subjects Division. Each subject was given 5 min of practice with the robot, and a total of 210 grasps were collected with the robot. Nine everyday objects were used in the experiment: three small objects, three medium-sized objects, and three large objects (see Fig. 3).

Since these everyday day objects had straightforward geometry, the experimenter used the objects's edge features to carefully position the objects at the required location and orientation with respect to a known world coordinate frame marked in Table 1.[4] Since the experiment was to study the details of human-planned grasping of everyday objects in natural configurations, the objects were placed in the vertical orientation only (as shown in Fig. 3). Since the robot base's position and location was also known accurately with respect to the world coordinate frame and the robot's forward kinematics available from the manufacturer, the robot hand's pose could be computed in the object's frame of reference.[5]

---

[4] In this particular experiment (see Fig. 2), a white rectangular box on which the objects were placed was used to align the object. This was only incidental to this experimental set-up, and any means of repeated accurate positioning of the object will suffice.

[5] We also placed the objects randomly in three different locations on the table (left, right, and center with respect to the robot base) to ensure that the human-planned grasps were not unduly influenced by the specificity of the arm posture required for a particular location. Since we did not find any significant differences between the grasps from different locations in terms of the robot wrist and finger posture relative to the object, we combined all the human-planned grasps from the different locations into one set to be tested by the stationary robot.

Fig. 3 Objects used in the experiment fall into three size categories: small, medium, and large

Table 1 Functional tasks

Object	Functional task
Wine glass	Lift to pour
1-l bottle	Lift to pour
Soda can	Lift to pour
Cereal box	Lift to pour
Coil of wire	Lift to remove wire
Phone	Lift to answer
Pitcher	Lift to pour
Soap dispenser	Lift to dispense
CD pouch	Lift to open

Each subject was asked to perform three different tasks for an object, namely, lifting the object, handing the object over, and performing a function with the object. For the handing over task, the subject was asked to grasp the object such that there was space left for someone else to grasp it. The functional tasks depended on the object. For example, the functional task for the wine glass was pouring and for the phone, the task was picking up to make a phone call (see Table 1 for complete list).

For each object-task pair, the subject was asked to provide two grasps, providing a total of six grasps per object. The subjects were asked to vary the grasps if they could so as to obtain some variety in the grasps collected. Each subject was randomly assigned to five objects, while ensuring an even distribution of grasps for each of the objects (each object was selected four times except for the soda can which was selected three times). After each human-planned grasp, the object was again placed carefully by the experimenter in the required position and orientation and the experiment was repeated.

The human subjects also responded to a questionnaire to help identify the heuristics they believed they used to perform the grasping task. Specifically, we wanted to find out what geometric and force-related aspects of the grasp the subjects thought they used to perform the grasp. For example, one specific question asked if the subjects paid attention to wrist orientation and finger posture (geometric) and wrist position relative to object center of mass (force-related). More details are presented in Sect. 5.

3 Grasp Testing on Physical Robot

After collecting the human-planned grasps, we wanted to test how well each grasp performed on average on those same objects. Several past works have tried to infer grasp quality simply from simulation models [3, 15, 25, 28, 31, 32, 35, 41] with mixed results (see Sect. 5). In this paper, the human-planned grasps were validated on a physical robot rather than in simulation. From the eight human-planned grasps for each object-task pair (six for the soda can), our protocol was to choose three grasps randomly for testing on a physical robot. Thus, we expected to test a total of 27 grasps for each task (3 candidate grasps $\times$ 9 objects). However, it turned out that some human-planned grasps for the lifting and functional task which were performed when the objects were placed to the left and right of the robot could not be tested on the stationary robot when the grasps were mapped to the center location (in particular, grasps from the object's front) due to the lack of an inverse kinematics solution. Thus, all the human-planned grasps for the lifting and functional tasks which could be tested (25) were tested.

The testing procedure was intentionally kept simple. The object was placed by the experimenter in a known position and orientation (similar to the procedure outlined in Sect. 2.3). Since each grasp was represented as an 11-dimensional vector of robot arm and hand joint angles relative to the object, the robot was simply commanded to the grasp posture as follows. The robot arm was commanded to the recorded arm joint angles with the fingers fully opened. The robot hand was then commanded to the required spread angle. Finally, the fingers were commanded to close in quickly on the object, and the robot lifted the object and then executed a shaking procedure, where the object was shaken by the robot four times in a continuous circular motion (see Table 2 for peak and mean velocities and accelerations). Note that this automated shaking by the robot was different and significantly more vigorous than the light shaking that the users performed after they planned the grasp (see Step 4 in Sect. 2.2). Also, the users did not know that the grasp would be tested in this manner.

If the object stayed in the hand after the shaking, it was considered a success (scored 1). All other situations (object pushed away during acquisition or object falls down during shaking) were considered a failure (scored 0). This testing process was repeated for each grasp five times.

Note that this simple grasp testing procedure helps maintain the focus of our research on grasp generation rather than elaborate grasp testing methods that include, say, feedback control. The success rate was computed for each grasp by averaging over the five trials. Hypothesis testing was performed with a significance level of 0.05, and standard errors were reported for all mean values.

Table 2 Shaking trajectory details

	Peak	Mean
Angular velocity (rad/s)	4.62	2.74
Linear velocity (m/s)	0.62	0.39
Angular acceleration (rad/s^2)	4.39	2.22
Linear acceleration (m/s^2)	0.63	0.33

4 Grasp Analysis

4.1 Analysis Using a Grasp Measure Set

Given our goal of identifying the principles behind human-planned grasping, we needed a grasp measure space which identified the properties of a grasp. Specifically, the 11-dimensional finger and wrist posture of the grasp alone does not provide insights into grasp quality, since the way the fingers are placed relative to the object is critical for the grasp. Several grasp measures have been proposed in prior literature to infer grasp quality [2, 7, 15, 31, 34, 37, 40]. After a detailed survey, we chose a set of 11 grasp measures from the literature (see Table 3; the citations correspond to all the features in each section of the table). The "grasp volume" measure is a three-dimensional version of the "grasp area" suggested for planar grasps in [7].

Each grasp measure is associated with a heuristic. For example, the epsilon metric in row 1 of Table 3 measures the minimum disturbance force and moments that a grasp can resist in all directions, and results from simulation show that a grasp is better if it has a larger epsilon score [15]. But few grasp measures have been rigorously evaluated using experiments on a physical robot.

Table 3 Grasp measure set

Grasp measure	Description	Minimum	Maximum	Citation
Epsilon[a]	Minimum disturbance wrench that can be resisted	0	1	[15, 31]
Wrench space volume[a]	Volume of grasp wrench space	0	2^6	
Grasp energy[b]	Hand-object proximity	Negative infinity	Infinity	
Point arrangement[a]	Proximity of fingertips being in a plane parallel to palm.	0	1	[7]
Grasp volume[a]	Volume enclosed by hand	0	$\sim$669 cm^3	
Hand flexion[b]	Similarity of finger flexion	0	1	
Hand spread[b]	Proximity of the finger spread to equilateral triangle	0	1	
Finger limit[c]	Extent of finger extensions	0	1	
Volume of object enclosed[a]	Object volume enclosed by hand normalized by object volume	0	1	[39, 40]
Parallel symmetry[b]	Distance between center of mass and contact point centroid along object principal axis	0	0.5	
Perpendicular symmetry[b]	Distance between center of mass and contact point centroid perpendicular to object principal axis	0	0.5	
Skewness	See Sect. 4.1	0	$\pi/4$	

[a] Larger $\Longrightarrow$ better grasp
[b] Smaller $\Longrightarrow$ better grasp
[c] Mid-range $\Longrightarrow$ better grasp

The last row of Table 3 also proposes a new grasp measure that was suggested by the data during the human-subject experiments and that we hypothesize may be broadly useful. The new measure, called skewness, measures the robot wrist orientation relative to the object's principal axis. Suppose the object's principal axis (axis of longest dimension) is represented by unit vector u, and the axis pointing out of the palm of the BarrettHand by unit vector v (see Fig. 4). The angle d between u and v may be computed as $\delta = \arccos(u \cdot v)$. Then the skewness measure α is defined as:

$$\alpha = \begin{cases} \delta, \text{ if } \delta < \pi/4 \\ \pi/2 - \delta, \text{ if } \pi/4 < \delta < \pi/2 \\ \delta - \pi/2, \text{ if } \pi/2 < \delta < 3\pi/4 \\ \pi - \delta, \text{ if } \delta > 3\pi/4 \end{cases} \quad (1)$$

In the human-planned lifting grasp for the bottle in Fig. 5, robot's wrist orientation in the bottle-lifting task is approximately parallel to the bottle's principal axis (vertical), and the grasp's skewness measure a is near zero. In contrast, the computer generated GraspIt! grasp for the bottle would have a skewness measure a close to 30°.

Note that it was easy to notice the peculiarity of wrist placement in the human-planned grasps only because we used the Physical Human Interactive Guidance method. This is because the robot hand geometry is simple and known explicitly and the subjects were comfortable with the guidance process. Thus, the subjects were able to use their natural grasping heuristics, and we could identify the new skewness measure. In contrast, if we were studying the human hand directly, it would have been significantly more difficult to identify a grasp measure such as skewness due to the complexity of the human hand geometry.

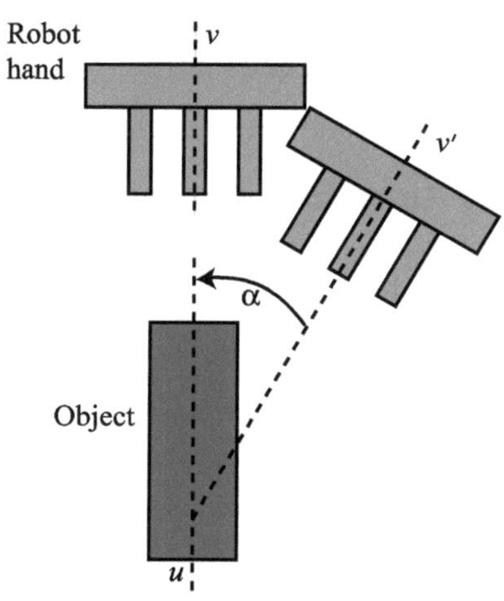

Fig. 4 Relative orientation of the object and robot hand: The object's principal axis is u. In pose v, the robot hand has skewness of zero, while in pose v, the robot hand has skewness close to 30°

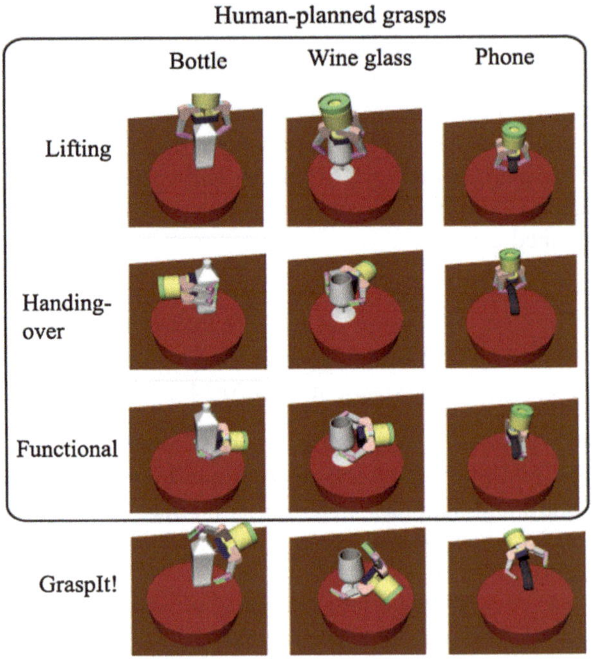

Fig. 5 Example grasp postures generated by human subjects (for a lifting tasks) and GraspIt! for three objects. Note that the human subjects manually specified the grasps on the physical Barrett robotic hand, which were then visualized using the OpenRAVE program [13]

We computed values for all the grasp measures for each human-planned grasp, and thus the grasp could now be evaluated in the chosen twelve dimensional grasp-measure space. This will help understand the parameters that humans optimize for when performing grasps. Note that this is in addition to testing on a physical robot, which provides a true measure of grasp quality.

4.2 Comparison with Automated Grasping Methods

Instead of a stand-alone analysis of human-planned grasping, we wanted to compare the human-planned grasping technique against existing grasp synthesis methods both in terms of average success rate as well as the heuristics optimized for during grasp generation. The most common and standardized procedure in the robotics community is automated grasp synthesis for robots using a set of grasp measures. We used an open-source state-of-the-art grasp planning software called GraspIt! developed by Columbia University [31] for grasp generation.[6] Note that we could have used other software such as openRAVE [13] as well for grasp

[6] http://grasping.cs.columbia.edu/

generation. However, many of these programs use the same force-closure metrics standard in robotics community [15] developed by for evaluating grasp quality. We chose GraspIt! since it had been well-tested for several years and their team helped us understand the code.

Given an object's three-dimensional model, GraspIt! finds grasps for an object by searching the high-dimensional hand-configuration space and then focuses the search on the best grasps by using a variety of grasp measures. Combined with a compliant contact (soft-contact) simulation, GraspIt! uses grasp measures that are popular in the robotics community: (1) wrench space computations (epsilon and volume [15]) that estimate a grasp's ability to provide force closure based on the minimum disturbance that the grasp can resist in all directions (first and second rows of Table 3); and (2) the shortest distance between the object and pre-defined grasp points on the hand (defined as grasp energy in the third row of Table 3).

Using the same procedure that was used to generate grasps for the Columbia Grasp Database [18], we ran GraspIt! for 30 min with the intention of generating multiple top grasps for each object according to its grasp heuristics. In 30 min, GraspIt! explored a large set (135,000) of varied wrist and finger configurations to generate six top grasps for most of the objects, but for three objects, GraspIt! generated only four or five grasps (wine glass: 4, coil of wire: 5, 1-l bottle: 4). This was partly due to search complexity as well as the lack of an inverse kinematics solution when implemented on the robot (since the robot was stationary relative to the table and object in the set-up). Thus, the automated grasp search provided a total of 49 grasps across the nine objects after exploring 1.2 million wrist and finger configurations. Since we collected a sufficient number of automated grasps from GraspIt! we did not feel the few grasps that we lost to the search complexity were significant in our results. Note again that each grasp is represented as the 11-dimensional vector containing robot arm and hand joint angles.

Note that GraspIt! cannot provide task-specific grasps, and its grasps are intended for lifting tasks only. So the performance of only the human-planned lifting grasps and GraspIt! grasps will be directly compared. The GraspIt! grasps also were validated using the same process as the human-planned grasps (see Sect. 3).

5 Results

5.1 Human-Planned Grasps Versus Automated Grasps

Figure 5 shows a sample of grasps generated by GraspIt! and through the Physical Human Interactive Guidance method for the different tasks. All the fingers were used in every grasp, whether human-planned or from GraspIt!.

Table 4 presents the success rates for each object (averaged over five trials) for the human-planned grasps and for the GraspIt! grasps (a total of $(25 + 27 + 25 + 49) \times 5 = 630$ testing trials). Across objects, the human-planned lifting strategy

Table 4 Mean success rates for human-planned grasping and GraspIt!

Object	Human-planned			GraspIt!
	Lifting	Handing-over	Functional	
Wine glass	93 (7)	33 (13)	100 (0)	100 (0)
1-l bottle	40 (13)	67 (13)	93 (7)	65 (26)
Soda can	93 (7)	87 (9)	100 (0)	90 (5)
Cereal box	93 (7)	87 (10)	100 (0)	90 (4)
Coil of wire	100 (0)	100 (0)	60 (13)	32 (11)
Phone	100 (0)	100 (0)	50 (13)	70 (6)
Pitcher	100 (0)	100 (0)	100 (0)	83 (5)
Soap dispenser	100 (0)	87 (9)	100 (0)	67 (11)
CD pouch	100 (0)	100 (0)	67 (13)	100 (0)
Overall	91 (3)*	84 (3)	86 (3)	77 (3)*
Number of grasps	25	27	25	49

*$p < 0.05$ when comparing the human-planned lifting and GraspIt! grasps

yielded a 91(3) % success rate while GraspIt! yielded 77(3) %. An outlier for the human lifting grasps was the 1-l bottle. If these grasps were removed, the success rate for human-planned lifting grasps would be 97(1) %. Interestingly, while the human-planned grasps for the handing-over task and the functional task did not perform as well as the human-planned grasps for lifting, they still outperformed on average the GraspIt! grasps which are meant for lifting only.

Table 5 shows the range of values for the grasp measures for human-planned grasps and the GraspIt! grasps. Looking first at the human-planned lifting grasps

Table 5 Grasp measure values for human-planned grasping and GraspIt!

Grasp measure	Mean (standard error)			GraspIt!
	Human-planned			
	Lifting	Handing-over	Functional	
Epsilon	0.1 (0.02)*	0.10 (0.01)	0.09 (0.01)	0.19 (0.01)*
Wrench space volume	0.15 (0.05)*	0.14 (0.03)	0.19 (0.06)	0.42 (0.04)*
Grasp energy	−1.33 (0.09)	−1.46 (0.07)	−1.2 (0.09)	3.95 (2.57)
Point arrangement	0.78 (0.02)	0.79 (0.02)	0.75 (0.02)	0.76 (0.02)
Grasp volume (cm^3)	281 (29)	271 (24)	238 (35)	259 (33)
Hand flexion	0.05 (0.01)*[a]	0.05 (0.01)[b]	0.29 (0.08)[a,b]	0.19 (0.04)*
Hand spread	0.39 (0.02)	0.38 (0.02)	0.43 (0.02)	0.37 (0.02)
Finger limit	0.70 (0.05)	0.71 (0.02)	0.73 (0.03)	0.76 (0.02)
Volume of object enclosed	0.06 (0.01)	0.06 (0.01)	0.05 (0.01)	0.05 (0.01)
Parallel symmetry	0.30 (0.05)	0.35 (0.05)	0.28 (0.04)	0.39 (0.03)
Perpendicular symmetry	0.33 (0.03)	0.30 (0.03)	0.33 (0.04)	0.28 (0.03)
Skewness	5.2 (1.3)*	6.09 (1.90)	4.79 (1.01)	23.2 (1.86)*
Number of grasps	25	27	25	49

*$p < 0.05$ when comparing the human-planned lifting and GraspIt! grasps
[a,b] $p < 0.05$ when comparing the human-planned grasps for different tasks

and the GraspIt! grasps, we notice that only four grasp measures, namely epsilon, grasp wrench-space volume, hand flexion, and skewness, were significantly different between the two grasp sets. The energy measure showed a borderline significant difference ($p = 0.05$) between human-planned lifting and GraspIt!, but that was due to outliers.

While larger epsilon and volume indicated better grasp quality theoretically, we noticed from the experiment that epsilon and volume were lower for the human-planned grasps when compared with the GraspIt! grasps even though the human guided grasps had a higher success rate than the GraspIt! grasps. The hand-flexion measure indicated that humans used lifting grasps which had significantly different finger flexion values when compared with the GraspIt! grasps. The hand-spread values for the human-planned grasps indicated that the humans used largely pinch grasps with low spread. This also led to small volumes of the object enclosed by a grasp.

The stand-out grasp measure however was skewness. The skewness measure for the human lifting grasps was significantly smaller than for the GraspIt! grasps, indicating that wrist orientation in the human-planned lifting grasps are much closer to the object's principal axis or its perpendiculars (see Fig. 5; the principal axis for the bottle and wine glass was vertical and phone horizontal). Figure 6 shows a scatter plot of the the skewness measure for human-planned lifting grasps (mean 5.2 (1.3)°) and the GraspIt! grasps (mean 23.2 (2)°), indicating that the human-planned grasps used wrist orientation that deviated very little from the objects's principal axes, whereas the automated grasps's wrist orientations were scattered all over.

Focusing on the task-dependent human-planned grasps, Fig. 5 shows some examples of variation in grasping strategy for different task requirements. Grasps used by the handing-over and functional tasks were not statistically different from the lifting-task grasps as measured by these grasp measures except for the hand flexion measure ($p < 0.05$). The hand flexion measure showed differences between the functional human-planned grasps and the lifting and handing-over human-planned grasps. This indicated that the functional task caused the human subjects to change the hand-flexion significantly.

There were near-significant differences ($0.05 < p < 0.1$) between the handing-over and functional human-planned grasps for the finger spread, parallel symmetry, and energy grasp measures and near-significant differences between the lifting and

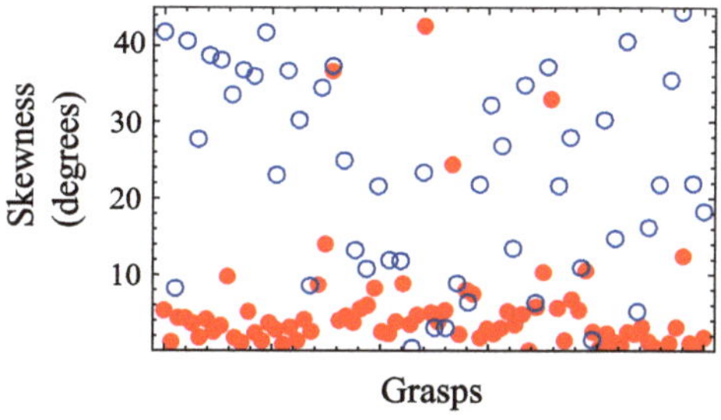

Fig. 6 A scatter plot of the skewness measure of the human-planned lifting grasps (*red dots*) and the GraspIt! grasps (*blue circles*)

functional human-planned grasps for the finger-spread grasp measure. Interestingly, the skewness measure was low for the human-planned handing-over (6.3 (1.8)°) and functional tasks (4.8 (1.0)°) also.

5.2 *GraspIt! Performance Improvement with Low Skewness*

Each grasp, whether from GraspIt! or planned by a human, was stored as a 11 dimensional vector containing the seven robot arm angles and four hand joint angles. All the grasps were divided into two groups: Group 1 was the set of grasps obtained by merging the set of human-planned lifting grasps and the set of grasps from GraspIt!. Group 2 consisted of GraspIt! grasps only. Figure 7 shows the variation in success rates for the two groups of grasps, each split by a skewness threshold of 13°. This result showed that the success rate of low-skewness grasps from GraspIt! was significantly higher than high-skewness grasps from GraspIt! (93(5) % compared with 77(3) %, p value = 0.01). In contrast, when investigating the significance of the hand-flexion measure for grasping, we did not see a significant difference in grasp success for grasps with small hand-flexion measures when compared with grasps with large hand-flexion measures. This indicated that a low hand-flexion measure was likely not a reason for a better grasp.

5.3 *Questionnaire*

Table 6 provides the results of the short questionnaire provided to the human subjects, the rows ordered in decreasing importance of the heuristic (according to the subject's belief) in generating a grasp. The results indicate that the human subjects consciously tried to use finger spread, the object's curves, and location of

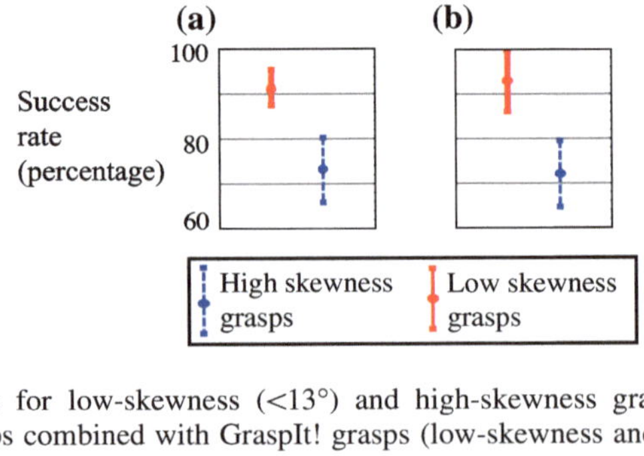

Fig. 7 Success rates for low-skewness (<13°) and high-skewness grasps from two groups: **a** human lifting grasps combined with GraspIt! grasps (low-skewness and high-skewness grasps $n = 37$ each; p-value = 0.01), and **b** GraspIt! grasps only (low-skewness grasps $n = 14$, high-skewness grasps $n = 35$; p-value = 0.01)

Table 6 Human subject questionnaire results

Heuristic	Positive response (%)
Object curves influence grasp	100
Use finger spread for stability	100
Grasp close to center of gravity	86
Object weight influence grasp	71
Keep robot wrist vertical/horizontal	71
Ridges influence grasp	71
Use palm of hand for grasp	57
Grasp strategy change with practice	43

the robot hand relative the object center of gravity to generate a good grasp. The role of object weight, and the robot wrist being vertical or horizontal, the ridges on object surface in determining a grasp were also strong, but the subjects were not unanimous in using that heuristic. The subjects did not feel that they used the robot hand's palm heavily in generating the grasp. Finally, the subjects felt that their grasp strategy did not change during the experiment with practice.

6 Discussion

The methods proposed by this paper (Physical Human Interactive Guidance, analysis using grasp measures, comparison with automated techniques) provides an exciting integration of the human physical experience and the human ability to extrapolate that experience to understand physical interaction in new scenarios along with the exhaustiveness of computer-based logic and simulation speed. Given the goals mentioned in Sect. 1, our work has provided interesting results. It was clearly shown that the human-planned grasps performed well with a high success rate (91 %). While not near-perfect like the human hand's grasping performance, the human-planned grasps were significantly better than the success rate of the state-of-the-art automated grasp planners (77 %, see Table 4).

Simultaneously, the physical interaction method showed that the human grasping method was similar in most aspects to a state-of-the-art automated grasp planner that exhaustively searched the entire configuration space for the best grasps. While the automated planner required hours across all objects to compute (suboptimal) grasps, the human subjects required only 5 min of practice to find the best grasps. This showed that humans excel at using their internal models to prune away large regions of the search space to exponentially speed up the search process. More work is also required in identifying why the existing grasping heuristics in the robotics community do not perform well when implemented on a physical robot and also develop better heuristics for automated grasping.

This experiment also showed that the human subjects used grasp measures different from those used by automated grasp planners. The strong preference of humans to exploit an object's principal axes to perform a grasp even with a robotic hand prompted us to create a new grasp measure called skewness, which to our knowledge has not been mentioned previously in the literature. Our robot experiments showed that when skewness was used to filter the grasps from automated grasp planning, low-skewness GraspIt! grasps performed significantly better than high-skewness GraspIt! grasps.

Note that while we used the Physical Human Interactive Guidance method in a specific experimental setting with a chosen robot arm and hand combination and with objects placed in their natural configurations without clutter, our approach can easily be extended to other scenarios, such as using a different robot arm and hand, using cluttered environments where a direct approach to the object is unavailable, and using objects placed in non-natural orientations. It would be interesting to see how the human-planned grasps and their performance would vary under other conditions.

6.1 The Robustness of Human-Planned Grasps

Humans have a strong sense of causal physicality, or how objects in the physical world interact. Humans use this sense everyday when they interact with the world, specifically when they use tools to perform various tasks. Indeed, the human subjects may have considered the robotic hands that they used in the experiment as tools to perform the required task. The models of physical interaction that the subjects have internalized through their daily interactions would certainly have been used in conceiving the grasps to performing the various tasks [21], which would explain the higher quality of the human-planned grasps.

Even though the humans did not have the opportunity in this experiment to dynamically control grasp forces or finger location during the disturbance, which they typically do when performing grasps with their own hands [20, 44], just the geometry of low skewness provided significant performance benefits over automated grasp planning. The low skewness of wrist orientation in the human-planned grasps might seem obvious in hindsight considering how a majority of everyday objects are designed with Cartesian coordinate frames. A grasp with low skewness provides two advantages with such “Cartesian” objects: (1) With Cartesian objects, palm contact and finger placement might be improved when the wrist orientation is parallel to or perpendicular to the object’s principal axis. Since the BarrettHand had a flat palm, grasp with low skewness would likely generate more palm contact which created a more robust grasp. (2) The contacts used in low-skewness grasps are more robust to small variations in contact location. For example, a grasp where the palm approached the soap bottle perpendicularly from the side is robust to small perturbations in position in the approach direction—all fingers will still have contact. In contrast, a grasp with high skewness that approached from the same side may lose contact at one finger due to a positioning error in the approach direction. This would cause the grasp to change from a three-finger contact to a weaker two-finger contact.

Indeed, humans might have a natural preference for grasps with low skewness, since human motor control literature has shown that many motor neurons encode human movements in extrinsic Cartesian coordinate frames rather than intrinsic (muscle or joint) coordinate frames [23]. A deeper analysis of how skewness influence grasping performance, particularly in different environment contexts (uncommon objects placed non-vertically in the presence of obstacles and clutter) will offer interesting insights into its effectiveness.

6.2 Implications for Automated Grasp Synthesis

While GraspIt! likely produced some of the best automated grasps, the mismatch between simulation models and the real world (in terms of, say, contact friction coefficients, unmodeled movement of the target object, and inaccurate soft-contact models) may have produced uncertainty in the grasping process and hurt the success rate of automated grasps. Also, it could be that the automated grasp planners did not have the optimal grasp measures to narrow down on the best grasps.

One of goals of this work is to use human skill to identify key grasp measures that can speed up automated grasp synthesis and improve real-world grasp quality. Table 5 shows that the skewness feature has significantly different values for human-planned grasps and GraspIt! grasps. Furthermore, Fig. 7 shows that low-skewness grasps have significantly higher success rate than high-skewness grasps. These results indicate that an automated search process can focus on grasps with low skewness values before exploring grasps with higher skewness values. This will likely result in better grasps faster for GraspIt! and other automated grasp synthesis methods.

This work did not further analyze the grasp measures that produced similar results between human-guided grasps and GraspIt!. This is because our data only contained highly successful grasps and thus it could not be used to identify good and bad grasp measures, unless significant differences were found between human-guided and GraspIt! grasps. Also, the lack of correlation between epsilon and grasp wrench space volume with the high human-planned grasp success rates is worth investigating further to validate the grasp measures used by the grasping research community. In particular, a more rigorous experimental testing of these grasping heuristics is necessary.

6.3 Task-Dependency of Grasps

An advantage of Physical Human Interactive Guidance is the simplicity with which the grasps that the human subjects specified for various tasks can be mapped into the robotic hand space. Previous studies have shown through human-subject experiments with datagloves such as the Cyberglove[7] that humans varied finger position carefully based on the task [16]. Indeed, it was also shown how finger posture influenced grasp force capability and stiffness. However, it is difficult to map the human-planned grasps to robot grasps.

In our work, the human subjects had an opportunity to control only finger and wrist placement (and not force and stiffness), and indeed we saw some variability between tasks in finger posture (hand flexion measure; see Table 5). Specifically, for the coil of wire the functional task was to "lift the object to remove a wire". It was noticed that the human subjects held the coil of wire by the rim, rather along its length as was the case in the lifting or handing-over task. Similarly, for the CD pouch, the functional task was to hold the object so that it may be opened. These differences in grasps have been captured by the hand flexion measure. However, we did expect to see more differences between the grasps for different tasks. We possibly need more appropriate grasp measures (than those measures listed in Table 3) and object-task pairs that are suitable for differentiating task-specific human-planned grasping strategies. Also, the large size of the robot hand relative to the object size could have influenced the human subjects to use similar grasps for the different tasks.

Interestingly, the human subjects chose grasps with low skewness independent of the task, indicating that humans valued a wrist configuration aligned with the object's principal axes significantly for grasping tasks. More work is required to understand wrist usage in grasping using the human hand.

6.4 The Human Grasping Heuristics

Table 6 presents a summary of the responses of the human subjects after performing the experiment. These responses provide insight into how the human subjects perceived their own actions and then enable us to compare the human subject's perception with a ground-truth measurement of their actions. It is clear that the subjects believed that object curves and using a spread-out finger configuration were critical aspects of the grasp. However, from Table 5, we noticed that the human subjects tended to use reasonably small hand-spread values (20°, compared to 60° for a equilateral-triangle grasp), indicating that they used the fingers closer to a parallel gripper form rather than a equilateral triangular grip.

Interestingly, even though low skewness was an important characteristic of the human-planned grasps, line 5 of Table 6 showed that only 71 % of the human subjects were conscious that the grasps that they performed had low skewness (we had expected a higher percentage). However, a more detailed study of the human-planned grasps with more subjects, more grasp measures, and machine learning techniques is necessary to derive insight into how humans plan grasps and manipulate objects in everyday life.

[7] http://www.cyberglovesystems.com/

6.5 Improving Human-Subject Experiment Protocol

When we were designing this experiment, we expected the human-planned grasps to have a success rate near 100 %; however, the human-planned grasps had a success rate of only 91 %. Why did the human-planned grasps have a success rate of only 91 %?

There might have been at least a few reasons related to the experiment protocol why the human-planned grasps had a lower success rate. First, we collected data from subjects who had never seen or interacted with a robotic arm/hand before. It is possible that with more practice with the robot, a subject would provide better grasping strategies.

Second, we asked human subjects to vary the grasping strategy every trial, if they could. In retrospect, we should not have forced the subjects to devise different grasping strategies as we do not believe that there are always multiple optimal solutions. Note however that there was one outlier in the human-planned lifting grasps success rates—the success rate for the one-liter bottle (only 40(13) %, see Table 4). If this outlier is removed, the human grasping success rate is 97(1) % even with vigorous shaking. As seen in Fig. 2, subjects chose to grasp the bottle from the top, when most humans with their own hand would not grasp a filled bottle this way. This strategy was chosen when we instructed subjects to vary the grasps when they could. This technique did not work well on the bottle's slippery surface and large mass.

Third, the subjects were not informed of the vigorous shaking used in the robustness test, and they only specified grasps for the various tasks. If the subjects had known about the shaking, they might have optimized their grasps for the shaking procedure. In contrast, the epsilon metric that GraspIt! uses actually optimizes the grasps for disturbances in all directions, similar to the disturbances in the shaking procedure. Thus, given that the human subjects did not know about the testing procedure, the human subjects were at a disadvantage compared with the GraspIt! grasps. However, the years of real-world experience still enabled the human subjects to perform better overall.

7 Conclusion

In this paper, we have shown that a novel experimental method called Physical Human Interactive Guidance can be used to obtain high-quality grasps planned by humans. The human-planned grasps were shown to be significantly better than grasps generated by state-of-the-art grasp planning algorithms (included in a program called GraspIt!). An elaborate grasp-measure set was also used to show that the human-planned grasps with the GraspIt! grasps were similar; however, a key contribution of this paper was finding a new grasp measure called skewness which explained why the GraspIt! grasps performed poorer than the human-planned grasps.

Finally, it was difficult to compare the performance of the physical interactive guidance method directly with other human-planned grasping methods, because of the lack of available data in the literature. In this paper, we performed extensive experiments with a physical robot arm and hand to evaluate the grasp performance of the human-planned grasps and also compared it with the state-of-the-art automated grasp planner. We look forward to comparing our results with results from other groups using different human-subject experiments for grasping.

Acknowledgments The authors thank Brian Mayton for help with the robot experiment set-up and Louis LeGrand for interesting discussions on grasp metrics. Gratitude is also due to Matei Ciocarlie and Peter Allen of the GraspIt! team for helping the authors use the GraspIt! code.

References

1. R. Balasubramanian, L. Xu, P. Brook, J. R. Smith, Y. Matsuoka, Human-guided grasp measures improve grasp robustness on physical robot, in *Proceedings of IEEE International Conference on Robotics and Automation*, pp. 2294–2301 (2010)
2. Y. Bekiroglu, J. Laaksonen, J. Jorgensen, V. Kyrki, D. Kragic, Assessing grasp stability based on learning and haptic data. IEEE Trans. Robot. **27**(3), 619–629 (2011)
3. G.M. Bone, Y. Du, Multi-metric comparison of optimal 2d grasp planning algorithms, in *Proceedings of IEEE International Conference on Robotics and Automation* (2001)
4. L.Y. Chang, R.L. Klatzky, N.S. Pollard, Selection criteria for preparatory object rotation in manual lifting actions. J. Mot. Behav. **42**(1), 11–27 (2010)
5. L.Y. Chang, Y. Matsuoka, A kinematic thumb model for the act hand, in *Proceedings of the 2006 IEEE International Conference on Robotics and Automation* (2006)
6. L.Y. Chang, N.S. Pollard, Constrained least-squares optimization for robust estimation of center of rotation. J. Biomech. **40**(6), 1392–1400 (2007)
7. E. Chinellato, A. Morales, R.B. Fisher, A.P. del Pobil, Visual quality measures for characterizing planar robot grasps. IEEE Trans. Syst. Man Cybernet. **35**(1), 30–41 (2005)
8. A. Churchill, B. Hopkins, L. Ronnqvist, S. Vogt, Vision of the hand and environmental context in human prehension. Exp. Brain Res. **134**, 81–89 (2000)
9. M.T. Ciocarlie, P.K. Allen, On-line interactive dexterous grasping, in *Proceedings of Eurohaptics* (2008)
10. S.T. Clanton, D.C. Wang, V.S. Chib, Y. Matsuoka, G.D. Stetten, Optical merger of direct vision with virtual images for scaled teleoperation. IEEE Trans. Vis. Comput. Graph. **12**(2), 277–285 (2006)
11. R.G. Cohen, D.A. Rosenbaum, Where grasps are made reveals how grasps are planned: generation and recall of motor plans. Exp. Brain Res. **157**, 486–495 (2004). doi:10.1007/s00221-004-1862-9
12. M.R. Cutkosky, On grasp choice, grasp models, and the design of hands for manufacturing tasks. IEEE Trans. Robot. Autom. **5**(3), 269–279 (1989)
13. R. Diankov, J. Kuffner, OpenRAVE: A planning architecture for autonomous robotics. Technical Report CMU-RI-TR-08-34, The Robotics Institute, Pittsburgh, PA, July 2008
14. C. Fernández, M.A. Vicente, C. Pérez, O. Reinoso, R. Aracil, Learning to grasp from examples in telerobotics, in *Proceeding Conference on Artificial Intelligence and Applications* (2003)
15. C. Ferrari, J. Canny, Planning optimal grasps, in *Proceeding of the IEEE International Conference on Robotics and Automation*, pp. 2290–2295 (1992)
16. J. Friedman, T. Flash, Task-dependent selection of grasp kinematics and stiffness in human object manipulation. Cortex **43**, 444–460 (2007)

17. S.S.H.U. Gamage, J. Lasenby, New least squares solutions for estimating the average centre of rotation and the axis of rotation. J. Biomech. **35**(1), 87–93, 01 (2002)
18. C. Goldfeder, M. Ciocarlie, H. Dang, P. Allen, The columbia grasp database, in *Proceedings of International Conference on Robotics and Automation*, pp. 1710–1716 (2009). doi:10.1109/ROBOT.2009.5152709
19. W.B. Griffin, R.P. Findley, M.L. Turner, M.R. Cutkosky, Calibration and mapping of a human hand for dexterous telemanipulation, in *Proceedings of ASME IMECE Conference on Haptic Interfaces for Virtual Environments and Teleoperator System Symposium* (2000)
20. R.S. Johansson, G. Westling, Roles of glabrous skin receptors and sensorimotor memory in automatic control of precision grip when lifting rougher or more slippery objects. Exp. Brain Res. **56**(3), 550–564 (1984)
21. S.H. Johnson-Frey, Whats so special about human tool use? *Neuron* (2003)
22. L.A. Jones, S.J. Lederman, *Human Hand Function* (Oxford University Press, Oxford, 2006)
23. S. Kakei, D.S. Hoffman, P.L. Strick, Muscle and movement representations in the primary motor cortex. Science **285**(5436), 2136–2139 (1999)
24. U. Kartoun, H. Stern, Y. Edan, *Advances in e-engineering and digital enterprise technology-1: Proc. Int. Conf. on e-Engineering and Digital Enterprise Technology*, chapter Virtual Reality Telerobotic System (John Wiley and Sons, New York, 2004)
25. D. Kirkpatrick, B. Mishra, C.K. Yap, Quantitative Steinitz's theorms with applications to multifingered grasping, in *ACM Symposium on Theory of Computing*, pp. 341–351 (1990)
26. C. Lee, Learning Reduced-Dimension Models of Human Actions. PhD thesis, The Robotics Institute, Carnegie Mellon University, 2000
27. C. Lee, Y. Xu, Reduced-dimension representations of human performance data for human-to-robot skill transfer, in *Proceedings of IEEE International Conference on Robotics and Automation*, pp. 84–90 (1998)
28. Z. Li, S.S. Sastry, Task-oriented optimal grasping by multifingered robot hands. IEEE J. Robot. Autom. **4**(1), 32–44 (1988)
29. J. Lloyd, J. Beis, D. Pai, D. Lowe, Model-based telerobotics with vision, in *Proceedings of IEEE International Conference on Robotics and Automation*, vol. 2, pp. 1297–1304 (1997)
30. J. Lukos, C. Ansuini, M. Santello, Choice of contact points during multidigit grasping: effect of predictability of object center of mass location. J. Neurosci. **27**(4), 3894–3903 (2007). doi:10.1523/JNEUROSCI.4693-06.2007
31. A. Miller, P.K. Allen, Graspit!: a versatile simulator for robotic grasping, in *IEEE Robotics and Automation Magazine* (2004)
32. B. Mirtich, J. Canny, Easily computable optimum grasps in 2-D and 3-D, in *Proceedings of IEEE International Conference on Robotics and Automation*, pp. 739–747 (1994)
33. N. Miyata, M. Kouchi, T. Kurihara, M. Mochimaru, Modeling of human hand link structure from optical motion capture data, in *Proceedings of IEEE/RSJ International Conference on Intelligent Robots and Systems*, pp. 2129–2135 (2004)
34. A. Morales, E. Chinellato, A. Fagg, A. del Pobil, An active learning approach for assessing robot grasp reliability, in *Proceedings of IEEE/RSJ International Conference on Intelligent Robots and Systems 2004 (IROS 2004),* vol. 1, pp. 485–490 (2004)
35. J. Ponce, B. Faveqon, On computing three-finger force-closure grasps of polygonal objects. IEEE Trans. Robot. Autom. **11**, 868–881 (1995)
36. M. Ralph, M. Moussa, An integrated system for user-adaptive robotic grasping. IEEE Trans. Robot. **26**(4), 698–709 (2010)
37. J. Romano, K. Hsiao, G. Niemeyer, S. Chitta, K. Kuchenbecker, Human-inspired robotic grasp control with tactile sensing. IEEE Trans. Robot. **27**(6), 1067–1079 (2011)
38. M. Santello, M. Flanders, J.F. Soechting, Postural hand synergies for tool use. J. Neurosci. **18**(23), 10105–10115 (1998)
39. A. Saxena, J. Driemeyer, A.Y. Ng, Robotic grasping of novel objects using vision. Int. J. Robotics Res. **27**(2), 157–173 (2008)
40. A. Saxena, L.L.S. Wong, A. Ng, Learning grasp strategies with partial shape information, in *Proceedings of AAAI Conference on Artificial Intelligence* (2008)

41. K.B. Shimoga, Robot grasp synthesis algorithms: a survey. Int. J. Robot. Res. (1996). doi:10.1177/027836499601500302
42. W.T. Townsend, The BarrettHand grasper—programmably flexible part handling and assembly. Ind. Robot Int. J. **27**(3), 181–188 (2000)
43. M. Veber, T. Bajd, Assessment of human hand kinematics, in *Proceedings of 2006 IEEE International Conference on Robotics and Automation, ICRA 2006*, pp. 2966–2971 (2006)
44. G. Westling, R. Johansson, Factors influencing the force control during precision grip. Exp. Brain Res. **53**, 277–284 (1984)
45. R. Wistort, J.R. Smith, Electric field servoing for robotic manipulation, in *Proceedings of IEEE/RSJ International Conference on Intelligent Robots and Systems* (2008)
46. D.M. Wolpert, Z. Ghahramani, M.I. Jordan, Perceptual distortion contributes to the curvature of human reaching movements. Exp. Brain Res. **98**, 153–156 (1994)
47. V.M. Zatsiorsky, M.L. Latash, Multifinger prehension: an overview. J. Motor Behav. **40**(5), 446–475 (2008)

Chapter 23
Pre-grasp Interaction for Object Acquisition in Difficult Tasks

Lillian Chang and Nancy Pollard

Abstract In natural manipulation activities of daily living, actions for object grasping must respect several constraints for successful task completion. For example, grasping actions must satisfy at a minimum the reachability of grasp contacts on the object surface, collision avoidance with obstacles, and kinematic as well as strength limits of the hand. In challenging manipulation scenarios with high constraints, direct reaching actions to grasp the object in place may not be sufficient for object acquisition. We have observed that humans use *pre-grasp interaction* to adjust the object placement during the grasping process. For example, an object may be slid or tumbled on its support surface before the final grasp contacts are achieved. In this chapter we provide an overview of the variety of pre-grasp actions that we have observed from a video survey of human manipulation activities in natural home and occupational environments. We then present our studies of object reorientation by rotation, as a particular type of human pre-grasp interaction. Finally we examine the utility of pre-grasp rotation for increasing object reachability and grasp reuse for a robot manipulator.

Manuscript received March 16, 2011. This work was supported in part by the National Science Foundation (CCF-0702443). L. Chang also received support from a NASA Harriet G. Jenkins Pre-Doctoral Fellowship. L. Chang was with the Robotics Institute at Carnegie Mellon University, Pittsburgh, PA 15213 USA, where this research was completed. L. Chang is a 2010–2012 NSF Computing Innovation Fellow hosted by Intel Corporation and received support from the National Science Foundation under Grant #1019343 to the Computing Research Association for the CI Fellows Project.

L. Chang (✉)
Robotics Institute at Carnegie Mellon University, Pittsburgh, PA 15213, USA
e-mail: lillian.chang@alumni.cmu.edu

N. Pollard
The School of Computer Science at Carnegie Mellon University, Pittsburgh, PA 15213, USA
e-mail: nsp@cs.cmu.edu

R. Balasubramanian and V. J. Santos (eds.), *The Human Hand as an Inspiration for Robot Hand Development*, Springer Tracts in Advanced Robotics 95,
DOI: 10.1007/978-3-319-03017-3_23,

Keywords Pre-grasp interaction · Manipulation · Rotation · Pushing · Daily activities · Task difficulty

1 Introduction

Humans typically use a few prototypical reaching and grasping actions to pick up objects during manipulation tasks. However, in daily life, humans must grasp objects from a variety of initial configurations, including many that may not be well-matched to canonical grasps and the arm approach directions. Observation of human grasping reveals that humans do not always grasp an object directly from its presented placement in the environment. Instead, humans often manipulate the object to adjust its configuration prior to grasping. For example, a person might slide a mug on a table closer to the body by pulling on the handle with unidirectional or non-grasping contact. When grasping a pen from a table surface, the fingers may quickly pivot the pen to orient the tip for the subsequent writing task.

These are examples of what we refer to as *pre-grasp interaction*. Pre-grasp interaction occurs whenever the manipulation first adjusts the object configuration prior to the final grasp. In several cases, the adjustment may occur while the object is partially supported by the environment, such as a table surface (Fig. 1a). This approach takes advantage of the object's movability on the supporting structure to effectively change the intermediate task parameters. Thus the anticipation of the grasping task includes object reconfiguration as a motion in addition to the arm reaching and hand pre-shaping movements.

This chapter includes a description of typical types of pre-grasp interaction from our survey of human hand activity in natural settings. In the observational survey in Sect. 3, we recorded video of people performing manipulation tasks in natural settings such as the home or place of occupation. The activities included tasks such as sorting office supplies, washing dishes, and moving furniture. We organized the survey examples into a framework that describes a pre-grasp interaction by the object reconfiguration and the task constraint that is modified. This framework provides the broad context for pre-grasp interaction strategies in which specific example of pre-grasp object rotation is situated.

In Sect. 4 and 5, we present studies of pre-grasp object rotation as a specific pre-grasp interaction strategy used by humans for manipulating objects by their handles. In our experiments on grasping familiar household objects in a kitchen setting, the action in anticipation of a grasp not only consisted of changes in the manipulator's posture, such as arm reaching movement and hand pre-shaping, but it also included re-orientation of the object to reach the handle with similar grasp postures (Fig. 1b). However, the pre-grasp object rotation occurred primarily when the task constraints were strict. That is, pre-grasp rotation was observed when the actions were restricted to one-handed object grasps, the object was heavy, and the

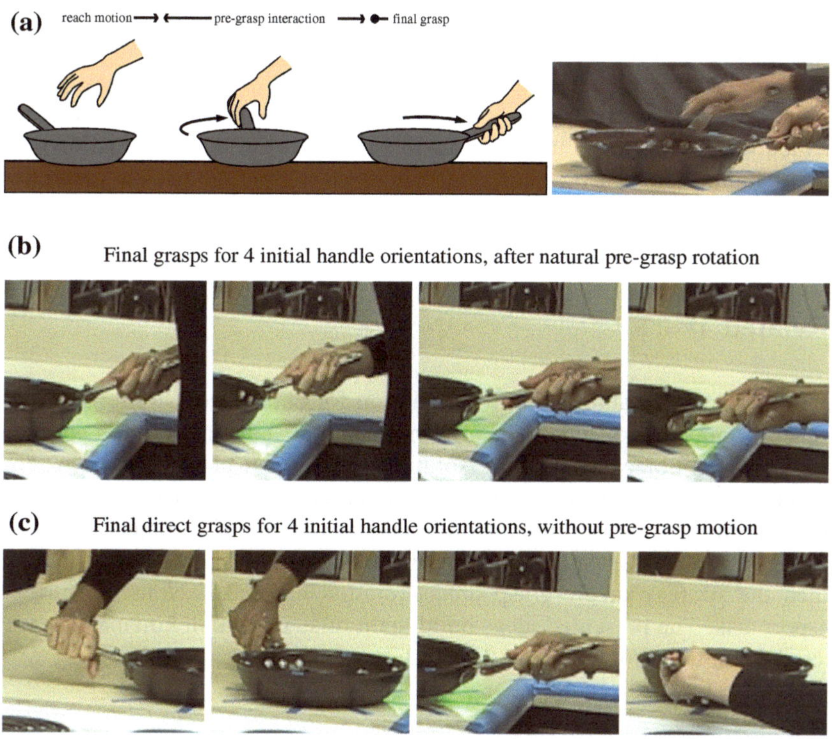

Fig. 1 **a** Pre-grasp rotation of a cooking pan adjusts its orientation before completing the final grasp of the handle. **b** Pan grasps at lift-off time using natural pre-grasp rotation strategy. The final grasps of the pan handle are similar when rotation has been used to adjust the object orientation. **c** Pan grasps at lift-off time using only direct grasping. Without object adjustment, the final direct grasps are much more varied than the grasp set that is reused after pre-grasp interaction

post-grasp transport task required precision constraints on the object orientation. In a second study of pre-grasp rotation, increasing the task difficulty factors of object mass and task precision were shown to decrease the variation in choice of object lifting orientations after pre-grasp rotation.

We observed in our human studies that it was often feasible to complete the instructed task by direct grasping without pre-grasp object adjustment, especially when the path to the hand was unobstructed by clutter in the environment. However, we also found that the hand grasp on the object changed when using the direct grasping strategy in these restricted maneuvers (Fig. 1c). The change in grasp may have accommodated the increased difficulty of reaching the object handle in the orientation outside a "comfort zone" suited to canonical reach-to-grasp actions while simultaneously avoiding the object as an obstacle. When a specific grasp of the object handle is required for the primary task, pre-grasp interaction such as object rotation may be preferred in order to achieve the desired grasp. In Sect. 5, we present a kinematic analysis and experimental validation of the utility of pre-grasp rotation for re-using robust grasps on a robot manipulator.

The studies reviewed in this chapter represent an initial investigation of pre-grasp interaction strategies, with a specific focus on pre-grasp rotation. We conclude with a discussion in Sect. 6 on directions for further examination of other types of human pre-grasp interaction that could inspire new robot manipulation strategies.

2 Related Work

Many approaches to automating robot motion for object acquisition have focused on reach-to-grasp tasks, where the arm motion and hand configuration are planned for grasping an object. With these solutions, the object placement often remains fixed in the environment until the object is carefully grasped from its presented configuration. In contrast, humans often take advantage of an object's movability to reorient and regrasp an object during the acquisition process. Specifically we studied the strategy of pre-grasp object rotation for grasp acquisition prior to a transport task. The presented ideas build upon related work in kinematic synergies (Sect. 2.1) and control strategies (Sect. 2.2), which we review here.

2.1 Kinematic Synergies in Human Manipulation

One contribution of this work is a framework for describing the types of pre-grasp interactions naturally used by humans in everyday tasks. The intent of the study is to provide insight and structure for understanding the complexity and variety of hand skills.

The dexterity of human manipulation has motivated many researchers to understand the structure and motion of the human upper limb. The structural complexity of the human upper limb is both a source of its flexibility and an obstacle for understanding the biological mechanisms for control. Often the human arm is kinematically modeled with 7 degrees of freedom (DoFs) describing the shoulder (3-DoF), elbow (1-DoF), and wrist (3-DoF) joints. The redundancy allows multiple arm configuration solutions for a single configuration of the hand segment. In addition, the hand itself exhibits more than 20 DoFs for the palm, thumb, and finger joints. An overview of basic hand anatomy and function can be found in Napier [1].

The high-dimensional kinematics allow for a wide variety of possible hand configurations. However, biomechanic and neuromuscular constraints suggest there are coordinated patterns underlying the apparent complexity. One approach to identifying such patterns has focused on categorizing hand poses as discrete grasp types. Napier [1] proposed a basic taxonomy of power grasps and precision grasps, which are distinguished by contact with the finger or palm surfaces versus the finger tips. More detailed taxonomies have been used to describe functional grasps for activities of daily living [2, 3] as well as for skilled machining tasks [4]. Although a large variety of possible hand postures exists given the different joint configurations of the fingers, a smaller subset may be sufficient to complete classes of daily or occupational tasks [5].

In another approach, studies from the motor control community provide evidence that hand motion can be described as combinations of low-dimensional synergies or components. Work by Santello et al. [6] and Mason et al. [7] found that mimed reach-to-grasp hand shapes can be represented by just a few principal components in the joint angle space. This reduction of the hand posture description has also been used to efficiently plan grasp postures for artificial robot hands with high-dimensional kinematics by Ciocarlie and Allen [8].

The survey of hand skills in Sect. 3 provides a novel framework for describing pre-grasp manipulation as a set of manipulation skills beyond direct grasping. Pre-grasp manipulation involves object reconfiguration in the environment and thus cannot be described only by the hand pose as classified in grasp taxonomies.

2.2 Motor Control Strategies

A portion of this chapter (Sects. 4 and 5) investigates the pre-grasp rotation strategy as a human behavior for manipulation tasks. The work contributes a study of pre-grasp manipulation to the field of human motor behavior and control.

Pre-grasp manipulation can be viewed as a coordinated pattern of movement for a certain class of manipulation tasks. Previous literature in the motor control community has already identified consistent patterns of human motion for the actions of reaching and grasping. Jeannerod [9] has investigated the coordination between arm reaching motion and hand pre-shaping. The experiments analyzed timing correlation of the hand trajectory toward the object with the hand shape and also investigated the effects of sensory input during the reaching task. Lacquaniti and Soechting [10] investigated the degree of coupling between the joints of the upper limb, and they found that shoulder and elbow but not wrist in particular were highly coupled during the tested reaching tasks. The research on grasp synergies by Santello et al. [6] and Mason et al. [7] demonstrated the low-dimensional variation in hand shape in response to different object geometries. Furthermore, Lukos et al. [11] showed that hand grasp shape, as measured by fingertip contact points on the object, not only responded to object geometry but was also modulated in anticipation of asymmetries in the location of the object's center of mass.

Coordination of the hand for grasping also involves the regulation of force and compliance in addition to kinematic configuration. Johansson [12] investigated how finger grip force magnitude responds to object surface friction. The experiments of Li et al. [13] and Latash et al. [14] suggest the existence of motor synergies in the force output of individual fingers, where force sharing patterns between digits in a multi-finger grasp address the motor redundancy of an overactuated system. The patterns of anticipatory contact point modulation in the hand grasp shape [11, 15] may be chosen to better reject force or torque disturbances during the grasp [16].

A consistent pattern of coordination represents a particular subset of the full set of movements possible by a redundant and over-actuated system. One approach to understanding the selection of particular strategies is to describe manipulation actions by the optimization of some cost criteria. Previous studies of possible optimization criteria of arm control have investigated minimum jerk [17] and minimum torque change [18] for arm motion trajectories. In addition, work in the biomechanics community has examined how static postures for lifting tasks may be predicted by energy or effort costs [19, 20]. For certain tasks such as one-handed lifting motions [21], humans may select between multiple movement strategies according to individual preference or capabilities. In two-handed lifting tasks, the selection between either a stooping or squatting whole-body lifting posture may depend on factors such as individual height or strength [22].

The work discussed in Sects. 4 and 5 contributes to the literature by demonstrating the consistent pattern of pre-grasp rotation in humans and examining factors which affect the strategy selection. In particular, our work extends the understanding of human behavior to task domains with more complicated grasping actions. For tasks where the object is easy to grasp, a direct reach-to-grasp action may be sufficient. However, more difficult tasks may require or encourage the use of pre-grasp interaction as a more successful manipulation strategy. We show that humans use pre-grasp object rotation for demanding tasks involving heavy objects and strict angular precision requirements when the object is presented with the handle in a non-canonical orientation relative to the person's body.

3 Pre-grasp Interaction in Natural Activities

For single objects placed in uncluttered environments, the object acquisition process may consist of only a direct reaching motion by the arm coordinated with the closing motion of the fingers that achieves a final stable grasp. However, the task conditions encountered in daily life can render a manipulation task more difficult, such as when the target object is near obstacles or requires specific grasping points due to its weight, size, or shape. Figure 1a illustrated one example of how *pre-grasp rotation* re-orients a cooking pan as part of the acquisition process such that the handle is easier to reach.

This section describes further examples of *pre-grasp interaction* observed in a video survey of human hand activity in natural settings. We discuss the survey examples by two main attributes, the object reconfiguration and the task constraint that is modified, to illustrate the richness of how pre-grasp interaction is exhibited in routine manipulation tasks. These examples provide a broad context for pre-grasp interaction strategies. The remainder of the chapter examines the specific example of pre-grasp object rotation.

3.1 Video Survey of Natural Pre-grasp Interactions

Our goal was to identify attributes for classifying the variety of pre-grasp action primitives which are integrated into complex reach-to-grasp tasks. We were specifically interested in surveying human hand activity in natural settings in contrast to instructed tasks within a laboratory environment. In this way, we could capture the richness of pre-grasp interactions beyond the direct reach-to-grasp actions studied previously in the literature.

In the video survey of human hand activity, we filmed people performing manipulation tasks in natural settings such as the home or place of occupation. All participants provided informed consent. In all observations, the participants performed manipulation skills which had been practiced previously as part of their regular occupation. There were a total of 10 sessions of both individual and group manipulation activities, such that overall 38 people were filmed. The sessions covered a total of 165 min of filmed activities for housekeeping, food preparation, office work, and mechanical repair. Specific tasks include sorting office supplies, washing dishes, and moving furniture.

The video footage was reviewed specifically for pre-grasp interaction motions. In this context, we considered pre-grasp interaction to be any manipulation that reconfigures the object for a subsequent interaction when the task does not require the reconfiguration. This definition allows for a broad concept of pre-grasp interaction as any "preparatory action" on the object prior to the final task phase. In this sense, many multi-step operations or complex manipulation processes inherently include pre-grasp interaction between the initial presented task conditions and the final primary interaction required for task completion. All actions that do not include pre-grasp interaction are referred to as "direct" manipulation strategies. Direct interactions consist of only the primary interaction that achieves the task goal.

We found that there is indeed a broad class of pre-grasp interactions where the object is not grasped directly from its presented placement in the environment. Our survey included observations over a range of manipulation scales from in-hand manipulation to whole-body grasps. Examples of the different interaction scales observed include:

- Unimanual grasp of a writing utensil,
- Bimanual grasp of a large pot,
- Whole-body grasp of large box making contact with the upper arm and torso, and
- Cooperative manipulation between multiple persons for lifting a piece of furniture.

The scale of interaction is partially related to the object size and mass, because larger and heavier objects are likely to require wider grasp apertures and contact forces for manipulation. For example, objects manipulated purely by unimanual contact are expected to be smaller or lighter than objects handled in cooperative manipulation between two people. In addition, we observed that larger scale interactions also included additional contact with the environment that effectively was an extension of the manipulator. This scenario often occurs when there is non-prehensile, or non-grasping, contact with the object, as in the case of pushing an object that is supported against gravity by a table.

In the next sections, we describe the observed examples according to two aspects: (1) the object reconfiguration and (2) the pre-grasp intent relative to the primary task goal. Without either component, the interaction is instead a direct manipulation where there is only manipulator reconfiguration prior to the primary interaction. Note that the primary interaction may include object reconfiguration (such as in lifting or pushing), which is distinct from pre-grasp reconfiguration when the task explicitly specifies the motion.

3.2 Categorization by Object Reconfiguration

First, pre-grasp object reconfiguration can be described according to the degrees of freedom which are changed by the pre-grasp interaction. For example, the object motion may be completely comprised by planar displacement. This case is common in instances of non-prehensile pre-grasp interaction where the object is primarily supported on a horizontal surface. Alternatively, for a bulky piece of furniture, pre-grasp tumbling interaction may result in general 6-DoF rigid displacement. In more complex cases, the pre-grasp interaction may cause a morphological reconfiguration of a deformable or articulated object, such as a bucket with a hinged handle.

Figure 2 presents a taxonomy for classifying pre-grasp interaction instances by the degrees of freedom which were adjusted by the reconfiguration. Typical objects for each sub-class are listed based on the observed examples from the video survey. Note that there were several examples of non-rigid pre-grasp reconfigurations for deformable objects or sets of multiple objects. This suggests that the pre-grasp interaction strategy is particularly relevant for describing more difficult object acquisition tasks that involve more than grasping a single rigid object.

The main patterns of object reconfiguration observed in our survey were:

- Planar displacement—Planar 3-DoF rigid displacement consisting of 1-DoF in-plane rotation with 2-DoF in-plane translation was observed for interactions where the object was supported on a flat surface such as a table. Pre-grasp planar displacement also occurred in the manipulation of stacked objects, such as removing the top stacked book by sliding (Fig. 3a).
- Rigid displacement—General 6-DoF rigid displacements included pivoting or tumbling out of the horizontal support plane. For example, a piece of furniture may be tumbled in order to reach a particular handhold before transporting (Fig. 3b).

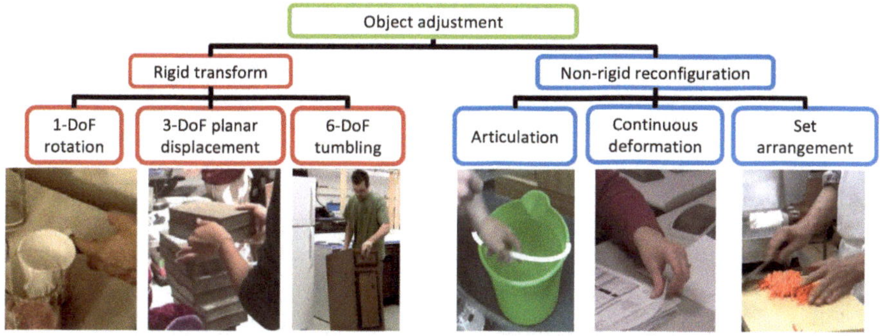

Objects in observed examples

mug	books	cabinet	bucket	paper	peelings
stapler	CD cases	pan	ladder	rug	bolts
pan	pitcher	wrench		foil	paper stack
	large box	paperclip		plastic bag	set of jars
	plates	spatula			chopped produce
	food carton	chopsticks			
		books			

Fig. 2 Taxonomy for the object reconfiguration aspect of pre-grasp interaction. Examples of rigid planar transformations included rotation of a cup by its handle and sliding books off the top of a stack. General rigid tumbling was used to achieve a whole body grasp of a bulky piece of furniture. Pre-grasp interaction was also observed for non-rigid objects. A hinged bucket handle was rotated to achieve a hook grasp, and a piece of paper was curled to achieve a pinch grasp. Multiple objects were also rearranged as a set, such as in the scooping interaction with a pile of carrot peelings

- Morphological reconfiguration—Pre-grasp interaction also changed the shape of an articulated or deformable object. The change in shape may create a better hand-hold for a desired grasp, as in the case of reorienting the handle of a bucket before lifting (Fig. 3c). Morphological reconfiguration included the manipulation of a set of objects which are separate but conceptually linked to each other, as in a pile of homogeneous objects such as game pieces or an assembly of multiple objects such as a container with a matching lid.

It is assumed that the reconfiguration did not violate any specified task constraints. In fact, the motion should have improved a presented object configuration that is suboptimal. This assumed improvement in the object configuration is related to the intent of the pre-grasp manipulation.

3.3 *Categorization by Intended Change in Task Constraints*

Second, a pre-grasp interaction can be described by the intent of the object adjustment in relation to the final grasp. The intent of a pre-grasp interaction is described as the benefit of the object reconfiguration relative to the primary

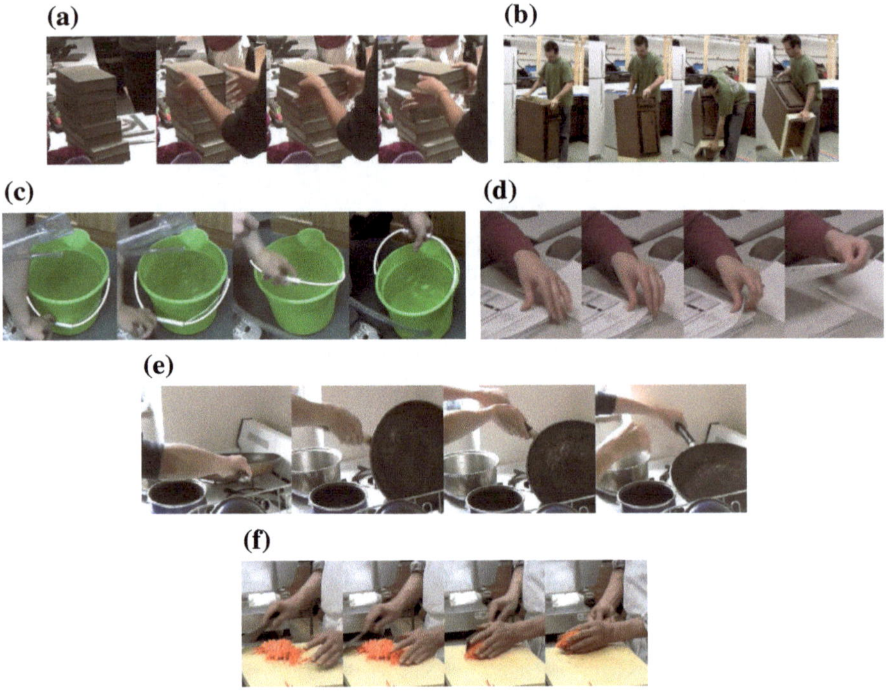

Fig. 3 (a, b, c) Examples of object reconfiguration classes for pre-grasp manipulation. **a** *Planar displacement*: Books are slid off the top of a stack to grasp the bottom surface. **b** *Rigid discplacement*: A cabinet is tumbled to reach a handhold on the bottom in order to carry it sideways. **c** *Articulated motion*: The bucket handle (*highlighted in white*) is lifted from the side to achieve a whole-hand grasp of the handle; (d, e, f) Constraints improved by pre-grasp interaction. **d** *Environment constraints*: The top sheet of paper is lifted from the stack to expose the underside for grasp contact. **e** *Manipulator constraints*: The pan is tumbled by the *right hand* so that the *left hand* can grasp the handle. **f** *Object constraints*: The pile of peelings is reshaped between two hands before lifting from the cutting board

interaction. The presented object configuration in the initial task conditions may be suboptimal with respect to the environment, manipulator, or object. These constraints are described by of how they limit the final grasp if only direct grasping without pre-grasp interaction is used.

- Environment constraints—In the initial configuration, the desired contact surface on the object could be unexposed or blocked by an obstacle in the environment. The obstacle in the environment could be the supporting structure if the desired grasp requires contact with the object's bottom surface. This case occurred when a sheet of paper resting on a surface was to be grasped on both faces (Fig. 3d). Clutter in the environment may also block the approach to a desired contact area.
- Manipulator constraints—Given the initial configuration of the manipulator relative to the object, a desired grasp of the object could be limited by a

suboptimal manipulator posture with direct reaching. For example, a direct grasp may require an arm posture associated with low kinematic manipulability. A person may prefer to reorient an object using bimanual regrasping while maintaining manipulability in both arms rather than reaching across the body with one hand if it requires a near-singular arm posture (Fig. 3e). Other factors which might affect the preference for a particular manipulator posture include torque capability constraints and grasp stability.

- Object constraints—In some situations, the presented object shape may not afford the desired grasp contacts or may have less desirable physical properties. The reshaping of an articulated object for grasping, such as the previous example of swinging the bucket handle (Fig. 3c), is similar in intent to overcoming environmental constraints. The ungrasped object subparts are obstacles to the grasped subpart. In the bucket example, the reconfiguration also changed the grasp location relative to the object's center of mass. Pre-grasp manipulation may also gather deformable materials or a set of multiple objects into a compact shape more suitable for grasping. An example of this case is the scooping of food scraps into a cupped hand (Fig. 3f).

The different constraints addressed by the pre-grasp intent are not necessarily mutually exclusive. In the example of bimanual regrasp of a pan by its handle (Fig. 3e), the intent of the pre-grasp action could be a combination of avoiding environment clutter and using an arm posture with improved manipulability.

The presented configuration of the object in the environment could be suboptimal for direct reach-to-grasp object acquisition due to preferences for a particular body posture and/or grasp. When the handle on a cooking pan is oriented away from the person, a direct grasp of the handle may be feasible but could require lifting the heavy pan from an uncomfortable body posture with limited lifting capability. In other scenarios, the intent of the pre-grasp interaction may be to improve the grasp quality rather the posture quality. This grasp improvement is especially relevant to situations where environmental clutter occludes the desired grasp contact surfaces, as in the case of a shelved book where only the spine is exposed in the initial task condition. The observed examples suggest that the intent of pre-grasp interaction was often a combination of preferences for both posture quality and grasp quality, and potentially other optimization metrics that influence the constraints described above.

3.4 Discussion

The presented pre-grasp interaction framework suggests several approaches for improving the dexterity of robotic manipulators. Taking advantage of object movability may extend the effective workspace by changing the environmental constraints when direct reach-to-grasp actions have insufficient posture and/or grasp quality. Non-prehensile pre-grasp interaction could reduce the load on the manipulator by using shared support with the work surface during the initial interaction with the object. Moreover, the expense of tuning control parameters for complex robot manipulators can be reduced if pre-grasp object reconfiguration enables the reuse of a single well-tuned grasp action for multiple initial placements. Finally, because pre-grasp strategies are part of natural human manipulation, incorporating them in the repertoire of assistive or teleoperated manipulators could facilitate more intuitive control for human operators.

4 Human Pre-grasp Rotation

The previous section presented examples of a broad set of human pre-grasp interaction strategies. In this section we describe a series of studies on human performance of pre-grasp object rotation. Pre-grasp object rotation is a specific example of planar pre-grap interaction, where the object displacement occurs in the plane of the support surface. The first set of studies examines the consistency of pre-grasp rotation patterns for human interaction with familiar household objects. The second set of studies demonstrate how the selected object orientation after pre-grasp rotation depends on the task difficulty, as measured by object mass and balance constraints.

4.1 Action Adaptation to Changes in Object Orientation

Movable objects often can be pushed or dragged along a tabletop surface using hand contacts that are less constrained than those for lifting grasps. For example, a person can use the fingers without the thumb to hook and drag a mug by its handle or to push a box to one side, and the table surface supports the weight of the object. In contrast, the hand interaction used to lift the same objects must use both thumbs and fingers in grasping, or prehensile, contact in order for the hand to fully support the load against gravity.

In general, planar sliding actions involve 3-DoF displacements of the object: two freedoms of translation in the plane of the surface, and one freedom of rotation around the axis normal to the plane. One subset of sliding actions are 1-DoF rotations around a pivot axis of the object. In this chapter, pre-grasp rotation refers to pre-grasp interactions where the sliding is dominated by planar re-orientation of the object with little translational movement. In particular, pre-grasp rotation is useful for describing interactions with household objects that have a single handle, such as mugs, pitchers, and pans.

In this section we highlight our observations of pre-grasp rotation that motivated the studies described later in Sects. 4.2 and 5. Pre-grasp interaction strategies such as pre-grasp rotation appear to be most relevant to tasks that are more constrained or more difficult. In such tasks, pre-grasp interaction appears to enable reuse of canonical grasping actions for successful task completion, as described in Sect. 4.1.1. Task constraints that affect the difficulty level can include the available freedoms of the manipulator, as well as object-centric task parameters such as weight and balance precision, described in Sect. 4.1.2.

4.1.1 Grasp Choice Changes Without Pre-grasp Rotation

This study investigated how humans would adapt their manipulation actions in response to changes in object orientation and under different strategies. The primary results from the study suggest that pre-grasp rotation is used to adjust object placement in the world for reuse of "canonical" grasping postures at the time of object lift-off from the surface. In contrast, when participants were restricted to a direct grasping strategy without pre-grasp object adjustment, there was larger variation in the body postures and hand grasp shapes used to lift the objects.

The experiments recorded how people complete an object transport task that included choices in both upper body posture and lower body position relative to the task space. Participants started and remained standing during the transport task. In the starting position, participants faced a kitchen countertop structure and an object placed on the right side of the counter. The transport goal location was a marked area on the left side of the counter, which required participants to move the object laterally from right to left. All 10 adult participants (5 male, 5 female) were right-handed and provided informed consent. The protocol encouraged as natural behavior as possible by allowing the participants to select the standing position(s) before the counter, the movement speed, the grasp points on the object, and the orientation of the object at the goal location. Body posture, hand shape, and object pose were recorded during the transport task using a marker-based camera system. Here we describe the main responses to the experiment variables of initial object orientation and grasping strategy. Additional details of the participant pool and experimental protocol can be found in Chang et al. [23] and Chang and Pollard [24].

The two objects tested in the experiments were a large plastic water jug and a cast iron frying pan. The objects were filled with water to a total mass of 3.4 kg for the jug and 1.5 kg for the pan. Within a set of 8 trials, the objects were placed at the same start location on the right of the counter, in one of 8 orientations for a uniform discretization of possible handle directions. The region for "canonical object orientations" within a participant's comfort zone was hypothesized to contain the orientations where the handle was pointing toward or to the right of the participant.

Two strategy constraints were tested on the transport task: unimanual lifting, and unimanual lifting without sliding. In the first case, participants were instructed to complete the transport task using only their right hand to contact the object. Except for this unimanual constraint, there were no restrictions on the task performance. We hypothesized that participants would use some amount of pre-grasp interaction in response to object orientations outside the comfort zone. However, the verbal instructions did not suggest pre-grasp object motion as a strategy, in order to capture the participants' natural unimanual strategy. Two sets of 8 trials were completed per participant and per object. In the second set of trials, participants were instructed to transport the object with the right hand and without any lateral sliding on the surface prior to lifting the object from the start position. Participants were able to abort the trial at any time if lifting the object from its presented orientation was perceived to be too difficult or uncomfortable.

For each trial, the time of object lift was determined by the frame where the upward object displacement first exceeded 1 cm. The participant upper body posture and hand shape at this lift-off time point are together referred to as the *grasping posture*, where the object load is fully supported by the grasp and not the table surface. The object orientation at this lift-off point is referred to as the *lift-off object orientation.* The lift-off object orientation is used to compute the amount of pre-grasp object rotation in a trial relative to the starting object orientation (Fig. 4).

The results compared the participants' grasping postures between the two strategy constraint cases. Under both constraints, there was no or little pre-grasp rotation when the object handle started in an orientation facing or to the right of the participant. The grasping postures used for these orientations in the comfort zone represent the canonical grasping postures that are used when direct grasping is sufficient and preferred.

In the unimanual case where pre-grasp interaction was implicitly allowed, the grasping postures were similar for different starting object orientations due to pre-grasp object rotation that adjusted the handle direction to be within the comfort zone. In particular, the type of hand shapes used to grasp the object handle were similar within trials for a single participant. For the jug, participants had individual preferences for grasping the upright handle with the palm facing either to the left side or toward the jug body. For the pan, participants had individual preferences for using either an oblique underhand grasp of the handle (Fig. 1b) or a straight overhand grasp.

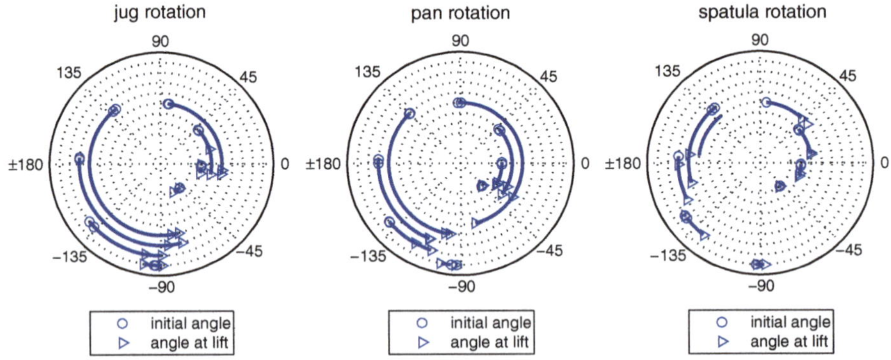

Fig. 4 Pre-grasp rotation for different household objects. The plots show the difference between the initial handle orientation and the lift-off orienation for one example participant. **a** Jug object. **b** Pan object. **c** Spatula object, tested only for pilot participants. There is less rotation for the lightweight spatula compared to the heavier jug and pan

In contrast, under the unimanual constraint without pre-grasp rotation, participant completed the task with new grasping postures for direct grasping in response to the 8 object orientations. Only one participant chose to abort a lifting trial, and otherwise the transport task was completed for all object orientation conditions. To reach the object handle when it faced to the side or away, several participants leaned their upper torso over the object and/or held their elbow out. Furthermore, the hand grasps changed for these direct grasping trials (Fig. 1c), unlike the grasp reuse observed in the cases with pre-grasp rotation (Fig. 1b). Quantitative results describing the degree of grasp changes can be found in Chang et al. [23].

The results from this study suggested that grasp reuse was preferred to new grasping plans because the new grasping plans only appeared when the participants were instructed to eliminate their natural pre-grasp interaction for object adjustment. The pan and jug objects tested in this study would be considered heavy or difficult to lift for one-handed manipulation. The next section briefly describes task situations where there was diminished pre-grasp rotation relative to direct grasping. The examples observed on these typical household objects motivated the studies described in Sect. 4.2.

4.1.2 Response to Different Objects and Task Constraints

Here we highlight some informal observations about how task difficulty affects pre-grasp rotation choices. In the study described above in Sect. 4.1.1, the difficulty of the object transport task was characterized by multiple aspects. First, the available degrees of freedom from the arm and/or body joints determine the range of reachable grasping postures. Second, the object weight limits which grasping postures are sufficient for lifting off the table surface. Third, the task constraints on the primary task after lifting, such as the object transport to the goal, further limit which initial grasping postures are preferred after pre-grasp interaction. Our observations suggest that pre-grasp interaction strategies have more utility, and thus would be selected more frequently, when a task has greater difficulty.

The available degrees of freedom for the transport task in Sect. 4.1.1 consisted of the standing position, the torso orientation, and the right arm and hand. The transport task difficulty is lower when there are more available degrees of freedom that allow for a wider range of grasps. In the experiments with the jug and pan transport tasks (Sect. 4.1.1), participants were first familiarized with the task space during a few practice trials where there was neither a unimanual constraint nor a non-sliding interaction constraint. In these practice trials, several participants lifted the objects from the start position using two-handed grasps, where the right hand grasped the handle while the left hand supported an opposing surface. With these bimanual grasps, participants appeared to use less pre-grasp sliding and instead used more direct grasping of the object from the surface.

The object properties can also affect the difficulty of a transport task, for example by the object weight or the location and shape of the handle. In the above experiments, pre-grasp rotation was observed for two objects with different handle characteristics, as the jug's handle axis was near vertical while the pan's handle was near horizontal. Further investigation is required to determine whether task difficulty could be quantified from object shape properties that affect allowable grasps. Object weight also can affect the allowable grasps for completing a manipulation task. A grasp that is sufficient for a lightweight object may not be able to support a heavier object. The weight of the jug and pan could have made the transport task more difficult such that pre-grasp interaction was necessary to obtain a desired grasp of the handle. In contrast, when a few participants completed the transport task in Sect. 4.1.1 for a lightweight spatula tool, we observed more direct grasping and less pre-grasp rotation. Figure 4 shows an example of the pre-grasp rotation results for the jug, pan, and spatula objects for a pilot study participant.

Another aspect of task difficulty is constraints on the primary manipulation task that occur after the initial object acquisition from the support surface. One constraint that is common to the transport of container objects is the requirement to maintain the object in an upright position to avoid spilling its contents. This constraint can be cast as a tolerance or precision requirement on the angular deviation from the upright orientation. A task with lower tolerance would be considered more difficult. This constraint can affect the selected grasping postures after pre-grasp rotation because the grasping posture not only must support the object at the time of lift-off but must also be able to maintain a stable grasp orientation over the duration of the following transport motion. For example, the transport of an empty jug has a high orientation tolerance, and the decreased task difficulty may allow a wider range of grasping postures that are reachable without pre-grasp object rotation.

The following sections reviews the results of an experiment which quantifies formally how the task difficulty factors of object weight and required angular precision affect the usage of pre-grasp object rotation.

4.2 Effects of Task Difficulty Constraints

This experiment studied the effect of object weight and balance precision constraints on the grasping postures following pre-grasp rotation. In particular, we examine the effect of these task difficulty factors on the selected object orientations comprising the "comfort zone" of a person's grasping postures. The previous study [23] illustrated that pre-grasp rotation enabled reuse of similar grasping postures, and these selected postures corresponded to object orientations where the handle faced toward participant's right side. These lift-off object orientations represent the preferred "comfort zone" that is the goal of the pre-grasp object interaction.

As a preview of the results, this study showed that increased task difficulty resulted in a smaller comfort zone for preferred object orientations. In this study, the range of the comfort zone is quantified by the variation in a set of lift-off object orientations. A large variation in object orientations corresponds to large set of object placements where direct grasping would be sufficient for task completion. In contrast, a small variation in selected orientation corresponds to a more constrained comfort zone that may require pre-grasp interaction to achieve.

The comfort zone for pre-grasp rotation was measured for a two-handed token retrieval task. Participants lifted a canister by their right hand to uncover a token that was then retrieved by their left hand. The token retrieval required that the grasping posture support the object weight by lifting the canister. Additionally, an upright orientation constraint was imposed with a ball balance apparatus on the canister lid. As in the previous study described in Sect. 4.1.1, the object position and orientation was tracked using a marker-based system. The time of object lift-off from the surface was extracted as the key time-point for measuring the amount of pre-grasp rotation from the initial object orientation.

During the experiment, participants performed the token retrieval task with geometrically-identical canisters with different weight and balance precision levels. The canisters were color-coded by weight and the precision level was visible by the balance apparatus size. These visual codings were reviewed before the task retrieval trials such that participants could perceive the task difficulty level before physical object interaction. There were four versions of the canister to test two levels of object weight and two levels of upright orientation precision. The two weight levels were 0.40 kg (light) and 1.20 kg (heavy). The two precision levels were controlled by using two diameters, 3.6 cm (wide) and 0.8 cm (thin), of the ball support ring for the balance apparatus. The wide ring allowed low precision in the upright orientation while still supporting the balanced ball, and the thin support required high precision in the orientation.

Twelve adults (6 male, 6 female) volunteered for the study and provided informed consent. All participants were right-handed by self-report. Each participant performed the token retrieval task for all four canisters. A single task difficulty condition is defined by the combination of the object weight and the precision level. For each of the four task difficulty conditions, multiple trials were tested where the canister was presented with its handle in each of eight initial orientations. The comfort zone was measured for a single task difficulty condition by the average absolute deviation (AAD) from the mean of the 8 lift-off orientations. Further details of the experimental procedure and data analysis can be found in Chang et al. [25].

The object weight and precision level were both significant main effects on the comfort zone variation measured by the lift-off angle AAD. For the least difficult task condition with the light object weight and wide balance support for high tolerance, the AAD from the mean lift-off orientation was 43.0°. This least difficult task condition was considered the baseline condition in the regression model. Compared to the baseline condition, the lift-off orientation variability decreased by 4.6° for increased object mass ($t(34) = -2.12$, $p = 0.0414$). The variability also decreased 12.9° for increased angular precision ($t(34) = -5.90$, $p < 0.001$). The results show that the two task difficulty factors of object weight and precision level both affect the goal states of pre-grasp rotation in the token retrieval task.

In additional experiments described in Chang et al. [25], it was found that the selection of the grasping postures in the comfort zone was correlated with the maximum lifting capability of the whole body posture. The effects of object weight and precision level may depend on this correlation between lifting capability and comfort zone. The weight of the heavy canister was closer to the maximum weight liftable by the reachable grasping postures for the initial object orientation. Pre-grasp rotation adjusted the object to a new orientation where there was greater lifting capability for robust task completion. Similarly, postures with increased strength for supporting a static load may be similar to the postures with increased control ability for maintaining the grasp orientation that holds the object upright.

Together these results suggest that pre-grasp interaction has the most utility in difficult manipulation tasks because the object adjustment enables strong grasping postures to satisfy the task constraints. In less difficult tasks involving lightweight objects or low precision motion, task completion may be possible with a smaller amount of pre-grasp interaction.

5 Robot Grasp Reuse Using Pre-grasp Rotation

The studies of human pre-grasp rotation suggested that one utility of object adjustment is the reuse of preferred grasping postures. In particular, the comfort zone of liftable object orientations was more restricted when the manipulation task was subject to constraints on allowable grasps. These conditions influencing human pre-grasp interaction are also relevant to the actions of a robot manipulator.

In this section, we present a workspace analysis comparison between a human model and a robot model. Even though the robot manipulator has large range of motion in several of its joints, there may still be a limited range of object orientations that are reachable with a specific desired grasp. In these cases where no direct grasping posture can complete the task, pre-grasp rotation can extend the effective workspace of the robot's grasping action.

In addition, a complete grasping action for object acquisition includes, besides the target grasping posture itself, the reaching motion to achieve the desired grasping configuration. Pre-grasp rotation can also be applied for reuse of such grasping actions. This is demonstrated on an example object acquisition task for a anthropomorphic robot manipulator.

5.1 Grasp Workspace Comparison

A workspace analysis provides an estimate of the reachable object positions based on the kinematic degrees of freedom and joint limits of a manipulator. The goal of the analysis in this section is the comparison between an example human model and a robot manipulator. The reachable object positions from a purely kinematic analysis represent the maximum feasible comfort zone of graspable object poses. The observed comfort zone for performed actions may be smaller due to additional constraints such as environmental obstacles, strength restrictions, or neuromuscular preferences. For a robot manipulator, we refer to the possible graspable poses as the capture region rather than the comfort zone.

The workspace results for an example grasping task show that the capture regions for the modeled human grasps is qualitatively similar to that observed in the motion studies. The capture regions for the robot are of similar size but may be shifted in orientation. In particular, even though the robot manipulator has larger kinematic limits for some joints, there are still workspace areas with small capture regions for direct grasping. It is in these regions where pre-grasp interaction strategies such as object rotation offer gains in the manipulator performance.

5.1.1 Manipulator Kinematic Models

For both the human and robot manipulators, we consider the reachable grasps for only a single, right hand. Each manipulator is modeled as a serial chain of single-axis joints using the Denavit-Hartenberg (DH) convention for describing kinematic chains (see, e.g., [26]). Specific details of the kinematic models' parameters can be found in Chang [27].

The kinematic model of the human right arm consists of 10-DoFs representing the kinematics of a chain consisting of a 3-DoF trunk, 3-DoF shoulder, 1-DoF elbow, and 3-DOF wrist joints. The model limits the joint rotation to anatomic ranges of motion based on the average maximum voluntary range of motion for male and female adults (see Chang [27]).

The robot manipulator model is based on a system consisting of a Mitsubishi PA-10 7-DoF arm with a 24-DoF Shadow Hand C3 end effector (Shadow Robot Company, London, UK). The kinematic model for this workspace analysis only considers the degrees of freedom required to the position the hand segment without modeling the finger joints. There are a total of 9-DoFs for the robot system from the 7-DoF PA-10 arm combined with the 2 wrist DoFs of the Shadow Hand Robot.

5.1.2 Example Task Scenario

The example scenario for the workspace analysis is a pan grasping task similar to the pan transport in the human studies described in Sect. 4.1.1. The task is to achieve a particular grasp of a cooking pan object on a tabletop surface, where the grasp is defined by the relative transform of the hand palm segment and the pan handle. The goal of the workspace analysis is to determine the initial pan configurations on the tabletop where such a grasp is kinematically feasible. The boundary of the workspace will indicate potential areas where a preparatory manipulation strategy such as object rotation can extend the reachability of the grasp.

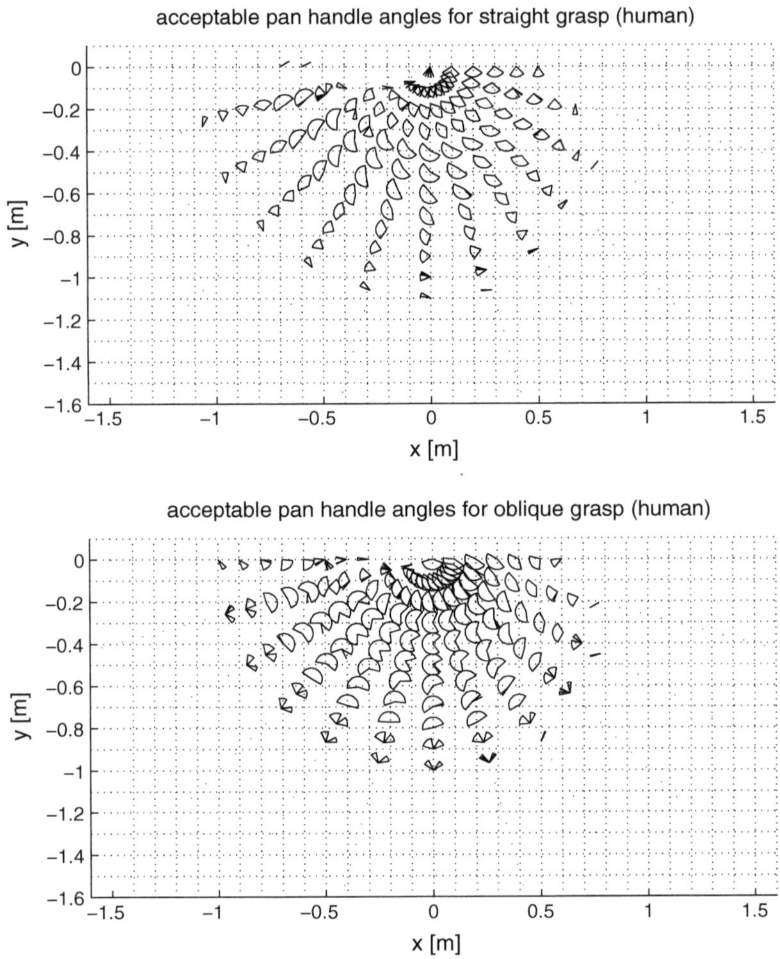

Fig. 5 Visualization of the pan orientations reachable by the human manipulator. The sectors at each selected tabletop position denote the range of the graspable handle directions

In the kinematic analysis, the manipulator's base is fixed at the origin, and the $+z$ axis points upwards such that the table surface spans the x–y plane (Figs. 5 and 6). For the example tasks, the table surface is set at 0.765 m, based on a physical table used with the actual robot manipulator in the experiments described later in Sect. 5.2.

The cooking pan is an typical object relevant to pre-grasp interaction because desired grasps are limited to its handle area in many tasks. The two specific grasps considered in this analysis are (a) an overhand, straight cylindrical grasp where the thumb wraps under the handle and (b) an underhand, oblique cylindrical grasp, where the fingers wrap underneath the handle with the thumb above. These grasps are based on the two main styles of cooking pan grasps observed of the participants

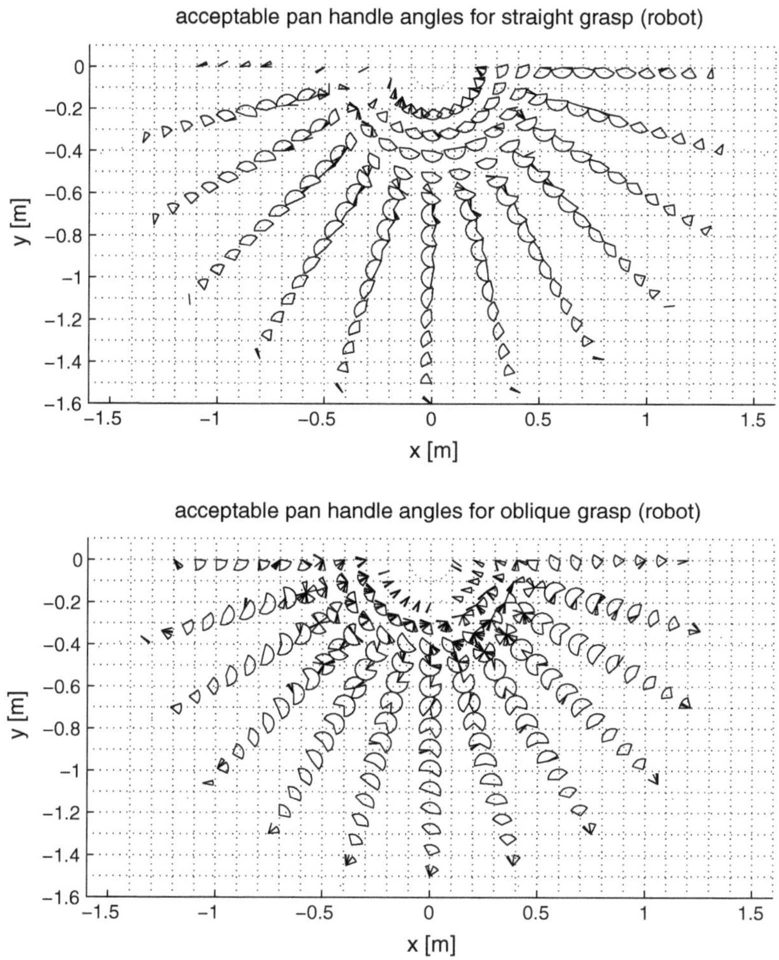

Fig. 6 Visualization of the pan orientations reachable by the robot manipulator. The sectors at each selected tabletop position denote the range of the graspable handle directions

in the previous study (Sect. 4.1.1). The kinematic analysis focused on achieving the relative transformation between the pan handle and the hand palm. No inter-joint dependencies on joint limits are considered. For example, it is assumed that the finger configuration of the grasp is reachable regardless of the wrist configuration.

5.1.3 Computation of Grasping Postures

The grasp capture region is determined by the set of object configurations for which there exists a grasping posture satisfying the grasp constraints. The desired grasp is specified by a 6-DoF description of the hand palm pose relative to the object pose in the world. The grasping posture is the manipulator configuration, which either has 10-DoFs for the human upper body model or 9-DoFs for the anthropomorphic robot model. The redundancy of the 9/10-DoF systems relative to the 6-DoF task space results in the possibility that a subspace of multiple grasping postures exist as solutions to the 6-DoF grasp constraint.

In general it is non-trivial to solve for a complete inverse kinematics (IK) solution, especially for high-dimensional redundant systems. The analysis presented here addresses the IK problem with a combination of sampling the manipulator joint-space configurations and a local search using the samples as initialization points. First, the configuration space is sampled at discretized joint angle values to pre-compute the forward kinematics (FK) of a representative set of possible end-effector configurations in the workspace. Then, given a desired hand configuration for the pan grasp, we search for an inverse kinematics (IK) solution for the arm configuration using the nearby pre-computed samples as initial guesses for a gradient-based search.

For high-dimensional systems it is prohibitive to compute and store a fine sampling of the configuration space for possible grasping postures. Our solution reduces the size of the FK pre-computation by decomposition of the 3 wrist DoFs from the remaining proximal arm joints. This provides a separation of the end-effector position and orientation. Details of the sampling and adjustment for non-intersecting wrist axes are described in Chang [27].

Given a specific desired end-effector configuration for the palm, we determine a set of IK initialization points by choosing samples which approximately match the desired position and exactly match the desired orientation of the forearm. For the position, the candidate samples of the neutral-wrist configuration must lie within a 10 cm cube centered around the desired end-effector position. For the orientation, the candidate samples of neutral-wrist orientations must be within the allowable set of neutral-wrist orientations which could reach the desired end-effector orientation with respect to the three wrist DoF joint limits. This allowable set of neutral-wrist orientations is computed by transforming the desired end-effector orientation by the inverse transformations of the pre-computed wrist orientation contributions. For an admitted IK initialization point which satisfies both the position and orientation criteria, the three wrist angles can be computed directly such that the initialization point is the full 9 or 10-DoF configuration rather than the 6 or 7-DoF neutral-wrist sample. We found that initializing with the wrist angles which achieved the desired end-effector orientation was important for successfully finding an IK solution which satisfies the wrist joint range of motion.

From this initialization set, the IK solution for the grasping posture was computed using an iterative pseudo-inverse Jacobian method, implemented by Corke [28]. The implementation of the pseudo-inverse Jacobian search does not respect joint limits, such that it was possible for the returned solution to violate the joint ranges of motion. This limitation is one reason why the success of finding a feasible IK solution that satisfies the joint angle limits is highly-dependent on the initial guess for the iterative search. Any IK solution which did not satisfy all the joint limits was discarded. All configurations in the initialization set were tested such that the overall search may generate multiple IK solutions which reach a single desired end-effector pose.

5.1.4 Reachable Object Configurations in Workspace

For both manipulator systems, the reachable pan orientations were computed for both the straight and oblique cylindrical grasps for a set of sampled pan positions.

For the human manipulator, there are few pan positions for which more than 75 % of the possible handle orientations can be reached by either grasp (Fig. 5). In particular, near the boundary of the workspace, only half or less of the pan orientations are graspable. For most pan positions, there is a larger capture region for the oblique grasp compared to the straight cylindrical grasp. The pan positions near the coordinates (−0.3 m, −0.6 m) are of particular interest since this region corresponds to the approximate x–y pan position relative to the waist for the lifting postures from the human motion capture experiments described in Sect. 4. The limited reachability of the straight grasp compared to the oblique grasp for this region may be a reason that there were fewer examples of this grasp observed from informal inspection of the human pan grasping. The capture region size of about 210° and handle direction toward the body for the oblique grasp in this area is consistent with the large variability in the lift angles observed from the human examples.

The robot manipulator has a larger workspace than the human manipulator due to the longer limb lengths (Fig. 6). Even though the robot has fewer degrees of freedom (9 vs. 10-DoFs), the large range of motion of the PA-10 arm joints results in reachable pan configurations which are qualitatively similar in capture region direction as that for the human manipulator. Similar to the results for the human manipulator, the capture region size for the straight cylindrical grasp is generally smaller than that for the oblique grasp for radii greater than 0.5 m. The area of interest for the lab demonstration setting are the pan locations about 1 m in front of the robot $((x, y) = (0\,\mathrm{m}, -1.0\,\mathrm{m}))$ due to the position of the table. In this area, the capture region points toward the manipulator's right side ($-x$) for the straight grasp and faces toward the manipulator (+y) for the oblique grasp, which is similar to the human manipulator. In terms of the potential for workspace extension by pre-grasp rotation, there are also several pan positions for which only about half of the pan orientations are graspable. This case occurs particularly for the outer regions of the workspace at greater than 0.7 m radial length from the base torso position.

For example, consider the initial object position in the robot workspace at (0, −0.9 m), which is centered in front of the base at radial length of 0.9 m. Both the oblique grasp and the straight cylindrical grasp can reach a capture region which is larger than 180°. However, at least one-third of the possible handle orientations are still unreachable by the manipulator. Objects starting in orientations outside the direct grasping capture region could be adjusted using pre-grasp rotation to complete an acquisition task over a greater range of initial task conditions. This example is considered in an empirical demonstration described in the following section.

5.2 Robot Grasp Reuse

The workspace analysis in the previous section (Sect. 5.1) illustrated regions where pre-grasp rotation may expand the grasp capture region beyond the direct-grasping boundaries. The analysis computed individual grasping postures that satisfied a desired grasp of the object handle without considering the motion required to reach a grasping posture.

In this section we examine the concept of pre-grasp object rotation as a strategy for reusing a single well-tuned grasp routine programmed for reaching a particular grasping posture on a robotic manipulator system. Here the grasp routine that is reused consists of the complete arm reaching action and the grasping motion of the fingers after the palm is placed in the desired grasp frame relative to the object. A grasp routine may only successfully lift the object from a small set of initial poses if the finger closing action is small relative to the possible object orientations. Pre-grasp object rotation can extend the effective workspace of such a grasp prototype. Reusing grasping routines for larger grasp regions can reduce the number of motor action primitives necessary for robot manipulation. This reuse can save programming time for manual actions or search time for automated sequencing of action primitives.

The manipulation actions were implemented on the system, previously described in Sect. 5.1, consisting of the 7-DoF PA-10 arm and 24-DOF Shadow Hand C3 end effector, shown in Fig. 7. The grasping routine completed a pan transport task similar to that in the human studies (Sect. 4.1.1), where the object is grasped from a position in front of the robot base and then laterally transported to the left side. The object starting position on the table was located 0.9 m in front of the manipulator base. The object goal position was located on the table 0.35 m to the left of the start position. Twenty-four handle directions were selected to sample initial object orientations at 15° intervals. The object was an empty cooking pan with a handle that had a mass of 0.46 kg. The pan pose was tracked by attached markers and a Vicon camera system.

5.2.1 Open-Loop Routines for Grasp Reuse

The grasping strategy using pre-grasp rotation was implemented as two manually-programmed open-loop routines. One action is the pre-grasp rotation routine for reconfiguring the handle orientation prior to grasping. The other action is the grasping routine for lifting and transporting the pan by its handle. The two actions are executed sequentially for a complete manipulation action which transports the pan from any initial handle orientation.

The grasping routine is a sequence of three manually-programmed motion components. The robot arm with an open hand grasp shape is moved from an initial configuration on the right toward the pan position in the approach or reaching motion. In the grasp motion, the PA-10 arm configuration remains fixed while the hand's finger joints close around the handle. Finally, the hand maintains

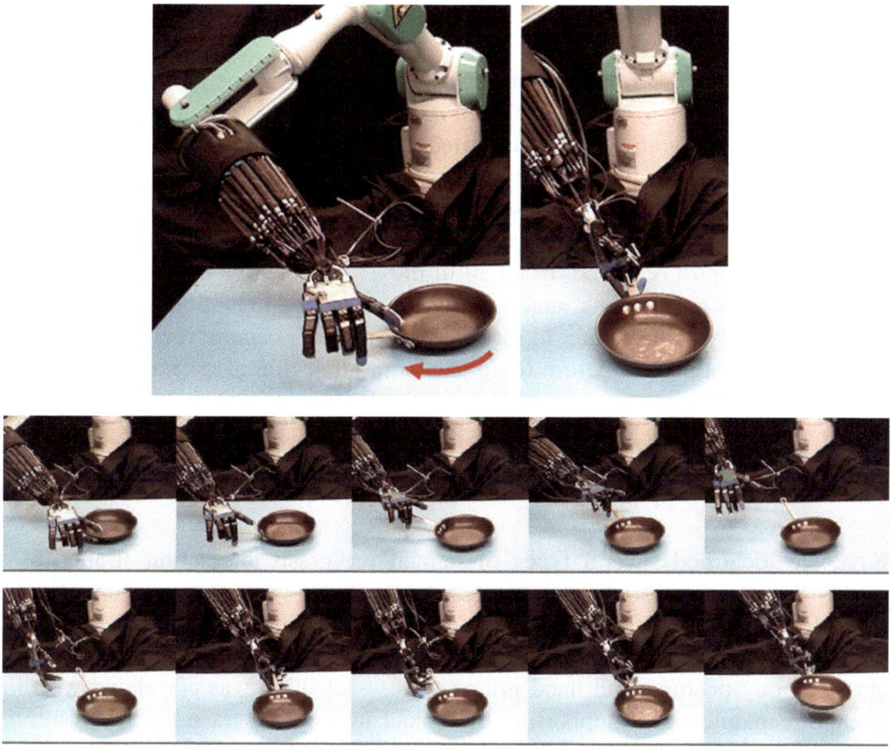

Fig. 7 Pre-grasp interaction improves the capture region of a pan-grasping routine for an anthopomophic robot manipulator. The pre-grasp rotation adjusts the object orientation in the environment prior to grasping, without requiring a new reach-to-grasp motion plan for each new orientation outside the reference grasping capture region

a closed grasp shape around the handle during the transport motion where the PA-10 arm lifts and moves left to the goal position.

The pre-grasp rotation routine was implemented as a pushing motion using single-finger contact with the object to turn the cooking pan around its natural pivot point. The index finger was flexed 90°, normal to the palm, while the thumb and other three fingers remained extended in the plane of the palm and parallel to the table surface. With this fixed hand shape, the arm moved in an arc such that the index fingertip traced a circular arc of 315° in a clockwise direction around the object perimeter and ended within the intended grasp capture region. Please see Chang et al. [23] for illustrations of the pre-grasp rotation and grasping routines.

5.2.2 Empirical Evaluation of Workspace Extension

We evaluated the effective capture region of the grasping routine alone in comparison to the sequence of the pre-grasp rotation followed by grasping action.

The two methods were each tested on the different initial handle orientations in a set of 24 consecutive trials.

The empirical capture region of the grasping routine alone covered 4 of the 24 initial handle angles, for a region size of 45°. Note that this capture region is much smaller than the capture region reported previously in the workspace analysis Sect. 5.1 because the grasping routine is based on a single reachable grasping posture rather than the complete set of grasping postures for the object location.

In contrast, when the pre-grasp rotation routine preceded the grasping action, successful transport occurred for all 24 of the 24 consecutive trials. The pre-grasp rotation routine reduced the uncertainty in the object orientation by consistently rotating the pan into the center of the grasping routine's capture region. The handle angles after rotation and before the grasp were all within a 15° range.

6 Discussion

First, we identified several types of human pre-grasp interaction strategies. A further examination of the specific example of pre-grasp rotation investigated possible factors that drive object adjustment over a direct reaching action in difficult grasping tasks. These factors include a preference for a reuse of a particular class of body postures and hand grasps, whose selection was correlated with the strength capabilities for supporting the object in the post-grasp task. Second, we performed a workspace analysis of both the human upper body and a robot manipulator. This demonstrated the potential advantages for a robot manipulator to use pre-grasp interaction strategies to increase the effective reachability of objects for specific desired grasps.

This work has investigated pre-grasp interaction as a manipulation strategy for successfully grasping objects during a difficult task. In our survey of human manipulation actions in natural settings (Sect. 3), we found several examples where humans do not or cannot grasp an object at the desired surface contacts without first re-configuring the object in the workspace. Regrasping is one method of reconfiguration, but pre-grasp interaction also occurs in the form of non-prehensile manipulation such as pushing or tumbling. We suggest that pre-grasp interaction may be especially useful for difficult tasks involving heavy objects and/or cluttered environments where the object is not conveniently presented for reaching the desired grasp.

Overall, we promote a broad view of the richness of manipulation skills for object interaction. We have shown that the seemingly simple skill of object acquisition can involve a sophisticated process of pre-grasp manipulation to achieve the desired grasp for the subsequent task. Here we review the highlights and limitations of our findings. We also suggest avenues for future research on human manipulation and commonalities with motion planning for robotic manipulation.

6.1 Human Manipulation Strategies

Pre-grasp object interaction is a broad description of one approach humans use to complete manipulation tasks robustly. Our experiments on human pre-grasp rotation represent an initial investigation of how object manipulation actions such as pushing or pivoting interact with the grasp acquisition.

Within the pre-grasp rotation category, we investigated the effect of particular task parameters such as initial object orientation, object weight, and lifting precision on the choice of object rotation. We also found that aspects of the manipulator, such as an individual's lifting capability or a robot's payload capacity, may drive the selection of a preferred grasping posture enabled by pre-grasp rotation. Other factors that plausibly influence the choice of preferred object orientation include the direction of transport to the goal location, the surface friction between the object and support surface, and obstacles in the environment. Learning how these factors change the performance of pre-grasp rotation may further contribute to the understanding of why and when it is chosen as a motor action.

At a higher level, there remain several interesting questions about how humans choose among multiple strategies rather than the parameters for a single strategy. The taxonomy of pre-grasp interaction strategies presented in Sect. 3 categorizes interaction examples based on the degrees of freedom in the object reconfiguration. The categorization provides a guide for classes of pre-grasp strategies, but it would benefit from a model of how the task specifications determine the choice between two possible strategies. For example, for articulated pliers, how does the primary manipulation task (for example, either transport, hand-off to another person, or bending a wire) determine whether the tool is pivoted, tumbled, or folded during the pre-grasp interaction?

We have shown for pre-grasp rotation that the choice of whether and how much to rotate the object is in part a response to the task difficulty factors of object weight and upright precision for our canister lifting task experiment. Along these lines, it would be useful to develop a formal, generalizable representation of the primary manipulation task difficulty. Descriptors of task difficulty in such a representation may provide a mapping for predicting the probable resulting pre-grasp interaction. Example factors that we speculate contribute to difficulty and thus more sophisticated strategies include the number and occlusion of possible object surface contacts for grasps, the object inertial properties, and the friction or resistance to movement of the object or environment in the possible reconfiguration freedoms. Manipulator constraints of the human body—e.g., strength, bimanual availability, or position mobility—are additional considerations.

The uncertainty in and/or the human perception of these difficulty factors may be more critical than the actual values, such that the cognitive load in planning a difficult maneuver is a driving factor. For example, the strategy choice will be limited by the perception of allowable object affordances, such as whether an object can be tumbled or whether a set of objects is stackable.

Finally, we have implicitly referred to pre-grasp interaction as an intentional, planned strategy that may optimize some aspect of the primary task. It is possible that some pre-grasp interactions arise from exploratory interactions with an object, such as in the haptic exploration of an unfamiliar surface for a grasp point, or from recovery responses to unintentional disturbances, such as when an object slips during an imprecise grasp action.

6.2 Generalization of Manipulation Planning Optimization Functions

While one approach to achieving robust and dexterous manipulation is to build a repertoire of task-specific behaviors, another avenue toward general manipulation capabilities is the identification of an explanatory optimization function for planning any action. In addition to providing insight into human cognition and motor skills, a unifying model for motor actions would avoid the need for manual observation and extraction of promising strategies for mimicry. Other behaviors of gathering, separation, and bracing are similar to pre-grasp interaction in that they involve some preparatory action to "set-up" a task.

Below is a non-exhaustive list of possible criteria from whose optimization could emerge the natural motor strategies of pre-grasp interaction and other behaviors:

- Time or speed:
  - Time for task completion
  - Planning or decision time
  - Cognitive retrieval time or load for new versus repeated/similar actions
- Torque, load, or effort:
  - Total joint torque magnitudes
  - Torque of weakest joint
  - Margin from multiple joint torque limits
  - Margin from weakest joint torque limit (payload margin, lifting capability, or available strength)
  - Energy or perceived exertion
- Posture or form:
  - Margin from joint limits (available travel)
  - Manipulability or margin from singularities
  - Balance or stance stability
  - Grasp stability
- Confidence, robustness, or flexibility:
  - Reachability of target from multiple postures
  - Reliability of sensory information: visibility, tactile redundancy
  - Reliability of action execution or remaining within actuation capabilities

  - Availability of alternative exit strategies: setting an object down, or changing a grip mid-task
  - Collision avoidance or distance to obstacles
- Comfort in action execution, or avoidance of pain
- Interaction predictability or social acceptance: avoidance of awkward or unlikely postures.

In this work we have examined a small set of metrics in the context of pre-grasp interaction. In the human motion studies, we found that time for task completion was not an explanatory metric for the pre-grasp rotation, since task completion took longer for pre-grasp rotation compared to direct grasping. However, time for planning initial movement was shorter and suggests that decision making or cognitive load could be a driving factor. We also tested metrics based on physical load at the time of object lift-off. These included the magnitude of joint torques over multiple joints.

Furthermore we surveyed subject's self report of comfort but did not find a clear preference for the pre-grasp rotation method. An objective measure of discomfort, such as local muscle fatigue [29], may be a more reliable metric if it could be practically recorded.

Future research is required to determine if there is indeed an underlying optimization criteria driving general human manipulation actions. A holistic criteria such as robustness or success rate that includes multiple factors may be necessary for modeling complex human motion over a wide range of behaviors. In addition, human motion may not necessarily be optimal but instead merely sufficient to complete the task objective using any one of multiple similar strategies. It also remains to be seen whether identification and implementation of such an optimization criteria on a robotic system would result in the emergence of similar manipulation behaviors observed in humans, despite the differences in kinematic, sensory, actuation, and computation capabilities.

Acknowledgments The authors would like to thank Roberta Klatzky, Howard Seltmann, Garth Zeglin, and Justin Macey for their contributions to the studies presented in this chapter.

References

1. J.R. Napier, *Hands*, ed. by R. H. Tuttle (Princeton University Press, Princeton, 1993)
2. S.J. Edwards, D.J. Buckland, J.D. McCoy-Powlen, *Developmental & Functional Hand Grasps* (Slack Incorporated, Thorofare, 2002)
3. N. Kamakura, M. Matsuo, H. Ishii, F. Mitsuboshi, Y. Miura, Patterns of static prehension in normal hands. Am. J. Occup. Ther. **34**(7), 437–445 (1980)
4. M.R. Cutkosky, On grasp choice, grasp models, and the design of hands for manufacturing tasks. IEEE J. Robot. Autom. **5**(3), 269–279 (1989)
5. J.Z. Zheng, S.D.L. Rosa, A.M. Dollar, An investigation of grasp type and frequency in daily household and machine shop tasks, in *Proceedings of the IEEE International Conference on Robotics and Automation (ICRA)*, 2011

6. M. Santello, M. Flanders, J. Soechting, Postural hand synergies for tool use. J. Neurosci. **18,** 10105–10115 (1998)
7. C. Mason, J. Gomez, T. Ebner, Hand synergies during reach-to-grasp. J. Neurophysiol. **86**, 2896–2910 (2001)
8. M. Ciocarlie, P.K. Allen, Hand posture subspaces for dexterous robotic grasping. Int. J. Robot. Res. **28**, 851–867 (2009)
9. M. Jeannerod, in *Attention and Performance IX*, ed. by J. Long, A. Baddeley. Intersegmental coordination during reaching at natural visual objects (Lawrence Erlbaum Associates, Hillsdale, 1981), pp. 153–169
10. F. Lacquaniti, J.F. Soechting, Coordination of arm and wrist motion during a reaching task. J. Neurosci. **2**(4), 399–408 (1982)
11. J. Lukos, C. Ansuini, M. Santello, Choice of contact points during multidigit grasping: effect of predictability of object center of mass location. J. Neurosci. **27**(14), 3894–3903 (2007)
12. R. Johansson, Sensory control of dexterous manipulation in humans, in *Hand and Brain: The Neurophysiology and Psychology of Hand Movements*, ed. by A. Wing, P. Haggard, J.J.R. Flangan (Academic Press, New York, 1996), pp. 381–414
13. Z.M. Li, M.L. Latash, V.M. Zatsiorsky, Force sharing among fingers as a model of the redundancy problem. Exp. Brain Res. **119**(3), 276–286 (1998)
14. M.L. Latash, I.M. Gelfand, Z.M. Li, V.M. Zatsiorsky, Changes in the force-sharing pattern induced by modifications of visual feedback during force production by a set of fingers. Exp. Brain Res. **123**(3), 255–262 (1998)
15. J.R. Lukos, C. Ansuini, M. Santello, Anticipatory control of grasping: independence of sensorimotor memories for kinematics and kinetics. J. Neurosci. **28**(48), 12765–12774 (2008)
16. M. Ciocarlie, P.K. Allen, Hand posture subspaces for dexterous robotic grasping. Int. J. Robot. Res. **28**, 851–867 (2009)
17. T. Flash, N. Hogan, The coordination of arm movements: an experimentally confirmed mathematical model. J. Neurosci. **5**(7), 1688–1703 (1985)
18. Y. Uno, M. Kawato, R. Suzuki, Formation and control of optimal trajectory in human multi joint arm movement minimum torque-change model. Biol. Cybern. **61**(2), 89–101 (1989)
19. M.J. Dysart, J.C. Woldstad, Posture prediction for static sagittal-plane lifting. J. Biomech. **29**(10), 1393–1397 (1996)
20. C.C. Chang, D.R. Brown, D.S. Bloswick, S.M. Hsiang, Biomechanical simulation of manual lifting using spacetime optimization. J. Biomech. **34**(4), 527–532 (2001)
21. W. Park, B.J. Martin, S. Choe, D.B. Chaffin, M.P. Reed, Representing and identifying alternative movement techniques for goal-directed manual tasks. J. Biomech. **38**(3), 519–527 (2005)
22. R. Burgess-Limerick, B. Abernethy, Qualitatively different modes of lifting. Int. J. Ind. Ergon. **19**, 413–417 (1997)
23. L.Y. Chang, G.J. Zeglin, N.S. Pollard, Preparatory object rotation as a human-inspired grasping strategy, in *IEEE-RAS International Conference on Humanoid Robots (Humanoids)*, Daejeon, Korea, Dec 2008, pp. 527–534
24. L.Y. Chang, N.S. Pollard, On preparatory object rotation to adjust handle orientation for grasping, Robotics Institute, Carnegie Mellon University, Pittsburgh, PA, Tech. Rep. CMU-RI-TR-08-10, April 2008
25. L.Y. Chang, R.L. Klatzky, N.S. Pollard, Selection criteria for preparatory object rotation in manual lifting actions. J. Mot. Behav. **42**(1), 11–27 (2010)
26. M.W. Spong, S. Hutchinson, M. Vidyasagar, *Robot Modeling and Control* (Wiley, Hoboken, 2006) (Chap. 3.2)
27. L.Y. Chang, Pre-grasp interaction as a manipulation strategy for movable objects, Ph.D. dissertation, Robotics Institute, Carnegie Mellon University, Pittsburgh, PA, December 2010
28. P. Corke, A robotics toolbox for MATLAB. IEEE Robot. Autom. Mag. **3**(1), 24–32 (1996)
29. D.B. Chaffin, Localized muscle fatigue–definiton and measurement. J. Occup. Med. **15**(4), 346–354 (1973)

Chapter 24
Grasp Planning Using Low Dimensional Subspaces

Peter K. Allen, Matei Ciocarlie and Corey Goldfeder

Abstract Recent advances in neuroscience research have shown that posture variation of the human hand during grasping is dominated by movement in a configuration space of highly reduced dimensionality. In this chapter we explore how robot and artificial hands may take advantage of similar subspaces to reduce the complexity of dexterous grasping. We first describe our method for grasp synthesis using a low-dimensional posture subspace, and apply it to a set of hand models with different kinematics and numbers of degrees of freedom. We then discuss two applications of the method: online interactive grasp planning and data-driven grasp planning using a pre-computed database of stable grasps.

Keywords Grasping · Low-dimensional subspaces · Robotic hands

---

This work has been supported by NSF grant IIS-0904514, NIH BRP grant 1RO1 NS 050256-01A2 and a Google Research Grant.

P. K. Allen (✉)
Department of Computer Science, Columbia University, New York, NY, USA
e-mail: allen@cs.columbia.edu

M. Ciocarlie
Willow Garage, Inc., Menlo Park, CA, USA
e-mail: cmatei@cs.columbia.edu

C. Goldfeder
Google Inc., New York, NY, USA
e-mail: goldfeder@gmail.com

R. Balasubramanian and V. J. Santos (eds.), *The Human Hand as an Inspiration for Robot Hand Development*, Springer Tracts in Advanced Robotics 95,
DOI: 10.1007/978-3-319-03017-3_24,

1 Introduction

In this chapter we explore the concept of low-dimensional posture subspaces for artificial hands. Recent advances in neuroscience research have shown that control of the human hand during grasping is dominated by movement in a configuration space of highly reduced dimensionality. This has led our group to explore how artificial hands may take advantage of similar subspaces to perform complex grasping tasks.

Subspaces are important not only because they are biologically motivated but because they allow us to create computational frameworks that are tractable for difficult problems, characterized by a large number of degrees of freedom, such as dexterous grasping. The work described in this chapter allows us to compute metrics on thousands of grasps quite efficiently, leading to a sampling-based approach that can adequately cover the space of grasps for a number of robotic hands as well as a human hand model.

This approach is based on a hand posture subspace defined by a small number of basis vectors which we call *eigengrasps*. The implied dimensionality reduction has allowed us to perform *online dexterous grasp planning* both for robots needing to find a correct grasp for an object and for prosthetic devices in which the human provides a subset of the necessary Degrees of Freedom (DOFs), allowing the planner to work in real-time to find a stable grasp. Further, the ability to pre-compute thousands and thousands of stable grasps for dexterous hands over a large class of objects has motivated a new direction in grasp synthesis, which we call *data driven grasping*. If the number of objects to be grasped in the database is very large and comprehensive then robotic grasping becomes a pre-computed indexed database lookup which is extensible to grasping novel objects not in the database.

2 Low-Dimensional Kinematics

As the result of complex, longterm evolutionary adaptations at the structural and neural level [1], biological versatility is challenging to replicate in robotics. By using experimental approaches to study human hand control and generation of hand postures we can however gain new insights into how humans so effortlessly grasp objects.

A common thread in the study of human grasping and its applications to robotics builds on the observation that humans simplify the huge space of possible grasps through learning and experience, enabling them to quickly choose good grasps for a wide variety of objects. The attempt to simplify the configuration space of dexterous artificial hands has resulted in the concept of "grasp taxonomy", widely discussed in the robotics literature (for reviews, see [2, 3]). When focusing on continuous subspaces rather than discrete taxonomies, previous studies have also shown that control of grasping in humans occurs in a smaller space of possible hand configurations [4–7]. Simultaneous motion of the fingers is characterized by coordination patterns that reduce the number of independent degrees of freedom to be controlled: despite the large number of DOFs of the hand, *humans utilize linear combinations of relatively few eigenvectors of the DOF space to match hand shape to object geometry*. This behavior results from biomechanically constrained interactions among joints of the digits as well as complex coordination patterns of hand muscle activity [1].

Numerical analysis of human hand postures can thus not only reveal the intrinsic low-dimensional nature of the data, but also suggest new directions for robotic hands. Based on inter-digit coordination caused by mechanical constraints in the anatomy of the hand, robotic hands can be built with highly interconnected finger actuation mechanisms [8, 9]. The assumption that motor control synergies also take place at a higher level in the Central Nervous System [10, 11] implies the use of low-dimensional control algorithms for dexterous robotic hands, such as the ones presented in this chapter.

Another important aspect of grasping concerns the relationship between low dimensional subspaces and the task being performed. Previous work [12] has shown that the execution of different manipulation tasks (such as flipping pages or crumpling paper) is characterized by different linear combinations. Interestingly, a posture subspace can be found even in the less constrained setting of haptic exploration [13]. Hand posture during the reach phase of a complete reach-to-grasp action is also described by a different (and lower-dimensional) principal component spectrum than the grasp phase [5, 11]. These results show that, when using a low-dimensional control space for robotic hands, the choice of the subspace has to be correlated with the proposed task. Finally, all the studies discussed so far have used Principal Component Analysis, and thus have addressed only linear subspaces that can be extracted from hand posture data. Linear decomposition has been successfully used in the past on different types of biometric data, ranging from face appearance [14] to dynamics of arm motion [15]. However, nonlinear dimensionality reduction methods can potentially reveal different manifold structures of the same data [16].

2.1 Eigengrasps

When performing human user studies, the usefulness of a hand posture subspace can be quantified by how well it approximates a given set of input data. In this chapter, we present a constructive approach, oriented towards application for artificial hands: given a hand posture subspace, we will use it to synthesize new hand postures for accomplishing a particular grasping task.

Any hand posture is fully specified by its joint values, and can therefore be thought of as a point in a high-dimensional joint space. If d is the number of DOFs of the hand, than a posture $\mathbf{p}$ can be defined as

$$\mathbf{p} = [\theta_1\, \theta_2 \ldots \theta_\mathrm{d}] \in \mathcal{R}^\mathrm{d} \tag{1}$$

where θ_i is the value of i-th degree of freedom.

As we have discussed above, previous research suggests that most human grasping postures derive from a relatively small set of discrete pre-grasp shapes. In particular, Santello et al. [4] collected and analyzed a large set of data containing grasping poses from subjects that were asked to shape their hands as if they were grasping a familiar object. Principal Component Analysis of this data revealed that *the first two principal components account for more than 80 % of the variance*, suggesting that a very good characterization of the recorded data can be obtained using a much lower dimensionality approximation of the joint space.

In our work, we will refer to the principal components of these postures as *eigengrasps*. The implication is that they form a **low-dimensional basis for grasp postures**, and can be linearly combined to closely approximate most common grasping positions. By choosing a basis comprising b eigengrasps, a hand posture placed in the subspace defined by this basis can be expressed as a function of the amplitudes a_i along each eigengrasp direction e_i:

$$\mathbf{p} = \mathbf{p}_m + \sum_{i=1}^{b} a_i \mathbf{e}_i \tag{2}$$

$$\mathbf{e}_i = \left[e_{i,1}\, e_{i,2} \ldots e_{i,d}\right] \tag{3}$$

where $\mathbf{p}_m$ is the "mean" posture that describes the origin of the eigengrasp subspace. When computing eigengrasps based on recorded human data, $\mathbf{p}_m$ is simply the average of all recorded hand postures. Each eigengrasp $\mathbf{e}_i$ is a d-dimensional vector and can also be thought of as direction of motion in joint space. Motion along one eigengrasp direction will usually imply motion along all (or most) degrees of freedom of the hand. We note that the hand posture is now defined by the amplitudes vector $\mathbf{a} = [a_1 \ldots a_b] \in \mathcal{R}^b$.

2.2 Extending Eigengrasps to Robotic Hands

Although the work of Santello et al. is centered on the study of the human hand, we have found this approach to be extremely useful for robotic hands as well. In our study, we have applied the eigengrasp concept to a total of 4 hand models: the BarrettHand, the DLR hand [17], the Robonaut hand [18] and finally a human hand model. All our hand models, as well as the eigengrasps used in each case, are presented in Table 1.

For the human hand we have chosen eigengrasp directions based on the published results in [4], taking advantage of the fact that they have been derived through rigorous study over a large number of recorded samples. Since such data is not available for robotic hand models, we have derived eigengrasps attempting to

Table 1 Eigengrasps defined for the robotic hand models used in this chapter

Model	DOFs	Eigengrasp 1			Eigengrasp 2		
		Description	Minimum	Maximum	Description	Minimum	Maximum
Barrett	4	Spread angle opening	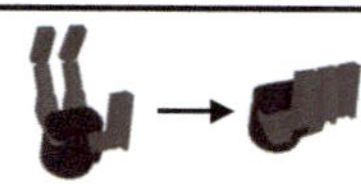		Finger flexion	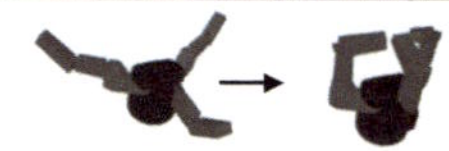	
DLR	12	Prox. joints flexion Finger abduction thumb flexion			Dist. joints flexion Prox. joints extension thumb flexion	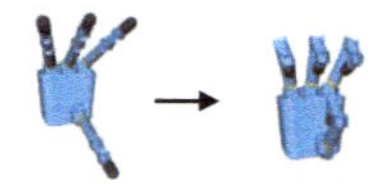	
Robonaut	14	Thumb flexion MCP flexion index abduction	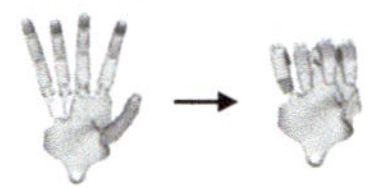		Thumb flexion MCP extension PIP flexion	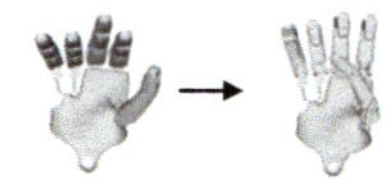	
Human	20	Thumb rotation thumb flexion MCP flexion index abduction	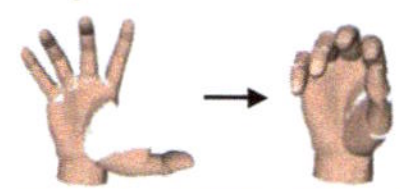		Thumb flexion MCP extension PIP flexion	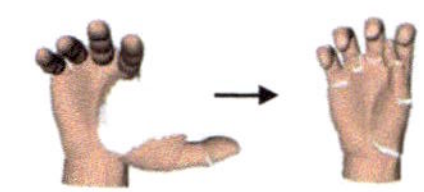	

define grasp subspaces similar to the one obtained using human hand eigengrasps. In most cases, such decisions could be made based directly on the similarities with the human hand. For example, the human metacarpophalangeal (MCP) and interphalangeal (IP) joints can be mapped to the proximal and distal joints of robotic fingers. In the case of the BarrettHand, changes in the spread angle DOF were mapped to human finger abduction. All eigengrasps were defined such that they satisfy the orthogonality constraint that naturally occurs when using Principal Component Analysis. While we found our choices to produce good results, the optimal choice of eigengrasps for non-human hands, as well as the choice of which eigengrasps to use for a particular task, are open questions and interesting directions for future research.

The eigengrasp concept allows us to design flexible control algorithms that operate identically across all the presented hand models. The key to our approach is that the eigengrasps encapsulate the kinematic characteristics of each hand design. This enables control algorithms that operate on eigengrasp amplitudes to ignore low-level operations and concentrate on the high-level task. Another advantage is the significant dimensionality reduction (by as much as a factor of 10 for complex hands) obtained by operating in the reduced basis eigengrasp space as opposed to the full joint space. In the next section we will derive a grasp planning algorithm that makes use of both these concepts.

3 Grasp Planning Using Eigengrasps

In essence, the grasp planning task can be thought of as an optimization problem in a high-dimensional space that describes both hand posture (intrinsic DOFs) and position (extrinsic DOFs). Consider the goal of maximizing a quality function of the form:

$$Q = f(\boldsymbol{p}, \boldsymbol{w}) \tag{4}$$

If $\mathbf{d}$ is the number of intrinsic hand DOFs then $\mathbf{p} \in \mathcal{R}^d$ represents the hand posture and $\mathbf{w} \in \mathcal{R}^6$ contains the position and orientation of the wrist.

Intuitively, this function has to be related to the quality of the grasp. However, most formulations pose a number of problems. First, it is very difficult, or even impossible, to compute an analytical gradient. Second, such functions are highly non-linear, as small changes in both finger posture and wrist position can drastically alter the quality of the resulting grasp. Finally, the legal parameter space is complex, having to satisfy multiple constraints: prevent inter-penetration with the object to be grasped as well as potential obstacles, and maintain joint values within their acceptable ranges.

3.1 Optimization Algorithm

We directly address all of these problems by using **simulated annealing** as the preferred optimization method (for a general review of the simulated annealing algorithm we refer the reader to [19]). Its stochastic nature makes it a particularly good choice for our task: since new states are generated as random neighbors of the current state, computation of the energy gradient is not necessary, and the algorithm works well on non-linear functions. Furthermore, the possibility of an "uphill move" to a state of higher energy allows it to escape local minima which can trap greedier methods such as gradient descent. However, the random exploration of the input domain means that high dimensionality of the parameter space will severely affect the computational efficiency of this algorithm.

We therefore propose performing the optimization **in eigengrasp space**, as opposed to DOF space. The energy function takes the form

$$Q = f(\mathbf{a}, \mathbf{w}) \tag{5}$$

where $\mathbf{a} \in \mathcal{R}^2$ is the vector of eigengrasp amplitudes. This effectively reduces the parameter space to 8 dimensions (2 eigengrasp amplitudes plus 6 extrinsic DOFs) from as high as 26 dimensions in the case of the human hand.

3.2 Energy Function

Most grasp quality metrics that have been proposed in the literature are based on the locations of the contacts between the hand and the target object. Our context is somewhat different: we need a quality metric that can also assess the quality of a *pre-grasp*, where the hand is very close, but not in contact with the target. For each hand model, we pre-define a number of *expected* contact locations by sampling each link of the fingers as well as the palm, as shown in Fig. 1. The value of the quality function is maximized for those hand postures that bring each expected contact location as close as possible to the target object. We are therefore searching for postures where the hand is wrapped around the object, generating a large contact area using all the fingers as well as the palm. As shown in Fig. 1, for each desired contact location on the hand, identified by the index i, we define the local surface normal $\hat{\mathbf{n}}_i$ as well as the distance $\mathbf{o}_i$ between the desired contact location and the target object. We then compute a measure δ_i of the distance (both linear and angular) between the desired contact and the surface of the object:

$$\delta_i = \frac{|\mathbf{o}_i|}{\alpha} + \left(1 - \frac{\hat{\mathbf{n}}_i \cdot \mathbf{o}_i}{|\mathbf{o}_i|}\right) \tag{6}$$

where α is a scaling parameter required to bring the range of useful linear distances (measured in mm) in the same range as the normalized dot product between $\hat{n}_i$ and

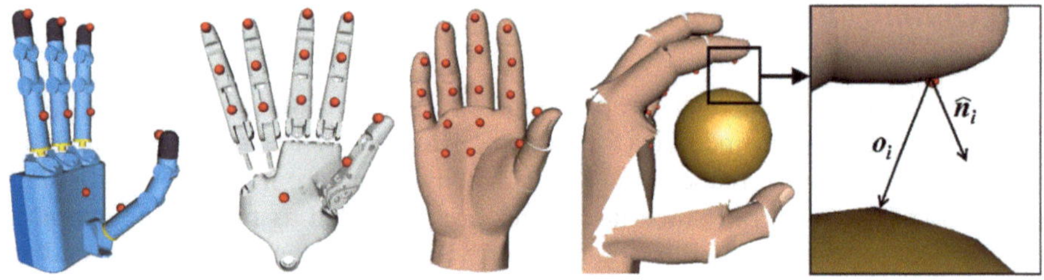

Fig. 1 *Left* complete set of pre-defined desired contact locations for the DLR, Robonaut and Human hands. *Right* for a desired contact with index i, we define the surface normal $\hat{n}_i$ and the current distance to the target object o_i

o_i (in our study we use a value of $\alpha = 50$). For a given hand posture, the total value of the quality function is then computed as:

$$Q = \sum_{\substack{all\, desired \\ contacts}} (1 - \delta_i) \tag{7}$$

In most cases, the hand postures that maximize the value of Q create an enveloping grasp of the object, especially for complex models grasping objects similar in size to the hand. The optimized value of this function can be seen as a measure of how well the hand shape can be set in order to match a given object while operating in a low-dimensional subspace. In Sect. 4 we will also present an alternative quality function formulation that includes a built-in notion of grasp wrench space analysis.

3.3 Simulated Annealing Example

We have implemented the simulated annealing approach using our publicly available *GraspIt!* simulation engine [20]. For each state generated during the annealing schedule, *GraspIt!* uses forward kinematics to place the robotic hand model in the correct posture and checks for collisions against the object to be grasped as well as other obstacles. If the state is found to be legal, the corresponding energy function is computed and the annealing algorithm continues. This process can be repeated until a satisfactory energy level has been reached, or a pre-specified number of iterations has been exceeded.

Before presenting an extensive test of this optimization method, involving multiple hand models as well as different objects, we will discuss a typical example, in order to analyze the behavior of the simulated annealing algorithm in more detail. This example involves the Robonaut hand grasping a glass. The optimization was performed over 100,000 iterations. Figure 2 shows the temporary solution (best state found so far) at various points during the optimization. We can observe what is considered typical behavior for a simulated annealing

$k = 5000$ $k = 15000$ $k = 25000$ $k = 29000$ $k = 30000$ $k = 40000$ $k = 60000$ $k = 10000$

Fig. 2 Simulated annealing example over 100,000 iterations. Each image shows the best state found until iteration k

implementation: at first, the search goes through random states, accepting bad positions as well as good positions. As the annealing schedule progresses, the search space is sampled more often in the vicinity of the good states, while bad states are no longer accepted. Finally, in the later stages, the search is confined in a small neighborhood around the best state, which is progressively refined. The total time required for the optimization was 173 s, or 1.73 ms/iteration. The most significant amount of computation was spent checking the feasibility of each generated state (i.e. checking for collisions and inter-penetrations).

3.4 Grasp Planning Examples

In order to test the effectiveness of our framework for the task of dexterous grasp planning, we have applied the 2-dimensional eigengrasp optimization using all four previously discussed robotic hand models on a set of six objects. Figure 3 shows the result of the annealing search for each hand-object combination. Our focus in this section is to evaluate the best hand postures that can be found in eigengrasp space. Therefore, Fig. 3 presents the best hand posture found by the optimization algorithm without any additional refinements, allowing a direct assessment of the optimization method through visual inspection of its output.

These results show that, when the search is confined to a low-dimensional eigengrasp space, it does not reach a global optimum of the quality function where all the desired contact location touch the target object. However, the local optimum found in eigengrasp space can be used as a *pre-grasp* by performing an additional adjustment where the hand leaves the planning subspace in order to conform to the surface of the object: execution of a binary "close all fingers" command, that simply closes all fingers until contact with either the object or another finger prevents further motion (Fig. 4). We use form-closure as the analysis criterion for the resulting grasps, as our goal is the synthesis of stable grasps with no weak points. Following [21], we label a grasp as form-closed if and only if it can resist an infinitesimal disturbance regardless of its direction.

In order to perform a quantitative analysis of the pre-grasps obtained through posture optimization, we have applied this adjustment to the 20 distinct postures with the highest quality values found by one execution of the optimization algorithm for each hand-object combination. On average, we obtained 6 form-closure

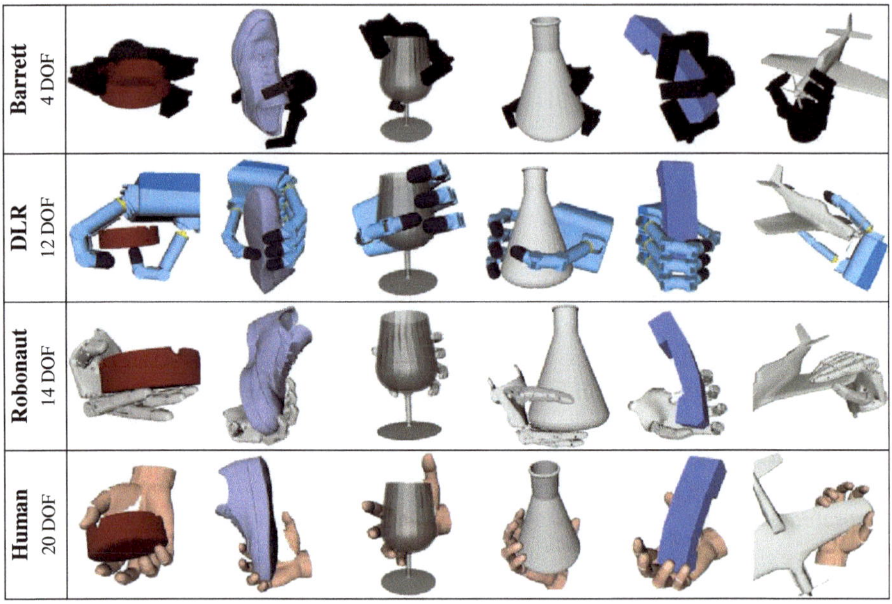

Fig. 3 Best hand postures found in a 2-dimensional eigengrasp space using simulated annealing optimization

Fig. 4 Examples of final grasps obtained from optimized postures by closing each finger until motion is stopped by contact with the object

grasps for each case; the detailed results for all the hand models and objects are presented in Table 2. Each optimization was performed over 70,000 iterations, with an average running time of 162 s. In the case of the human hand, Fig. 4 also shows all the final grasps obtained when using as pre-grasps the corresponding postures from Fig. 3.

These findings confirm our expectations of eigengrasp space as a *pre-grasp* or *grasp planning space*: in general, closing the fingers of a dexterous hand starting from a random configuration is not enough to obtain a stable grasp. However, the presented results show that if the starting position is the result of the eigengrasp optimization algorithm, we can obtain multiple form-closure grasps even when using highly dexterous hands.

Table 2 Number of form-closed grasps obtained from 20 pre-grasps found in a 2D eigengrasp space (average and standard deviation over 5 executions for each hand and object)

	Ashtray		Shoe		Glass		Flask		Phone		Plane	
	Average	Standard deviation	Average	Standard deviation	Average	Standard deviation	Average	Standard deviation	Average	Standard deviation	Average	Standard deviation
BarrettHand	2.8	2.2	1.0	1.0	5.8	5.3	3.4	1.7	1.4	2.1	4.0	3.1
DLR	11.0	3.6	6.0	3.4	0.8	0.8	3.2	2.7	2.8	3.6	1.8	0.8
Robonaut	7.0	2.3	9.0	1.7	14.4	3.7	14.4	3.7	10.0	2.8	3.6	2.3
Human	14.6	2.3	11.0	2.5	11.2	1.6	13.4	3.5	8.4	2.3	1.8	1.3

4 Online Interactive Dexterous Grasping

In the previous section we have presented an optimization algorithm that uses a low-dimensional subspace when searching for hand postures that match the shape of a grasped object. However, a significant amount of the computational effort was dedicated to optimizing extrinsic DOFs (wrist position and orientation, 6 variables) versus intrinsic DOFs (eigengrasp amplitudes, 2 variables). As the focus has been on dimensionality reduction for the intrinsic DOFs domain, no attempt has been made to simplify the extrinsic DOFs search domain. For fully autonomous grasp synthesis this is a necessary component: a correct finger posture is only relevant when combined with an appropriate wrist position relative to the target object.

An important category of grasp planning applications that do not require complete autonomy stems from the field of neural hand prosthetics. Such devices combine a degree of human control with artificial hardware and algorithms. Formalizing this concept using our framework means that an external operator can specify desired values for some, but not all of, the variables that define a grasp. For example, previous research [22] has enabled a primate to directly control the linear velocity of the endpoint of a robot arm through 3 DOFs in real time. This control was achieved by measuring the activity of individual cortical neurons that correspond to individual preferred directions of each neuron in space. The vector sum of preferred directions of a population of neurons, each scaled by their individual unit activity, provided the velocity of robotic end-effector movement.

In contrast, controlling finger posture has proven to be significantly more difficult. A number of possible approaches are described in the literature, including electromyography [23] and cortical implants [24]. These studies have shown success in decoding a limited number of information channels, therefore controlling a highly dexterous hand for interactive grasping remains an open and challenging problem. In this study we propose a grasp planning method that combines the eigengrasp framework for reducing the dimensionality of the hand configuration space with real-time operator input simplifying the spatial components of the search. Our goal is to enable an operator to complete dexterous grasping tasks with limited direct control over finger posture.

4.1 System Overview

In our current implementation, the user provides on-line information on the position and orientation of the wrist. This data is currently provided using a 6-DOF magnetic tracker. While we have not yet integrated this component in a real prosthetic system, it is our directional goal; we envision that hand position information will by extracted from cortical activity in a similar fashion to the primate study described above. In contrast, finger posture is entirely under control of the automatic component, which selects an appropriate hand shape by combining information about the geometry and pose of the target object with the position input provided by the operator. The only additional information needed from the user is a binary "click" command for completing a grasp, which we will describe below.

It is important to note that our approach must compensate for the lack of complete user grasp data by using knowledge of the target object geometry, as well as its initial position relative to the hand. In previous work [25], we have shown that it is possible to perform grasp planning using a vision-based system for object recognition and localization. Compared to the optimization method presented in the previous section, this system also has to satisfy two important criteria: first, the output has to be in the form of explicit form-closure grasps rather than optimized pre-grasps; second, solution grasps must be found at a fast enough rate to enable on-line interaction with the operator and usage of real-time input. The execution of a grasping task must therefore be performed at a speed that approaches natural human behavior, on the order of seconds as opposed to minutes.

A high-level overview of the complete system and the interaction with the operator is provided in Fig. 5. The planning algorithm runs on the *GraspIt!* simulator platform. Even though the grasp planner runs in a simulated environment, the results can be applied to a real robotic hand, allowing the user to interact with the hand directly and use it to pick up surrounding objects. The simulator receives user input and sends it to the grasp planner which processes it and outputs potential grasps, which are in turn used to generate commands for the robotic hand. The operator can hold the hand and approach the target object; the position of the hand relative to the target is tracked using a Flock of Birds (Ascension Corp., VA) magnetic tracker. We have also applied our method on a range of more complex hand designs (including the DLR hand, the Robonaut hand as well as the human hand model) using the virtual environment in *GraspIt!*; the operator can change the position of the virtual wrist by directly manipulating the magnetic tracker. In both

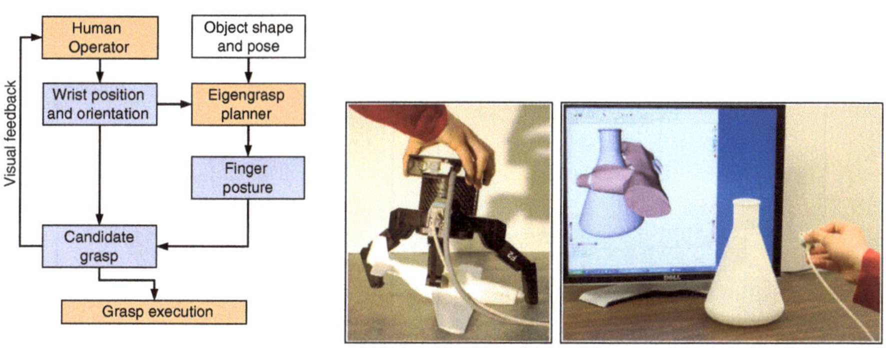

Fig. 5 Interactive grasp planning using wrist position input from a human operator. *Left* system overview; *Right* applied examples using a real BarrettHand and a dexterous hand in a simulated environment

cases, wrist position is supplied as input to the grasp planner, but the operator has no direct control over finger posture.

4.2 User Interaction with the Grasp Planner

In general, in order to uniquely identify a grasp, three variables are needed to specify the position of the wrist and three more for its orientation. In the context of our application, we expect the user to specify a desired approach direction to the target; however, the presence of such external input does not fully eliminate the spatial component of the grasp planning search. First, it is not practical to wait until the user has brought the wrist into a final grasping position before starting the search for an appropriate finger posture, as this behavior would decrease the interactivity of the system. Rather, it is preferable to start the search early, and attempt to predict where the user intends to place the wrist. Second, this prediction allows the system to offer feedback to the user: as soon as an anticipated grasp is found, the grasp planner can shape the fingers accordingly. The user can then decide if the grasp is satisfactory and either continue towards the target or choose another approach direction if the system is unable to find an acceptable solution.

This behavior can be implemented efficiently by re-parameterizing the spatial component of the grasp planner as shown in Fig. 6. For each hand model, we define a preferred search direction **d** based on the kinematics of the hand, usually normal to the palm. Then, starting from a hand position specified by the operator, we search for good grasps in a conical region around the search direction using 3

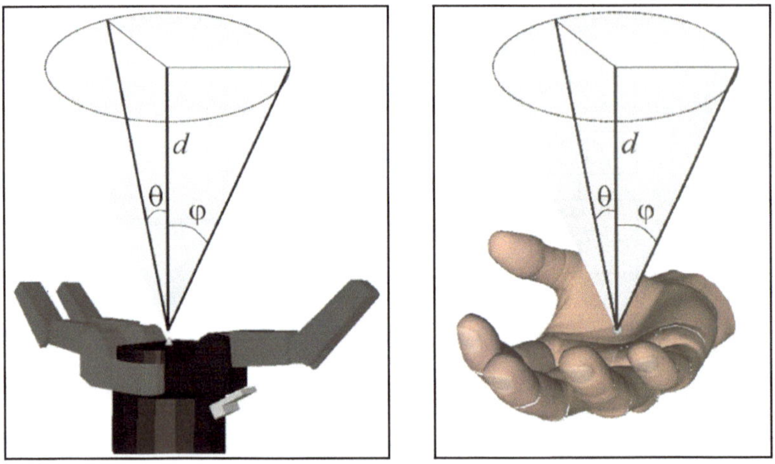

Fig. 6 Search directions defined for the BarrettHand and human hand models. The direction of the vector d is predefined relative to the palm. Its magnitude, as well as the values of the angles θ and ϕ are variables defining a conical search area

variables: the distance $|d|$ along the approach direction, as well as two angular variables, θ and ϕ. The operator is instructed to approach the object along a direction that is generally similar to the search cone; however, the search directions are defined in order to make this a natural choice. In the examples in Fig. 6 this means that the user is asked to keep the palm approximately facing the target, as opposed to other possibilities like a sideways or backwards approach.

The role of this parameterization is to reduce the number of extrinsic DOFs that are used for grasp planning, focusing on areas where good grasps are most likely to be found. Using this heuristic, the search will automatically ignore states where, for example, the hand is facing away from the target object. However, the user is not expected to specify an exact wrist position for a good grasp; by searching along the approach direction d the planner attempts to anticipate the intended final grasp. The angular variables θ and ϕ allow the planner to compensate for noisy measurements in the intended wrist position, and allow for more flexibility in the search for solution grasps. By adding these three variables to the eigengrasp amplitudes describing hand posture, we obtain a low-dimensional domain that can be searched fast enough to respond to on-line changes in the wrist position input provided by the human operator.

4.3 Quality Function Formulation Using Scaled Contact Wrench Spaces

When the posture optimization algorithm is used for on-line grasping tasks, we use a formulation of the quality function Q that is better adapted for interactive operation. Recall that, in the form presented in Sect. 3, our formulation rewards hand postures that bring all the fingers, as well as the palm, as close to the surface of the object as possible. For the application presented here, it is necessary to also reward hand postures that create stable, but not necessarily enveloping grasps (consider as an example the case of a fingertip pinch grasp applied on a thin object). We therefore propose an alternative quality function which is fast to compute and can assess the *potential* quality of a pre-grasp posture using a modified version of the Grasp Wrench Space (GWS) ϵ-metric [26]. A detailed description of this metric in its original form is beyond the scope of this paper; we provide a brief overview below, and for further details we also refer the reader to [27].

For each contact i, we assume that the space of forces and torques that can be transmitted is bounded by the convex hull of a finite set of 6D wrenches $\mathbf{w}_{i,j}$ where $1 \leq j \leq k$. The convex hull of these wrenches forms the Contact Wrench Space (CWS). We note that this approach imposes a linear form for what are normally quadratic friction constraints; for example, in the case of Coulomb friction, the torque components of all $\mathbf{w}_{i,j}$ are null, and the force components sample a polygonal approximation to the contact friction cone. In order to define the GWS, the contact wrenches from all contacts are first expressed relative to a common

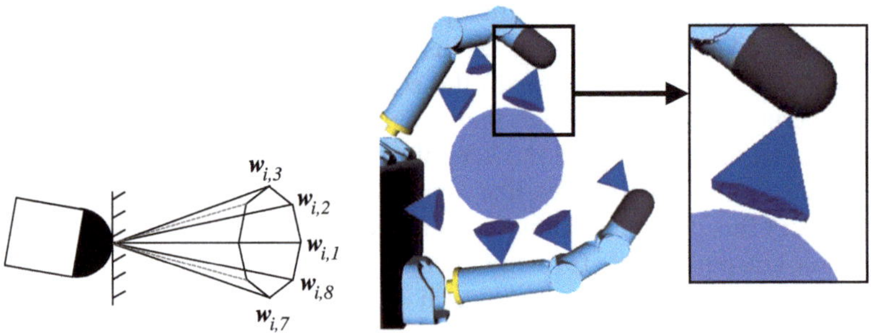

Fig. 7 *Left* contact wrench space example using a Coulomb friction cone; *Right* multiple contact wrench spaces, each scaled based on the contact distance metric δ_i

coordinate system. This coordinate system is usually anchored at the center of mass of the object and the choice of axes directions is arbitrary. We denote the matrix that transforms a wrench from the local coordinate system of contact i to the global object coordinate system by $\mathbf{R}_i \in \mathcal{R}^{6\times6}$.

In our implementation, we are usually assessing the quality of a pre-grasp shape where the fingers are not in contact with the target. Therefore, we assume that the hand can apply *potential* contact wrenches using the desired contact locations shown in Fig. 1. When computing the GWS, we scale the potential wrenches at each desired contact proportional to the inverse of the distance metric δ_i computed as in (6):

$$GWS = ConvexHull\left\{ \bigcup_{\substack{\text{all desired}\\ \text{contacts}}} (1-\delta_i)R_i \bigcup_{j=1}^{k} w_{i,j} \right\} \tag{8}$$

Thus, if the value of δ_i is small, the contact will have a significant contribution to the GWS, and states that bring it closer to the object surface will be rewarded by a higher quality value. If, on the contrary, the desired contact is far from the object, it will not significantly affect the grasp quality measurement. If the contact is far enough from the object so that its corresponding weight of $1 - \delta_i$ is negative, it is completely excluded from the computation.

After building the GWS, we compute the ϵ-quality measure as described in [26, 27]. The quality of the grasp is considered equal to the radius ϵ of the largest 6D ball, centered at the wrench space origin, that can be enclosed within the GWS. If $\epsilon = 0$ then the origin itself is not contained in the hull, and the grasp does not have form closure. For $\epsilon > 0$, the grasp can resist any disturbance, and the maximum magnitude of the contact forces needed to resist a disturbance is inversely proportional with ϵ.

The process is illustrated in Fig. 7 for the DLR hand grasping a disc. In this example, each contact is modeled by a friction cone, approximating Coulomb friction for rigid bodies. However, other local contact models can also be used. For example, the ability to create stable, encompassing grasps with subsets of fingers is increased by using soft fingertips that deform during contact. In addition to tangential friction, such contacts can also apply frictional torque. The friction cone is thus replaced by a four-dimensional "friction ellipsoid" which constrains the relationship between tangential force and frictional torque [28]. This effect can be captured by linearizing the friction ellipsoid and using the appropriate contact wrenches $w_{i,j}$, as shown in [29]. This method enables the use of rubber-coated fingertips for our robotic hands, without compromising the accuracy of the grasp quality computations.

4.4 Computation of Form-Closure Grasps

The automated grasp planner searches for solution grasps in two stages. The first stage is the posture-optimization algorithm presented in Sect. 3, using the quality function formulation described above. For interactive tasks, each run of the simulated annealing algorithm is performed over 2,000 iterations, taking advantage of the fact that the search domain is 5-dimensional (2 eigengrasp amplitudes and 3 wrist position/orientation variables), as opposed to the 8-dimensional domain used for fully autonomous searches. After reaching this number of iterations, the search is restarted by resetting the annealing temperature. As a result, the planner does not get stuck if one particular search fails; rather, the search is restarted and takes advantage of any changes in the approach direction provided by the operator.

The user-specified reference wrist position is updated continuously during the search. The results of the optimization are therefore always relative to the current position of the wrist. However, we recall that the low-dimensional optimization procedure can still only produce *pre-grasp* shapes; in order for the system to allow successful completion of the task, *final* grasping postures satisfying the form-closure requirement are necessary. In order to achieve interactive rates, this expensive computation is only performed using the best pre-grasps found during each run of the annealing optimization, which are queued and sent to the second stage of the planning process.

For each candidate pre-grasp resulting from the first stage, we use the contact detection engine within *GraspIt!* to compute the final grasp that results by closing the fingers on the object. Once the contacts between the hand and the object have been determined, we compute the exact quality value of the final grasp by applying the Ferrari-Canny metric in the original form presented in [26]. If the grasp is found to have form-closure, it is saved, along with its associated quality value, as a potential solution, and used by the next component of the system, which is responsible for interaction with the human user.

When computing the final grasping posture resulting from a candidate pre-grasp, we take into account specific mechanical properties of the hand, such as passive mechanical adaptation to the shape of the target. A number of robotic hands, such as the BarrettHand, the SDM Hand [30] and the CyberHand [31] rely on passive mechanical adaptation, as it significantly increases grasp stability without increasing the complexity of the control mechanisms. All of the results involving the BarrettHand presented in this paper take into account its adaptive actuation mechanism which allows distal joints to close even when proximal joints controlled by the same motor have been stopped due to contact.

In our implementation, the two planning phases described in this section (simulated annealing search for pre-grasps and final grasp testing for form-closure) run in separate threads. As soon as a candidate pre-grasp is found, it is queued for testing, but the search for new candidates continues independently of the testing phase. Also, candidate pre-grasps are independent of each other, and can be tested simultaneously. This parallelism allows us to take advantage of the current evolution in multi-core architectures, largely available on standard desktop computers.

We can now provide a complete step-by-step walk-through of a grasping task that combines user input and automated grasp planning. Figure 8 shows the execution of a grasp proceeding through the following stages:

- as the user approaches the target object, the grasp planner searches for a good grasp in a cone-shaped area around the given approach direction; when a solution is found, it is used to set the hand posture, allowing the user to react. If multiple solutions are found, the one that is closest to the current user approach direction is chosen for presentation (i.e., the solution with the lowest values for the angular variables θ and ϕ).
- the planner continuously attempts to improve the current result, by finding new grasps that either have a better quality metric or are closer to the current position established by the user.
- if the planner is unable to find a grasp in the current search area, or the user is not satisfied with the resulting hand posture, the user can reposition the hand and attempt to grasp a different part of the target object.
- if the user is satisfied with the hand posture, he/she continues along the current approach direction. As the real hand position approaches the target grasp, the fingers are gradually closed around the object. The user can therefore predict where the object will be touched and finally issue a "close all fingers" command which completes the grasping task.

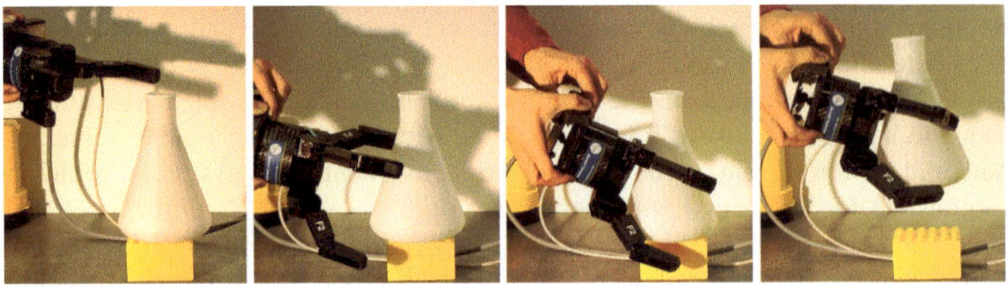

Fig. 8 Example of a complete grasping task: initial approach, finger-preshaping using grasp planning result, continued approach and final grasp execution

4.5 Results

Figure 9 presents the application of our method using the BarrettHand in a real environment, while Fig. 10 shows interactive grasps performed in a simulated environment using the DLR hand, the Robonaut hand and the human hand model. In most cases, the images show only the final grasp applied by the user; due to space constraints we are unable to include images showing the evolution of the grasping task from approach direction, pre-grasp and final grasp. In order to better evaluate the interactive nature of our application, a video showing a number of complete examples is available at http://www.cs.columbia.edu/~cmatei/interactive.

Fig. 9 Examples of interactive grasping tasks; each image shows the grasp found for a different approach direction or target object. In all cases the object was successfully grasped and lifted off the table

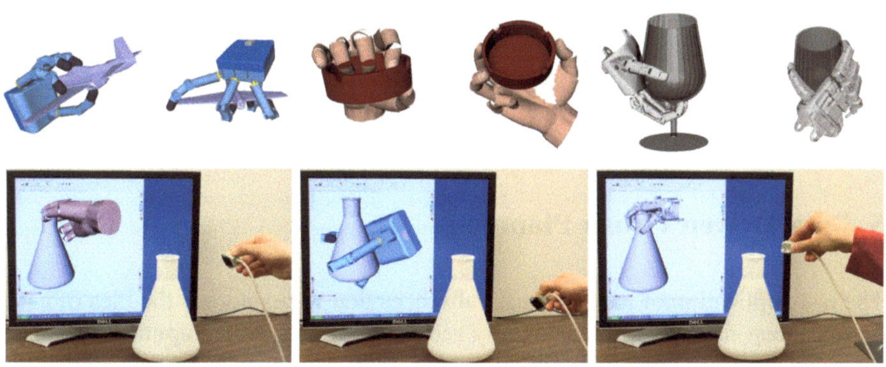

Fig. 10 Examples of grasping tasks executed in simulated environments. *Bottom row* images also show the user providing the approach direction via a magnetic tracker. All of the presented grasps have form-closure

For any given grasping task, the exact computational effort required to find a stable grasp depends on the complexity of the hand and target object, as well as the approach direction chosen by the user. On average, the first stage of the grasp planning algorithm processes approx. 1,000 hand postures per second, while the second testing phase, running in parallel, can evaluate approx. 20 candidate pre-grasps per second. In most cases, solution grasps are found at interactive rates: in the example presented in Fig. 8, the grasp planner found 8 stable grasps in 13.6 s of computation. These are representative numbers for the behavior of the system, which generally requires less than 2 s to find a solution grasp for a new approach direction. All of our tests were performed using a commodity desktop computer equipped with a 2.13 GHz Intel Core2 CPU.

The ability of the system to allow for successful task completion in a short time is more difficult to quantify, as it also depends on how well the user reacts to the behavior of the automated components. All the results presented in Figs. 9 and 10, as well as in the video referenced above, were obtained at interactive rates, usually requiring between 5 and 15 s from first approach to final grasp execution. For the more difficult tasks, taking up to 30 s to complete, we found two main reasons that led to the increased execution time: either the planner repeatedly failed to find form-closure grasps for selected approach directions, or the human user could not interpret some of the finger postures selected by the planner and had to attempt different grasps. These cases represent a small minority of our tests and examples; however, the tests were performed by well-trained users familiar with the inner workings of the planning algorithm.

In other work, we have used this online grasp planner for biomimetic grasp planning. Our idea is a shared control paradigm where incomplete reach and grasp control information from the cortex is supplemented with control synthesized by the automatic grasp planning system described above. A key aspect of this task is the presence of on-line user input for hand posture, which will ultimately be obtained by identifying and extracting the relevant signals from brain activity. Our grasping system can combine partial or noisy user input and autonomous planning to enable the robot to perform stable grasping tasks. We have tested two applications of this algorithm, using data collected from both primate and human subjects during grasping, to demonstrate its ability to synthesize stable grasps using partial control input in real or near-real time [32].

5 Data Driven Grasp Planning

The last application of hand posture subspaces that we explore is the idea of a data driven grasp planner. Dimensionality reduction allows pre-computation of many thousands of stable grasps on thousands of everyday objects. Using this data, we can create a database of known grasps. To grasp a novel object, we can index into a database of known 3D models and use a pre-computed grasp for a similarly shaped object to suggest a new grasp. We refer to this idea as data driven grasping.

To support this, we have built the Columbia Grasp Data Base (CGDB) [33], a database of pre-computed grasps, which is described below. The database is freely available at grasping.cs.columbia.edu both as a PostgreSQL database and as flat text files.

Our primary interest is in using an object's 3D geometry as an index into the database. Given a new 3D object, we can find geometric neighbors in the database, and the accompanying stable grasps for these similar objects. If the number of objects to be grasped in the database is very large and comprehensive then robotic grasping becomes a pre-computed database lookup.

We are not aware of any previous attempt to construct a large scale grasp database, or of any commonly used benchmarks for evaluating robotic grasping. However, researchers have investigated grasp planning approaches that assume such a database already exists. Bowers and Lumia [34] collected grasps for a small number of planar objects and used fuzzy logic to extrapolate grasping rules. Morales et al. [35] used *GraspIt!* to compute offline grasps for a small database of graspable objects, and successfully identified and executed those grasps on real objects in a complex environment. Unlike the planner we present in this chapter, their approach requires an exact model of every possible graspable object.

Other researchers have experimented with different forms of precomputed grasp knowledge. Li and Pollard [36] collected a database of 17 hand poses, and used shape matching to match poses to graspable objects. Their work highlighted the difficulty of automatically generating grasps that are both stable and plausibly humanlike. Aleotti and Caselli demonstrated grasp synthesis using examples acquired with a dataglove [37]. Saxena et al. [38] generated 2D renderings of a large set of example objects, and learned a model-free mapping from images to graspable features. Their work uses a two fingered gripper and does not have an obvious extension to dexterous hands.

5.1 Building a Grasp Database

Although the approach of building a large grasp database by direct data collection from human users has difficulty scaling to thousands of objects, the basic strategy of performing many grasps and recording their poses is still valid. However, it requires replacing human input with an automated procedure that does not need user attention. In our own previous work [39, 40] we have demonstrated the use of a grasp planner running in a simulation engine. Grasp planning can be considered an optimization task that can be performed in simulation without user supervision. However, even when simulations are performed on a powerful computer, the space of possible grasps is too large to sample directly. This is particularly true in the case of dexterous hands with many intrinsic degrees of freedom (DOFs).

5.2 Enhanced Eigengrasp Planner

In Sect. 2 we presented a grasp planning algorithm that optimizes hand posture in a low dimensional eigengrasp space to find effective pre-grasp postures. These pre-grasps are then heuristically developed into stable, form closure grasps. To build our database of grasps, we apply this planner to a very large set of objects, and obtain multiple results for each object. We use a slightly modified multi-threaded version which takes advantage of multi-core architectures widely available today on commodity computers. A single parent thread searches the eigengrasp space for likely pre-grasps; for each pre-grasp position that crosses a quality threshold, a separate child thread is created to refine it. The child thread performs a quick local simulated annealing search with finer step values, attempting to develop the pre-grasp into form closure. If the resulting grasp indeed has form closure, the pre-grasp and grasp are both saved in the result list.

After creating a child thread for a pre-grasp state, the parent thread's state generation function rejects states close to the child thread's search area, forcing it to look elsewhere in the state space for new grasps. The planner can be restarted at any point by resetting the annealing temperature. The process continues until either the desired number of grasps are found, or a pre-set time limit is exceeded.

5.3 Constructing the Database

A crucial aspect of the database construction is the running time required for each hand-object combination. As the planner has no intrinsic time limit, one has to be imposed based on the size of the object set and the computing power available. When building the database described in the next section, each execution of the planner ran until we found 15 form closure grasps of the target object. In general, this required about 10 min of run time. When dealing with large datasets and different hand models, a subset of objects will inevitably prove difficult to grasp using our algorithm. To prevent these from dominating the computation time we also set an upper time limit of 20 min per model.

The total number of planner executions used to build the database was approximately 22,000, as described in following section. The total computation time was approximately 1 month on 6 multi-core computers.

5.4 3D Models and Scaling

The first requirement for a grasp database is a set of objects to grasp. Rather than distributing a new set of 3D models, we chose to reuse the models from the Princeton Shape Benchmark (PSB) [41], which is already in common use in the shape matching community. We hope that a shared dataset will encourage increased collaboration between shape researchers and roboticists.

The disadvantage of this choice is that the PSB models were not originally selected with an eye towards robotic grasping, and so some of the models are not obvious choices for grasping experiments. For example, the database contains both airplanes and insects, all of which are outside the normal grasping range of a human-sized hand. We chose to treat all such "ungraspable" objects as toys, and rescaled them accordingly. The rescaling factor for each model, as compared with the "original" scale in the standard PSB, is included in the database.

Even with all of the models at graspable size, the issue of scale required further attention. Grasping is inherently scale-dependent, but most of the models in the PSB might plausibly exist anywhere within a small range of scales. This was particularly true for the models rescaled to "toy" size. To soften the impact of scale, we cloned each object at four distinct scales, 0.75, 1.0, 1.25 and 1.5, where 1.0 represents the rescaled size from above. As the PSB contains 1,814 distinct models, this left us with a total of 7,256 models to grasp.

For each model we store the radius of a ball that approximately contains the model. We use an approximately enclosing ball rather than an absolutely enclosing ball to avoid outlier sensitivity. We assume that all of the points on a model's surface are normally distributed around some fixed radius from the center of mass. The fixed radius is simply the mean distance from the surface to the center of mass, and the "approximate radius" stored in the database is the mean distance to the surface plus two standard deviations of the distance to the center once the mean has been subtracted.

Along with shape and scale, the space of possible grasps is influenced by the frictional properties and deformability of both the hand and the object. Both in the construction the database and in our experiments we treated all models as being made of rigid plastic. The properties of the different hands are described below.

5.5 Robotic Hands

Grasping is strongly hand-dependent, and so we need to specify the hands used in our database. For the first version of the database, we chose to focus on two hands; a human hand model in order to emphasize the "humanlike" nature of the grasp selection, and the three-fingered BarrettHand, which is ubiquitous in robotics research due to its durability and relatively low cost. The human hand model has 20 DoF. The BarrettHand has 4 DoF, plus a disengaging clutch mechanism which allows conformance even when the proximal link of a finger is blocked. Both models are available for download with *GraspIt!*.

Frictional forces play an important role in grasping, and so we must specify the materials for each of our hands. There is no exact consensus on the friction coefficient of human skin and so we chose $\mu = 1.0$ as a plausible value for the friction between the human hand and plastic [42]. The ability to create stable, encompassing grasps with subsets of fingers is also increased by using soft fingertips that deform during contact and apply a larger space of frictional forces and moments than their rigid counterparts. In order to take into account such effects, we use a fast analytical model for soft finger contacts that we have introduced in previous work [29]. The BarrettHand is made of metal, but can be coated with a higher friction material. We created two versions of the BarrettHand, one uncoated and one with rubberized fingers, and computed grasps for then independently, in effect giving us three hand models. For the metal BarrettHand we used $\mu = 0.4$ and for the rubber coated version we used $\mu = 1.0$. As the kinematic models are identical, grasps computed for either BarrettHand model can be executed using the other, making it possible to evaluate the advantage afforded by using the higher friction material. We note that grasps from the regular hand can be assumed to be form closure for the rubberized hand as well, but that this guarantee does not hold in reverse.

5.6 Grasps and Pre-Grasps

Our grasp database is intended to be used in conjunction with *GraspIt!* or a similar grasp simulation tool. As such, we provide the necessary data to recreate each grasp, in the form of joint angles and hand position, and the contact points between hand and object, which can be used as a check to ensure that the grasp was simulated correctly. We also provide the two measures of grasp quality mentioned in Sect. 5.3.

Each grasp entry consists of a *pre-grasp* and the final grasp pose. A pre-grasp is a pose from the instance before the hand contacts the object; it represents "pure" grasp information untainted by conformance to an exact object. Each of our pre-grasps lies within the two dimensional Eigengrasp subspace, as described above. In contrast, the grasp poses represent final positions for form closure. This first version of our database contains 238,737 distinct form closure grasps, each with an associated pre-grasp. The breakdown of these grasps is shown in Table 3.

5.7 Caveats

Since the grasps in the database were found using an automated planner, not all of the grasps are truly humanlike or reliable. There can be cases where a grasp satisfies our quality metrics, but would require a degree of precision that cannot be obtained

Table 3 Precomputed grasps for each hand over 7,256 scaled models

Metal Barrett	25,585
Rubber Barrett	132,421
Human	80,731
Total	238,737

in real-life execution. Recent work has underlined the fact that quality metrics used in simulation do not always translate to stable or human-like real-world grasps [44, 45]. Aside from the intrinsic limitations of grasp quality metrics, for which there is as of yet no firm consensus on which to use, our approach to grasp planning is purely geometric. This presents problems for objects that do not match our assumptions. For example, our assumption that all objects are rigid plastic results in geometrically correct but unrealistic grasps on objects such as flowers or leaves.

Furthermore, the lack of domain-specific knowledge means that some of our grasps are semantically incorrect, such as a mug grasped by placing the fingers inside, although they are still form closed.

Finally, all of our grasps were obtained from pre-grasps that sample a low-dimensional subspace of the hand DOF space. This is for the moment a necessary simplification, without which the planning problem for dexterous hands is intractable at this scale. While our choice of subspace is theoretically justified and shown to be effective [43], we cannot reasonably claim that the database covers the entire space of possible grasps. The choice of optimal subspace is one of our directions for future research.

6 Databased-Backed Grasp Planning

One of our primary motivations of building a grasp database was to collect enough grasping data to build new grasp planners based on learning. In this section we present a grasp planner that uses a k-Nearest-Neighbors approach to find candidate grasps for a model not in the database. In general, the relation between hand pose and grasp quality for a given object is both nonlinear and discontinuous, and more sophisticated learning methods such as SVMs have so far been shown to work only for simple objects [34, 46]. We hope that the data we have collected will facilitate further research in this direction.

6.1 Algorithm

Our grasp planning algorithm is based on the intuition that similar objects are likely to have similar grasps. Therefore, if we wish to grasp an object not in our database, it makes sense to look at the most similar objects that *are* in the database and to attempt to grasp the new object in the same way.

Given a model to grasp α, we use a shape matching algorithm to find $N = \{n_1 \ldots n_k\}$, the k models in the database most similar to α under some shape similarity metric. In this study we use L^2 distances between Zernike descriptors [47], which we have previously shown to be scalable to very large libraries of 3D models [48]. Zernike matching is scale-normalized, but as detailed in Sect. 5.4, each PSB model exists in our database at 4 distinct scales. For each n_i we consider up to 2 models, $n_i^<$, the largest neighbor smaller than α and $n_i^>$, the smallest neighbor larger than α, using the scaled approximate radius described in Sect. 5.4. In the case of α smaller or larger than all 4 versions of the neighbor we only used one model for n_i.

We present our grasp planning algorithm here. For simplicity, we have ignored the issue of scale and treated each n_i as a single model.

Algorithm 1 DATABASE-BACKED GRASP PLANNING

Require: Model α to grasp with hand H, using k neighbors and quality threshold τ.

```
N ← NearestNeighbors(α, k)
R ← {} {The resulting form closure grasps.}
for all n_i ∈ N do
    Align α and n_i using PCA.
    Co-locate the centers of mass of α and n_i.
    G ← the precomputed pre-grasps on n_i.
    for all g ∈ G do
        Transform g to α's coordinate system.
        Place H in the position and configuration of g.
        repeat
            Move H backwards.
        until NoCollisions(Palm(H), α)
        for all finger ∈ Fingers(H) do
            repeat
                Open finger.
            until NoCollisions(finger, α)
        end for
        repeat
            Move H forwards.
        until InContact(H, α)
        CloseFingers(H)
        if EpsilonQuality(H, α) > τ then
            R ← R ∪ g
        end if
    end for
end for
return R
```

The entire process, from shape matching through final output, takes approximately 20 s. To illustrate the behavior of this algorithm, we provide a number of examples in Fig. 11.

6.2 Experiments

For our experiments, we removed each PSB model at scale 1.0 from the database one at a time and attempted to grasp it using only the known grasps from the remaining models. To isolate the effects of shape matching, we used three methods of choosing similar models. In each case we used $k = 5$ neighbors for every model.

Our first method used the ground-truth labels provided with the PSB. For each model, we chose neighbors within the same shape category, starting with the finest

Fig. 11 Three example models and their grasps, using the database-backed planner with Zernike neighbors. For each model α (*left*), the *top row* of images shows a neighbor n_k from the database, the value of k, and a pre-computed grasp on that neighbor. Directly below each neighbor is the same grasp executed on α, along with its GWS epsilon quality measure

categorization and moving up to coarser categories if fewer than k neighbors were available. Within the same category the choice of neighbors was arbitrary. We designated the chosen models as the '*PSB classes*' neighbors. This method of indexing, while not usable for arbitrary unclassified models, approximates the performance of a theoretical ideal[1] shape matching algorithm that has perfect precision and recall over the PSB.

Our second method used L^2 distances between Zernike descriptors [47]. For each model, we designated the k models with the smallest L^2 distance in Zernike descriptor space as the '*Zernike*' neighbors. These descriptors are computed on voxel grids and are quite robust, making them suitable for use in matching newly acquired objects into the database.

For our third method, we randomly selected k models from the database and designated them as '*unordered*' neighbors. We emphasize that these are *not* random grasps in any sense; α has been translated and axis-aligned with some model of a similar scale. Furthermore, the pre-grasps applied to it are pointed in the right direction, with joint angles drawn from a high quality eigengrasp subspace and known to produce form closure on another model with aligned principal axes. We therefore expect that some of the pre-grasps taken from unordered neighbors will result in form closure grasps. Our aim in using the unordered neighbors is to isolate the performance gains based on shape matching while holding constant the performance due to the overall high quality of all of the grasps in the database.

We ran the experiment separately for each type of neighbor selection and averaged the grasp quality of the nth best grasp on each model over all 1,814 models in the database at scale 1.0. As mentioned in the introduction, we believe this to be one of the most comprehensive grasp planning experiments in the literature, as it consists of thousands of runs on a highly varying set of objects. We can analyze these results from a number of different perspectives: the absolute performance of database-backed grasp planning, the relative behavior of different neighbor selection methods and finally the performance compared to running the eigengrasp planner of Sect. 5.2 directly on the target object. Figure 12 shows these results for the human hand and the BarrettHand.

Although Zernike descriptors do not have perfect precision and recall over the PSB, their performance for our grasp planner is nearly identical to the ground truth PSB classification. This is likely because the PSB classification is partially semantic, whereas our grasp quality measures are purely geometric; a 'poor' neighbor for semantic shape matching may still be geometrically close enough to share high quality grasps.

The performance of the 'unordered' neighbors is as expected; good, due to the general quality of grasps in the database and the axis-alignment between α and the unordered neighbor, but not as good as the shape matching methods. Although the improvement due to shape matching is small for the BarrettHand, for the human hand the difference is quite significant. We attribute this difference to the many additional DOFs of the human hand, which creates a need for careful pre-grasping.

---

[1] Even with perfect precision and recall, this theoretical algorithm may not truly be 'ideal', as the categories in the PSB are semantic rather than purely geometric. Nevertheless, since shape matching algorithms are regularly evaluated using these categories as a ground truth, we adopt the same convention.

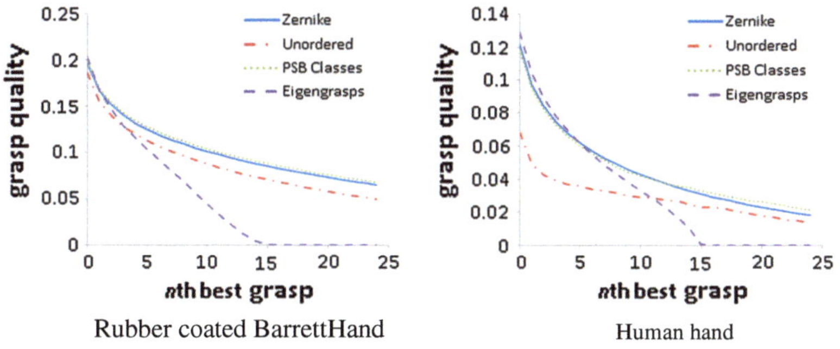

Fig. 12 The *n*th best grasp from database-backed grasping with three neighbor selection methods and from the eigengrasp planner, averaged over the 1,814 models in the database at scale 1.0. This figure is best viewed in color

The BarrettHand, with its 4 DoF, has a far simpler configuration space, and the importance of pre-grasping is correspondingly less.

Of special interest is the comparison between the database-backed methods and the eigengrasp planner. For the first few grasps, the performance of the shape matching methods is essentially identical to that of the eigengrasp planner. However, for subsequent grasps the quality quickly diverges, with the advantage going to the database-backed methods. This is even more impressive when we recall that the eigengrasp planner ran for approximately 10 min per model, whereas the database-backed planners ran for about 20 s. The database-backed approach can take advantage of pre-computed grasp data from multiple objects, essentially extracting the useful information obtained from several runs of the eigengrasp planner.

6.3 Planning for Real Objects

The ultimate goal of our database-backed planner is to grasp new objects that are not in our database, using sensor data. Realistically, sensors will only be able to acquire partial models for novel objects, which presents a problem for many planning algorithms. A data-driven planner, however, will be able to operate even on such incomplete data as long as the underlying 3D model database has been indexed to allow for partial matching.

While a comprehensive evaluation using such acquired data is beyond the scope of this chapter, we present here our preliminary results. Using a commodity desktop 3D scanner, we acquired a range image of a plastic toy. Due to the intrinsic limitations of the acquisition method, the range image was both noisy and incomplete, with several occlusions. We computed the Zernike descriptor of the scan, found the nearest neighbors in the database, and ran the planner as before for both the BarrettHand and the human hand. We found 88 form closure grasps using

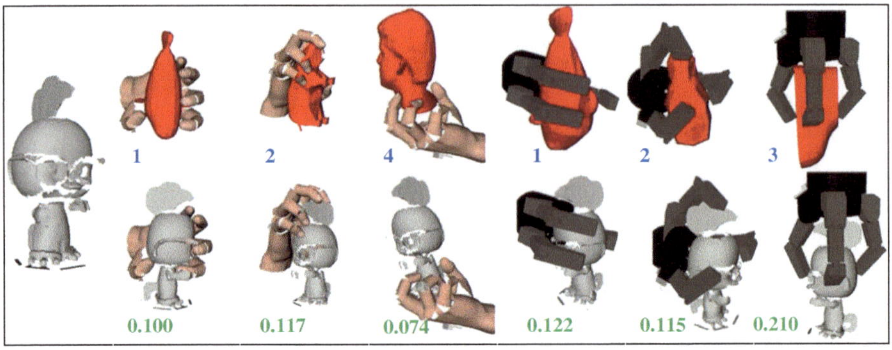

Fig. 13 Some of the grasps planned for an acquired object with holes and occlusions, using the database-backed planner with Zernike neighbors. In total 88 form closure grasps were found for the human hand and 112 for the BarrettHand

the human hand and 112 form closure grasps using the BarrettHand. Some of these results are shown in Fig. 13. For an in-depth treatment of data-driven grasping with partial sensor data, refer to [49].

7 Conclusion and Future Work

Our online eigengrasp planner and database planner each have limitations. In both cases, full knowledge of object geometry is needed to either compute a stable grasp or to match an object's shape in the database. We have performed initial experiments with the database planner to use only partial shape information to find the best grasp [49], [50].

In the case of the eigengrasp planner, choosing the correct couplings between the joints is somewhat ad-hoc. For anthropomorphic hands such as the Robonaut hand, these couplings can be directly transferred from human data observations. In non-anthropomorphic hands, determining the correct set of eigenvectors is a subject of further research.

For the database planner, the approach to grasping is entirely geometric. It leaves out important considerations such as non-uniform mass distribution in computing a stable grasp. It may be possible to add this information into the database, but it would require a scale-up of the database in size.

Overall, we have found the using these low-dimensional subspace ideas has been extremely fruitful in making robotic grasping systems that actually work in the physical world. We hope to push these ideas beyond static grasping and into manipulation in the future.

Acknowledgments The authors would like to thank Hao Dang for his help in building the Columbia Grasp Database.

References

1. M.H. Schieber, M. Santello, Hand function: Neural control and peripheral limits to performance, (invited review). J. Appl. Physiol. **96**, 2293–2300 (2004)
2. M.R. Cutkosky, On grasp choice, grasp models, and the design of hands for manufacturing tasks. IEEE Trans. Robot. Autom. **5**, 269–279 (1989)
3. T. Iberall, Human prehension and dexterous robot hands. Int. J. Robot. Res. **16**, 285–299 (1997)
4. M. Santello, M. Flanders, J.F. Soechting, Postural hand synergies for tool use. J. Neurosci. **18**(23), 10105–10115 (1998)
5. M. Santello, M. Flanders, J.F. Soechting, Patterns of hand motion during grasping and the influence of sensory guidance. J. Neurosci. **22**, 1426–1435 (2002)
6. M. Santello, J.F. Soechting, Matching object size by controlling finger span and hand shape. Somatosens. Mot. Res. **14**, 203–212 (1997)
7. M. Santello, J.F. Soechting, Gradual molding of the hand to object contours. J. Neurophysiol. **79**, 1307–1320 (1998)
8. C. Brown, H. Asada, Inter-finger coordination and postural synergies in robot hands via mechanical implementation of principal components analysis. In: *IEEE-RAS International Conference on Intelligent Robots and Systems*, pp. 2877–2882, 2007
9. C. Gosselin, F. Pelletier, T. Laliberte, An anthropomorphic underactuated robotic hand with 15 DoFs and a single actuator. In: *IEEE International Conference on Robotics and Automation*, 2008
10. V.C.K. Cheung, A. d'Avella, M.C. Tresch, E. Bizzi, Central and sensory contributions to the activation and organization of muscle synergies during natural motor behaviors. J. Neurosci. **25**(27), 6419–6434 (2005)
11. C.R. Mason, J.E. Gomez, T.J. Ebner, Hand synergies during reach-to-grasp. J. Neurophysiol. **86**, 2896–2910 (2001)
12. E. Todorov, Z. Ghahramani, Analysis of the synergies underlying complex hand manipulation. In: *26th Annual International Conference of the IEEE Engineering in Medicine and Biology Society*, pp. 4637–4640, 2004
13. P.H. Thakur, A.J. Bastian, S. Hsiao, Multidigit movement synergies of the human hand in an unconstrained haptic exploration task. J. Neurosci. **28**(6), 1271–1281 (2008)
14. M. Turk, A. Pentland, Eigenfaces for recognition. J. Cogn. Neurosci. **3**(1), 71–86 (1991)
15. A. Fod, M.J. Mataric, O.C. Jenkins, Automated derivation of primitives for movement classification. Autono. Robot. **12**(16), 39–54 (2002)
16. A. Tsoli, O.C. Jenkins, 2d subspaces for user-driven robot grasping. In: *Robotics, Science and Systems Conference: Workshop on Robot Manipulation*, 2007
17. C.S. Lovchik, M.A. Diftler, The robonaut hand: A dextrous robot hand for space. In: *IEEE International Conference on Robotics and Automation*, pp. 907–912, 1998
18. J. Butterfass, G. Hirzinger, S. Knoch, H. Liu. DLR's articulated hand, part I: Hard- and software architecture. In: *International Conference on Robotics and Automation*, pp. 2081–2086, 1998
19. L. Ingber, Very fast simulated re-annealing. J. Math. Comput. Model. **12**(8), 967–973 (1989)
20. A. Miller, P.K. Allen, Graspit!: a versatile simulator for robotic grasping. IEEE Robot. Autom. Mag. **11**(4), 110–122 (2004)
21. M. Mason, K. Salisbury, *Robot hands and the mechanics of manipulation* (MIT Press, Cambridge, 1985)
22. D.M. Taylor, S.H. Tillery, A.B. Schwartz, Direct cortical control of 3D neuroprosthetic devices. Science **296**(5574), 1829–1832 (2002)
23. M. Zecca, S. Micera, M.C. Carrozza, P. Dario, Control of multifunctional prosthetic hands by processing the electromyographic signal. Crit. Rev. Biomed. Eng. **30**, 459–485 (2002)
24. D. Taylor, S.H. Tillery, A. Schwartz, Information conveyed through brain control: Cursor versus robot. IEEE Trans. Neural Syst. Rehab. Eng. **1**(2), 195–199 (2003)

25. D. Kragic, A. Miller, P. Allen, Real-time tracking meets online planning. In: *IEEE International Conference on Robotics and Automation*, Seoul, 2001
26. C. Ferrari, J. Canny, Planning optimal grasps. In: *IEEE Internationl Conference on Robotics and Automation*, pp. 2290–2295, 1992
27. A. Miller, P. Allen, Examples of 3-d grasp quality computations. In: *IEEE International Conference on Robotics and Automation*, Detroit, MI, pp. 1240–1246, 1999
28. R.D. Howe, M.R. Cutkosky, Practical force-motion models for sliding manipulation. Intl. J. of Robot. Res. **15**(6), 557–572 (1996)
29. M. Ciocarlie, C. Lackner, P. Allen, Soft finger model with adaptive contact geometry for grasping and manipulation tasks. In: *Joint Eurohaptics Conference and IEEE Symposium on Haptic Interfaces*, 2007
30. A. Dollar, R. Howe, Simple, robust autonomous grasping in unstructured environments. In: *IEEE International Conference on Robotics and Automation*, 2007
31. M.C. Carrozza, G. Cappiello, S. Micera, B.B. Edin, L. Beccai, C. Cipriani, Design of a cybernetic hand for perception and action. Biol. Cybern. **95**(6), 629–644 (2006)
32. M.T. Ciocarlie, S.T. Clanton, M. C. Spalding, P.K. Allen, Biomimetic grasp planning for cortical control of a robotic hand. In: *IEEE/RSJ International Conference on Intelligent Robots and Systems*, 2008
33. C. Goldfeder, M. Ciocarlie, H. Dang, P.K. Allen, The Columbia grasp database. In: *IEEE International Conference on Robotics and Automation*, pp. 1710–1716, May 2009
34. D.L. Bowers, R. Lumia, Manipulation of unmodeled objects using intelligent grasping schemes. Trans. Fuzzy Syst. **11**(3), 320–329 (2003)
35. A. Morales, T. Asfour, P. Azad, S. Knoop, R. Dillmann, Integrated grasp planning and visual object localization for a humanoid robot with five-fingered hands. In: *International Conference on Intelligent Robots and Systems*, 2006
36. Y. Li, N.S. Pollard, A shape matching algorithm for synthesizing humanlike enveloping grasps. In: *IEEE-RAS International Conference on Humanoid Robots*, 2005
37. J. Aleotti, S. Caselli, Robot grasp synthesis from virtual demonstration and topology-preserving environment reconstruction. In: *International Conference on Intelligent Robots and Systems*, 2007
38. A. Saxena, J. Driemeyer, A.Y. Ng, Robotic grasping of novel objects using vision. Int. J. Robot. Res. **27**(2), 157–173 (2008)
39. C. Goldfeder, P.K. Allen, C. Lackner, R. Pelossof, Grasp planning via decomposition trees. In: *International Conference on Robotics and Automation*, 2007
40. A.T. Miller, S. Knoop, H.I. Christensen, P.K. Allen, Automatic grasp planning using shape primitives. In: *International Conference on Robotics and Automation*, 2003
41. P. Shilane, P. Min, M. Kazhdan, T. Funkhouser, The princeton shape benchmark. In: *Shape Modeling and Applications*, 2004
42. R.K. Sivamani, J. Goodman, N.V. Gitis, H.I. Maibach, Coefficient of friction: tribological studies in man an overview. Skin Res. Technol. **9**(3), 227–234 (2003)
43. M. Ciocarlie, C. Goldfeder, P.K. Allen, Dimensionality reduction for hand-independent dexterous robotic grasping. In: *International Conference on Intelligent Robots and Systems*, 2007
44. R. Balasubramanian, L. Xu, P. Brook, J.R. Smith, Y. Matsuoka, Human-guided grasp measures improve grasp robustness on physical robot. In: *International Conference on Robotics and Automation*, pp. 2294–2301, 2010
45. R. Balasubramanian, L. Xu, P. Brook, J.R. Smith, Y. Matsuoka, Physical human interactive guidance: Identifying grasping principles from human planned grasps. IEEE Trans. Robot. (2012) (in press)
46. R. Pelossof, A. Miller, P. Allen, T. Jebara, An SVM learning approach to robotic grasping. In: *International Conference on Robotics and Automation*, 2004
47. M. Novotni, R. Klein, 3D zernike descriptors for content based shape retrieval. In: *Solid Modeling and Applications*, June 2003

48. C. Goldfeder, P.K. Allen, Autotagging to improve text search for 3d models. In: *Joint Conference on Digital Libraries*, 2008
49. C. Goldfeder, P.K. Allen, Data-driven grasping. Auton. Robot. **31**(1), 1–20 (2011)
50. C. Goldfeder, M. Ciocarlie, J. Peretzman, H. Dang, P.K. Allen, Data-driven grasping with partial sensor data. In: *IEEE/RSJ International Conference on Intelligent Robots and Systems*, pp. 1278–1284, October 2009

Glossary

Active touch The process where objects are dynamically explored by a finger or hand as in object contour following.

Adaptive thresholding A procedure in which a stimulus is interactively increased and decreased in amplitude to determine a threshold.

Admittance A relation between force and resulting displacement.

Afferent or sensory input Information from sensory receptors that is generated in the peripheral parts of the body and transmitted to the brain.

Anatomical pulley An anatomical structure that is held to resemble a pulley; a soft tissue structure in which a tendon can run to change the direction of the pull.

Arc length parameter A parameter that keeps track of the distance along an arc.

Artificial neural network A machine learning technique that clusters information or approximates functions by transforming weighting and biasing of a number of inputs.

Biomimetic Utilizing principles of design or function found in nature.

Body representation Persistent information about an aspect of the body that is stored by the brain. Examples include information about body size or body position.

Capture region The set of initial poses from which an object can be successfully grasped with a particular grasping strategy or routine. See also comfort zone.

Catch-up response A characteristic pulse in grip (normal) force rate which is elicited by an unexpected perturbation of an object being grasped in a precision grip and results in a rapid initial increase in grip force.

Central nervous system The part of the nervous system that consists of the brain and the spinal cord.

Complementary metal oxide semiconductor (CMOS) A technology used commonly to make integrated circuits. Sensors made using the CMOS process

R. Balasubramanian and V. J. Santos (eds.), *The Human Hand as an Inspiration for Robot Hand Development*, Springer Tracts in Advanced Robotics 95,
DOI: 10.1007/978-3-319-03017-3,

can easily be integrated with electronics for signal processing, etc. by fabricating them together.

Comfort zone The set of object poses at the lift-off time point in a grasp acquisition action. See also capture region.

Convex contour A shape that bends or curves outward.

Corollary discharges Copies or corollaries of the motor command that originate in the motor areas of the cortex and are transmitted to the sensory centers in the brain that reflect the magnitude of the motor command generated.

Corticofugal Descending fibers from motor cortex.

Cortico-motoneuronal cell A neuron in the primary motor cortex with a descending axon that synapses directly on spinal motoneurons. Cortico-motoneuronal cells are found only in macaque monkeys, apes, humans.

Curvature of an object The reciprocal of the radius of curvature at a given point on the object.

Cutaneous Sensation derived from the skin or covering of a device or organism.

Cutaneous mechanoreceptors Sensory receptors found in glabrous and hairy skin that respond to mechanical inputs such as pressure vibration and force.

Damping The characteristic of matter or an arrangement of matter which resists velocity.

Degrees of freedom (DOF) The number of variables required to specify the configuration of a mechanism. Often equal to the number of joints.

Distal Near the end of a kinematic chain such as a finger.

Dynamic range A measure of the maximum measurable value of a quantity (e.g. pressure or shear stress for a tactile sensor) typically limited by saturation, divided by the minimum measurable value, typically limited by the noise floor.

Efference copy A copy of the motor command that is generated and received by the brain when it performs an action. The efference copy is used to process incoming sensory information by removing the component of the sensory information that is due to the action performed.

Elastomer A flexible compliant polymer having rubber-like properties.

Elbow flexors Muscles that are involved in flexing the elbow joint.

Encoder A device for digital measurement of rotary or linear displacement.

End-effector The rigid body segment or robot link of interest for task completion usually the most distal segment, e.g. the hand or palm segment for a manipulation task.

Euler-Lagrange equation A differential equation whose solutions are the functions for which a given functional (a function that inputs a vector and returns a scalar) is stationary. In Lagrangian mechanics, a physical system's evolution is described by the solution to the Euler–Lagrange equation for the system's action.

Exploration Motion of a user's limbs (usually fingers) for the purpose of haptic detection or recognition of a target.

Extensor hood A fibrous sheet on the posterior aspect of each finger; plexus of tendons with an aponeurotic sheet on the posterior aspect of the finger.

Extrinsic finger or hand muscles These are muscles that act directly on the fingers but have their muscle bellies in the forearm and tendons that cross the wrist and palm to insert ultimately on the bones of the fingers (phalanges). Contraction of any of these muscles thus acts not only on certain fingers, but also tends to rotate the wrist. Some extrinsic finger muscles send tendons to more than one digit. In contrast, the intrinsic finger muscles have their bellies in the palm of the hand, and each acts on a single digit.

Feasible force set The convex set in endpoint force space (normally 2-D or 3-D) that encloses all achievable force vectors based on the parameters of a tendon-driven manipulator.

Fingertip force vector angle The angle of the fingertip force vector with respect to the grip surface normal as viewed from the plane containing the grip axis.

Gaussian mixture model regression A machine learning technique that clusters information or approximates functions by combining a number of Gaussian functions over a number of dimensions.

Gaussian processes An advanced mathematical regression method to model nonlinear functions.

Golgi tendon organs Receptors found at the junction between muscle fibers and tendons that respond to the forces developed in series by the muscle fibers and so provide the CNS with information regarding muscle force.

Grasping posture The manipulator configuration at the lift-off time point.

Grip axis Line connecting the fingertips in a two-finger precision grasp.

Haptic inputs Sensory inputs arising from stimulation of receptors embedded in skin, muscles, tendons, and joints during active touch.

Hysteresis Having the property that when a material is loaded and unloaded it does not follow exactly the same curve but instead forms a loop on a force/displacement plot. The area of the loop is a measure of energy lost per loading/unloading cycle.

Impedance A measure of opposition to motion of a structure subjected to a force.

Individuated movements Movements in which the basic simultaneous motion of multiple body parts is modified such that one part moves more or differently than the others.

Inertia The characteristic of matter or an arrangement of matter which resists acceleration.

Intracortical microstimulation (ICMS) The delivery of small amounts of current (5–100 μA) to extremely localized targets in cortex with the goal of activating 100-1000s of neurons.

Intrinsic hand muscles Muscles that take their origin and insertion within the hand.

Jacobian matrix A matrix that maps values from one space to another space, such as joint angle velocities to manipulator endpoint velocities, or joint torques to manipulator endpoint forces and torques.

Kinematic redundancy The characteristic of a mechanism which has more degrees of freedom than required to specify its end effector position and orientation.

Kinesthesia Sense of limb movement that is derived from sensory receptors in muscle and skin.

Kinesthetic feedback Force feedback involving muscle stimulation.

Lagrange multipliers The method of Lagrange multipliers is a strategy for finding the local maxima and minima of a function subject to equality constraints.

Lateral pinch A grasp in which the pad of the thumb is in contact with the lateral surface of the middle phalanx of the index finger, as when holding a key.

Lift-off time point The first time when a grasped object is completely separated from the original support surface and completely supported by the manipulator.

Material properties of an object Properties including stiffness, texture, and weight.

Maximum isotropic value (MIV) The greatest possible force that a robotic finger or manipulator can exert in all directions (i.e., isotropically).

Mechanoreceptor Nerve endings that respond to mechanical phenomena such as vibrations or local tissue strain.

Mechanotransduction The mechanism by which mechanical stimulation and or deformation is converted to changes in membrane potential in mechanoreceptors.

Meissner corpuscles Mechanoreceptors that are located near the surface of the skin, have small well-defined receptive fields, and respond to dynamic variations in skin curvature with a frequency range of tens of cycles per second.

Minkowski sum In this the context of coordination patterns, it represents all possible positive or zero combinations of the collection of basis vectors for the tendons. Stated another way, it contains all possibilities of full or zero activation for all of the tendons.

Moment arm The perpendicular distance from line of force application to the axis of rotation. Also, the mechanical advantage of a tendon over a joint.

Moment arm matrix A matrix that contains information about all of the moment arms in a tendon-driven manipulator. Each row corresponds to a joint and each column to a specific tendon. The entries are the values of the moment arms.

Motor unit Motor neuron and skeletal muscle fibers innervated by the same motor neuron.

Motor-unit coherence Correlation in the discharge rate of pairs or groups of motor units within the same muscle or across different muscles.

Motor-unit synchrony Temporal correlation between action potentials from two concurrently active motor units within the same muscle or across different muscles.

Muscle synergy A group of muscles acting together concurrently in a fixed ratio.

Muscle thixotropy A property of skeletal muscle fibres that results in the resting muscle having a stiffness or resistance to length changes. This property extends to intrafusal muscle fibres and thus affects the firing rate of muscle spindles.

Musculotendon A unit composed of muscle and tendon.

Neuroprosthetics Interfaces between the nervous system and electronic systems for the purpose of restoring some element of lost or reduced function.

Nonprehensile Manual activities that do not involve grasping an object such as gestures made as part of normal speech or as a substitute for speech (sign language), and movements that involve depressing keys on a keyboard or musical instrument.

Nonprehensile interaction Non-grasping contact or interaction where the object is not necessarily fixtured rigidly to the hand or end-effector.

Object-moment External torque load imposed about the center of mass of a grasped object.

Pacinian corpuscles Mechanoreceptors located beneath the skin with large receptive fields, and responding strongly to vibrations with frequencies of tens or hundreds of cycles per second, up to 1000 Hz.

Palmar pinch A grasp in which the pulp surface of the thumb opposes the finger pads of the index and middle finger.

Passive joint impedance A resistive torque produced by joint rotation.

Passive touch The process where objects are brought in contact with a passive finger or hand.

Phantom limb The feeling that a body part exists when no sensory information about that body part is available to the brain. Many amputees experience a phantom of their amputated limb. That is, they continue to experience its presence despite its amputation. Experimental phantoms can be induced with anaesthesia. Here the subject feels the presence of the limb even when all sensory information is blocked from reaching the brain.

Piezoelectric Having the property that the material acquires an electrical charge when subject to strain.

Piezoresistive Having the property that the electrical resistance varies in proportion to strain.

Posture The relative deposition of the body parts (i.e., a collection of joint angles).

Power grip A grip that is used when force is the primary objective and there is a large area of contact between the object and the surfaces of the fingers and palm, such as when holding a hammer.

Pre-grasp or preparatory interaction Contact or interaction that adjusts object placement in the environment prior to grasping and a primary manipulation task.

Pre-grasp rotation Pre-grasp interaction where the object adjustment consists only of changes in the object orientation in the plane of the support surface.

Prehensile Related to actions that involve the thumb and one or more fingers in grasping an object.

Primary motor cortex (M1) The area of the cerebral cortex (Brodmann's cytoarchitectonic area 4) with the most direct projections to the motor centers of the brainstem and spinal cord, and hence the most direct control of body movement. In primates, this area lies directly in front of the central sulcus (Rolandic fissure). Axons descending from M1 provide the bulk of the cortical projection to the spinal cord, the corticospinal tract.

Primary task or primary interaction The action following or requiring object acquisition in a grasp (e.g., transport of the object while fixtured in a grasp).

Principle component analysis A mathematical method that linearly separates data along orthogonal dimensions according to variance.

Proprioception A sense of the position and movement of the body as well as the forces and efforts involved in movement. The sense is used by the brain for conscious knowledge of the body as well as for the control of movement. The sensations and perceptions associated with the posture of the body, including torso, limbs, and hands/feet.

Proximal Near the origin of a kinematic chain such as a finger.

Psychophysical tests Controlled human subjects tests meant to elucidate how subjects perceive or respond to physical phenomena.

PVDF Polyvinylidene fluoride. A polymer film that can be poled (placed under a strong electric field) making it piezoelectric.

Rotational slip Slip between a digit and grip surface caused by torques about the axis perpendicular to a grip surface.

Rubber hand illusion An illusion that manipulates the sense of ownership. By carefully controlling the location and timing of sensory information an experimenter can make a subject feel that a rubber hand is their own hand.

Shape properties of an object Properties including size orientation and curvature.

Size-weight illusion The size of an object influences the perceived weight of the same object.

Somatosensation The set of senses arising from the body including touch vibration, pressure, movement, body movement, temperature, and pain.

Static friction force The friction force between two solid objects that are not moving relative to each other.

Stiffness The characteristic of matter or an arrangement of matter which resists displacement or deformation as a result of an applied force.

Strain The stretching of a material defined as extension per unit length.

Tactile feedback Cutaneous feedback relying on skin stimulation.

Tactile sensor A device that measures physical effects arising from contact with objects and surfaces.

Taxel One element of an array of tactile sensing elements analogous to a pixel in graphics.

Tendon-driven system Any mechanical system that is actuated (either actively or passively) with tendons that can only pull and not push (e.g., wires, cables, strings).

Tendon routing The configuration in which the tendons are arranged in a tendon-driven system.

Texture of an object Properties of object surface including but not limited to roughness.

Thermistor An electronic device whose resistance varies with temperature.

Threshold A level of sensory amplitude at which a feature can just be detected.

Tip pinch A grip that involves holding small object between the pulps of the opposed thumb and index finger.

Vertical-horizontal illusion The tendency to overestimate the length of a vertical line compared to a horizontal line of the same length.

Vibrissae Whiskers of animals that can sense light contacts and vibrations.

Weber fraction For many sensory modalities the change in stimulus intensity that can be discriminated is a constant fraction of the intensity of the stimulus. This number is known as the Weber fraction.

MIX
Papier aus verantwortungsvollen Quellen
Paper from responsible sources
FSC® C105338
FSC
www.fsc.org

If you have any concerns about our products,
you can contact us on
ProductSafety@springernature.com

In case Publisher is established outside the EU,
the EU authorized representative is:
Springer Nature Customer Service Center GmbH
Europaplatz 3, 69115 Heidelberg, Germany

Printed by Libri Plureos GmbH
in Hamburg, Germany